BASIC BIOMECHANICS

SIXTH EDITION

Susan J. Hall, Ph.D.

College of Health Sciences
University of Delaware

Mc
Graw
Hill

Connect
Learn
Succeed™

The McGraw-Hill Companies

Connect
Learn
Succeed™

BASIC BIOMECHANICS, SIXTH EDITION

Published by McGraw-Hill, a business unit of The McGraw-Hill Companies, Inc., 1221
Avenue of the Americas, New York, NY 10020. Copyright © 2012 by The McGraw-Hill
Companies, Inc. All rights reserved. No part of this publication may be reproduced or
distributed in any form or by any means, or stored in a database or retrieval system,
without the prior written consent of The McGraw-Hill Companies, Inc., including, but not
limited to, in any network or other electronic storage or transmission, or broadcast for
distance learning.

Some ancillaries, including electronic and print components, may not be available to cus-
tomers outside the United States.

♻ This book is printed on recycled, acid-free paper containing 10% postconsumer waste.

1 2 3 4 5 6 7 8 9 0 QDB/QDB 1 0 9 8 7 6 5 4 3 2 1

ISBN 978-0-07-122151-1
MHID 0-07-122151-4

www.mhhe.com

BRIEF CONTENTS

CONTENTS

PREFACE

The sixth edition of *Basic Biomechanics* has been significantly updated and redesigned from the previous edition. As the interdisciplinary field of biomechanics grows in both breadth and depth, it is important that even introductory textbooks reflect the nature of the science. Accordingly, the text has been revised, expanded, and updated, with the objectives being to present relevant information from recent research findings and to prepare students to *analyze* human biomechanics.

The approach taken remains an integrated balance of qualitative and quantitative examples, applications, and problems designed to illustrate the principles discussed. The sixth edition also retains the important sensitivity to the fact that some beginning students of biomechanics possess weak backgrounds in mathematics. For this reason, it includes numerous sample problems and applications, along with practical advice on approaching quantitative problems.

ORGANIZATION

Each chapter follows a logical and readable format, with the introduction of new concepts consistently accompanied by practical human movement examples and applications from across the life span and across sport, clinical, and daily living activities.

NEW CONTENT HIGHLIGHTS

New content has been added to provide updated scientific information on relevant topics. All chapters have been revised to incorporate the latest information from the biomechanics research literature, and numerous new sport and clinical applications and examples are included. Topics added or expanded include fall prevention in older adults, vibration effects on the musculoskeletal system, the female athlete triad, exercise and bone, bone microarchitecture, articular cartilage repair, swimming biomechanics, and stretching protocols related to strength, power, and injury potential.

Balanced Coverage

The Biomechanics Academy of AAHPERD recommends that undergraduate students in biomechanics devote approximately one-third of study time to anatomical considerations, one-third to mechanical considerations, and the remainder to applications. The integrated approach to coverage of these areas taken in previous editions is continued in this sixth edition.

Applications Oriented

All chapters in this new edition contain discussion of a broad range of updated human movement applications, many of which are taken from the recent biomechanics research literature. Special emphasis has been placed on examples that span all ages and address clinical and daily living issues, as well as sport applications.

Laboratory Experiences

The integrated laboratory manual appears at the end of each chapter with references to simulations on the text's Online Learning Center. The soft-cover design with perforation allows laboratory manual pages to be completed and turned in to instructors.

Integrated Technology

Technology is integrated throughout the text, with an Online Learning Center box appearing on every chapter-opening page and directing students to resources online, while lists of related websites at the end of each chapter offer pertinent sources to students. Problems and laboratory experiences are incorporated throughout the text and updated to reference the Online Learning Center.

The Sixth Edition of *Basic Biomechanics* can be bundled (for a small additional price) with *MaxTRAQ*™ software. MaxTRAQ is a downloadable motion analysis software that offers an easy-to-use tool to track data and analyze various motion elected by the authors. The MaxTRAQ software includes video clips of motions such as golf swing and gait, 2D manual tracking, coverage of distance and angles—and more!

Visit www.mhhe.com/hall6e for more information about the MaxTRAQ software.

PEDAGOGICAL FEATURES

In addition to the sample problems, problem sets, laboratory experiences, Online Learning Center boxes, end-of-chapter key terms lists, and lists of websites, the book contains other pedagogical features from previous editions. These include **key concepts, marginal definitions, sample problems, chapter summaries, introductory and additional problems, references,** and **appendices.**

ANCILLARIES

Online Learning Center

www.mhhe.com/hall6e
This website offers resources to students and instructors. It includes downloadable ancillaries, Web links, student quizzing, additional information on topics of interest, and much more.

Resources for the instructor include:

- Downloadable PowerPoint presentation with annotated lecture notes
- Online instructor's manual, originally developed by Darla Smith, University of Texas at El Paso
- Test bank
- Interactive links

- Links to professional resources
- Online laboratory manual with simulations

Resources for the student include:

- Downloadable PowerPoint presentations
- Self-grading quizzes
- Online laboratory manual with simulations

ACKNOWLEDGMENTS

I would like to thank developmental editor Chris Narozny and managing editor Marley Magaziner for their quality work on the new edition of this book. Many thanks also to copy editor Sheryl Rose and production editor Erin Melloy for their very capable and professional work on this revision. I also wish to extend appreciation to the following reviewers:

Debra A. Allyn
University of Wisconsin–River Falls

Sarah Bingham
Longwood University

Michael G. Collins
Lewis-Clark State College

Shannon Crumpton
Auburn Montgomery

Barry A. Frishberg
South Carolina State University

Douglas Haladay
University of Scranton

Larry McDaniel
Dakota State University

Jan Prins
University of Hawaii

Andrea Woodson-Smith
Chicago State University

Finally, I also very much appreciate the excellent suggestions I have received over the six editions of this book from numerous students and colleagues.

Susan J. Hall

Deputy Dean,
College of Health Sciences
University of Delaware

CourseSmart
Learn Smart. Choose Smart.

COURSESMART eTEXTBOOKS

This text is available as an eTextbook from CourseSmart, a new way for faculty to find and review eTextbooks. It's also a great option for students who are interested in accessing their course materials digitally and saving money. CourseSmart offers thousands of the most commonly adopted textbooks across hundreds of courses from a wide variety of higher education publishers. It is the only place for faculty to review and compare the full text of a textbook online, providing immediate access without the environmental impact of requesting a print exam copy. At CourseSmart, students can save up to 50% off the cost of a print book, reduce their impact on the environment, and gain access to powerful Web tools for learning, including full text search, notes and highlighting, and e-mail tools for sharing notes between classmates. For further details contact your sales representative or go to www.coursesmart.com.

McGRAW-HILL CREATE

www.mcgrawhillcreate.com

Craft your teaching resources to match the way you teach! With McGraw-Hill Create you can easily rearrange chapters, combine material from other content sources, and quickly upload content you have written such as your course syllabus or teaching notes. Find the content you need in Create by searching through thousands of leading McGraw-Hill textbooks. Arrange your book to fit your teaching style. Create even allows you to personalize your book's appearance by selecting the cover and adding your name, school, and course information. Order a Create book and you'll receive a complimentary print review copy in 3–5 business days or a complimentary electronic review copy (eComp) via e-mail in about one hour. Go to www.mcgrawhillcreate.com today and register. Experience how McGraw-Hill Create empowers you to teach *your* students *your* way.

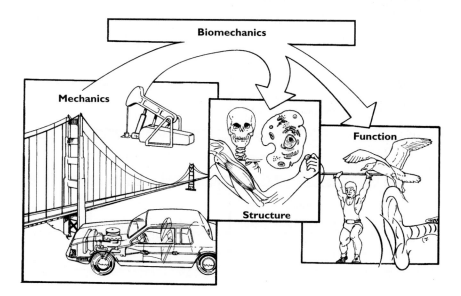

FIGURE 1-1

Biomechanics uses the principles of mechanics for solving problems related to the structure and function of living organisms.

Biomechanists use the tools of mechanics, the branch of physics involving analysis of the actions of forces, to study the anatomical and functional aspects of living organisms (Figure 1-1). Statics and dynamics are two major subbranches of mechanics. Statics is the study of systems that are in a state of constant motion, that is, either at rest (with no motion) or moving with a constant velocity. Dynamics is the study of systems in which acceleration is present.

Kinematics and kinetics are further subdivisions of biomechanical study. What we are able to observe visually when watching a body in motion is termed the *kinematics* of the movement. Kinematics involves the study of the size, sequencing, and timing of movement, without reference to the forces that cause or result from the motion. The kinematics of an exercise or a sport skill execution is also known, more commonly, as *form* or *technique*. Whereas kinematics describes the appearance of motion, kinetics is the study of the forces associated with motion. Force can be thought of as a push or pull acting on a body. The study of human biomechanics may include questions such as whether the amount of force the muscles are producing is optimal for the intended purpose of the movement.

Although biomechanics is relatively young as a recognized field of scientific inquiry, biomechanical considerations are of interest in several different scientific disciplines and professional fields. Biomechanists may have academic backgrounds in zoology; orthopedic, cardiac, or sports medicine; biomedical or biomechanical engineering; physical therapy; or kinesiology, with the commonality being an interest in the biomechanical aspects of the structure and function of living things.

The biomechanics of human movement is one of the subdisciplines of kinesiology, the study of human movement (Figure 1-2). Although some biomechanists study topics such as ostrich locomotion, blood flow through constricted arteries, or micromapping of dental cavities, this book focuses primarily on the biomechanics of human movement from the perspective of the movement analyst.

As shown in Figure 1-3, biomechanics is also a scientific branch of sports medicine. *Sports medicine* is an umbrella term that encompasses both clinical and scientific aspects of exercise and sport. The American College of Sports Medicine is an example of an organization that promotes interaction between scientists and clinicians with interests in sports medicine–related topics.

mechanics
branch of physics that analyzes the actions of forces on particles and mechanical systems

statics
branch of mechanics dealing with systems in a constant state of motion

dynamics
branch of mechanics dealing with systems subject to acceleration

kinematics
study of the description of motion, including considerations of space and time

kinetics
study of the action of forces

kinesiology
study of human movement

sports medicine
clinical and scientific aspects of sports and exercise

The subdisciplines of kinesiology.

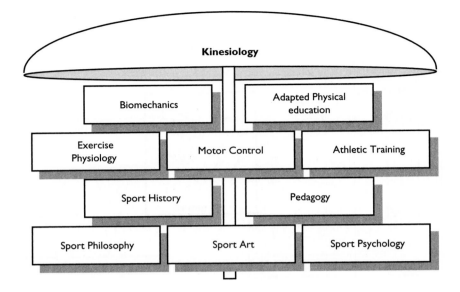

The branches of sports medicine.

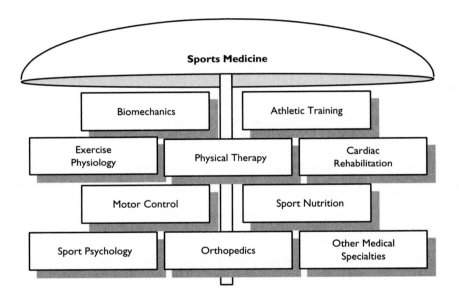

What Problems Are Studied by Biomechanists?

As expected given the different scientific and professional fields represented, biomechanists study questions or problems that are topically diverse. For example, zoologists have examined the locomotion patterns of dozens of species of animals walking, running, trotting, and galloping at controlled speeds on a treadmill to determine why animals choose a particular stride length and stride rate at a given speed. They have found that running actually consumes less energy than walking in small animals up to the size of dogs, but running is more costly than walking for larger animals such as horses (35). One of the challenges of this type of research is determining how to persuade a cat, a dog, or a turkey to run on a treadmill (Figure 1-4).

Among humans, although the energy cost of running increases with running speed, sizable differences in energy cost between individuals become even larger as running speed increases (21). Although some

•In research, each new study, investigation, or experiment is usually designed to address a particular question or problem.

FIGURE 1-4

Research on the biomechanics of animal gaits poses some interesting problems.

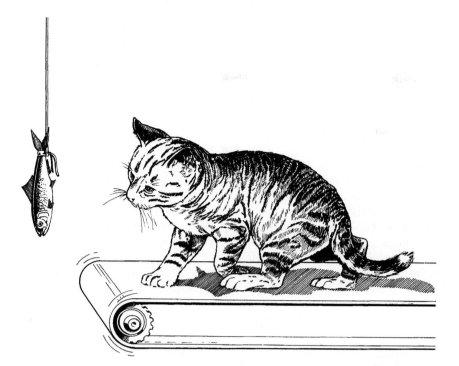

individuals appear to run more smoothly and comfortably than others, no particular biomechanical factors have been associated with either good or poor running economy (21). Differences in muscle fiber type composition appear to translate into differences in energy utilization during running (see Chapter 6) (22).

The U.S. National Aeronautics and Space Administration (NASA) sponsors another multidisciplinary line of biomechanics research to promote understanding of the effects of microgravity on the human musculoskeletal system. Of concern is the fact that astronauts who have been out of the earth's gravitational field for just a few days have returned with muscle atrophy, cardiovascular and immune system changes, and reduced bone density, mineralization, and strength, especially in the lower extremities (11). The issue of bone loss, in particular, is currently a limiting factor for long-term space flights, with bone lost at a rate of about 1% per month from the lumbar spine and 1.5% per month from the hips (26). Both increased bone resorption and decreased calcium absorption appear to be responsible (see Chapter 4) (39).

Since those early days of space flight, biomechanists have designed and built a number of exercise devices for use in space to take the place of normal bone-maintaining activities on earth. Some of this research has focused on the design of treadmills for use in space that load the bones of the lower extremity with deformations and strain rates that are optimal for stimulating new bone formation (8, 30). Another approach involves combining voluntary muscle contraction with electrical stimulation of the muscles to maintain muscle mass and tone (44). So far, however, no adequate substitute has been found for weight bearing for the prevention of bone and muscle loss in space (11).

Maintaining sufficient bone-mineral density is also a topic of concern here on earth. Osteoporosis is a condition in which bone mineral mass and strength are so severely compromised that daily activities can cause bone pain and fracturing (13). This condition is found in most elderly individuals, with earlier onset in women, and is becoming increasingly prevalent throughout the world with the increasing mean age of the population. Approximately 40% of women experience one or more osteoporotic fractures after age 50, and after age 60, about 90% of all fractures in both men and women are osteoporosis-related (23, 34). The most common fracture site is the vertebrae, with the presence of one fracture indicating increased risk for future vertebral and hip fractures (15). This topic is explored in depth in Chapter 4.

Another problem area challenging biomechanists who study the elderly is mobility impairment. Age is associated with decreased ability to balance, and older adults both sway more and fall more than young adults, although the reasons for these changes are not well understood. Falls, and particularly fall-related hip fractures, are extremely serious, common, and costly medical problems among the elderly. Each year, falls cause large percentages of the wrist fractures, head injuries, vertebral fractures, and lacerations, as well as over 90% of the hip fractures, occurring in the United States (37). Biomechanical research teams are investigating the biomechanical factors that enable individuals to avoid falling, the characteristics of safe landings from falls, the forces sustained by different parts of the body during falls, and the ability of protective clothing

Exercise in space is critically important for preventing loss of bone mass among astronauts. Photo courtesy of NASA.

and floors to prevent falling injuries (37). Promising work in the development of intervention strategies has shown that the key to preventing falls may be the ability to limit trunk motion (14). Older adults can quickly learn strategies for limiting trunk motion through task-specific training combined with whole-body exercise.

Research by clinical biomechanists has resulted in improved gait among children with cerebral palsy, a condition involving high levels of muscle tension and spasticity. The gait of the cerebral palsy individual is characterized by excessive knee flexion during stance. This problem is treated by surgical lengthening of the hamstring tendons to improve knee extension during stance. In some patients, however, the procedure also diminishes knee flexion during the swing phase of gait, resulting in dragging of the foot. After research showed that patients with this problem exhibited significant co-contraction of the rectus femoris with the hamstrings during the swing phase, orthopedists began treating the problem by surgically attaching the rectus femoris to the sartorius insertion. This creative, biomechanics research–based approach has enabled a major step toward gait normalization for children with cerebral palsy.

Research by biomedical engineers has also resulted in improved gait for children and adults with below-knee amputations. Ambulation with a prosthesis creates an added metabolic demand, which can be particularly significant for elderly amputees and for young active amputees who participate in sports requiring aerobic conditioning. In response to this problem, researchers have developed an array of lower-limb and foot prostheses that store and return mechanical energy during gait, thereby reducing the metabolic cost of locomotion (2). Studies have shown that the more compliant prostheses are better suited for active and fast walkers, whereas prostheses that provide a more stable base of support are generally preferred for the elderly population (3). Microchip-controlled "Intelligent Prostheses" show promise for reducing the energy cost of walking at a range of speeds (7). Researchers are currently developing a new class of "bionic" prosthetic feet that are designed to better imitate normal gait (41).

Occupational biomechanics is a field that focuses on the prevention of work-related injuries and the improvement of working conditions and worker performance. Researchers in this field have learned that work-related low back pain can derive not only from the handling of heavy materials but from unnatural postures, sudden and unexpected motions, and the characteristics of the individual worker (27). Sophisticated biomechanical models of the trunk are now being used in the design of materials-handling tasks in industry to enable minimizing potentially injurious stresses to the low back (4).

Biomechanists have also contributed to performance improvements in selected sports through the design of innovative equipment. One excellent example of this is the Klapskate, the speed skate equipped with a hinge near the toes that allows the skater to plantar flex at the ankle during push-off, resulting in up to 5% higher skating velocities than were obtainable with traditional skates (17). The Klapskate was designed by van Ingen Schenau and de Groot, based on study of the gliding push-off technique in speed skating by van Ingen Schenau and Baker, as well as work on the intermuscular coordination of vertical jumping by Bobbert and van Ingen Schenau (9). When the Klapskate was used for the first time by competitors in the 1998 Winter Olympic Games, speed records were shattered in every event.

Numerous innovations in sport equipment and apparel have also resulted from findings of experiments conducted in experimental chambers called *wind tunnels* that involved controlled simulation of the air

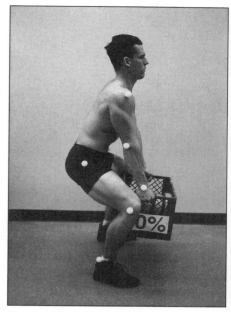

Occupational biomechanics involves study of safety factors in activities such as lifting.

carpal tunnel syndrome
overuse condition caused by compression of the median nerve in the carpal tunnel and involving numbness, tingling, and pain in the hand

Aerodynamic cycling equipment has contributed to new world records. Photo courtesy of Getty Images.

Biomechanists Develop a Revolutionary New Figure Skate

What do 1996 U.S. figure skating champion Rudy Galindo and 1998 Olympic gold medal winner Tara Lipinski have in common besides figure skating success? They have both had double hip replacements, Galindo at age 32 and Lipinski at age 18.

Overuse injuries among figure skaters are on the rise at an alarming rate, with most involving the lower extremities and lower back (4, 12). With skaters performing more and more technically demanding programs including multirotation jumps, on-ice training time for elite skaters now typically includes over 100 jumps per day, six days per week, year after year.

Yet, unlike most modern sports equipment, the figure skate has undergone only very minor modifications since 1900. The soft-leather, calf-high boots of the nineteenth century are now made of stiffer leather to promote ankle stability and are not quite as high to allow a small amount of ankle motion. However, the basic design of the rigid boot with a screwed-on steel blade has not changed.

The problem with the traditional figure skate is that when a skater lands after a jump, the rigid boot severely restricts motion at the ankle, forcing the skater to land nearly flat-footed and preventing motion at the ankle that could help attenuate the landing shock that gets translated upward through the musculoskeletal system. Not surprisingly, the incidence of overuse injuries in figure skating is mushrooming due to the increased emphasis on performing jumps, the increase in training time, and the continued use of outdated equipment.

To address this problem, biomechanist Jim Richards and graduate student Dustin Bruening, working at the University of Delaware's Human Performance Lab, have designed and tested a new figure skating boot. Following the design of modern-day Alpine skiing and in-line skating boots, the new boot incorporates an articulation at the ankle that permits flexion movement but restricts potentially injurious sideways movement.

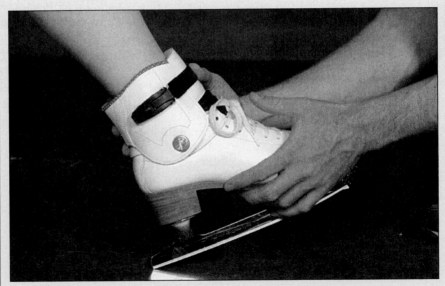

New figure skating boot with an articulation at the ankle designed by biomechanists at the University of Delaware.

The boot enables skaters to land toe-first, with the rest of the foot hitting the ice more slowly. This extends the landing time, thereby spreading the impact force over a longer time and dramatically diminishing the peak force translated up through the body. As shown in the graph, the new boot attenuates the peak landing force on the order of 30%.

Although the new figure skating boot design was motivated by a desire to reduce the incidence of stress injuries in skating, it may also promote performance. The ability to move through a larger range of motion at the ankle may well enable higher jump heights and concomitantly more rotations while the skater is in the air.

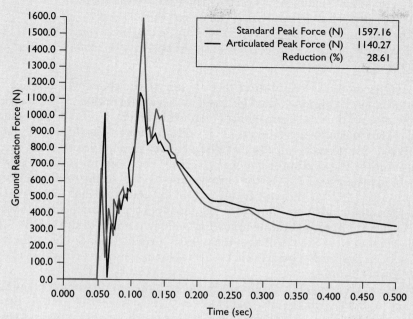

	Standard Peak Force (N)	1597.16
	Articulated Peak Force (N)	1140.27
	Reduction (%)	28.61

The new figure skating boot with an articulation at the ankle reduces peak impact forces during landing from a jump on the order of 30%.
Graph courtesy of D. Bruening and J. Richards.

Skaters who adopt the new boot are finding that using it effectively requires a period of acclimatization. Those who have been skating in the traditional boot for many years tend to have reduced strength in the musculature surrounding the ankle. Improving ankle strength is likely to be necessary for optimal use of a boot that now allows ankle motion.

resistance actually encountered during particular sports. Examples include the aerodynamic helmets, clothing, and cycle designs used in competitive cycling, and the ultrasmooth suits worn in other competitive speed-related events, such as swimming, track, skating, and skiing. Wind tunnel experiments have also been conducted to identify optimal body configuration during events such as ski jumping (42).

Sport biomechanists have also directed efforts at improving the biomechanical, or technique, components of athletic performance. They have learned, for example, that factors contributing to superior performance in the long jump, high jump, and pole vault include high horizontal velocity going into takeoff and a shortened last step that facilitates continued elevation of the total-body center of mass (6, 16). Study of baseball pitchers has determined that high-velocity pitchers display greater external rotation at the shoulder, more forward trunk tilt at ball release, higher-extension angular velocity at the lead knee, and greater angular velocity of the pelvis and upper torso than lower-velocity pitchers (25, 40).

One rather dramatic example of performance improvement partially attributable to biomechanical analysis is the case of four-time Olympic discus champion Al Oerter. Mechanical analysis of the discus throw requires precise evaluation of the major mechanical factors affecting the flight of the discus. These factors include the following:

1. The speed of the discus when it is released by the thrower
2. The projection angle at which the discus is released
3. The height above the ground at which the discus is released
4. The angle of attack (the orientation of the discus relative to the prevailing air current)

By using computer simulation techniques, researchers can predict the needed combination of values for these four variables that will result in a throw of maximum distance for a given athlete (18). High-speed cameras can record performances in great detail, and when the film or video is analyzed, the actual projection height, velocity, and angle of attack can be compared to the computer-generated values required for optimal performance. At the age of 43, Oerter bettered his best Olympic performance by 8.2 m. Although it is difficult to determine the contributions of motivation and training to such an improvement, some part of Oerter's success was a result of enhanced technique following biomechanical analysis (38). Most adjustments to skilled athletes' techniques produce relatively modest results because their performances are already characterized by above-average technique.

The USOC began funding sports medicine research in 1981. Other countries began sponsoring research to boost the performance of elite athletes in the early 1970s.

Some of the research produced by sport biomechanists has been done in conjunction with the Sports Medicine Division of the United States Olympic Committee (USOC). Typically, this work is done in direct cooperation with the national coach of the sport to ensure the practicality of results. USOC-sponsored research has yielded much new information about the mechanical characteristics of elite performance in various sports. Because of continuing advances in scientific analysis equipment, the role of sport biomechanists in contributing to performance improvements is likely to be increasingly important in the future.

The influence of biomechanics is also being felt in sports popular with both nonathletes and athletes, such as golf. Computerized video analyses of golf swings designed by biomechanists are commonly available at golf courses and equipment shops. The science of biomechanics can play a role in optimizing the distance and accuracy of all golf shots, including putting, through analysis of body angles, joint forces, and muscle activity patterns (19). A common technique recommendation is to maintain a single fixed center of rotation to impart force to the ball (19).

Other concerns of sport biomechanists relate to minimizing sport injuries through both identifying dangerous practices and designing safe equipment and apparel. In recreational runners, for example, research shows that the most serious risk factors for overuse injuries are training errors such as a sudden increase in running distance or intensity, excess cumulative mileage, and running on cambered surfaces (20). The complexity of safety-related issues increases when the sport is equipment-intensive. Evaluation of protective helmets involves ensuring not only that the impact characteristics offer reliable protection but also that the helmet does not overly restrict wearers' peripheral vision.

Impact testing of protective sport helmets is carried out scientifically in engineering laboratories.

An added complication is that equipment designed to protect one part of the body may actually contribute to the likelihood of injury in another part of the musculoskeletal system. Modern ski boots and bindings, while effective in protecting the ankle and lower leg against injury, unfortunately contribute to severe bending moments at the knee when the skier loses balance. Recreational Alpine skiers consequently experience a higher incidence of anterior cruciate ligament tears than participants in any other sport (33). Injuries in snowboarding are also more frequent with rigid, as compared to pliable, boots, although more than half of all snowboarding injuries are to the upper extremity (24, 32).

Another challenging area of research for biomechanists in the realm of sport safety is investigation of the efficacy of prophylactic knee braces (29). Approximately 60% of all sport injuries are to the knee (36). Research shows that knee braces can contribute 20–30% added resistance against lateral blows to the knee, with custom-fitted braces providing the best protection (1). A possible concern, however, is that knee braces act to change the pattern of lower-extremity muscle activity during gait, with less work performed at the knee and more at the hip (10). Other documented problems that appear to affect some athletes more than others and may be brace-specific include reduced sprinting speed and earlier onset of fatigue (1). The research literature is almost evenly divided on the efficacy of prophylactic knee braces in preventing knee ligament injuries in football players, with some studies showing decreases and others showing increases in injury incidence (31).

An area of biomechanics research with implications for both safety and performance is sport shoe design. Today sport shoes are designed both to prevent excessive loading and related injuries and to enhance performance. Because the ground or playing surface, the shoe, and the human body compose an interactive system, athletic shoes are specifically designed for particular sports, surfaces, and anatomical considerations. Aerobic dance shoes are constructed to cushion the metatarsal arch. Football shoes to be used on artificial turf are designed to minimize the risk of knee injury. Running shoes are available for training and racing on snow and ice. In fact, sport shoes today are so specifically designed for designated activities that wearing an inappropriate shoe can contribute to the likelihood of injury.

These examples illustrate the diversity of topics addressed in biomechanics research, including some examples of success and some areas of continuing challenge. Clearly, biomechanists are contributing to the knowledge base on the full gamut of human movement, from the gait of the physically challenged child to the technique of the elite athlete. Although varied, all of the research described is based on applications of mechanical principles in solving specific problems in living organisms. This book is designed to provide an introduction to many of those principles and to focus on some of the ways in which biomechanical principles may be applied in the analysis of human movement.

Why Study Biomechanics?

As is evident from the preceding section, biomechanical principles are applied by scientists and professionals in a number of fields to problems related to human health and performance. Knowledge of basic biomechanical concepts is also essential for the competent physical education teacher, physical therapist, physician, coach, personal trainer, or exercise instructor.

An introductory course in biomechanics provides foundational understanding of mechanical principles and their applications in analyzing movements of the human body. The knowledgeable human movement analyst should be able to answer the following types of questions related to biomechanics: Why is swimming *not* the best form of exercise for individuals with osteoporosis? What is the biomechanical principle behind variable-resistance exercise machines? What is the safest way to lift a heavy object? Is it possible to judge what movements are more/less economical from visual observation? At what angle should a ball be thrown for maximum distance? From what distance and angle is it best to observe a patient walk down a ramp or a volleyball player execute a serve? What strategies can an elderly person or a football lineman employ to maximize stability? Why are some individuals unable to float?

Perusing the objectives at the beginning of each chapter of this book is a good way to highlight the scope of biomechanical topics to be covered at the introductory level. For those planning careers that involve visual observation and analysis of human movement, knowledge of these topics will be invaluable.

PROBLEM-SOLVING APPROACH

Scientific research is usually aimed at providing a solution for a particular problem or answering a specific question. Even for the nonresearcher, however, the ability to solve problems is a practical necessity for functioning in modern society. The use of specific problems is also an effective approach for illustrating basic biomechanical concepts.

Quantitative versus Qualitative Problems

quantitative
involving the use of numbers

qualitative
involving nonnumeric description of quality

Analysis of human movement may be either quantitative or qualitative. *Quantitative* implies that numbers are involved, and *qualitative* refers to a description of quality without the use of numbers. After watching the performance of a standing long jump, an observer might qualitatively state, "That was a very good jump." Another observer might quantitatively announce that the same jump was 2.1 m in length. Other examples of qualitative and quantitative descriptors are displayed in Figures 1-5 and 1-6.

It is important to recognize that *qualitative* does not mean *general*. Qualitative descriptions may be general, but they may also be extremely detailed. It can be stated qualitatively and generally, for example, that a man is walking down the street. It might also be stated that the same man is walking very slowly, appears to be leaning to the left, and is bearing weight on his right leg for as short a time as possible. The second description is entirely qualitative but provides a more detailed picture of the movement.

Both qualitative and quantitative descriptions play important roles in the biomechanical analysis of human movement. Biomechanical researchers rely heavily on quantitative techniques in attempting to answer

FIGURE 1-5

Examples of qualitative and quantitative descriptors.

FIGURE 1-6

Quantitatively, the robot missed the coffee cup by 15 cm. Qualitatively, it malfunctioned.

specific questions related to the mechanics of living organisms. Clinicians, coaches, and teachers of physical activities regularly employ qualitative observations of their patients, athletes, or students to formulate opinions or give advice.

Solving Qualitative Problems

Qualitative problems commonly arise during daily activities. Questions such as what clothes to wear, whether to major in botany or English, and whether to study or watch television are all problems in the sense that they are uncertainties that may require resolution. Thus, a large portion of our daily lives is devoted to the solution of problems.

Analyzing human movement, whether to identify a gait anomaly or to refine a technique, is essentially a process of problem solving. Whether the analysis is qualitative or quantitative, this involves identifying, then studying or analyzing, and finally answering a question or problem of interest.

To effectively analyze a movement, it is essential first to formulate one or more questions regarding the movement. Depending on the specific purpose of the analysis, the questions to be framed may be general or specific. General questions, for example, might include the following:

1. Is the movement being performed with adequate (or optimal) force?
2. Is the movement being performed through an appropriate range of motion?
3. Is the sequencing of body movements appropriate (or optimal) for execution of the skill?
4. Why does this elderly woman have a tendency to fall?
5. Why is this shot putter not getting more distance?

More specific questions might include these:

1. Is there excessive pronation taking place during the stance phase of gait?
2. Is release of the ball taking place at the instant of full elbow extension?
3. Does selective strengthening of the vastus medialis obliquus alleviate mistracking of the patella for this person?

Coaches rely heavily on qualitative observations of athletes' performances in formulating advice about technique. Photo courtesy of Ken Karp for MMH.

Once one or more questions have been identified, the next step in analyzing a human movement is to collect data. The form of data most commonly collected by teachers, therapists, and coaches is qualitative visual observation data. That is, the movement analyst carefully observes the movement being performed and makes either written or mental notes. To acquire the best observational data possible, it is useful to plan ahead as to the optimal distance(s) and perspective(s) from which to make the observations. These and other important considerations for qualitatively analyzing human movement are discussed in detail in Chapter 2.

Formal versus Informal Problems

When confronted with a stated problem taken from an area of mathematics or science, many individuals believe they are not capable of finding a solution. Clearly, a stated math problem is different from a problem such as what to wear to a particular social gathering. In some ways, however, the informal type of problem is the more difficult one to solve. According to Wickelgren (43), a formal problem (such as a stated math problem) is characterized by three discrete components:

1. A set of given information
2. A particular goal, answer, or desired finding
3. A set of operations or processes that can be used to arrive at the answer from the given information

In dealing with informal problems, however, individuals may find the given information, the processes to be used, and even the goal itself to be unclear or not readily identifiable.

Solving Formal Quantitative Problems

Formal problems are effective vehicles for translating nebulous concepts into well-defined, specific principles that can be readily understood and applied in the analysis of human motion. People who believe themselves incapable of solving formal stated problems do not recognize that, to a large extent, problem-solving skills can be learned. Entire books on problem-solving approaches and techniques are available. However, most students are not exposed to coursework involving general strategies of the problem-solving process. A simple procedure for approaching and solving problems involves 11 sequential steps:

1. Read the problem *carefully*. It may be necessary to read the problem several times before proceeding to the next step. Only when you clearly understand the information given and the question(s) to be answered should you undertake step 2.
2. Write down the given information in list form. It is acceptable to use symbols (such as v for velocity) to represent physical quantities if the symbols are meaningful.
3. Write down what is wanted or what is to be determined, using list form if more than one quantity is to be solved for.
4. Draw a diagram representing the problem situation, clearly indicating all known quantities and representing those to be identified with question marks. (Although certain types of problems may not easily be represented diagrammatically, it is critically important to carry out this step whenever possible to accurately visualize the problem situation.)
5. Identify and write down the relationships or formulas that might be useful in solving the problem. (More than one formula may be useful and/or necessary.)

Summary of Steps for Solving Formal Problems

1. Read the problem carefully.
2. List the given information.
3. List the desired (unknown) information for which you are to solve.
4. Draw a diagram of the problem situation showing the known and unknown information.
5. Write down formulas that may be of use.
6. Identify the formula to use.
7. If necessary, reread the problem statement to determine whether any additional needed information can be inferred.
8. Carefully substitute the given information into the formula.
9. Solve the equation to identify the unknown variable (the desired information).
10. Check that the answer is both reasonable and complete.
11. Clearly box in the answer.

FIGURE 1-7

Using the systematic process helps simplify problem solving.

6. From the formulas that you wrote down in step 5, select the formula(s) containing both given variables (from step 2) and the variables that are desired unknowns (from step 3). If a formula contains only one unknown variable that is the variable to be determined, skip step 7 and proceed directly to step 8.
7. If you cannot identify a workable formula (in more difficult problems), certain essential information was probably not specifically stated but can be determined by inference and by further thought on and analysis of the given information. If this occurs, it may be necessary to repeat step 1 and review the pertinent information relating to the problem presented in the text.
8. Once you have identified the appropriate formula(s), write the formula(s) and carefully substitute the known quantities given in the problem for the variable symbols.
9. Using the simple algebraic techniques reviewed in Appendix A, solve for the unknown variable by (a) rewriting the equation so that the unknown variable is isolated on one side of the equals sign and (b) reducing the numbers on the other side of the equation to a single quantity.
10. Do a commonsense check of the answer derived. Does it seem too small or too large? If so, recheck the calculations. Also check to ensure that *all* questions originally posed in the statement of the problem have been answered.
11. Clearly box in the answer and include the correct units of measurement.

inference
process of forming deductions from available information

Figure 1-7 provides a summary of this procedure for solving formal quantitative problems. These steps should be carefully studied, referred to, and applied in working the quantitative problems included at the end of each chapter. Sample Problem 1.1 illustrates the use of this procedure.

Units of Measurement

Providing the correct units of measurement associated with the answer to a quantitative problem is important. Clearly, an answer of 2 cm is quite different from an answer of 2 km. It is also important to recognize the units of measurement associated with particular physical quantities. Ordering 10 km of gasoline for a car when traveling in a foreign country would clearly not be appropriate.

The predominant system of measurement still used in the United States is the English system. The English system of weights and measures

English system
system of weights and measures originally developed in England and used in the United States today

SAMPLE PROBLEM 1.1

A baseball player hits a triple to deep center field. As he is approaching third base, he notices that the incoming throw to the catcher is wild, and he decides to break for home plate. The catcher retrieves the ball 10 m from the plate and runs back toward the plate at a speed of 5 m/s. As the catcher starts running, the base runner, who is traveling at a speed of 9 m/s, is 15 m from the plate. Given that time = distance/speed, who will reach the plate first?

Solution

Step 1   Read the problem carefully.

Step 2   Write down the given information:

$$\text{base runner's speed} = 9 \text{ m/s}$$
$$\text{catcher's speed} = 5 \text{ m/s}$$
$$\text{distance of base runner from plate} = 15 \text{ m}$$
$$\text{distance of catcher from plate} = 10 \text{ m}$$

Step 3   Write down the variable to be identified: Find which player reaches home plate in the shortest time.

Step 4   Draw a diagram of the problem situation.

Step 5   Write down formulas of use:

$$\text{time} = \text{distance/speed}$$

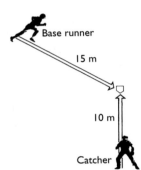

Base runner

15 m

10 m

Catcher

Step 6   Identify the formula to be used: It may be assumed that the formula provided is appropriate because no other information relevant to the solution has been presented.

Step 7   Reread the problem if all necessary information is not available: It may be determined that all information appears to be available.

Step 8   Substitute the given information into the formula:

$$\text{time} = \frac{\text{distance}}{\text{speed}}$$

Catcher:

$$\text{time} = \frac{10 \text{ m}}{5 \text{ m/s}}$$

Base runner:

$$\text{time} = \frac{15 \text{ m}}{9 \text{ m/s}}$$

Step 9   Solve the equations:
Catcher:

$$\text{time} = \frac{10 \text{ m}}{5 \text{ m/s}}$$
$$\text{time} = 2 \text{ s}$$

Base runner:

$$\text{time} = \frac{15 \text{ m}}{9 \text{ m/s}}$$
$$\text{time} = 1.67 \text{ s}$$

Step 10   Check that the answer is both reasonable and complete.

Step 11   Box in the answer:

The base runner arrives at home plate first, by 0.33 s.

arose over the course of several centuries primarily for purposes of commerce and land parceling. Specific units came largely from royal decrees. For example, a yard was originally defined as the distance from the end of the nose of King Henry I to the thumb of his extended arm. The English system of measurement displays little logic. There are 12 inches to the foot, 3 feet to the yard, 5280 feet to the mile, 16 ounces to the pound, and 2000 pounds to the ton.

The system of measurement that is presently used by every major country in the world except the United States is Le Système International d'Unites (the International System of Units), which is commonly known as the S.I. or the metric system. The metric system originated as the result of a request of King Louis XVI to the French Academy of Sciences in the 1790s. Although the system fell briefly from favor in France, it was readopted in 1837. In 1875, the Treaty of the Meter was signed by 17 countries agreeing to adopt the metric system.

metric system
system of weights and measures used internationally in scientific applications and adopted for daily use by every major country except the United States

Since that time the metric system has enjoyed worldwide popularity for several reasons. First, it entails only four base units—the meter, of length; the kilogram, of mass; the second, of time; and the degree Kelvin, of temperature. Second, the base units are precisely defined, reproducible quantities that are independent of factors such as gravitational force. Third, all units excepting those for time relate by factors of 10, in contrast to the numerous conversion factors necessary in converting English units of measurement. Last, the system is used internationally.

For these reasons, as well as the fact that the metric system is used almost exclusively by the scientific community, it is the system used in this book. For those who are not familiar with the metric system, it is useful to be able to recognize the approximate English system equivalents of metric quantities. Two conversion factors that are particularly valuable are 2.54 cm for every inch and approximately 4.45 N for every pound. All of the relevant units of measurement in both systems and common English-metric conversion factors are presented in Appendix C.

SUMMARY

Biomechanics is a multidisciplinary science involving the application of mechanical principles in the study of the structure and function of living organisms. Because biomechanists come from different academic backgrounds and professional fields, biomechanical research addresses a spectrum of problems and questions.

Basic knowledge of biomechanics is essential for competent professional analysts of human movement, including physical education teachers, physical therapists, physicians, coaches, personal trainers, and exercise instructors. The structured approach presented in this book is designed to facilitate the identification, analysis, and solution of problems or questions related to human movement.

INTRODUCTORY PROBLEMS

1. Locate and read three articles from the scientific literature that report the results of biomechanical investigations. (The *Journal of Biomechanics*, the *Journal of Applied Biomechanics*, and *Medicine and Science in Sports and Exercise* are possible sources.) Write a one-page summary of each article, and identify whether the investigation involved statics or dynamics and kinetics or kinematics.

2. List 8–10 websites that are related to biomechanics, and write a paragraph describing each site.
3. Write a brief discussion about how knowledge of biomechanics may be useful in your intended profession or career.
4. Choose three jobs or professions, and write a discussion about the ways in which each involves quantitative and qualitative work.
5. Write a summary list of the problem-solving steps identified in the chapter, using your own words.
6. Write a description of one informal problem and one formal problem.
7. Step by step, show how to arrive at a solution to one of the problems you described in Problem 6.
8. Solve for x in each of the equations below. Refer to Appendix A for help if necessary.

   a. $x = 5^3$
   b. $7 + 8 = x/3$
   c. $4 \times 3^2 = x \times 8$
   d. $-15/3 = x + 1$

   e. $x^2 = 27 + 35$
   f. $x = \sqrt{79}$
   g. $x + 3 = \sqrt{38}$

   h. $7 \times 5 = -40 + x$
   i. $3^3 = x/2$
   j. $15 - 28 = x \times 2$

   (Answers: a. 125; b. 45; c. 4.5; d. −6; e. 7.9; f. 8.9; g. 3.2; h. 75; i. 54; j. −6.5)

9. Two schoolchildren race across a playground for a ball. Tim starts running at a distance of 15 m from the ball, and Jan starts running at a distance of 12 m from the ball. If Tim's average speed is 4.2 m/s and Jan's average speed is 4.0 m/s, which child will reach the ball first? Show how you arrived at your answer. (See Sample Problem 1.1.) (Answer: Jan reaches the ball first.)
10. A 0.5 kg ball is kicked with a force of 40 N. What is the resulting acceleration of the ball? (Answer: 80 m/s^2)

ADDITIONAL PROBLEMS

1. Select a specific movement or sport skill of interest, and read two or three articles from the scientific literature that report the results of biomechanical investigations related to the topic. Write a short paper that integrates the information from your sources into a scientifically based description of your chosen movement.
2. When attempting to balance your checkbook, you discover that your figures show a different balance in your account than was calculated by the bank. List an ordered, logical set of procedures that you may use to discover the error. You may use list, outline, or block diagram format.
3. Sarah goes to the grocery store and spends half of her money. On the way home, she stops for an ice cream cone that costs $0.78. Then she stops and spends one-fourth of her remaining money to settle a $5.50 bill at the dry cleaners. How much money did Sarah have originally? (Answer: $45.56)
4. Wendell invests $10,000 in a stock portfolio made up of Petroleum Special at $30 per share, Newshoe at $12 per share, and Beans & Sprouts at $2.50 per share. He places 60% of the money in P.S., 30% in N, and 10% in B & S. With market values changing (P.S. down $3.12, N up 80%, and B & S up $0.20), what is his portfolio worth six months later? (Answer: $11,856)
5. The hypotenuse of right triangle ABC (shown here) is 4 cm long. What are the lengths of the other two sides? (Answer: $A = 2$ cm; $B = 3.5$ cm)
6. In triangle DEF, side E is 4 cm long and side F is 7 cm long. If the angle between sides E and F is 50 degrees, how long is side D? (Answer: 5.4 cm)

7. An orienteer runs 300 m north and then 400 m to the southeast (at a 45° angle to north). If he has run at a constant speed, how far away is he from the starting position? (Answer: 283.4 m)

8. John is out for his daily noontime run. He runs 2 km west, then 2 km south, and then runs on a path that takes him directly back to the place he started at.
   a. How far did John run?
   b. If he has run at an average speed of 4 m/s, how long did the entire run take?
   (Answers: a. 6.83 km; b. 28.5 min)

9. John and Al are in a 15 km race. John averages 4.4 m/s during the first half of the race and then runs at a speed of 4.2 m/s until the last 200 m, which he covers at 4.5 m/s. At what average speed must Al run to beat John? (Answer: > 4.3 m/s)

10. A sailboat heads north at 3 m/s for 1 hour and then tacks back to the southeast (at 45° to north) at 2 m/s for 45 minutes.
    a. How far has the boat sailed?
    b. How far is it from its starting location?
    (Answers: a. 16.2 km; b. 8.0 km)

LABORATORY EXPERIENCES

1. Working in a group of 3–5 students, choose three human movements or motor skills with which you are all familiar. (A vertical jump is an example.) For each movement, list at least three general questions and three specific questions that an analyst might choose to answer.

Movement/Skill 1: _____

General Questions

1. _____

   _____

2. _____

   _____

3. _____

   _____

Specific Questions

1. _____

   _____

2. _____

   _____

3. _____

   _____

Movement/Skill 2: _____

General Questions

1. _____

   _____

2. _____

   _____

3. _____

   _____

Specific Questions

1. _____

   _____

2. _____
   _____

3. _____
   _____

Movement/Skill 3: _____

General Questions

1. _____
   _____

2. _____
   _____

3. _____
   _____

Specific Questions

1. _____
   _____

2. _____
   _____

3. _____
   _____

2. Working in a group of 3–5 students, choose a human movement or motor skill with which you are all familiar, and have two members of the group simultaneously perform the movement several times as the group observes. Based on your comparative observations, list any differences and similarities that you can detect. Which of these are of potential importance and which are more a matter of personal style?

Movement Differences	**Important? (Y/N)**
_____	_____
_____	_____
_____	_____
_____	_____
_____	_____
_____	_____

Movement Similarities   **Important? (Y/N)**

_____   _____

_____   _____

_____   _____

_____   _____

_____   _____

_____   _____

3. Working in a group of 3–5 students, view a previously taken video or film of a human movement or motor skill performance. After viewing the movement several times, list at least three general questions and three specific questions that an analyst might choose to answer regarding the movement.

General Questions

1. _____

   _____

2. _____

   _____

3. _____

   _____

Specific Questions

1. _____

   _____

2. _____

   _____

3. _____

   _____

4. Having completed Laboratory Experiences 1–3, discuss in your group the relative advantages and disadvantages of each of the three exercises in terms of your ability to formulate meaningful questions.
5. Have one member of your group perform several trials of walking as the group observes from front, side, and rear views. The subject may walk either on a treadmill or across the floor. What observations can be made about the subject's gait from each view that are not visible or apparent from the other views?

Front View Observations

_____

_____

_____

_____

_____

Side View Observations

_____

_____

_____

_____

_____

Rear View Observations

_____

_____

_____

_____

_____

REFERENCES

1. Albright JP, Saterbak A, and Stokes J: Use of knee braces in sport. Current recommendations, *Sports Med* 20:281, 1995.
2. Block RM: Figure skating injuries, *Phys Med Rehabil Clin N Am* 10:177, 1999.
3. Casillas JM, Dulieu V, Cohen M, Marcer I, and Didier JP: Bioenergetic comparison of a new energy-storing foot and SACH foot in traumatic below-knee vascular amputations, *Arch Phys Med Rehabil* 76:39, 1995.
4. Chaffin DB: Primary prevention of low back pain through the application of biomechanics in manual materials handling tasks, *G Ital Med Lav Ergon* 27:40, 2005.
5. Chang YH, Hamerski CM, and Kram R: Applied horizontal force increases impact loading in reduced-gravity running, *J Biomech* 34:679, 2001.
6. Dapena J and Chung CS: Vertical and radial motions of the body during the take-off phase of high jumping, *Med Sci Sports Exerc* 20:290, 1988.
7. Datta D, Heller B, and Howitt J: A comparative evaluation of oxygen consumption and gait pattern in amputees using Intelligent Prostheses and conventionally damped knee swing-phase control, *Clin Rehabil* 19:398, 2005.
8. Davis BL, Cavanagh PR, Sommer HJ 3rd, and Wu, G: Ground reaction forces during locomotion in simulated microgravity, *Aviat Space Environ Med* 67:235, 1996.
9. De Koning JJ, Houdijk H, de Groot G, and Bobbert MF: From biomechanical theory to application in top sports: the Klapskate story, *J Biomech* 33:1225, 2000.
10. DeVita P, Torry M, Glover KL, and Speroni DL: A functional knee brace alters joint torque and power patterns during walking and running, *J Biomech* 29:583, 1996.
11. Doty SB: Space flight and bone formation, *Materwiss Werksttech* 35:951, 2004.
12. Dubravcic-Simunjak S, Pecina M, Kuipers H, Moran J, and Haspl M: The incidence of injuries in elite junior figure skaters, *Am J Sports Med* 31:511, 2003.
13. Frost HM: Osteoporosis: a rationale for further definitions? *Calcif Tissue Int* 62:89, 1998.
14. Grabiner MD, Donovan S, Bareither ML, Marone JR, Hamstra-Wright K, Gatts S, and Troy KL: Trunk kinematics and fall risk of older adults: translating biomechanical results to the clinic, *J Electromyogr Kinesiol* 18:197, 2007.
15. Greenblatt D: Treatment of postmenopausal osteoporosis, *Pharmacotherapy* 25:574, 2005.
16. Hay JG and Nohara H: The techniques used by elite long jumpers in preparation for take-off, *J Biomech* 23:229, 1990.
17. Houdijk H, de Koning JJ, de Groot G, Bobbert MF, and van Ingen Schenau GJ: Push-off mechanics in speed skating with conventional skates and klapskates, *Med Sci Sprt Exerc* 32:635, 2000.
18. Hubbard M, de Mestre NJ, and Scott J: Dependence of release variables in the shot put, *J Biomech* 34:449, 2001.
19. Hume PA, Keogh K, and Reid D: The role of biomechanics in maximizing distance and accuracy of golf shots, *Sports Med* 35:429, 2005.
20. Johnston CA, Taunton JE, Lloyd-Smith DR, and McKenzie DC: Preventing running injuries. Practical approach for family doctors, *Can Fam Physician* 49:1101, 2003.
21. Kyröläinen H, Belli A, and Komi P: Biomechanical factors affecting running economy, *Med Sci Sports Exer* 33:1330, 2001.
22. Kyröläinen H, Kivela R, Koskinen S, McBride J, Andersen JL, Takala T, Sipila S, and Komi PV: Interrelationships between muscle structure, muscle strength, and running economy, *Med Sci Sports Exerc* 35:45, 2003.
23. Lips P: Epidemiology and predictors of fractures associated with osteoporosis, *Am J Med* 103:3S, 1997.
24. Machold W, Kwansy O, Gässler P, Kolonja A, Reddy B, Bauer E, and Lehr S: Risk of injury through snowboarding, *J Trauma* 48:1109, 2000.
25. Matsuo T, Escamilla RF, Fleisig GS, Barrentine SW, and Andrews JR: Comparison of kinematic and temporal parameters between different pitch velocity groups, *J Appl Biomech* 17:1, 2001.

26. McCarthy ID: Fluid shifts due to microgravity and their effects on bone: a review of current knowledge, *Ann Biomed Eng* 33:95, 2005.
27. McGill SM: Evolving ergonomics? *Ergonomics* 52:80, 2009.
28. Nelson RC: Biomechanics: past and present. In Cooper JM and Haven B, eds: *Proceedings of the Biomechanics Symposium,* Bloomington, Ind, 1980.
29. Paluska SA and McKeag DB: Knee braces: current evidence and clinical recommendations for their use, *Am Fam Physician* 61:411, 2000.
30. Peterman MM, Hamel AJ, Cavanagh PR, Paizza SJ, and Shrakey NA: In vitro modeling of human tibial strains during exercise in micro-gravity, *J Biomech* 34:693, 2001.
31. Pietrosimone BG, Grindstaff TL, Linens SW, Uczekaj E, Hertel J: A systematic review of prophylactic braces in the prevention of knee ligament injuries in collegiate football players, *J Athl Train* 43:409, 2008.
32. Pigozzi F, Santori N, Di Salvo V, Parisi A, and Di-Luigi L: Snowboard traumatology: an epidemiological study, *Orthopedics* 20:505, 1997.
33. Prodromos CC, Han Y, Rogowski J, Joyce B, Shi K: A meta-analysis of the incidence of anterior cruciate ligament tears as a function of gender, sport, and a knee injury-reduction regimen, *Arthroscopy* 23:1320, 2007.
34. Recker RR: Osteoporosis, *Contemp Nutr* 8:1, 1983.
35. Reilly SM, McElroy EJ, and Biknevicius AR: Posture, gait and the ecological relevance of locomotor costs and energy-saving mechanisms in tetrapods, *Zoology (Jena)* 110:271, 2007.
36. Rishiraj N, Taunton JE, Lloyd-Smith R, Woollard R, Regan W, and Clement DB: The potential role of prophylactic/functional knee bracing in preventing knee ligament injury, *Sports Med* 39:937, 2009.
37. Robinovitch SN, Hsiao ET, Sandler R, Cortez J, Liu Q, and Paiement GD: Prevention of falls and fall-related fractures through biomechanics, *Exer Sprt Sci Rev* 28:74, 2000.
38. Ruby D: Biomechanics—how computers extend athletic performance to the body's far limits, *Popular Science* p 58, Jan 1982.
39. Smith SM, Wastney ME, O'Brien KO, Morukov BV, Larina IM, Abrams SA, Davis-Street JE, Oganov V, and Shackelford LC: Bone markers, calcium metabolism, and calcium kinetics during extended-duration space flight on the mir space station, *J Bone Miner Res* 20:208, 2004.
40. Stodden DF, Fleisig GS, McLean SP, Lyman SL, and Andrews JR: Relationship of pelvis and upper torso kinematics to pitched baseball velocity, *J Appl Biomech* 17:164, 2001.
41. Versluys R, Beyl P, Van Damme M, Desomer A, Van Ham R, and Lefeber D: Prosthetic feet: state-of-the-art review and the importance of mimicking human ankle-foot biomechanics, *Disabil Rehabil Assist Technol* 4:65, 2009.
42. Virmavirta M, Kivekäs J, and Komi P: Take-off aerodynamics in ski jumping, *J Biomech* 34:465, 2001.
43. Wickelgren WA: *How to solve problems,* San Francisco, 1974, WH Freeman.
44. Yoshimitsu K, Shiva N, Matsuse H, Takano Y, Matsugaki T, Inada T, Tagawa Y, and Nagata K: Development of a training method for weightless environment using both electrical stimulation and voluntary muscle contraction, *Tohoku J Exp Med* 220:83, 2010.

ANNOTATED READINGS

Chaffin DB, Andersson GBJ, and Martin BJ: *Occupational biomechanics* (3rd ed.), New York, 2006, John Wiley & Sons.
  Serves as a comprehensive text on the field of occupational biomechanics.
Chapman AE: *Biomechanical analysis of fundamental human movements,* Champaign, IL, 2008, Human Kinetics.
  Analyzes common fundamental movements such as walking, running, jumping, throwing, climbing, etc.

Winter DA: *Biomechanics and motor control of human movement* (4th ed.), New York, 2010, John Wiley & Sons.

    Serves as an advanced textbook for the study of human biomechanics.

Zeitz P: *The art and craft of problem solving* (2nd ed.), New York, 2007, John Wiley & Sons.

    Provides general strategies, as well as specific tools and techniques for solving quantitative problems.

RELATED WEBSITES

American College of Sports Medicine—Biomechanics Interest Group
http://www.acsm.org

    Provides a link to the American College of Sports Medicine Member Service Center, which links to the ACSM Interest Groups, including the Biomechanics Interest Group.

American Society of Biomechanics
http://asb-biomech.org/

    Home page of the American Society of Biomechanics. Provides information about the organization, conference abstracts, and a list of graduate programs in biomechanics.

The Biomch-L Newsgroup
http://www.biomch-l.org/

    Provides information about an e-mail discussion group for biomechanics and human/animal movement science.

Biomechanics Classes on the Web
http://www.uoregon.edu/~karduna/biomechanics/

    Contains links to over 100 biomechanics classes with web-based instructional components.

Biomechanics Yellow Pages
http://www.sciencecentral.com/site/433521

    Provides information on technology used in biomechanics-related work and includes a number of downloadable video clips.

Biomechanics World Wide
http://www.uni-due.de/~qpd800/WSITECOPY.html

    A comprehensive site with links to other websites for a wide spectrum of topics related to biomechanics.

International Society of Biomechanics
http://www.isbweb.org/

    Home page of the International Society of Biomechanics (ISB). Provides information on ISB, biomechanical software and data, and pointers to other sources of biomechanics-related information.

KEY TERMS

anthropometric	related to the dimensions and weights of body segments
biomechanics	application of mechanical principles in the study of living organisms
carpal tunnel syndrome	overuse condition caused by compression of the median nerve in the carpal tunnel and involving numbness, tingling, and pain in the hands
dynamics	branch of mechanics dealing with systems subject to acceleration
English system	system of weights and measures originally developed in England and used in the United States today
inference	process of forming deductions from available information
kinematics	study of the description of motion, including considerations of space and time
kinesiology	study of human movement

kinetics	study of the action of forces
mechanics	branch of physics that analyzes the actions of forces on particles and mechanical systems
metric system	system of weights and measures used internationally in scientific applications and adopted for daily use by every major country except the United States
qualitative	involving nonnumeric description of quality
quantitative	involving the use of numbers
sports medicine	clinical and scientific aspects of sports and exercise
statics	branch of mechanics dealing with systems in a constant state of motion

Kinematic Concepts for Analyzing Human Motion

2

After completing this chapter, you will be able to:

Provide examples of linear, angular, and general forms of motion.

Identify and describe the reference positions, planes, and axes associated with the human body.

Define and appropriately use directional terms and joint movement terminology.

Explain how to plan and conduct an effective qualitative human movement analysis.

Identify and describe the uses of available instrumentation for measuring kinematic quantities.

ONLINE LEARNING CENTER RESOURCES

www.mhhe.com/hall6e

Log on to our Online Learning Center (OLC) for access to these additional resources:

- Online Lab Manual
- Flashcards with definitions of chapter key terms
- Chapter objectives
- Chapter lecture PowerPoint presentation
- Self-scoring chapter quiz
- Additional chapter resources
- Web links for study and exploration of chapter-related topics

Is it best to observe walking gait from a side view, front view, or back view? From what distance can a coach best observe a pitcher's throwing style? What are the advantages and disadvantages of analyzing a movement captured on video? To the untrained observer, there may be no differences in the forms displayed by an elite hurdler and a novice hurdler or in the functioning of a normal knee and an injured, partially rehabilitated knee. What skills are necessary and what procedures are used for effective analysis of human movement kinematics?

One of the most important steps in learning a new subject is mastering the associated terminology. Likewise, learning a general analysis protocol that can be adapted to specific questions or problems within a field of study is invaluable. In this chapter, human movement terminology is introduced, and the problem-solving approach is adapted to provide a template for qualitative solving of human movement analysis problems.

FORMS OF MOTION

general motion
involving translation and rotation simultaneously

Most human movement is general motion, a complex combination of linear and angular motion components. Since linear and angular motion are "pure" forms of motion, it is sometimes useful to break complex movements down into their linear and angular components when performing an analysis.

linear
along a line that may be straight or curved, with all parts of the body moving in the same direction at the same speed

Linear Motion

angular
involving rotation around a central line or point

Pure linear motion involves uniform motion of the system of interest, with all system parts moving in the same direction at the same speed. Linear motion is also referred to as translatory motion, or translation. When a body experiences translation, it moves as a unit, and portions of the body do not move relative to each other. For example, a sleeping passenger on a smooth airplane flight is being translated through the air. If the passenger awakens and reaches for a magazine, however, pure translation is no longer occurring because the position of the arm relative to the body has changed.

translation
linear motion

Linear motion may also be thought of as motion along a line. If the line is straight, the motion is rectilinear; if the line is curved, the motion is curvilinear. A motorcyclist maintaining a motionless posture as the bike moves along a straight path is moving rectilinearly. If the motorcyclist jumps the bike and the frame of the bike does not rotate, both rider and bike (with the exception of the spinning wheels) are moving curvilinearly while airborne. Likewise, a Nordic skier coasting in a locked static position down a short hill is in rectilinear motion. If the skier jumps over a gully with all body parts moving in the same direction at the same speed along a curved path, the motion is curvilinear. When a motorcyclist or skier goes over the crest of a hill, the motion is *not* linear, because the top of the body is moving at a greater speed than lower body parts. Figure 2-1 displays a gymnast in rectilinear, curvilinear, and rotational motion.

rectilinear
along a straight line

curvilinear
along a curved line

Angular Motion

axis of rotation
imaginary line perpendicular to the plane of rotation and passing through the center of rotation

Angular motion is rotation around a central imaginary line known as the axis of rotation, which is oriented perpendicular to the plane in which the rotation occurs. When a gymnast performs a giant circle on

Rotation of a body segment at a joint occurs around an imaginary line known as the axis of rotation that passes through the joint center. Photo © Design Pics/PunchStock.

a bar, the entire body rotates, with the axis of rotation passing through the center of the bar. When a springboard diver executes a somersault in midair, the entire body is again rotating, this time around an imaginary axis of rotation that moves along with the body. Almost all volitional human movement involves rotation of a body segment around an imaginary axis of rotation that passes through the center of the joint to which the segment attaches. When angular motion or rotation occurs, portions of the body in motion are constantly moving relative to other portions of the body.

General Motion

When translation and rotation are combined, the resulting movement is general motion. A football kicked end over end translates through the air as it simultaneously rotates around a central axis (Figure 2-2). A runner is translated along by angular movements of body segments at the hip, knee, and ankle. Human movement usually consists of general motion rather than pure linear or angular motion.

• *Most human movement activities are categorized as general motion.*

Mechanical Systems

Before determining the nature of a movement, the mechanical system of interest must be defined. In many circumstances, the entire human body is chosen as the system to be analyzed. In other circumstances, however, the system might be defined as the right arm or perhaps even a ball being projected by the right arm. When an overhand throw is executed, the body as a whole displays general motion, the motion of the throwing arm is primarily angular, and the motion of the released ball is linear. The mechanical system to be analyzed is chosen by the movement analyst according to the focus of interest.

system
object or group of objects chosen by the analyst for study

FIGURE 2-1

Examples of rectilinear, curvilinear, and rotational motion.

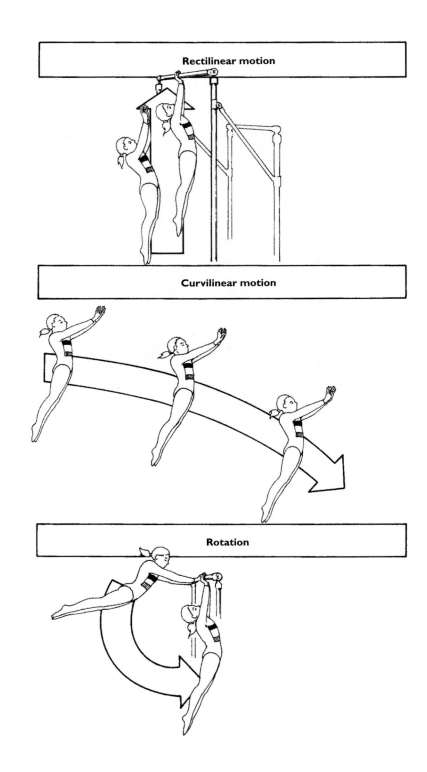

Rectilinear motion

Curvilinear motion

Rotation

STANDARD REFERENCE TERMINOLOGY

Communicating specific information about human movement requires specialized terminology that precisely identifies body positions and directions.

Anatomical Reference Position

Anatomical reference position is an erect standing position with the feet slightly separated and the arms hanging relaxed at the sides, with the

anatomical reference position

erect standing position with all body parts, including the palms of the hands, facing forward; considered the starting position for body segment movements

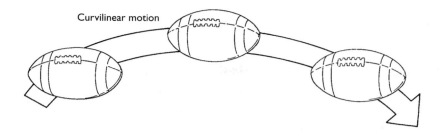

Curvilinear motion

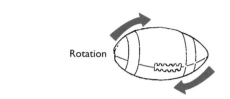

Rotation

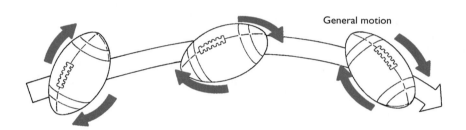

General motion

FIGURE 2-2

General motion is a
combination of linear and
angular motion.

palms of the hands facing forward. It is not a natural standing position,
but is the body orientation conventionally used as the reference position
or starting place when movement terms are defined.

Directional Terms

In describing the relationship of body parts or the location of an external
object with respect to the body, the use of directional terms is necessary.
The following are commonly used directional terms:

Superior: closer to the head (In zoology, the synonymous term is *cranial.*)

Inferior: farther away from the head (In zoology, the synonymous term is
caudal.)

Anterior: toward the front of the body (In zoology, the synonymous term
is *ventral.*)

Posterior: toward the back of the body (In zoology, the synonymous term
is *dorsal.*)

Medial: toward the midline of the body

Lateral: away from the midline of the body

Proximal: closer in proximity to the trunk (For example, the knee is proximal to the ankle.)

Distal: at a distance from the trunk (For example, the wrist is distal to
the elbow.)

Superficial: toward the surface of the body

Deep: inside the body and away from the body surface

Anatomical reference position.

●*Reference planes and axes are useful in describing gross body movements and in defining more specific movement terminology.*

cardinal planes
three imaginary perpendicular reference planes that divide the body in half by mass

sagittal plane
plane in which forward and backward movements of the body and body segments occur

frontal plane
plane in which lateral movements of the body and body segments occur

transverse plane
plane in which horizontal body and body segment movements occur when the body is in an erect standing position

●*Although most human movements are not strictly planar, the cardinal planes provide a useful way to describe movements that are primarily planar.*

mediolateral axis
imaginary line around which sagittal plane rotations occur

anteroposterior axis
imaginary line around which frontal plane rotations occur

longitudinal axis
imaginary line around which transverse plane rotations occur

All of these directional terms can be paired as antonyms—words having opposite meanings. Saying that the elbow is proximal to the wrist is as correct as saying that the wrist is distal to the elbow. Similarly, the nose is superior to the mouth and the mouth is inferior to the nose.

Anatomical Reference Planes

The three imaginary cardinal planes bisect the mass of the body in three dimensions. A *plane* is a two-dimensional surface with an orientation defined by the spatial coordinates of three discrete points not all contained in the same line. It may be thought of as an imaginary flat surface. The sagittal plane, also known as the anteroposterior (AP) plane, divides the body vertically into left and right halves, with each half containing the same mass. The frontal plane, also referred to as the coronal plane, splits the body vertically into front and back halves of equal mass. The horizontal or transverse plane separates the body into top and bottom halves of equal mass. For an individual standing in anatomical reference position, the three cardinal planes all intersect at a single point known as the body's center of mass or center of gravity (Figure 2-3). These imaginary reference planes exist only with respect to the human body. If a person turns at an angle to the right, the reference planes also turn at an angle to the right.

Although the entire body may move along or parallel to a cardinal plane, the movements of individual body segments may also be described as sagittal plane movements, frontal plane movements, and transverse plane movements. When this occurs, the movements being described are usually in a plane that is parallel to one of the cardinal planes. For example, movements that involve forward and backward motion are referred to as sagittal plane movements. When a forward roll is executed, the entire body moves parallel to the sagittal plane. During running in place, the motion of the arms and legs is generally forward and backward, although the planes of motion pass through the shoulder and hip joints rather than the center of the body. Marching, bowling, and cycling are all largely sagittal plane movements (Figure 2-4). Frontal plane movement is lateral (side-to-side) movement; an example of total-body frontal plane movement is the cartwheel. Jumping jacks, side stepping, and side kicks in soccer require frontal plane movement at certain body joints. Examples of total-body transverse plane movement include a twist executed by a diver, trampolinist, or airborne gymnast and a dancer's pirouette.

Although many of the movements conducted by the human body are not oriented sagittally, frontally, or transversely, or are not planar at all, the three major reference planes are still useful. Gross-body movements and specifically named movements that occur at joints are often described as primarily frontal, sagittal, or transverse plane movements.

Anatomical Reference Axes

When a segment of the human body moves, it rotates around an imaginary axis of rotation that passes through a joint to which it is attached. There are three reference axes for describing human motion, and each is oriented perpendicular to one of the three planes of motion. The mediolateral axis, also known as the frontal-horizontal axis, is perpendicular to the sagittal plane. Rotation in the frontal plane occurs around the anteroposterior axis, or sagittal-horizontal axis (Figure 2-5). Transverse plane rotation is around the longitudinal axis, or vertical axis. It is important to recognize that each of these three axes is always associated with the same single plane—the one to which the axis is perpendicular.

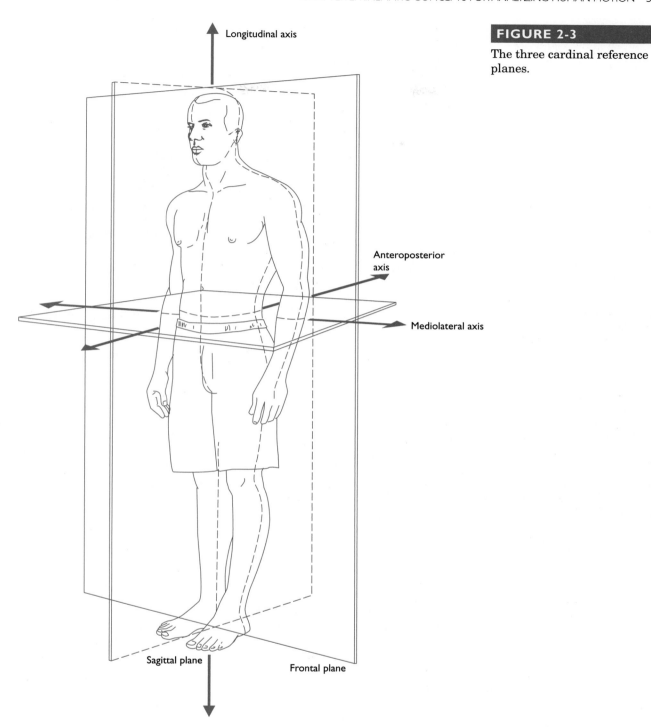

Longitudinal axis

Anteroposterior axis

Mediolateral axis

Sagittal plane

Frontal plane

FIGURE 2-3

The three cardinal reference planes.

FIGURE 2-4

Cycling requires sagittal plane movement of the legs.

Sagittal plane

FIGURE 2-5

For a jumping jack, the major axes of rotation are anteroposterior axes passing through the shoulders and hips.

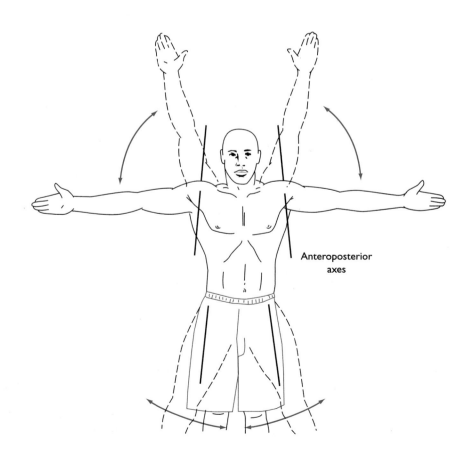

Anteroposterior axes

JOINT MOVEMENT TERMINOLOGY

When the human body is in anatomical reference position, all body segments are considered to be positioned at zero degrees. Rotation of a body segment away from anatomical position is named according to the direction of motion and is measured as the angle between the body segment's position and anatomical position.

Sagittal Plane Movements

From anatomical position, the three primary movements occurring in the sagittal plane are *flexion*, *extension*, and *hyperextension* (Figure 2-6). Flexion includes anteriorly directed sagittal plane rotations of the head, trunk, upper arm, forearm, hand, and hip, and posteriorly directed sagittal plane rotation of the lower leg. Extension is defined as the movement that returns a body segment to anatomical position from a position of flexion, and hyperextension is the rotation beyond anatomical position in the direction opposite the direction of flexion. If the arms or legs are internally or externally rotated from anatomical position, flexion, extension, and hyperextension at the knee and elbow may occur in a plane other than the sagittal.

Sagittal plane rotation at the ankle occurs both when the foot is moved relative to the lower leg and when the lower leg is moved relative to the foot. Motion bringing the top of the foot toward the lower leg is known as *dorsiflexion*, and the opposite motion, which can be visualized as "planting" the ball of the foot, is termed *plantar flexion* (Figure 2-7).

• *Sagittal plane movements include flexion, extension, and hyperextension, as well as dorsiflexion and plantar flexion.*

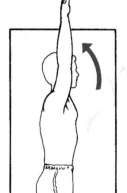
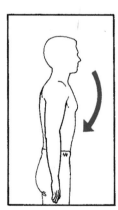
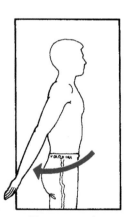

Flexion   **Extension**   **Hyperextension**

FIGURE 2-6

Sagittal plane movements at the shoulder.

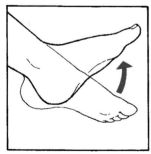
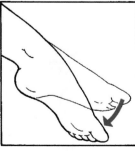

Dorsiflexion   **Plantar flexion**

FIGURE 2-7

Sagittal plane movements of the foot.

Frontal Plane Movements

•*Frontal plane movements include abduction and adduction, lateral flexion, elevation and depression, inversion and eversion, and radial and ulnar deviation.*

The major frontal plane rotational movements are *abduction* and *adduction*. Abduction (*abduct* meaning "to take away") moves a body segment away from the midline of the body; adduction (*add* meaning "to bring back") moves a body segment closer to the midline of the body (Figure 2-8).

Other frontal plane movements include sideways rotation of the trunk, which is termed right or left *lateral flexion* (Figure 2-9). *Elevation* and *depression* of the shoulder girdle refer to movement of the shoulder girdle in superior and inferior directions, respectively (Figure 2-10). Rotation of the hand at the wrist in the frontal plane toward the radius (thumb side) is referred to as *radial deviation*, and *ulnar deviation* is hand rotation toward the ulna (little finger side) (Figure 2-11).

Movements of the foot that occur largely in the frontal plane are eversion and inversion. Outward rotation of the sole of the foot is termed *eversion,* and inward rotation of the sole of the foot is called *inversion* (Figure 2-12). Abduction and adduction are also used to describe outward and inward rotation of the entire foot. *Pronation* and *supination* are often used to describe motion occurring at the subtalar joint. Pronation at the subtalar joint consists of a combination of eversion, abduction, and dorsiflexion, and supination involves inversion, adduction, and plantar flexion.

FIGURE 2-8

Frontal plane movements at the hip.

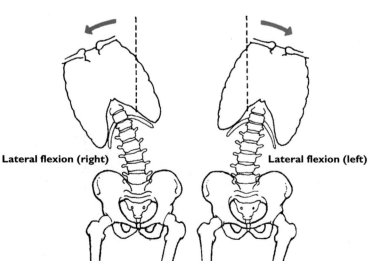

Abduction   **Adduction**

FIGURE 2-9

Frontal plane movements of the spinal column.

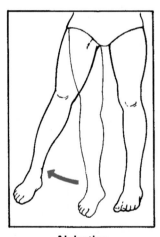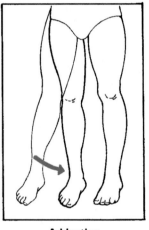

Lateral flexion (right)   **Lateral flexion (left)**

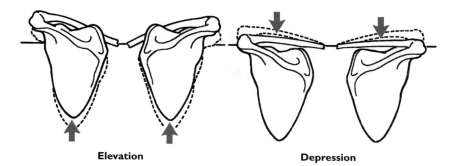

FIGURE 2-10

Frontal plane movements of the shoulder girdle.

Elevation   **Depression**

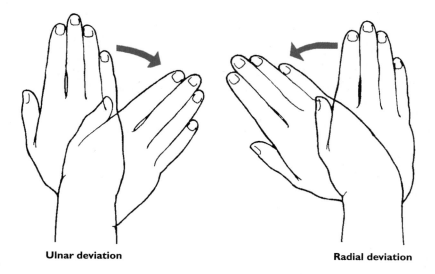

FIGURE 2-11

Frontal plane movements of the hand.

Ulnar deviation   **Radial deviation**

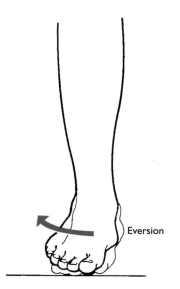
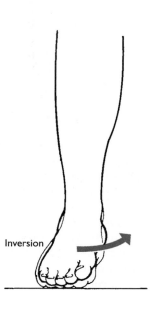

FIGURE 2-12

Frontal plane movements of the foot.

Eversion   Inversion

FIGURE 2-13

Transverse plane movements of the leg.

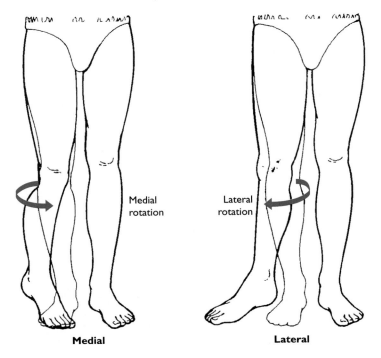

Medial rotation

Lateral rotation

Medial

Lateral

Transverse Plane Movements

• *Transverse plane movements include left and right rotation, medial and lateral rotation, supination and pronation, and horizontal abduction and adduction.*

Body movements in the transverse plane are rotational movements about a longitudinal axis. *Left rotation* and *right rotation* are used to describe transverse plane movements of the head, neck, and trunk. Rotation of an arm or leg as a unit in the transverse plane is called *medial rotation,* or internal rotation, when rotation is toward the midline of the body, and *lateral rotation,* or external rotation, when the rotation is away from the midline of the body (Figure 2-13).

Specific terms are used for rotational movements of the forearm. Outward and inward rotations of the forearm are respectively known as *supination* and *pronation* (Figure 2-14). In anatomical position the forearm is in a supinated position.

Although abduction and adduction are frontal plane movements, when the arm or thigh is flexed to a position, movement of these segments in the transverse plane from an anterior position to a lateral position is termed *horizontal abduction*, or horizontal extension (Figure 2-15). Movement in the transverse plane from a lateral to an anterior position is called *horizontal adduction*, or horizontal flexion.

Other Movements

Many movements of the body limbs take place in planes that are oriented diagonally to the three traditionally recognized cardinal planes. Because human movements are so complex, however, nominal identification of every plane of human movement is impractical.

One special case of general motion involving circular movement of a body segment is designated as *circumduction.* Tracing an imaginary circle in the air with a fingertip while the rest of the hand is stationary requires circumduction at the metacarpophalangeal joint (Figure 2-16). Circumduction combines flexion, extension, abduction, and adduction, resulting in a conical trajectory of the moving body segment.

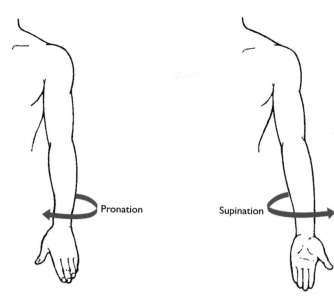

FIGURE 2-14

Transverse plane movements of the forearm.

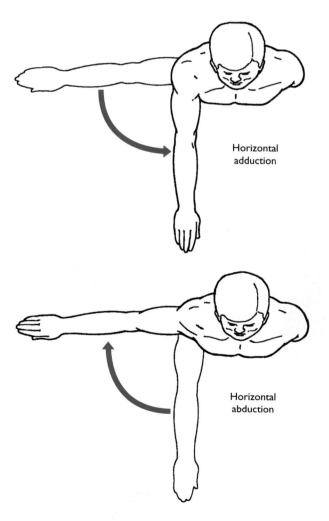

FIGURE 2-15

Transverse plane movements at the shoulder.

FIGURE 2-16

Circumduction of the index finger at the metacarpophalangeal joint.

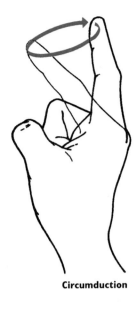

Circumduction

A tennis serve requires arm movement in a diagonal plane.

● *Qualitative analysis requires knowledge of the specific biomechanical purpose of the movement and the ability to detect the causes of errors.*

SPATIAL REFERENCE SYSTEMS

Whereas the three cardinal planes and their associated axes of rotation move along with the body, it is also often useful to make use of a fixed system of reference. When biomechanists quantitatively describe the movement of living organisms, they use a spatial reference system to standardize the measurements taken. The system most commonly used is a Cartesian coordinate system, in which units are measured in the directions of either two or three primary axes.

Movements that are primarily in a single direction, or planar, such as running, cycling, or jumping, can be analyzed using a two-dimensional Cartesian coordinate system (Figure 2-17). In two-dimensional Cartesian coordinate systems, points of interest are measured in units in the x, or horizontal, direction and in the y, or vertical, direction. When a biomechanist is analyzing the motion of the human body, the points of interest are usually the body's joints, which constitute the end points of the body segments. The location of each joint center can be measured with respect to the two axes and described as (x,y), where x is the number of horizontal units away from the y-axis and y is the number of vertical units away from the x-axis. These units can be measured in both positive and negative directions (Figure 2-18). When a movement of interest is three-dimensional, the analysis can be extended to the third dimension by adding a z-axis perpendicular to the x- and y-axes and measuring units away from the x,y plane in the z direction. With a two-dimensional coordinate system, the y-axis is normally vertical, and the x-axis horizontal. In the case of a three-dimensional coordinate system, it is usually the z-axis that is vertical, with the x- and y-axes representing the two horizontal directions.

QUALITATIVE ANALYSIS OF HUMAN MOVEMENT

A good command of the language associated with forms of motion, standard reference terminology, and joint movement terminology is essential for being able to describe a qualitative analysis of human movement. The ability to qualitatively assess human movement also requires both knowl-

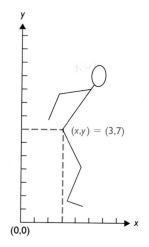

FIGURE 2-17

A Cartesian coordinate system showing the x and y coordinates of the hip.

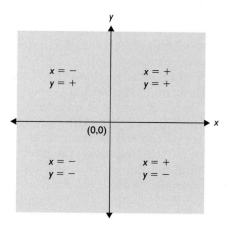

FIGURE 2-18

Coordinates can be both positive and negative in a Cartesian coordinate system.

edge of the movement characteristics desired and the ability to observe and analyze whether a given performance incorporates these characteristics. As introduced in Chapter 1, the word *qualitative* refers to a description of quality without the use of numbers. Visual observation is the most commonly used approach for qualitatively analyzing the mechanics of human movement. Based on information gained from watching an athlete perform a skill, a patient walk down a ramp, or a student attempt a novel task, coaches, clinicians, and teachers make judgments and recommendations on a daily basis. To be effective, however, a qualitative analysis cannot be conducted haphazardly, but must be carefully planned and conducted by an analyst with knowledge of the biomechanics of the movement.

Prerequisite Knowledge for a Qualitative Analysis

There are two main sources of information for the analyst diagnosing a motor skill. The first is the kinematics or technique exhibited by the performer, and the second is the performance outcome. Evaluating performance outcome is of limited value, since the root of optimal performance outcome is appropriate biomechanics.

To effectively analyze a motor skill, it is very helpful for the analyst to understand the specific purpose of the skill from a biomechanical perspective. The general goal of a volleyball player serving a ball is to legally project the ball over the net and into the opposite court. Specifically, this

Many jobs require conducting qualitative analyses of human movement daily. Photo courtesy of Digital Vision/Alamy.

•*Analysts should be able to distinguish the cause of a problem from symptoms of the problem or an unrelated movement idiosyncrasy.*

•*Experience in performing a motor skill does not necessarily translate to proficiency in analyzing the skill.*

requires a coordinated summation of forces produced by trunk rotation, shoulder extension, elbow extension, and forward translation of the total-body center of gravity, as well as contacting the ball at an appropriate height and angle. Whereas the ultimate purpose of a competitive sprint cyclist is to maximize speed while maintaining balance in order to cross the finish line first, biomechanically this requires factors such as maximizing perpendicular force production against the pedals and maintaining a low body profile to minimize air resistance.

Without knowledge of relevant biomechanical principles, analysts may have difficulty in identifying the factors that contribute to (or hinder) performance and may misinterpret the observations they make. More specifically, to effectively analyze a motor skill, the analyst must be able to identify the *cause* of a technique error, as opposed to a symptom of the error, or a performance idiosyncrasy. Inexperienced coaches of tennis or golf may focus on getting the performer to display an appropriate follow-through after hitting the ball. Inadequate follow-through, however, is merely a symptom of the underlying performance error, which may be failure to begin the stroke or swing with sufficient trunk rotation and backswing, or failure to swing the racquet or club with sufficient velocity. The ability to identify the cause of a performance error is dependent on an understanding of the biomechanics of the motor skill.

One potential source of knowledge about the biomechanics of a motor skill is experience in performing the skill. A person who performs a skill proficiently usually is better equipped to qualitatively analyze that skill than is a person less familiar with the skill. For example, advanced batters demonstrate greater perceptual decision making during a pitch than do intermediate batters, particularly when the pitch is a curve ball (4). In most cases, a high level of familiarity with the skill or movement being performed improves the analyst's ability to focus attention on the critical aspects of the event.

Direct experience in performing a motor skill, however, is not the only or necessarily the best way to acquire expertise in analyzing the skill. Skilled athletes often achieve success not because of the form or technique they display, but in spite of it! Furthermore, highly accomplished athletes do not always become the best coaches, and highly successful coaches may have had little or no participatory experience in the sports they coach.

The conscientious coach, teacher, or clinician typically uses several avenues to develop a knowledge base from which to evaluate a motor skill. One is to read available materials from textbooks, scientific journals, and lay (coaching) journals, despite the facts that not all movement patterns and skills have been researched and that some biomechanics literature is so esoteric that advanced training in biomechanics is required to understand it. However, when selecting reading material, it is important to distinguish between articles supported by research and those based primarily on opinion, as "commonsense" approaches to skill analyses may be flawed. There are also opportunities to interact directly with individuals who have expert knowledge of particular skills at conferences and workshops.

Planning a Qualitative Analysis

Even the simplest qualitative analysis may yield inadequate or faulty information if approached haphazardly. As the complexity of the skill and/or the level of desired analytical detail increases, so does the level of required planning.

The first step in any analysis is to identify the major question or questions of interest. Often, these questions have already been formulated by the analyst, or they serve as the original purpose for the observation. For example, has a post–knee surgery patient's gait returned to normal? Why is a volleyball player having difficulty hitting cross-court? What might be causing a secretary's wrist pain? Or simply, is a given skill being performed as effectively as possible? Having one or more particular questions or problems in mind helps to focus the analysis. Preparing a criteria sheet or a checklist prior to performing an analysis is a useful way to help focus attention on the critical elements of the movement being evaluated. Of course, the ability to identify appropriate analysis questions and formulate a checklist is dependent on the analyst's knowledge of the biomechanics of the movement. When an analyst is observing a skill that is less than familiar, it can be helpful to recall that many motor skills have commonalities. For example, serves in tennis and volleyball and the badminton overhead clear are all very similar to the overarm throw.

The analyst should next determine the optimal perspective(s) from which to view the movement. If the major movements are primarily planar, as with the legs during cycling or the pitching arm during a softball pitch, a single viewing perspective such as a side view or a rear view may be sufficient. If the movement occurs in more than one plane, as with the motions of the arms and legs during the breaststroke or the arm motion during a baseball batter's swing, the observer may need to view the movement from more than one perspective to see all critical aspects of interest. For example, a rear view, a side view, and a top view of a martial artist's kick all yield different information about the movement (Figure 2-19).

The analyst's viewing distance from the performer should also be selected thoughtfully (Figure 2-20). If the analyst wishes to observe subtalar pronation and supination in a patient walking on a treadmill, a close-up rear view of the lower legs and feet is necessary. Analyzing where a particular volleyball player moves on the court during a series of plays under rapidly changing game conditions is best accomplished from a reasonably distant, elevated position.

Another consideration is the number of trials or executions of the movement that should be observed in the course of formulating an analysis. A skilled athlete may display movement kinematics that deviate only slightly across performances, but a child learning to run may take no two steps alike. Basing an analysis on observation of a single performance is usually unwise. The greater the inconsistency in the performer's kinematics, the larger the number of observations that should be made.

Other factors that potentially influence the quality of observations of human movement are the performer's attire and the nature of the surrounding environment. When biomechanic researchers study the kinematics of a particular movement, the subjects typically wear minimal attire so that movements of body segments will not be obscured. Although there are many situations, such as instructional classes, competitive events, and team practices, for which this may not be practical, analysts should be aware that loose clothing can obscure subtle motions. Adequate lighting and a nondistracting background of contrasting color also improve the visibility of the observed movement.

A final consideration is whether to rely on visual observation alone or to use a video camera. As the speed of the movement of interest increases, it becomes progressively less practical to rely on visual observation. Consequently, even the most careful observer may miss important aspects of a rapidly executed movement. Video also enables the performer to view the movement, as well as allowing repeated viewing of the movement by analyst and performer, enabling performance feedback that can enhance the learning of a motor skill. Most playback units also enable slow-motion

A tennis player's eyes should follow the oncoming ball long enough to enable the player to contact the ball with the racquet.

•*Repeated observation of a motor skill is useful in helping the analyst to distinguish consistent performance errors from random errors.*

•*Use of a video camera provides both advantages and disadvantages to the movement analyst.*

FIGURE 2-19

Whereas skills that are
primarily planar may require
only one viewing perspective,
the movement analyst should
view multiplanar skills from
more than one direction.

**Primarily planar
skills**

**Multiplanar
skills**

FIGURE 2-20

The observation distance
between analyst and
performer should be selected
based on the specific
questions of interest.

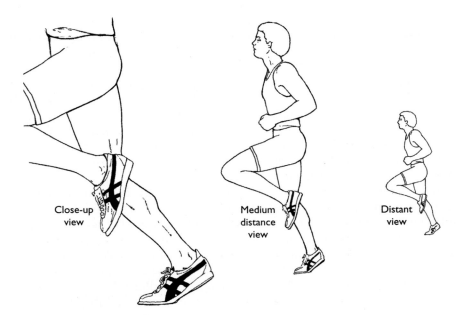

Close-up
view

Medium
distance
view

Distant
view

viewing and single-picture advance that facilitate isolation of the critical aspects of a movement.

The analyst should be aware, however, that there is a potential drawback to the use of video. The subject's awareness of the presence of a camera sometimes results in changes in performance. Movement analysts should be aware that subjects may be distracted or unconsciously modify their techniques when a recording device is used.

Conducting a Qualitative Analysis

Despite careful planning of a qualitative analysis, new questions occasionally emerge during the course of collecting observations. Movement modifications may be taking place with each performance as learning occurs, especially when the performer is unskilled. Even when this is not the case, the observations made may suggest new questions of interest. For example, what is causing the inconsistencies in a golfer's swing? What technique changes are occurring over the 30–40 m range in a 100 m sprint? A careful analysis is not strictly preprogrammed, but often involves identifying new questions to answer or problems to solve. The teacher, clinician, or coach often is involved in a continuous process of formulating an analysis, collecting additional observations, and formulating an updated analysis (Figure 2-21).

Answering questions that have been identified requires that the analyst be able to focus on the critical aspects of the movement. Once a biomechanical error has been generally identified, it is often useful for the analyst to watch the performer over several trials and to progressively zero in on the specific problem. Evaluating a softball pitcher's technique might begin with observation of insufficient ball speed, progress to an evaluation of upper-extremity kinematics, and end with an identification of insufficient wrist snap at ball release.

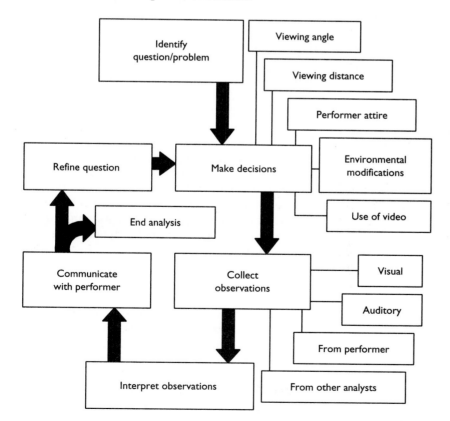

FIGURE 2-21

The qualitative analysis process is often cyclical, with observations leading to refinement of the original question.

The analyst should also be aware that every performance of a motor skill is affected by the characteristics of the performer. These include the performer's age, gender, and anthropometry; the developmental and skill levels at which the performer is operating; and any special physical or personality traits that may impact performance. Providing a novice, pre-school-aged performer with cues for a skilled, mature performance may be counterproductive, since young children do not have the same motor capabilities as adults. Likewise, although training can ameliorate loss of muscular strength and joint range of motion once thought to be inevitably associated with aging, human movement analysts need increased knowledge of and sensitivity to the special needs of older adults who wish to develop new motor skills. Analysts should also be aware that, although gender has traditionally been regarded as a basis for performance differences, research has shown that before puberty most gender-associated performance differences are probably culturally derived rather than biologically determined (3). Young girls are usually not expected to be as skilled or even as active as young boys. Unfortunately, in many settings, these expectations extend beyond childhood into adolescence and adulthood. The belief that an activity is not gender appropriate has been shown to negatively affect college-aged women's ability to learn a new motor skill (1). Analysts of female performers should not reinforce this cultural misunderstanding by lowering their expectations of girls or women based on gender. Analysts should also be sensitive to other factors that can influence performance. Has the performer experienced a recent emotional upset? Is the sun in his eyes? Is she tired? Being an effective observer requires full awareness of the surrounding environment.

● *Auditory information is often a valuable source in the analysis of human motor skills.*

To supplement visual observation, the analyst should be aware that nonvisual forms of information can also sometimes be useful during a movement analysis. For example, auditory information can provide clues about the way in which a movement was executed. Proper contact of a golf club with a ball sounds distinctly different from when a golfer "tops" the ball. Similarly, the crack of a baseball bat hitting a ball indicates that the contact was direct rather than glancing. The sound of a double contact of a volleyball player's arms with the ball may identify an illegal hit. The sound of a patient's gait usually reveals whether an asymmetry is present.

Another potential source of information is feedback from the performer (Sample Application 2.1). A performer who is experienced enough to recognize the way a particular movement feels as compared to the way a slight modification of the same movement feels is a useful source of information. However, not all performers are sufficiently kinesthetically attuned to provide meaningful subjective feedback of this nature. The performer being analyzed may also assist in other ways. Performance deficiencies may result from errors in technique, perception, or decision making. Identification of perceptual and decision-making errors by the performer often requires more than visual observation of the performance. In these cases, asking meaningful questions of the performer may be useful. However, the analyst should consider subjective input from the performer in conjunction with more objective observations.

Another potential way to enhance the thoroughness of an analysis is to involve more than one analyst. This reduces the likelihood of oversight. Students in the process of learning a new motor skill may also benefit from teaming up to analyze each other's performances under appropriate teacher direction.

● *The ability to effectively analyze human movement improves with practice.*

Finally, analysts must remember that observation skills improve with practice. As analysts gain experience, the analysis process becomes more natural, and the analyses conducted are likely to become more effective

SAMPLE APPLICATION 2.1

Problem: Sally, a powerful outside hitter on a high school volleyball team, has been out for two weeks with mild shoulder bursitis but has recently received her physician's clearance to return to practice. Joan, Sally's coach, notices that Sally's spikes are traveling at a slow speed and are being easily handled by the defensive players.

Planning the Analysis

1. What specific problems need to be solved or questions need to be answered regarding the movement? Joan first questions Sally to make sure that the shoulder is not painful. She then reasons that a technique error is present.
2. From what angle(s) and distance(s) can problematic aspects of the movement best be observed? Is more than one view needed? Although a volleyball spike involves transverse plane rotation of the trunk, the arm movement is primarily in the sagittal plane. Joan therefore decides to begin by observing a sagittal view from the side of Sally's hitting arm.
3. How many movement performances should be observed? Since Sally is a skilled player and her spikes are consistently being executed at reduced velocity, Joan reasons that only a few observations may be needed.
4. Is special subject attire, lighting, or background environment needed to facilitate observation? The gym where the team works out is well lit and the players wear sleeveless tops. Therefore, no special accommodations for the analysis seem necessary.
5. Will a video recording of the movement be necessary or useful? A volleyball spike is a relatively fast movement, but there are definite checkpoints that the knowledgeable observer can watch in real time. Is the jump primarily vertical, and is it high enough for the player to contact the ball above the net? Is the hitting arm positioned with the upper arm in maximal horizontal abduction prior to arm swing to allow a full range of arm motion? Is the hitting movement initiated by trunk rotation followed by shoulder flexion, then elbow extension, then snaplike wrist flexion? Is the movement being executed in a coordinated fashion to enable imparting a large force to the ball?

Conducting the Analysis

1. Review, and sometimes reformulate, specific questions of focus. After watching Sally execute two spikes, Joan observes that her arm range of motion appears to be relatively small.
2. Repeatedly view movements to gradually zero in on causes of performance errors. After watching Sally spike three more times, Joan suspects that Sally is not positioning her upper arm in maximal horizontal abduction in preparation for the hit.
3. Be aware of the influence of performer characteristics. Joan talks to Sally on the sideline and asks her to put her arm in the preparatory position for a hit. She asks Sally if this position is painful, and Sally responds that it is not.
4. Pay attention to nonvisual cues. (None are apparent in this situation.)
5. When appropriate, ask the performer to self-analyze. Joan tells Sally that she suspects Sally has been protecting the shoulder by not rotating her arm back far enough in preparation for spikes. She can correct the problem. Sally's next few spikes are executed at much faster velocity.
6. Consider involving other analysts to assist. Joan asks her assistant coach to watch Sally for the remainder of practice to determine whether the problem has been corrected.

Motion analysis software tracks joint markers in three-dimensional space.

and informative. The expert analyst is typically better able to both identify and diagnose errors than the novice. Novice analysts should take every opportunity to practice movement analysis in carefully planned and structured settings, as such practice has been shown to improve the ability to focus attention on the critical aspects of performance (2).

TOOLS FOR MEASURING KINEMATIC QUANTITIES

Biomechanics researchers have available a wide array of equipment for studying human movement kinematics. Knowledge gained through the use of this apparatus is often published in professional journals for teachers, clinicians, coaches, and others interested in human movement.

Video and Film

Photographers began employing cameras in the study of human and animal movement during the late nineteenth century. One famous early photographer was Eadweard Muybridge, a British landscape photographer and a rather colorful character who frequently published essays praising his own work. Muybridge used electronically controlled still cameras aligned in sequence with an electromagnetic tripping device to capture serial shots of trotting and galloping horses, thereby resolving the controversy about whether all four hooves are ever airborne simultaneously (they are). More importantly, however, he amassed three volumes of photographic work on human and animal motions that provided scientific documentation of some of the subtle differences between normal and pathological gait.

Movement analysts today have quite an array of camera types from which to choose. The type of movement and the requirements of the analysis largely determine the camera and analysis system of choice. Standard video provides 30 resolvable pictures per second, which is perfectly adequate for many human movement applications. Scientists and clinicians performing detailed quantitative study of the kinematics of human motion typically require a more sophisticated video camera and playback unit, with higher rates of picture capture. Digital video capture systems designed for human movement analysis are commercially available with frame rates of up to 2000 Hz. For both qualitative and quantitative analysis, however, a consideration often of greater importance than camera speed is the clarity of the captured images. It is the camera's shutter speed that allows user control of the exposure time, or length of time that the shutter is open when each picture in the video record is taken. The faster the movement being analyzed, the shorter the duration of the exposure time required to prevent blurring of the image captured.

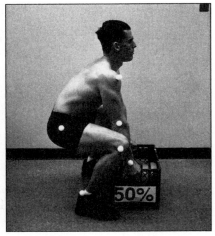

Reflective joint markers can be tracked by a camera for automatic digitizing of the movement.

Another important consideration when analyzing human movement with video is the number of cameras required to adequately capture the aspects of interest. Because most human movement is not constrained to a single plane, it is typically necessary to use multiple cameras to ensure that all of the movements can be viewed and recorded accurately for a detailed analysis. When practicality dictates that a single camera be used, thoughtful consideration should be given to camera positioning relative to the movement of interest. Only when human motion is occurring perpendicular to the optical axis of a camera are the angles present at joints viewed without distortion.

Biomechanists typically conduct quantitative analyses of human motion by adhering small, reflective markers over the subject's joint centers and other points of interest on the body, with marker locations depending on the purpose of the analysis. High-speed digital video cameras with infrared light rings encircling the lenses then capture high-contrast images of the reflective markers. Since human motion is rarely purely planar, researchers typically position six to eight and sometimes more cameras around the staging area in strategic locations to enable generation of three-dimensional representations of the movements of the markers. Much of today's biomechanical analysis software is capable of providing graphical outputs displaying kinematic and kinetic quantities of interest within minutes after a motion has been digitally captured by the cameras.

Reflective joint markers can be tracked by a camera for automatic digitizing of the movement.

Other Movement-Monitoring Systems

An *accelerometer* is a transducer used for the direct measurement of acceleration. The accelerometer is attached as rigidly as possible to the body segment or other object of interest, with electrical output channeled to a recording device. Three-dimensional accelerometers that incorporate multiple linear accelerometers are commercially available for monitoring acceleration during nonlinear movements.

A digital camera with infrared light ring is used for tracking reflective markers on a subject.

SUMMARY

Movements of the human body are referenced to the sagittal, frontal, and transverse planes, with their respectively associated mediolateral, anteroposterior, and longitudinal axes. Most human motion is general, with both linear and angular components. A set of specialized terminology is used to describe segment motions and joint actions of the human body.

Teachers of physical activities, clinicians, and coaches all routinely perform qualitative analyses to assess, correct, or improve human movements. Both knowledge of the specific biomechanical purpose of the movement and careful preplanning are necessary for an effective qualitative analysis. A number of special tools are available to assist researchers in collecting kinematic observations of human movement.

INTRODUCTORY PROBLEMS

1. Using appropriate movement terminology, write a qualitative description of the performance of a maximal vertical jump. Your description should be sufficiently detailed that the reader can completely and accurately visualize the movement.
2. Select a movement that occurs primarily in one of the three major reference planes. Qualitatively describe this movement in enough detail that the reader of your description can visualize the movement.
3. List five movements that occur primarily in each of the three cardinal planes. The movements may be either sport skills or activities of daily living.
4. Select a familiar animal. Does the animal move in the same major reference planes in which humans move? What are the major differences

in the movement patterns of this animal and the movement patterns of humans?

5. Select a familiar movement, and list the factors that contribute to skilled versus unskilled performance of that movement.

6. Test your observation skills by carefully observing the two photos shown on the top. List the differences that you are able to identify between these two photos.

7. Choose a familiar movement, and list aspects of that movement that are best observed from close up, from 2 to 3 m away, and from reasonably far away. Write a brief explanation of your choices.

8. Choose a familiar movement, and list aspects of the movement that are best observed from the side view, front view, rear view, and top view. Write a brief explanation of your choices.

9. Choose one of the instrumentation systems described and write a short paragraph explaining the way in which it might be used to study a question related to analysis of a human movement of interest to you.

ADDITIONAL PROBLEMS

1. Select a familiar movement and identify the ways in which performance of that movement is affected by strength, flexibility, and coordination.

2. List three human movement patterns or skills that are best observed from a side view, from a front or rear view, and from a top view.

3. Select a movement that is nonplanar and write a qualitative description of that movement sufficiently detailed to enable the reader of your description to picture the movement.

4. Select a nonplanar movement of interest and list the protocol you would employ in analyzing that movement.

5. What special expectations, if any, should the analyst have of movement performances if the performer is an older adult? An elementary school–aged girl? A novice? An obese high school–aged boy?

6. What are the advantages and disadvantages of collecting observational data on a sport skill during a competitive event as opposed to a practice session?

7. Select a movement with which you are familiar and list at least five questions that you, as a movement analyst, might ask the performer of the movement to gain additional knowledge about a performance.

8. List the auditory characteristics of five movements and explain in each case how these characteristics provide information about the nature of the movement performance.

9. List the advantages and disadvantages of using a video camera as compared to the human eye for collecting observational data.

10. Locate an article in a professional or research journal that involves kinematic description of a movement of interest to you. What instrumentation was used by the researchers? What viewing distances and perspectives were used? How might the analysis described have been improved?

LABORATORY EXPERIENCES

1. Observe and analyze a single performer executing two similar but different versions of a particular movement—for example, two pitching styles or two gait styles. Explain what viewing perspectives and distances you selected for collecting observational data on each movement. Write a paragraph comparing the kinematics of the two movements.

Movement selected: _____

Viewing perspectives: _____

Reasons for selection of viewing perspectives: _____

_____

Viewing distances: _____

Reasons for selection of viewing distances: _____

_____

Kinematic comparison: _____

_____

_____

_____

_____

_____

_____

_____

2. Observe a single sport skill as performed by a highly skilled individual, a moderately skilled individual, and an unskilled individual. Qualitatively describe the differences observed.

Sport skill selected: _____

Highly Skilled Performer	Moderately Skilled Performer	Unskilled Performer
_____	_____	_____
_____	_____	_____
_____	_____	_____
_____	_____	_____
_____	_____	_____
_____	_____	_____
_____	_____	_____
_____	_____	_____

_____   _____   _____
_____   _____   _____
_____   _____   _____
_____   _____   _____
_____   _____   _____
_____   _____   _____
_____   _____   _____

3. Select a movement at which you are reasonably skilled. Plan and carry out observations of a less-skilled individual performing the movement, and provide verbal learning cues for that individual, if appropriate. Write a short description of the cues provided, with a rationale for each cue.

Movement selected: _____

Cues Provided   **Rationale**

_____   _____
_____   _____
_____   _____
_____   _____
_____   _____
_____   _____
_____   _____
_____   _____
_____   _____
_____   _____
_____   _____
_____   _____

4. Select a partner, and plan and carry out an observational analysis of a movement of interest. Write a composite summary analysis of the movement performance. Write a paragraph identifying in what ways the analysis process was changed by the inclusion of a partner.

Movement selected: _____

Analysis of Performance

_____

_____

_____

_____

_____

_____

_____

_____

_____

_____

_____

How the analysis process was different when working with a partner: _____

_____

_____

_____

_____

_____

5. Plan and carry out a video session of a slow movement of interest as performed by two different subjects. Write a comparative analysis of the subjects' performances.

Subject 1 Performance   **Subject 2 Performance**

_____   _____

_____   _____

_____   _____

_____   _____

_____   _____

_____   _____

_____   _____

_____   _____

_____   _____

_____   _____

_____   _____
_____   _____
_____   _____
_____   _____
_____   _____
_____   _____
_____   _____
_____   _____

REFERENCES

1. Belcher D, Lee AM, Solmon MA, and Harrison L Jr: The influence of gender-related beliefs and conceptions of ability on women learning the hockey wrist shot, *Res Q Exerc Sport* 74:183, 2005.
2. Jenkins JM, Garn A, and Jenkins P: Preservice teacher observations in peer coaching, *J Teach Phys Educ* 24:2, 2005.
3. Lorson KM and Goodway JD: Gender differences in throwing form of children ages 6–8 years during a throwing game, *Res Q Exerc Sport* 79:174, 2008.
4. Radlo SJ, Janelle CM, Barba DA, and Frehlich SG: Perceptial decision making for baseball pitch recognition: using P300 latency and amplitude to index attentional processing, *Res Q Exerc Sport* 72:22, 2001.

ANNOTATED READINGS

Burkett B: *Sport mechanics for coaches* (3rd ed.), Champaign, IL, 2010, Human Kinetics.
Provides an introductory look at the mechanics of sport to help readers understand and incorporate technology to enhance training, identify errors in technique, and improve performance.
Hudson JL: Applied biomechanics in an instructional setting, *JOPERD* 77:25, 2006.
Describes application of biomechanics in analyzing sport skills in a practical context.
Payton C and Bartlett R (eds.): *Biomechanical evaluation of movement in sport and exercise,* New York, 2008, Routledge.
Practical guide to using the range of biomechanics movement analysis equipment and software available today, including detailed explanations of the theory underlying biomechanics testing along with advice concerning choice of equipment and how to use laboratory equipment most effectively.
Reiman M and Manske R: *Functional testing in human performance,* Champaign, IL, 2009, Human Kinetics.
Serves as a comprehensive reference on functional testing for assessment of physical activities in sport, recreation, work, and daily living.

RELATED WEBSITES

Mikromak
http://www.mikromak.com
Advertises video hardware and software for sports, medicine, and product research.
Motion Analysis Corporation
http://www.motionanalysis.com
Offers an optical motion capture system utilizing reflective markers for entertainment, biomechanics, character animation, and motion analysis.
Northern Digital, Inc.
http://www.ndigital.com
Presents optoelectronic 3-D motion measurement systems that track light-emitting diodes for real-time analysis.
Qualisys, Inc.
http://www.qualisys.com
Presents a system in which cameras track reflective markers, enabling real-time calculations; applications described for research, clinical, industry, and animation.
Redlake Imaging
http://www.redlake.com/imaging
Advertises high-speed video products for scientific and clinical applications.
SIMI Reality Motion Systems
http://www.simi.com
Describes computer-based video analysis for the human body and cellular applications; includes demo of gait analysis, among others.

KEY TERMS

anatomical reference position	erect standing position with all body parts, including the palms of the hands, facing forward; considered the starting position for body segment movements
angular	involving rotation around a central line or point
anteroposterior axis	imaginary line around which frontal plane rotations occur
axis of rotation	imaginary line perpendicular to the plane of rotation and passing through the center of rotation
cardinal planes	three imaginary perpendicular reference planes that divide the body in half by mass
curvilinear	along a curved line
frontal plane	plane in which lateral movements of the body and body segments occur
general motion	motion involving translation and rotation simultaneously
linear	along a line that may be straight or curved, with all parts of the body moving in the same direction at the same speed
longitudinal axis	imaginary line around which transverse plane rotations occur
mediolateral axis	imaginary line around which sagittal plane rotations occur
rectilinear	along a straight line
sagittal plane	plane in which forward and backward movements of the body and body segments occur
system	mechanical system chosen by the analyst for study
translation	linear motion
transverse plane	plane in which horizontal body and body segment movements occur when the body is in an erect standing position

Kinetic Concepts for Analyzing Human Motion

After completing this chapter, you will be able to:

Define and identify common units of measurement for mass, force, weight, pressure, volume, density, specific weight, torque, and impulse.

Identify and describe the different types of mechanical loads that act on the human body.

Identify and describe the uses of available instrumentation for measuring kinetic quantities.

Distinguish between vector and scalar quantities.

Solve quantitative problems involving vector quantities using both graphic and trigonometric procedures.

ONLINE LEARNING CENTER RESOURCES

www.mhhe.com/hall6e

Log on to our Online Learning Center (OLC) for access to these additional resources:

- Online Lab Manual
- Flashcards with definitions of chapter key terms
- Chapter objectives
- Chapter lecture PowerPoint presentation
- Self-scoring chapter quiz
- Additional chapter resources
- Web links for study and exploration of chapter-related topics

A skater has a tendency to continue gliding with constant speed and direction due to inertia.

inertia

tendency of a body to resist a change in its state of motion

W hen muscles on opposite sides of a joint develop tension, what determines the direction of joint motion? In which direction will a swimmer swimming perpendicular to a river current actually travel? What determines whether a push can move a heavy piece of furniture? The answers to these questions are rooted in kinetics, the study of forces.

The human body both generates and resists forces during the course of daily activities. The forces of gravity and friction enable walking and manipulation of objects in predictable ways when internal forces are produced by muscles. Sport participation involves application of forces to balls, bats, racquets, and clubs, and absorption of forces from impacts with balls, the ground or floor, and opponents in contact sports. This chapter introduces basic kinetic concepts that form the basis for understanding these activities.

BASIC CONCEPTS RELATED TO KINETICS

Understanding the concepts of inertia, mass, weight, pressure, volume, density, specific weight, torque, and impulse provides a useful foundation for understanding the effects of forces.

Inertia

In common usage, inertia means resistance to action or to change (Figure 3-1). Similarly, the mechanical definition is resistance to acceleration. Inertia is the tendency of a body to maintain its current state of motion, whether motionless or moving with a constant velocity. For example, a 150 kg weight bar lying motionless on the floor has a tendency to remain motionless. A skater gliding on a smooth surface of ice has a tendency to continue gliding in a straight line with a constant speed.

Although inertia has no units of measurement, the amount of inertia a body possesses is directly proportional to its mass. The more massive an object is, the more it tends to maintain its current state of motion and the more difficult it is to disrupt that state.

FIGURE 3-1

A static object tends to maintain its motionless state because of inertia.

Mass

Mass (m) is the quantity of matter composing a body. The common unit of mass in the metric system is the kilogram (kg), with the English unit of mass being the *slug,* which is much larger than a kg.

mass
quantity of matter contained in an object

Force

A force (F) can be thought of as a push or a pull acting on a body. Each force is characterized by its magnitude, direction, and point of application to a given body. Body weight, friction, and air or water resistance are all forces that commonly act on the human body. The action of a force causes a body's mass to accelerate:

force
push or pull; the product of mass and acceleration

$$F = ma$$

Units of force are units of mass multiplied by units of acceleration (a). In the metric system, the most common unit of force is the Newton (N), which is the amount of force required to accelerate 1 kg of mass at 1 m/s^2:

$$1 \text{ N} = (1 \text{ kg})(1 \text{ m/s}^2)$$

In the English system, the most common unit of force is the pound (lb). A pound of force is the amount of force necessary to accelerate a mass of 1 slug at 1 ft/s^2, and 1 lb is equal to 4.45 N:

$$1 \text{ lb} = (1 \text{ slug})(1 \text{ ft/s}^2)$$

Because a number of forces act simultaneously in most situations, constructing a free body diagram is usually the first step when analyzing the effects of forces on a body or system of interest. A *free body* is any object, body, or body part that is being focused upon for analysis. A free body diagram consists of a sketch of the system being analyzed and vector representations of the acting forces (Figure 3-2). Even though a hand must be applying force to a tennis racquet in order for the racquet to forcefully contact a ball, if the racquet is the free body of interest, the hand is

free body diagram
sketch that shows a defined system in isolation with all of the force vectors acting on the system

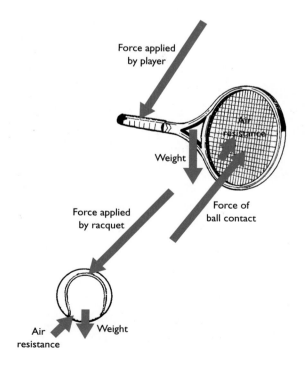

Force applied by player

Air resistance

Weight

Force applied by racquet

Force of ball contact

Air resistance   Weight

FIGURE 3-2

Two free body diagrams showing the acting forces.

represented in the free body diagram of the racquet only as a force vector. Similarly, if the tennis ball constitutes the free body being studied, the force of the racquet acting on the ball is displayed as a vector.

Since a force rarely acts in isolation, it is important to recognize that the overall effect of many forces acting on a system or free body is a function of the net force, which is the vector sum of all of the acting forces. When all acting forces are balanced, or cancel each other out, the net force is zero, and the body remains in its original state of motion, either motionless or moving with a constant velocity. When a net force is present, the body moves in the direction of the net force and with an acceleration that is proportional to the magnitude of the net force.

net force
resultant force derived from the composition of two or more forces

Center of Gravity

A body's center of gravity, or center of mass, is the point around which the body's weight is equally balanced, no matter how the body is positioned, (see Chapter 13). In motion analyses, the motion of the center of gravity serves as an index of total body motion. From a kinetic perspective, the location of the center of mass determines the way in which the body responds to external forces.

center of gravity
point around which a body's weight is equally balanced, no matter how the body is positioned

Weight

Weight is defined as the amount of gravitational force exerted on a body. Algebraically, its definition is a modification of the general definition of a force, with weight (wt) being equal to mass (m) multiplied by the acceleration of gravity (a_g):

$$wt = ma_g$$

Since weight is a force, units of weight are units of force—either N or lb.

As the mass of a body increases, its weight increases proportionally. The factor of proportionality is the acceleration of gravity, which is $-9.81 \ m/s^2$.

weight
gravitational force that the earth exerts on a body

Although a body's mass remains unchanged on the moon, its weight is less due to smaller gravitational acceleration. **Photo courtesy of NASA.**

The negative sign indicates that the acceleration of gravity is directed downward, or toward the center of the earth. On the moon or another planet with a different gravitational acceleration, a body's weight would be different, although its mass would remain the same.

Because weight is a force, it is also characterized by magnitude, direction, and point of application. The direction in which weight acts is always toward the center of the earth. Because the point at which weight is assumed to act on a body is the body's center of gravity, the center of gravity is the point where the weight vector is shown to act in free body diagrams.

Although body weights are often reported in kilograms, the kilogram is actually a unit of mass. To be technically correct, weights should be identified in Newtons and masses reported in kilograms. Sample Problem 3.1 illustrates the relationship between mass and weight.

SAMPLE PROBLEM 3.1

1. If a scale shows that an individual has a mass of 68 kg, what is that individual's weight?

Known

$$m = 68 \text{ kg}$$

Solution

Wanted: weight

Formulas: $wt = ma_g$

$1 \text{ kg} = 2.2 \text{ lb}$ (English/metric conversion factor)

(Mass may be multiplied by the acceleration of gravity to convert to weight within either the English or the metric system.)

$$wt = ma_g$$
$$wt = (68 \text{ kg})(9 .81) \text{ m/s}^2$$
$$\boxed{wt = 667 \text{ N}}$$

Mass in kg may be multiplied by the conversion factor 2.2 lb/kg to convert to weight in pounds:

$$(68 \text{ kg})(2.2 \text{ lb/kg}) = \boxed{150 \text{ lb}}$$

2. What is the mass of an object weighing 1200 N?

Known

$$wt = 1200 \text{ N}$$

Solution

Wanted: mass

Formula: $wt = ma_g$

(Weight may be divided by the acceleration of gravity within a given system of measurement to convert to mass.)

$$wt = ma_g$$
$$1200 \text{ N} = m(9.81 \text{ m/s}^2)$$
$$\frac{1200 \text{ N}}{9.81 \text{ m/s}^2} = m$$

$$\boxed{m = 122.32 \text{ kg}}$$

Pressure

pressure
force per unit of area over which force acts

Pressure (P) is defined as force (F) distributed over a given area (A):

$$P = \frac{F}{A}$$

Units of pressure are units of force divided by units of area. Common units of pressure in the metric system are N per square centimeter (N/cm^2) and Pascals (Pa). One Pascal represents one Newton per square meter ($Pa = N/m^2$). In the English system, the most common unit of pressure is pounds per square inch (psi or lb/in^2).

The pressure exerted by the sole of a shoe on the floor beneath it is the body weight resting on the shoe divided by the surface area between the sole of the shoe and the floor. As illustrated in Sample Problem 3.2, the smaller amount of surface area on the bottom of a spike heel as compared to a flat sole results in a much larger amount of pressure being exerted.

Volume

volume
amount of three-dimensional space occupied by a body

A body's volume is the amount of space that it occupies. Because space is considered to have three dimensions (width, height, depth), a unit of volume is a unit of length multiplied by a unit of length multiplied by a unit of length. In mathematical shorthand, this is a unit of length raised to the exponential power of three, or a unit of length *cubed*. In the metric

Pairs of balls that are similar in volume but markedly different in weight, including a solid metal shot and a softball (photo 1) and a table tennis ball and golf ball (photo 2).

SAMPLE PROBLEM 3.2

Is it better to be stepped on by a woman wearing a spike heel or by the same woman wearing a smooth-soled court shoe? If a woman's weight is 556 N, the surface area of the spike heel is 4 cm², and the surface area of the court shoe is 175 cm², how much pressure is exerted by each shoe?

Known

$$wt = 556 \text{ N}$$
$$A_s = 4 \text{ cm}^2$$
$$A_c = 175 \text{ cm}^2$$

Solution

Wanted: pressure exerted by the spike heel
            pressure exerted by the court shoe

Formula: $P = F/A$

Deduction: It is necessary to recall that weight is a force.

For the spike heel:

$$P = \frac{556 \text{ N}}{4 \text{ cm}^2}$$

$$P = 139 \text{ N/cm}^2$$

For the court shoe:

$$P = \frac{556 \text{ N}}{175 \text{ cm}^2}$$

$$P = 3.18 \text{ N/cm}^2$$

Comparison of the amounts of pressure exerted by the two shoes:

$$\frac{P_{\text{spike heel}}}{P_{\text{court shoe}}} = \frac{139 \text{ N/cm}^2}{3.18 \text{ N/cm}^2} = 43.75$$

Therefore, 43.75 times more pressure is exerted by the spike heel than by the court shoe worn by the same woman.

system, common units of volume are cubic centimeters (cm^3), cubic meters (m^3), and liters (l):

$$1 \, l = 1000 \text{ cm}^3$$

In the English system of measurement, common units of volume are cubic inches (in^3) and cubic feet (ft^3). Another unit of volume in the English system is the quart (qt):

$$1 \text{ qt} = 57.75 \text{ in}^3$$

Volume should not be confused with weight or mass. An 8 kg shot and a softball occupy approximately the same volume of space, but the weight of the shot is much greater than that of the softball.

Density

density
mass per unit of volume

The concept of density combines the mass of a body with the body volume. Density is defined as mass per unit of volume. The conventional symbol for density is the Greek letter rho (ρ).

$$\text{density } (\rho) = \text{mass/volume}$$

Units of density are units of mass divided by units of volume. In the metric system, a common unit of density is the kilogram per cubic meter (kg/m^3). In the English system of measurement, units of density are not commonly used. Instead, units of specific weight (weight density) are employed.

specific weight
weight per unit of volume

Specific weight is defined as weight per unit of volume. Because weight is proportional to mass, specific weight is proportional to density. Units of specific weight are units of weight divided by units of volume. The metric unit for specific weight is Newtons per cubic meter (N/m^3), and the English system uses pounds per cubic foot (lb/ft^3).

Although a golf ball and a ping-pong ball occupy approximately the same volume, the golf ball has a greater density and specific weight than the ping-pong ball because the golf ball has more mass and more weight. Similarly, a lean person with the same body volume as an obese person has a higher total body density because muscle is denser than fat. Thus, percent body fat is inversely related to body density.

Torque

When a force is applied to an object such as a pencil lying on a desk, either translation or general motion may result. If the applied force is directed parallel to the desktop and through the center of the pencil (a *centric force*), the pencil will be translated in the direction of the applied force. If the force is applied parallel to the desktop but directed through a point other than the center of the pencil (an *eccentric force*), the pencil will undergo both translation and rotation (Figure 3-3).

torque
rotary effect of a force

The rotary effect created by an eccentric force is known as torque (T), or moment of force. Torque, which may be thought of as *rotary force,* is the angular equivalent of linear force. Algebraically, torque is the product of force (F) and the perpendicular distance ($d_\perp$) from the force's line of action to the axis of rotation:

$$T = Fd_\perp$$

The greater the amount of torque acting at the axis of rotation, the greater the tendency for rotation to occur. Units of torque in both the metric and the English systems follow the algebraic definition. They are units of force multiplied by units of distance: Newton-meters (N-m) or foot-pounds (ft-lb).

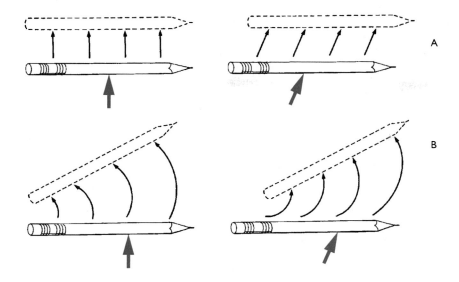

FIGURE 3-3

A. Centric forces produce translation. **B.** Eccentric forces produce translation and rotation.

Impulse

When a force is applied to a body, the resulting motion of the body is dependent not only on the magnitude of the applied force but also on the duration of force application. The product of force (F) and time (t) is known as impulse (J):

$$J = Ft$$

A large change in an object's state of motion may result from a small force acting for a relatively long time or from a large force acting for a relatively short time. A golf ball rolling across a green gradually loses speed because of the small force of rolling friction. The speed of a baseball struck vigorously by a bat changes because of the large force exerted by the bat during the fraction of a second it is in contact with the ball. When a vertical jump is executed, the larger the impulse generated against the floor, the greater the jumper's takeoff velocity and the higher the resulting jump.

Units of physical quantities commonly used in biomechanics are shown in Table 3-1.

impulse
product of force and the time over which the force acts

QUANTITY	SYMBOL	FORMULA	METRIC UNIT	ENGLISH UNIT
Mass	m		kg	slug
Force	F	F = ma	N	lb
Pressure	P	P = F/A	Pa	psi
Volume (solids)	V		m³	ft³
(liquids)	V		liter	gallon
Density	ρ	ρ = m/V	kg/m³	slugs/ft³
Specific weight	γ	γ = wt/V	N/m³	lb/ft³
Torque	T	T = Fd	N-m	ft-lb
Impulse	J	J = Ft	N · s	lb · s

TABLE 3-1

Common Units for Kinetic Quantities

MECHANICAL LOADS ON THE HUMAN BODY

Muscle forces, gravitational force, and bone-breaking force such as that encountered in a skiing accident all affect the human body differently. The effect of a given force depends on its direction and duration as well as its magnitude, as described in the following section.

Compression, Tension, and Shear

compression
pressing or squeezing force directed axially through a body

Compressive force, or compression, can be thought of as a squeezing force (Figure 3-4). An effective way to press wildflowers is to place them inside the pages of a book and to stack other books on top of that book. The weight of the books creates a compressive force on the flowers. Similarly, the weight of the body acts as a compressive force on the bones that support it. When the trunk is erect, each vertebra in the spinal column must support the weight of that portion of the body above it.

tension
pulling or stretching force directed axially through a body

The opposite of compressive force is tensile force, or tension (Figure 3-4). Tensile force is a pulling force that creates tension in the object to which it is applied. When a child sits in a playground swing, the child's weight creates tension in the chains supporting the swing. A heavier child creates even more tension in the supports of the swing. Muscles produce tensile force that pulls on the attached bones.

shear
force directed parallel to a surface

A third category of force is termed shear. Whereas compressive and tensile forces act along the longitudinal axis of a bone or other structure to which they are applied, shear force acts parallel or tangent to a surface. Shear force tends to cause one portion of the object to slide, displace, or shear with respect to another portion of the object (Figure 3-4). For example, a force acting at the knee joint in a direction parallel to the tibial plateau is a shearing force at the knee. During the landing from a ski jump the impact force includes a component of anteriorly directed shear on the tibial plateau, elevating stress on the anterior cruciate ligament (1). (Figure 3-5).

FIGURE 3-4

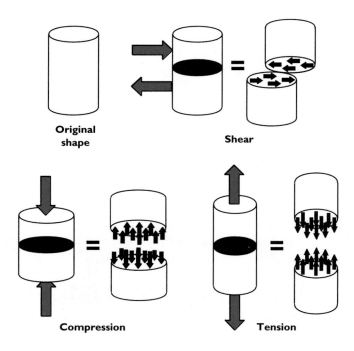

Original shape   Shear

Compression   Tension

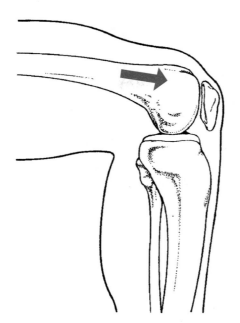

FIGURE 3-5

During the landing from a ski jump the axial impact force at the knee includes a component of anteriorly directed shear on the tibial plateau.

Mechanical Stress

Another factor affecting the outcome of the action of forces on the human body is the way in which the force is distributed. Whereas pressure represents the distribution of force external to a solid body, stress represents the resulting force distribution inside a solid body when an external force acts. Stress is quantified in the same way as pressure: force per unit of area over which the force acts. As shown in Figure 3-6, a given force acting on a small surface produces greater stress than the same force acting over a larger surface. When a blow is sustained by the human body, the likelihood of injury to body tissue is related to the magnitude and direction of the stress created by the blow. Compressive stress, tensile stress, and shear stress are terms that indicate the direction of the acting stress.

Because the lumbar vertebrae bear more of the weight of the body than the thoracic vertebrae when a person is in an upright position, the compressive stress in the lumbar region should logically be greater. However, the amount of stress present is not directly proportional to the amount of weight borne, because the load-bearing surface areas of the lumbar vertebrae are greater than those of the vertebrae higher in the spinal column (Figure 3-7). This increased surface area reduces the amount of compressive stress present.

stress
distribution of force within a body, quantified as force divided by the area over which the force acts

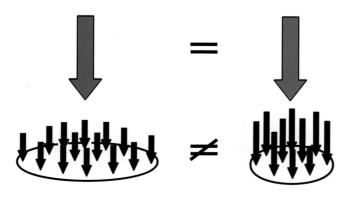

FIGURE 3-6

The amount of mechanical stress created by a force is inversely related to the size of the area over which the force is spread.

The surfaces of the vertebral bodies increase in surface area as more weight is supported.

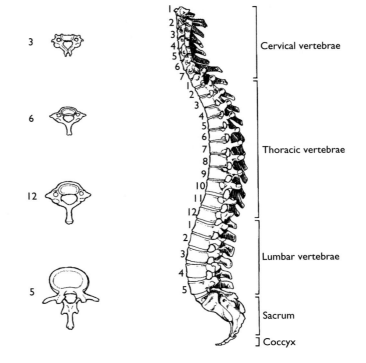

Cervical vertebrae

Thoracic vertebrae

Lumbar vertebrae

Sacrum

Coccyx

Nevertheless, the L5-S1 intervertebral disc (at the bottom of the lumbar spine) is the most common site of disc herniations, although other factors also play a role (see Chapter 9). Quantification of mechanical stress is demonstrated in Sample Problem 3.3.

Torsion, Bending, and Combined Loads

bending
asymmetric loading that produces tension on one side of a body's longitudinal axis and compression on the other side

axial
directed along the longitudinal axis of a body

torsion
load-producing twisting of a body around its longitudinal axis

combined loading
simultaneous action of more than one of the pure forms of loading

A somewhat more complicated type of loading is called bending. Pure compression and tension are both axial forces—that is, they are directed along the longitudinal axis of the affected structure. When an eccentric (or nonaxial) force is applied to a structure, the structure bends, creating compressive stress on one side and tensile stress on the opposite side (Figure 3-8).

Torsion occurs when a structure is caused to twist about its longitudinal axis, typically when one end of the structure is fixed. Torsional fractures of the tibia are not uncommon in football injuries and skiing accidents in which the foot is held in a fixed position while the rest of the body undergoes a twist.

The presence of more than one form of loading is known as combined loading. Because the human body is subjected to a myriad of simultaneously acting forces during daily activities, this is the most common type of loading on the body.

THE EFFECTS OF LOADING

deformation
change in shape

When a force acts on an object, there are two potential effects. The first is acceleration and the second is deformation, or change in shape. When a diver applies force to the end of a springboard, the board both accelerates and deforms. The amount of deformation that occurs in response to a given force depends on the stiffness of the object acted upon.

When an external force is applied to the human body, several factors influence whether an injury occurs. Among these are the magnitude and

SAMPLE PROBLEM 3.3

How much compressive stress is present on the L1, L2 vertebral disc of a 625 N woman, given that approximately 45% of body weight is supported by the disc (a) when she stands in anatomical position and (b) when she stands erect holding a 222 N suitcase? (Assume that the disc is oriented horizontally and that its surface area is 20 cm^2.)

Solution

1. Given:

$$F = (625 \text{ N})(0.45)$$

$$A = 20 \text{ cm}^2$$

Formula:

$$\text{stress} = F/A$$

$$\text{stress} = \frac{(625 \text{ N})(0.45)}{20 \text{ cm}^2}$$

$$\boxed{\text{stress} = 14 \text{ N/cm}^2}$$

2. Given:

$$F = (625 \text{ N})(0.45) + 222 \text{ N}$$

Formula:

$$\text{stress} = F/A$$

$$\text{stress} = \frac{(625 \text{ N})(0.45) + 222 \text{ N}}{20 \text{ cm}^2}$$

$$\boxed{\text{stress} = 25.2 \text{ N/cm}^2}$$

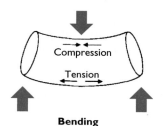

Bending

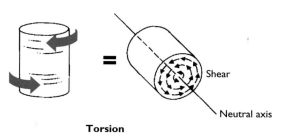

Torsion

FIGURE 3-8

Objects loaded in bending are subject to compression on one side and tension on the other. Objects loaded in torsion develop internal shear stress, with maximal stress at the periphery and no stress at the neutral axis.

direction of the force, and the area over which the force is distributed. Also important, however, are the material properties of the loaded body tissues.

The relationship between the amount of force applied to a structure and the structure's response is illustrated by a load deformation curve (Figure 3-9). With relatively small loads, deformation occurs, but the response is elastic, meaning that when the force is removed the structure returns to its original size and shape. Since stiffer materials display less deformation in response to a given load, greater stiffness translates

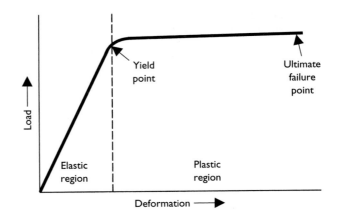

When a structure is loaded, it deforms, or changes shape. The deformation is temporary within the elastic region and permanent in the plastic region. Structural integrity is lost at the ultimate failure point.

yield point (elastic limit)
point on the load deformation curve past which deformation is permanent

failure
loss of mechanical continuity

repetitive loading
repeated application of a subacute load that is usually of relatively low magnitude

acute loading
application of a single force of sufficient magnitude to cause injury to a biological tissue

to a steeper slope of the load deformation curve in the elastic region. If the force applied causes the deformation to exceed the structure's yield point or elastic limit, however, the response is plastic, meaning that some amount of deformation is permanent. Deformations exceeding the ultimate failure point produce mechanical failure of the structure, which in the human body means fracturing of bone or rupturing of soft tissues.

Repetitive versus Acute Loads

The distinction between repetitive and acute loading is also important. When a single force large enough to cause injury acts on biological tissues, the injury is termed *acute* and the causative force is termed *macrotrauma*. The force produced by a fall, a rugby tackle, or an automobile accident may be sufficient to fracture a bone.

Injury can also result from the repeated sustenance of relatively small forces. For example, each time a foot hits the pavement during running, a force of approximately two to three times body weight is sustained. Although a single force of this magnitude is not likely to result in a fracture of healthy bone, numerous repetitions of such a force may cause a fracture of an otherwise healthy bone somewhere in the lower extremity. When repeated or chronic loading over a period produces an injury, the injury is called a *chronic injury* or a *stress injury*, and the causative mechanism is termed *microtrauma*. The relationship between the magnitude of the load sustained, the frequency of loading, and the likelihood of injury is shown in Figure 3-10.

The general pattern of injury likelihood as a function of load magnitude and repetition. Injury can be sustained, but is less likely, with a single large load and with a repeated small load.

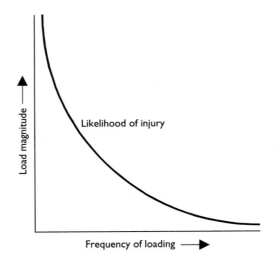

TOOLS FOR MEASURING KINETIC QUANTITIES

Biomechanics researchers use equipment for studying both muscle forces and forces generated by the feet against the ground during gait and other activities. Knowledge gained through the use of these tools is often published in professional journals for teachers, clinicians, coaches, and others interested in human movement.

Electromyography

Eighteenth-century Italian scientist Galvani made two interesting discoveries about skeletal muscle: (a) It develops tension when electrically stimulated, and (b) it produces a detectable current or voltage when developing tension, even when the stimulus is a nerve impulse. The latter discovery was of little practical value until the twentieth century, when technology became available for the detection and recording of extremely small electrical charges. The technique of recording electrical activity produced by muscle, or myoelectric activity, is known today as *electromyography* (EMG).

Electromyography is used to study neuromuscular function, including identification of which muscles develop tension throughout a movement and which movements elicit more or less tension from a particular muscle or muscle group. It is also used clinically to assess nerve conduction velocities and muscle response in conjunction with the diagnosis and tracking of pathological conditions of the neuromuscular system. Scientists also employ electromyographic techniques to study the ways in which individual motor units respond to central nervous system commands.

The process of electromyography involves the use of transducers known as *electrodes* that sense the level of myoelectric activity present at a particular site over time. Depending on the questions of interest, either surface electrodes or fine wire electrodes are used. Surface electrodes, consisting of small discs of conductive material, are positioned on the surface of the skin over a muscle or muscle group to pick up global myoelectric activity. When more localized pickup is desired, indwelling, fine-wire electrodes are injected directly into a muscle. Output from the electrodes is amplified and graphically displayed or mathematically processed and stored by a computer.

Dynamography

Scientists have devised several types of platforms and portable systems for the measurement of forces and pressure on the plantar surface of the foot. These systems have been employed primarily in gait research, but have also been used to study phenomena such as starts, takeoffs, landings, baseball and golf swings, and balance.

Both commercially available and homemade *force platforms* and *pressure platforms* are typically built rigidly into a floor flush with the surface and are interfaced to a computer that calculates kinetic quantities of interest. Force platforms are usually designed to transduce ground reaction forces in vertical, lateral, and anteroposterior directions with respect to the platform itself; pressure platforms provide graphical or digital maps of pressures across the plantar surfaces of the feet. The force platform is a relatively sophisticated instrument, but its limitations include the restrictions of a laboratory setting and

Myoelectric signal traces displayed on a computer monitor.

myoelectric activity
electric current or voltage produced by a muscle developing tension

transducers
devices that detect signals

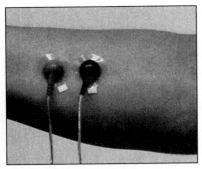

Surface electromyographic electrodes are small discs that attach directly to the skin over a muscle or muscle group of interest to transduce electrical activity in the underlying tissue.

vector
physical quantity that possesses both magnitude and direction

scalar
physical quantity that is completely described by its magnitude

vector composition
process of determining a single vector from two or more vectors by vector addition

resultant
single vector that results from vector composition

potential difficulties associated with the subject's consciously targeting the platform.

Portable systems for measuring plantar forces and pressures are also available in commercial and homemade models as instrumented shoes, shoe inserts, and thin transducers that adhere to the plantar surfaces of the feet. These systems provide the advantage of data collection outside the laboratory but lack the precision of the built-in platforms.

VECTOR ALGEBRA

A vector is a quantity that has both magnitude and direction. Vectors are represented by arrow-shaped symbols. The magnitude of a vector is its size; for example, the number 12 is of greater magnitude than the number 10. A vector symbol's orientation on paper represents direction, and its length represents magnitude. Force, weight, pressure, specific weight, and torque are kinetic vector quantities; displacement, velocity, and acceleration (see Chapter 10) are kinematic vector quantities. No vector is fully defined without the identification of both its magnitude and its direction. Scalar quantities possess magnitude but have no particular direction associated with them. Mass, volume, length, and speed are examples of scalar quantities.

Vector Composition

When vectors are added together, the operation is called vector composition. The composition of two or more vectors that have exactly the same direction results in a single vector that has a magnitude equal to the sum of the magnitudes of the vectors being added (Figure 3-11). The single vector resulting from a composition of two or more vectors is known as the resultant vector, or the resultant. If two vectors that are oriented in exactly opposite directions are composed, the resultant has the direction of the longer vector and a magnitude equal to the difference in the magnitudes of the two original vectors (Figure 3-12).

It is also possible to add vectors that are not oriented in the same or opposite directions. When the vectors are coplanar, that is, contained in the same plane, a procedure that may be used is the *"tip-to-tail"* method, in which the tail of the second vector is placed on the tip of the first vector, and the resultant is then drawn with its tail on the tail of the first vector

FIGURE 3-11

The composition of vectors with the same direction requires adding their magnitudes.

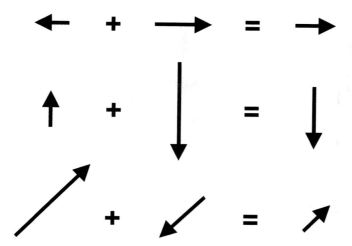

FIGURE 3-12

Composition of vectors with opposite directions requires subtracting their magnitudes.

and its tip on the tip of the second vector. This procedure may be used for combining any number of vectors if each successive vector is positioned with its tail on the tip of the immediately preceding vector and the resultant connects the tail of the first vector to the tip of the previous vector (Figure 3-13).

Through the laws of vector combination, we often can calculate or better visualize the resultant effect of combined vector quantities. For example, a canoe floating down a river is subject to both the force of the current and the force of the wind. If the magnitudes and directions of these two forces are known, the single resultant or *net force* can be derived through the process of vector composition (Figure 3-14). The canoe travels in the direction of the net force.

Vector Resolution

Determining the perpendicular components of a vector quantity relative to a particular plane or structure is often useful. For example, when a ball is thrown into the air, the horizontal component of its velocity determines the distance it travels, and the vertical component of its velocity determines the height it reaches (see Chapter 10). When a vector is resolved

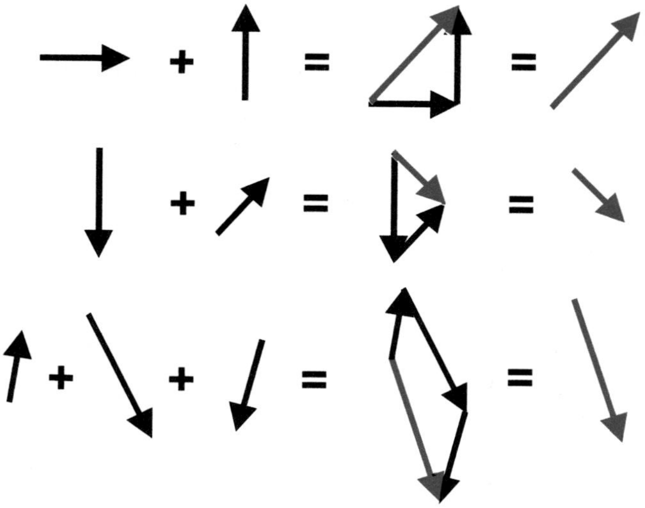

FIGURE 3-13

The "tip-to-tail" method of vector composition.

FIGURE 3-14

The net force is the resultant of all acting forces.

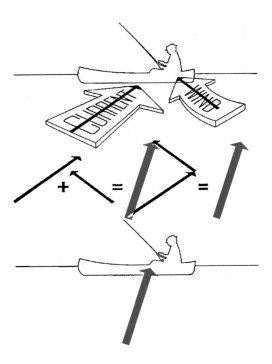

vector resolution

operation that replaces a single vector with two perpendicular vectors such that the vector composition of the two perpendicular vectors yields the original vector

into perpendicular components—a process known as vector resolution— the vector sum of the components always yields a resultant that is equal to the original vector (Figure 3-15). The two perpendicular components, therefore, are a different but equal representation of the original vector.

Graphic Solution of Vector Problems

When vector quantities are uniplanar (contained in a single plane), vector manipulations may be done graphically to yield approximate results. Graphic solution of vector problems requires the careful measurement of vector orientations and lengths to minimize error. Vector lengths, which represent the magnitudes of vector quantities, must be drawn to scale. For example, 1 cm of vector length could represent 10 N of force. A force of 30 N would then be represented by a vector 3 cm in length, and a force of 45 N would be represented by a vector of 4.5 cm length.

Trigonometric Solution of Vector Problems

A more accurate procedure for quantitatively dealing with vector problems involves the application of trigonometric principles. Through the use of trigonometric relationships, the tedious process of measuring and

FIGURE 3-15

Vectors may be resolved into perpendicular components. The vector composition of each perpendicular pair of components yields the original vector.

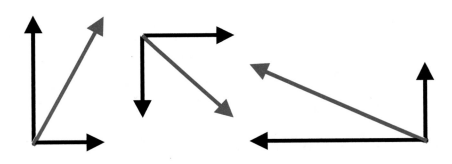

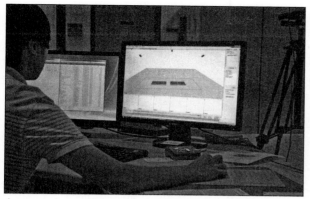

A researcher calibrates force plates in a laboratory in preparation for a motion analysis data capture.

drawing vectors to scale can be eliminated (see Appendix B). Sample Problem 3.4 provides an example of the processes of both graphic and trigonometric solutions using vector quantities.

SAMPLE PROBLEM 3.4

Terry and Charlie must move a refrigerator to a new location. They both push parallel to the floor, Terry with a force of 350 N and Charlie with a force of 400 N, as shown in the diagram below. (a) What is the magnitude of the resultant of the forces produced by Terry and Charlie? (b) If the amount of friction force that directly opposes the direction of motion of the refrigerator is 700 N, will they be able to move the refrigerator?

Graphic Solution

1. Use the scale 1 cm = 100 N to measure the length of the resultant.

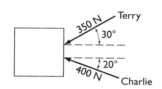

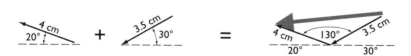

The length of the resultant is approximately 6.75 cm, or   675 N.

2. Since 675 N < 700 N, they will not be able to move the refrigerator.

Trigonometric Solution

Given: $F_T = 350$ N
                                 $F_C = 400$ N

Wanted: magnitude of the resultant force

Horizontal plane free body diagram:

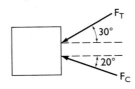

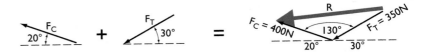

Formula:

$$C^2 = A^2 + B^2 - 2(A)(B)\cos \gamma \text{ (the law of cosines)}$$
$$R^2 = 400^2 + 350^2 - 2(400)(350) \cos 130$$
$$R = 680 \text{ N}$$

3. Since 680 N < 700 N, they will not be able to move the refrigerator unless they exert more collective force while pushing at these particular angles. (If both Terry and Charlie pushed at a 90° angle to the refrigerator, their combined force would be sufficient to move it.)

SUMMARY

Basic concepts related to kinetics include mass, the quantity of matter composing an object; inertia, the tendency of a body to maintain its current state of motion; force, a push or pull that alters or tends to alter a body's state of motion; center of gravity, the point around which a body's weight is balanced; weight, the gravitational force exerted on a body; pressure, the amount of force distributed over a given area; volume, the space occupied by a body; density, the mass or weight per unit of body volume; and torque, the rotational effect of a force.

Several types of mechanical loads act on the human body. These include compression, tension, shear, bending, and torsion. Generally, some combination of these loading modes is present. The distribution of force within a body structure is known as mechanical stress. The nature and magnitude of stress determine the likelihood of injury to biological tissues.

Vector quantities have magnitude and direction; scalar quantities possess magnitude only. Problems with vector quantities can be solved using either a graphic or a trigonometric approach. Of the two procedures, the use of trigonometric relationships is more accurate and less tedious.

INTRODUCTORY PROBLEMS

1. William Perry, defensive tackle and part-time running back better known as "The Refrigerator," weighed in at 1352 N during his 1985 rookie season with the Chicago Bears. What was Perry's mass? (Answer: 138 kg)
2. How much force must be applied to a 0.5 kg hockey puck to give it an acceleration of 30 m/s²? (Answer: 15 N)

3. A rugby player is contacted simultaneously by three opponents who exert forces of the magnitudes and directions shown in the diagram at right. Using a graphic solution, show the magnitude and direction of the resultant force.

4. Using a graphic solution, compose the muscle force vectors to find the net force acting on the scapula shown below.

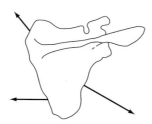

5. Draw the horizontal and vertical components of the vectors shown below.

6. A gymnastics floor mat weighing 220 N has dimensions of 3 m × 4 m × 0.04 m. How much pressure is exerted by the mat against the floor? (Answer: 18.33 Pa)

7. What is the volume of a milk crate with sides of 25 cm, 40 cm, and 30 cm? (Answer: 30,000 cm^3 or 30 l)

8. Choose three objects that are within your field of view, and estimate the volume of each. List the approximate dimensions you used in formulating your estimates.

9. If the contents of the crate described in Problem 7 weigh 120 N, what are the average density and specific weight of the box and contents? (Answer: 0.0004 kg/cm^3; 0.004 N/cm^3)

10. Two children sit on opposite sides of a playground seesaw. Joey, who weighs 220 N, sits 1.5 m from the axis of the seesaw, and Suzy, who weighs 200 N, sits 1.7 m from the axis of the seesaw. How much torque is created at the axis by each child? In which direction will the seesaw tip? (Answer: Joey, 330 N-m; Suzy, 340 N-m; Suzy's end)

ADDITIONAL PROBLEMS

1. What is your own body mass in kg?

2. Gravitational force on planet X is 40% of that found on the earth. If a person weighs 667.5 N on earth, what is the person's weight on planet X? What is the person's mass on the earth and on planet X? (Answer: weight on planet X = 267 N; mass = 68 kg on either planet)

3. A football player is contacted by two tacklers simultaneously. Tackler A exerts a force of 400 N, and tackler B exerts a force of 375 N. If the forces are coplanar and directed perpendicular to each other, what is the magnitude and direction of the resultant force acting on the player? (Answer: 548 N at an angle of 43° to the line of action of tackler A)

4. A 75 kg skydiver in free fall is subjected to a crosswind exerting a force of 60 N and to a vertical air resistance force of 100 N. Describe the resultant force acting on the skydiver. (Answer: 638.6 N at an angle of 5.4° to vertical)

5. Use a trigonometric solution to find the magnitude of the resultant of the following coplanar forces: 60 N at 90°, 80 N at 120°, and 100 N at 270°. (Answer: 49.57 N)

6. If 37% of body weight is distributed above the superior surface of the L5 intervertebral disc and the area of the superior surface of the disc is 25 cm^2, how much pressure exerted on the disc is attributable to body weight for a 930 N man? (Answer: 13.8 N/cm^2)

7. In the nucleus pulposus of an intervertebral disc, the compressive load is 1.5 times the externally applied load. In the annulus fibrosus, the compressive force is 0.5 times the external load. What are the compressive loads on the nucleus pulposus and annulus fibrosus of the L5-S1 intervertebral disc of a 930 N man holding a 445 N weight bar across his shoulders, given that 37% of body weight is distributed above the disc? (Answer: 1183.7 N acts on the nucleus pulposus; 394.5 N acts on the annulus fibrosus.)

8. Estimate the volume of your own body. Construct a table that shows the approximate body dimensions you used in formulating your estimate.

9. Given the mass or weight and the volume of each of the following objects, rank them in the order of their densities.

OBJECT	WEIGHT OR MASS	VOLUME
A	50 kg	15.00 in^3
B	90 lb	12.00 cm^3
C	3 slugs	1.50 ft^3
D	450 N	0.14 m^3
E	45 kg	30.00 cm^3

10. Two muscles develop tension simultaneously on opposite sides of a joint. Muscle A, attaching 3 cm from the axis of rotation at the joint, exerts 250 N of force. Muscle B, attaching 2.5 cm from the joint axis, exerts 260 N of force. How much torque is created at the joint by each muscle? What is the net torque created at the joint? In which direction will motion at the joint occur? (Answer: A, 7.5 N-m; B, 6.5 N-m; net torque equals 1 N-m in the direction of A)

LABORATORY EXPERIENCES

1. Use a ruler to measure the dimensions of the sole of one of your shoes in centimeters. Being as accurate as possible, calculate an estimate of the surface area of the sole. (If a planimeter is available, use it to more accurately assess surface area by tracing around the perimeter of the sole.) Knowing your own body weight, calculate the amount of pressure exerted over the sole of one shoe. How much change in pressure would result if your body weight changed by 22 N (5 lb)?

Surface area calculation:

Surface area: _____

Body weight: _____

Pressure calculation:

Pressure: _____

Pressure calculation with 22 N (5 lb) change in body weight:

Pressure: _____

2. Place a large container filled three-quarters full of water on a scale and record its weight. To assess the volume of an object of interest, completely submerge the object in the container, holding it just below the surface of the water. Record the *change* in weight on the scale. Remove the object from the container. Carefully pour water from the container into a measuring cup until the container weighs its original weight less the change in weight recorded. The volume of water in the measuring cup is the volume of the submerged object. (Be sure to use correct units when recording your measured values.)

Weight of container of water: _____

Change in weight with object submerged: _____

Volume of object: _____

3. Secure one end of a pencil by firmly clamping it in a vise. Grip the other end of the pencil with an adjustable wrench and slowly apply a bending load to the pencil until it begins to break. Observe the nature of the break.

On which side of the pencil did the break begin? _____

Is the pencil stronger in resisting compression or tension? _____

Repeat the exercise using another pencil and applying a torsional (twisting) load. What does the nature of the initial break indicate about the distribution of shear stress within the pencil?

4. Experiment with pushing open a door by applying force with one finger. Apply force at distances of 10 cm, 20 cm, 30 cm, and 40 cm from the hinges. Write a brief paragraph explaining at which force application distance it is easiest/hardest to open the door.

5. Stand on a bathroom scale and perform a vertical jump as a partner carefully observes the pattern of change in weight registered on the scale. Repeat the jump several times, as needed for your partner to determine the pattern. Trade positions and observe the pattern of weight change as your partner performs a jump. In consultation with your partner, sketch a graph of the change in exerted force (vertical axis) across time (horizontal axis) during the performance of a vertical jump.

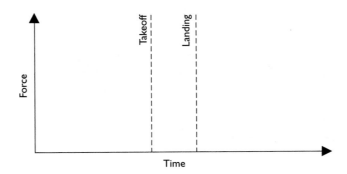

What does the area under the curve represent? _____

REFERENCES

1. Yeow CH, Lee PV, and Goh JC: Direct contribution of axial impact compressive load to anterior tibial load during simulated ski landing impact, *J Biomech* 43:242, 2010.

ANNOTATED READINGS

Caldwell GE, Hamill J, Kmen G, Whittlesey SN, and Robertson DGE: *Research methods in biomechanics,* Champaign, IL, 2004, Human Kinetics.
  Includes chapters on kinetics, forces and their measurement, inverse dynamics, and electromyography, among others.
Kamen G and Gabriel D: *Essentials of electromyography,* Champaign, IL, 2010, Human Kinetics.
  Presents both fundamental and advanced concepts related to collection, analysis, and interpretation of electromyography data.
LeVeau BF: *Biomechanics of human motion: Basics and beyond for the health professions,* Thorofare, NJ, 2010, SLACK.
  Discusses mechanical loading on the human body in addition to other topics.
Winter DA: *Biomechanics and motor control of human movement* (4th ed.), Hoboken, NJ, 2009, John Wiley and Sons.
  Describes a wide spectrum of biomechanical movement analysis techniques, among other topics.

RELATED WEBSITES

Advanced Medical Technology, Inc.
http://www.amtiweb.com
  Provides information on the AMTI force platforms, with reference to force and torque sensors, gait analysis, balance and posture, and other topics.
B & L Engineering
http://www.bleng.com/
  Describes electromyography systems and footswitches for gait analysis.
Biokinetics and Associates, Ltd.
http://www.biokinetics.com/
  Markets products designed to prevent injury.
Bortec Biomedical Ltd.
http://www.bortec.ca/pages/home.htm
  Describes a multichannel telemetered electromyography system.
Delsys, Inc.
http://www.delsys.com/
  Provides a description of surface electromyography equipment.
Kistler
http://www.kistler.com
  Describes a series of force platforms.
RSscan
http://www.rsscan.co.uk/users/university.php
  Describes a within-shoe pressure measurement system.

KEY TERMS

acute loading	application of a single force of sufficient magnitude to cause injury to a biological tissue
axial	directed along the longitudinal axis of a body
bending	asymmetric loading that produces tension on one side of a body's longitudinal axis and compression on the other side

center of gravity	point around which a body's weight is equally balanced, no matter how the body is positioned
combined loading	simultaneous action of more than one of the pure forms of loading
compression	pressing or squeezing force directed axially through a body
deformation	change in shape
density	mass per unit of volume
failure	loss of mechanical continuity
force	push or pull; the product of mass and acceleration
free body diagram	sketch that shows a defined system in isolation with all of the force vectors acting on the system
impulse	product of force and the time over which the force acts
inertia	tendency of a body to resist a change in its state of motion
mass	quantity of matter contained in an object
myoelectric activity	electric current or voltage produced by a muscle developing tension
net force	resultant force derived from the composition of two or more forces
pressure	force per unit of area over which a force acts
repetitive loading	repeated application of a subacute load that is usually of relatively low magnitude
resultant	single vector that results from vector composition
scalar	physical quantity that is completely described by its magnitude
shear	force directed parallel to a surface
specific weight	weight per unit of volume
stress	distribution of force within a body, quantified as force divided by the area over which the force acts
tension	pulling or stretching force directed axially through a body
torque	rotary effect of a force
torsion	load-producing twisting of a body around its longitudinal axis
transducers	devices that detect signals
vector	physical quantity that possesses both magnitude and direction
vector composition	process of determining a single vector from two or more vectors by vector addition
vector resolution	operation that replaces a single vector with two perpendicular vectors such that the vector composition of the two perpendicular vectors yields the original vector
volume	space occupied by a body
weight	attractive force that the earth exerts on a body
yield point (elastic limit)	point on the load deformation curve past which deformation is permanent

The Biomechanics of Human Bone Growth and Development

4

After completing this chapter, you will be able to:

Explain how the material constituents and structural organization of bone affect its ability to withstand mechanical loads.

Describe the processes involved in the normal growth and maturation of bone.

Describe the effects of exercise and of weightlessness on bone mineralization.

Explain the significance of osteoporosis and discuss current theories on its prevention.

Explain the relationship between different forms of mechanical loading and common bone injuries.

ONLINE LEARNING CENTER RESOURCES

www.mhhe.com/hall6e

Log on to our Online Learning Center (OLC) for access to these additional resources:

- Online Lab Manual
- Flashcards with definitions of chapter key terms
- Chapter objectives
- Chapter lecture PowerPoint presentation
- Self-scoring chapter quiz
- Additional chapter resources
- Web links for study and exploration of chapter-related topics

W hat determines when a bone stops growing? How are stress fractures caused? Why does space travel cause reduced bone mineral density in astronauts? What is osteoporosis and how can it be prevented?

The word *bone* typically conjures up a mental image of a dead bone—a dry, brittle chunk of mineral that a dog would enjoy chewing. Given this picture, it is difficult to realize that living bone is an extremely dynamic tissue that is continually modeled and remodeled by the forces acting on it. Bone fulfills two important mechanical functions for human beings: (a) It provides a rigid skeletal framework that supports and protects other body tissues, and (b) it forms a system of rigid levers that can be moved by forces from the attaching muscles (see Chapter 12). This chapter discusses the biomechanical aspects of bone composition and structure, bone growth and development, bone response to stress, osteoporosis, and common bone injuries.

COMPOSITION AND STRUCTURE OF BONE TISSUE

The material constituents and structural organization of bone influence the ways in which bone responds to mechanical loading. The composition and structure of bone yield a material that is strong for its relatively light weight.

Material Constituents

The major building blocks of bone are calcium carbonate, calcium phosphate, collagen, and water. The relative percentages of these materials vary with the age and health of the bone. Calcium carbonate and calcium phosphate generally constitute approximately 60–70% of dry bone weight. These minerals give bone its stiffness and are the primary determiners of its compressive strength. Other minerals, including magnesium, sodium, and fluoride, also have vital structural and metabolic roles in bone growth and development. Collagen is a protein that provides bone with flexibility and contributes to its tensile strength.

The water content of bone makes up approximately 25–30% of total bone weight. The water present in bone tissue is an important contributor to bone strength. For this reason, scientists and engineers studying the material properties of different types of bone tissue must ensure that the bone specimens they are testing do not become dehydrated. The flow of water through bones also carries nutrients to and waste products away from the living bone cells within the mineralized matrix. In addition, water transports mineral ions to and from bone for storage and subsequent use by the body tissues when needed.

Structural Organization

The relative percentage of bone mineralization varies not only with the age of the individual but also with the specific bone in the body. Some bones are more porous than others. The more porous the bone, the smaller the proportion of calcium phosphate and calcium carbonate, and the greater the proportion of nonmineralized tissue. Bone tissue has been classified into two categories based on porosity (Figure 4-1). If the porosity is low, with 5–30% of bone volume occupied by nonmineralized tissue, the tissue is termed cortical bone. Bone tissue with a relatively high porosity, with 30% to greater than 90% of bone volume occupied by nonmineralized tissue, is known as spongy, cancellous, or

lever
a relatively rigid object that may be made to rotate about an axis by the application of force

stiffness
ratio of stress to strain in a loaded material—that is, the stress divided by the relative amount of change in the structure's shape

compressive strength
ability to resist pressing or squeezing force

tensile strength
ability to resist pulling or stretching force

•*Collagen resists tension and provides flexibility to bone.*

porous
containing pores or cavities

cortical bone
compact mineralized connective tissue with low porosity that is found in the shafts of long bones

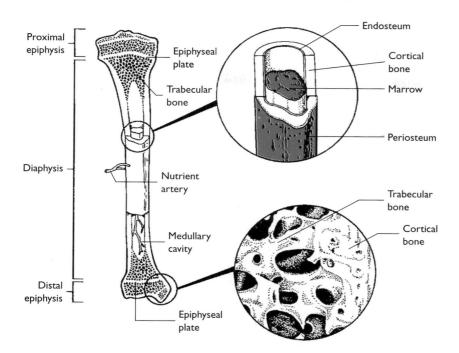

FIGURE 4-1

Structures of cortical and trabecular bone.

trabecular bone. Trabecular bone has a honeycomb structure with mineralized vertical and horizontal bars, called *trabeculae,* forming cells filled with marrow and fat.

The porosity of bone is of interest because it directly affects the mechanical characteristics of the tissue. With its higher mineral content, cortical bone is stiffer, so that it can withstand greater stress, but less strain or relative deformation, than trabecular bone. Because trabecular bone is spongier than cortical bone, it can undergo more strain before fracturing.

The function of a given bone determines its structure. The shafts of the long bones are composed of strong cortical bone. The relatively high

trabecular bone
less compact mineralized connective tissue with high porosity that is found in the ends of long bones and in the vertebrae

strain
amount of deformation divided by the original length of the structure or by the original angular orientation of the structure

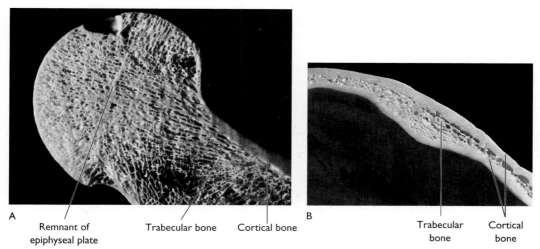

(**A**) In the femur, trabecular bone is encased by a thin layer of cortical bone.
(**B**) In the skull, trabecular bone is sandwiched between plates of cortical bone.
From Shier, Butler, and Lewis. *Hole's Human Anatomy and Physiology,* © 1996. Reprinted by permission of The McGraw-Hill Companies, Inc.

FIGURE 4-2

Relative bone strength in resisting compression, tension, and shear.

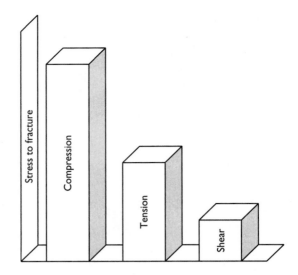

●*Because cortical bone is stiffer than trabecular bone, it can withstand greater stress but less strain.*

anisotropic
exhibiting different mechanical properties in response to loads from different directions

●*Bone is strongest in resisting compression and weakest in resisting shear.*

axial skeleton
the skull, vertebrae, sternum, and ribs

appendicular skeleton
bones composing the body appendages

short bones
small, cubical skeletal structures, including the carpals and tarsals

flat bones
skeletal structures that are largely flat in shape—for example, the scapula

irregular bones
skeletal structures of irregular shape—for example, the sacrum

long bones
skeletal structures consisting of a long shaft with bulbous ends—for example, the femur

articular cartilage
protective layer of firm, flexible connective tissue over the articulating ends of long bones

trabecular bone content of the vertebrae contributes to their shock-absorbing capability.

Both cortical and trabecular bone are anisotropic; that is, they exhibit different strength and stiffness in response to forces applied from different directions. Bone is strongest in resisting compressive stress and weakest in resisting shear stress (Figure 4-2).

Types of Bones

The structures and shapes of the 206 bones of the human body enable them to fulfill specific functions. The skeletal system is nominally subdivided into the central or axial skeleton and the peripheral or appendicular skeleton (Figure 4-3). The axial skeleton includes the bones that form the axis of the body, which are the skull, the vertebrae, the sternum, and the ribs. The other bones form the body appendages, or the appendicular skeleton. Bones are also categorized according to their general shapes and functions.

Short bones, which are approximately cubical, include only the carpals and the tarsals (Figure 4-4). These bones provide limited gliding motions and serve as shock absorbers.

Flat bones are also described by their name (Figure 4-4). These bones protect underlying organs and soft tissues and also provide large areas for muscle and ligament attachments. The flat bones include the scapulae, sternum, ribs, patellae, and some of the bones of the skull.

Irregular bones have different shapes to fulfill special functions in the human body (Figure 4-4). For example, the vertebrae provide a bony, protective tunnel for the spinal cord; offer several processes for muscle and ligament attachments; and support the weight of the superior body parts while enabling movement of the trunk in all three cardinal planes. The sacrum, coccyx, and maxilla are other examples of irregular bones.

Long bones form the framework of the appendicular skeleton (Figure 4-4). They consist of a long, roughly cylindrical shaft (also called the *body*, or *diaphysis*) of cortical bone, with bulbous ends known as condyles, tubercles, or tuberosities. A self-lubricating articular cartilage protects the

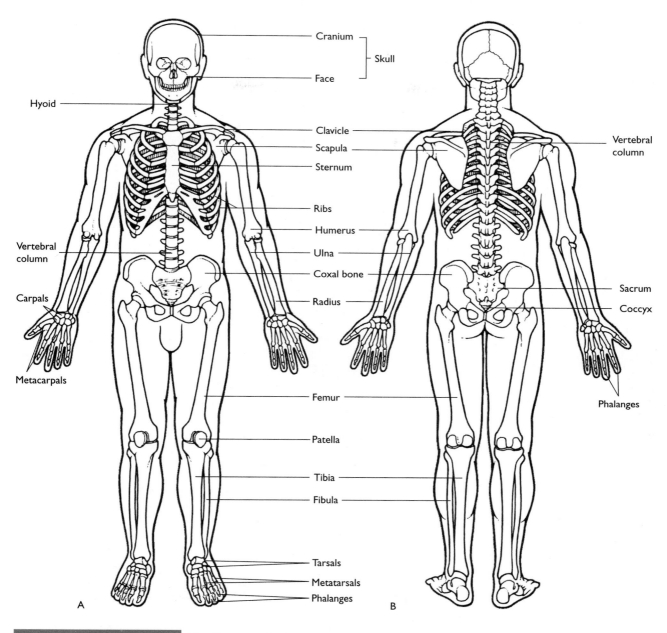

FIGURE 4-3   **The human skeleton.**

ends of long bones from wear at points of contact with other bones. Long bones also contain a central hollow area known as the *medullary cavity* or *canal*.

The long bones are adapted in size and weight for specific biomechanical functions. The tibia and femur are large and massive to support the weight of the body. The long bones of the upper extremity, including the humerus, radius, and ulna, are smaller and lighter to promote ease of movement. Other long bones include the clavicle, fibula, metatarsals, metacarpals, and phalanges.

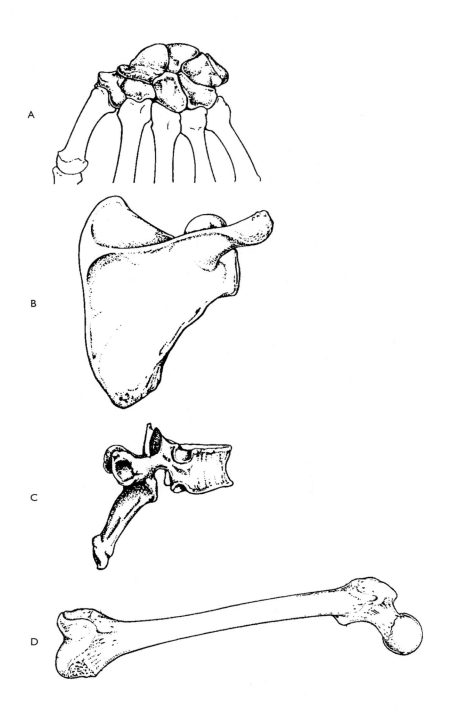

FIGURE 4-4

A. The carpals are categorized as short bones. **B.** The scapula is categorized as a flat bone. **C.** The vertebrae are examples of irregular bones. **D.** The femur represents the long bones.

A

B

C

D

BONE GROWTH AND DEVELOPMENT

Bone growth begins early in fetal development, and living bone is continually changing in composition and structure during the life span. Many of these changes represent normal growth and maturation of bone.

Longitudinal Growth

Longitudinal growth of a bone occurs at the epiphyses, or epiphyseal plates (Figure 4-5). The epiphyses are cartilaginous discs found near the ends of the long bones. The diaphysis (central) side of each epiphysis

●*Most epiphyses close around age 18, although some may be present until about age 25.*

epiphysis
growth center of a bone that produces new bone tissue as part of the normal growth process until it closes during adolescence or early adulthood

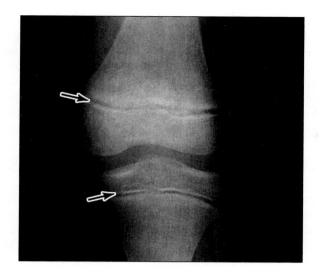

FIGURE 4-5

Epiphyseal plates are the sites of longitudinal growth in immature bone. From Shier, Butler, and Lewis. *Hole's Human Anatomy and Physiology,* © 1996. Reprinted by permission of The McGraw-Hill Companies, Inc.

continually produces new bone cells. During or shortly following adolescence, the plate disappears and the bone fuses, terminating longitudinal growth. Most epiphyses close around age 18, although some may be present until about age 25.

Circumferential Growth

Long bones grow in diameter throughout most of the life span, although the most rapid bone growth occurs before adulthood. The internal layer of the periosteum builds concentric layers of new bone tissue on top of existing ones. At the same time, bone is resorbed or eliminated around the circumference of the medullary cavity, so that the diameter of the cavity is continually enlarged. This occurs in such a way that both bending stresses and torsional stresses on the bones remain relatively constant (65).

These changes in bone size and shape are the work of specialized cells called osteoblasts and osteoclasts, which respectively form and resorb bone tissue. In healthy adult bone, the activity of osteoblasts and osteoclasts is largely balanced.

Adult Bone Development

There is a progressive loss of collagen and increase in bone brittleness with aging. Thus, the bones of children are more pliable than the bones of adults.

Bone mineral normally accumulates throughout childhood and adolescence, reaching a peak at about age 25–28 in women and age 30–35 in men (55). Following this peak, researchers disagree as to the length of time that bone density remains constant (62). However, an age-related, progressive decline in bone density and bone strength in both men and women may begin as soon as the early twenties (46). This involves a progressive diminishment in the mechanical properties and general toughness of bone, with increasing loss of bone substance and increasing porosity (15). Trabecular bone is particularly affected, with progressive disconnection and disintegration of trabeculae compromising the integrity of the bone's structure and seriously diminishing bone strength (41).

These changes are much more pronounced in women than in men, however. In women, there is a notable decrease in both volume and density of cortical bone, and a decrease in the density of trabecular bone

periosteum
double-layered membrane covering bone; muscle tendons attach to the outside layer, and the internal layer is a site of osteoblast activity

osteoblasts
specialized bone cells that build new bone tissue

osteoclasts
specialized bone cells that resorb bone tissue

with aging (67). Approximately 0.5–1.0% of bone mass is lost each year, until women reach about age 50 or menopause (62). Following menopause, there appears to be an increased rate of bone loss, with values as high as 6.5% per year reported during the first five to eight years (36). Although similar changes occur in men, they do not become significant before a more advanced age. Women at all ages tend to have smaller bones and less cortical bone area than do men (65), although volumetric bone mineral density is similar for both genders (69).

BONE RESPONSE TO STRESS

Other changes that occur in living bone throughout the life span are unrelated to normal growth and development. Bone responds dynamically to the presence or absence of different forces with changes in size, shape, and density. This phenomenon was originally described by the German scientist Julius Wolff in 1892:

> The form of a bone being given, the bone elements place or displace themselves in the direction of functional forces and increase or decrease their mass to reflect the amount of the functional forces (79).

Bone Modeling and Remodeling

● *Wolff's law indicates that bone strength increases and decreases as the functional forces on the bone increase and decrease.*

According to Wolff's law, the densities and, to a much lesser extent, the shapes and sizes of the bones of a given human being are a function of the magnitude and direction of the mechanical stresses that act on the bones. Dynamic mechanical loading causes bones to deform or strain, with larger loads producing higher levels of strain. These strains are translated into changes in bone shape and strength through a process known as

The structure of a long bone.
From Shier, Butler, and Lewis, *Hole's Human Anatomy and Physiology,* © 1996. Reprinted by permission of The McGraw-Hill Companies, Inc.

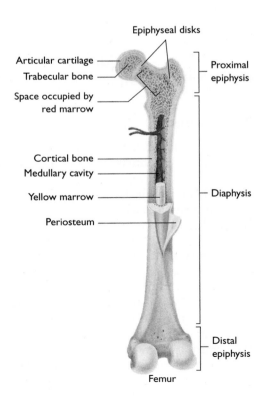

Epiphyseal disks

Articular cartilage

Trabecular bone

Space occupied by red marrow

Proximal epiphysis

Cortical bone

Medullary cavity

Yellow marrow

Periosteum

Diaphysis

Distal epiphysis

Femur

remodeling. Remodeling involves resorption of fatigue-damaged older bone and subsequent formation of new bone (43). *Bone modeling* is the term given to formation of new bone that is not preceded by resorption, and is the process by which immature bones grow.

Adult bones gain or lose mass in accordance with Wolff's law. When strain on a bone exceeds a certain threshold, new bone is laid down at the strain sites, and overall bone mass and density are increased. When strain magnitudes stay below a lower threshold, bone remodeling occurs, with bone removed close to the marrow (18). Strain magnitudes in between these two thresholds occur in what is termed the "lazy zone" and do not trigger bone adaptation (73). Remodeling can occur in either "conservation mode," with no change in bone mass, or "disuse mode," with a net loss of bone mass characterized by an enlarged marrow cavity and thinned cortex (18). Bone is a very dynamic tissue, with the modeling and remodeling processes continuously acting to increase, decrease, or reshape bone.

The modeling and remodeling processes are directed by osteocytes, cells embedded in bone that are sensitive to changes in the flow of interstitial fluid through the pores resulting from strain on the bone (68). Dynamic loading resulting from high-level impact produces a high rate of deformation that best pushes fluid through the bone matrix (61). It is for this reason that activities involving high levels of impact are best at stimulating bone formation. In response to the motion of fluid within the bone matrix, osteocytes trigger the actions of osteoblasts and osteoclasts, the cells that respectively form and resorb bone (61). A predominance of osteoblast activity produces bone modeling, with a net gain in bone mass. Bone remodeling involves a balance of osteoblast and osteoclast action or a predominance of osteoclast activity, with associated maintenance or loss of bone mass. Approximately 25% of the body's trabecular bone is remodeled each year through this process (29). Strains resulting from an activity such as walking are sufficient to provoke bone turnover and new bone formation (80).

Thus, bone mineralization and bone strength in both children and adults are a function of stresses producing strains on the skeleton. Since body weight provides the most constant mechanical stress to bones, bone mineral density generally parallels body weight, with heavier individuals having more massive bones. Adults who gain or lose weight tend to also gain or lose bone mineral density (17). However, a given individual's physical activity profile, diet, lifestyle, and genetics can also dramatically influence bone density. Factors such as lean body mass, muscle strength, and regular participation in weight-bearing exercise have been shown to exert stronger influences on bone density than weight, height, and race (20, 38, 76). Dynamic loading during participation in gymnastics has been shown to affect bone size and strength more than muscle mass (16). Even in young, nonathletic children, bone appears to remodel in response to the presence or absence of physical activity (31).

The malleability of bone is dramatically exemplified by the case of an infant who was born in normal physical condition but missing one tibia, the major weight-bearing bone of the lower extremity. After the child was walking for a time, X-rays revealed that modeling of the fibula in the abnormal leg had occurred to the extent that it could not be distinguished from the tibia of the other leg (1).

Another interesting case is that of a construction worker who had lost all but the fifth finger of one hand in a war injury. After 32 years, the

● *The processes causing bone remodeling are not fully understood and continue to be researched by scientists.*

metacarpal and phalanx of the remaining finger had been modeled to resemble the third finger of the other hand (58).

Bone Hypertrophy

bone hypertrophy
increase in bone mass resulting from a predominance of osteoblast activity

Although cases of complete changes in bone shape and size are unusual, there are many examples of bone modeling, or bone hypertrophy, in response to regular physical activity. The bones of physically active individuals tend to be denser and therefore more mineralized than those of sedentary individuals of the same age and gender. Moreover, the results of several studies indicate that occupations and sports particularly stressing a certain limb or region of the body produce accentuated bone hypertrophy in the stressed area. For example, professional tennis players display not only muscular hypertrophy in the tennis arm but also hypertrophy of that arm's radius (35). Similar bone hypertrophy has been observed in the dominant humerus of baseball players (49).

It also appears that the greater the forces or loads habitually encountered, the more dramatic the increased mineralization of the bone. In one study involving collegiate female athletes in basketball, volleyball, soccer, track, and swimming, the athletes participating in high-impact sports (basketball and volleyball) were found to have higher bone mineral densities and bone formation values than the swimmers (12). In another investigation, the bone mineral densities of trained runners and cyclists were compared to those of sedentary individuals of the same age (65). Compared to the nonexercisers, the runners were found to have increased bone density, although this was not true for the cyclists. Among older women, both yard work and weight training have been found to be strong predictors for bone density, with jogging, swimming, and calisthenics being weak predictors (70). On the whole, the research evidence suggests that physical activity involving impact forces is necessary for maintaining or increasing bone mass (51). Competitive swimmers, who spend a lot of time in the water where the buoyant force counteracts gravity, may have bone mineral densities lower than those of sedentary individuals (56).

Bone Atrophy

bone atrophy
decrease in bone mass resulting from a predominance of osteoclast activity

Whereas bone hypertrophies in response to increased mechanical stress, it displays the opposite response to reduced stress. When the normal stresses exerted on bone by muscle contractions, weight bearing, or impact forces are reduced, bone tissue atrophies through remodeling. When bone atrophy occurs, the amount of calcium contained in the bone diminishes, and both the weight and the strength of the bone decrease. Loss of bone mass due to reduced mechanical stress has been found in bedridden patients, sedentary senior citizens, and astronauts. Four to six weeks of bed rest can result in significant decrements in bone mineral density that are not fully reversed after six months of normal weight-bearing activity (5).

Bone demineralization is a potentially serious problem. From a biomechanical standpoint, as bone mass diminishes, strength and thus resistance to fracture also decrease, particularly in trabecular bone.

The results of calcium loss studies conducted during the Skylab flights indicate that urinary calcium loss is related to time spent out of the earth's gravitational field. The pattern of bone loss observed is highly similar to that documented among patients during periods of bed rest, with greater bone loss in the weight-bearing bones of the

Loss of bone mass during periods of time spent outside of the earth's gravitational field is a problem for astronauts. Photo courtesy of NASA.

lumbar spine and lower extremity than in other parts of the skeleton (64). During one month in space, astronauts lose 1–3% of bone mass, or approximately as much bone mass as postmenopausal women lose in a year (8, 23).

It is not yet clear what specific mechanism or mechanisms are responsible for bone loss outside of the gravitational field. Research has consistently documented a negative calcium balance in astronauts and experimental animals during space flight, with reduced intestinal absorption of calcium and increased excretion of calcium (71). It is not known, however, whether this is caused by an increase in bone remodeling, a decrease in bone remodeling, or an imbalance between osteoblast and osteoclast activity (71). It appears that the normal balance between formation and resorption of bone becomes disturbed, with an initial increase in osteoclast activity followed by a prolonged decrease in osteoblast activity (40). One hypothesis is that these changes in bone remodeling are precipitated by changes in bone blood flow related to being outside of the gravitational field (10). More research on this topic is clearly needed.

It remains to be seen if measures other than the artificial creation of gravity can effectively prevent bone loss during space travel. Astronauts' current exercise programs during flights in space are designed to prevent bone loss by increasing the mechanical stress and strain placed on bones using muscular force. However, the muscles of the body exert mainly tensile forces on bone, whereas gravity provides a compressive force. Therefore, it may be that no amount of physical exercise alone can completely compensate for the absence of gravitational force.

Recent research shows that resistive exercise combined with whole-body vibration may be an effective countermeasure for preventing muscle atrophy and bone loss during space flight (57). Researchers hypothesize that low-amplitude, high-frequency vibration stimulates muscle spindles and alpha-motoneurons (see Chapter 6), which initiate muscle contraction (34, 53). The effects of several months' intervention treatment with whole-body vibration appear to include improved bone mineral density resulting from increased bone deposition coupled with decreased bone resorption, with bone density particularly improved in the femur and tibia (30, 45).

Since joints support the body weight positioned above them, the magnitude of skeletal loading varies from joint to joint during both resistance exercise and vibration. Sample Problem 4.1 illustrates this point.

OSTEOPOROSIS

osteoporosis
disorder involving decreased bone mass and strength with one or more resulting fractures

osteopenia
condition of reduced bone mineral density that predisposes the individual to fractures

Bone atrophy is a problem not only for astronauts and bedridden patients but also for a growing number of senior citizens and female athletes. Osteoporosis is found in most elderly individuals, with earlier onset in women, and is becoming increasingly prevalent with the increasing mean age of the population (37). The condition begins as osteopenia, reduced bone mass without the presence of a fracture, but often progresses to osteoporosis, a condition in which bone mineral mass and strength are so severely compromised that daily activities can cause bone pain and fracturing (50).

SAMPLE PROBLEM 4.1

The tibia is the major weight-bearing bone in the lower extremity. If 88% of body mass is proximal to the knee joint, how much compressive force acts on each tibia when a 600 N person stands in anatomical position? How much compressive force acts on each tibia if the person holds a 20 N sack of groceries?

Solution

Given: wt = 600 N
(It may be deduced that weight = compressive force, F_c.)

Formula: F_c on knees $= (600 \, N)(0.88)$

$$F_c \text{ on one knee} = \frac{(600 \, N)(0.88)}{2}$$

$$F_c \text{ on one knee} = 264 \, N$$

$$F_c \text{ with groceries} = \frac{(600 \, N)(0.88) + 20 \, N}{2}$$

$$F_c \text{ with groceries} = 274 \, N$$

Postmenopausal and Age-Associated Osteoporosis

The majority of those affected by osteoporosis are postmenopausal and elderly women, although elderly men are also susceptible, with more than half of all women and about one-third of men developing fractures related to osteoporosis (32). Although it was once regarded as primarily a health concern for women, with the increasing age of the population, osteoporosis is now also emerging as a serious health-related concern for men (69). Risk factors for osteoporosis include being female, white or Asian ethnicity, older age, small stature or frame size, and family history of osteoporosis (32).

●*Osteoporosis is a serious health problem for most elderly individuals, with women affected more dramatically than men.*

Type I osteoporosis, or postmenopausal osteoporosis, affects approximately 40% of women after age 50 (39). The first osteoporotic fractures usually begin to occur about 15 years postmenopause, with women suffering approximately three times as many femoral neck fractures, three times as many vertebral fractures, and six times as many wrist fractures as men of the same age (39).

This discrepancy occurs partially because men reach a higher peak of bone mass and strength than women in early adulthood, and partially because of a greater prevalence of disconnections in the trabecular network among postmenopausal women than among men (47).

Type II osteoporosis, or age-associated osteoporosis, affects most women and also affects men after age 70 (69). After age 60, about 90% of all fractures in both men and women are osteoporosis-related, and these fractures are one of the leading causes of death in the elderly population (50).

Although the radius and ulna, femoral neck, and spine are all common sites of osteoporotic fractures, the most common symptom of osteoporosis is back pain derived from fractures of the weakened trabecular bone of the vertebral bodies. Crush fractures of the lumbar vertebrae resulting from compressive loads created by weight bearing during activities of daily living frequently cause reduction of body height. Because most body weight is anterior to the spine, the resulting fractures often leave the vertebral bodies wedge-shaped, accentuating thoracic kyphosis (see Chapter 9). This disabling deformity is known as *dowager's hump*. Vertebral compression fractures are extremely painful and debilitating and affect physical, functional, and psychosocial aspects of the person's life. As spinal height is lost, there is added discomfort from the rib cage pressing on the pelvis.

●*Painful, deforming, and debilitating crush fractures of the vertebrae are the most common symptom of osteoporosis.*

As the skeleton ages in men, there is an increase in vertebral diameter that serves to reduce compressive stress during weight bearing (47). Thus, although osteoporotic changes may be taking place, the structural strength of the vertebrae is not reduced. Why the same compensatory change does not occur in women is unknown.

Female Athlete Triad

The desire to excel at competitive sports causes some young female athletes to strive to achieve an undesirably low body weight. This dangerous practice commonly involves a combination of disordered eating, amenorrhea, and osteoporosis, a combination that has come to be known as the "female athlete triad." This condition often goes unrecognized, but because the triad can result in negative consequences ranging from irreversible bone loss to death, friends, parents, coaches, and physicians need to be alert to the signs and symptoms.

amenorrhea
cessation of menses

●*Disordered eating, amenorrhea, and osteoporosis constitute a dangerous and potentially lethal triad for young female athletes.*

As many as 62% of female athletes in certain sports display disordered eating behaviors, with those participating in endurance or artistic sports such as gymnastics and figure skating most likely to be involved (48). Prolonged disordered eating can lead to anorexia nervosa or bulimia nervosa, illnesses that affect 1–10% of all adolescent and college-age women (22).

●*Anorexia nervosa and bulimia nervosa are life-threatening eating disorders.*

Competitive female athletes in endurance and appearance-related sports are particularly at risk for developing the dangerous female athlete triad. Photo courtesy of Royalty-Free/ CORBIS.

●*Regular exercise has been shown to be effective to some extent in mediating age-related bone loss.*

Symptoms of anorexia nervosa in girls and women include body weight 15% or more below minimal normal weight for age and height, an intense fear of gaining weight, a disturbed body image, and amenorrhea. Symptoms of bulimia nervosa are a minimum of two eating binges a week for at least three months, a feeling of lack of control during binges, regular use of self-induced vomiting, laxatives, diuretics, strict dieting, or exercise to prevent weight gain, and excessive concern with body image and weight (22). Disordered eating behavior has been found to be strongly associated with both menstrual irregularity and low bone mineral density (9).

The relationship between disordered eating and amenorrhea appears to be related to a decrease in hypothalamus secretion of gonadotrophin-releasing hormone, which in turn decreases the secretion of luteinizing hormone and follicle-stimulating hormone, with subsequent shutting down of stimulation of the ovary (75). The prevalence of primary amenorrhea, with menarche delayed beyond 16 years of age, is less than 1% in the general population, but as high as 22% in sports such as cheerleading, diving, and gymnastics (48). Secondary amenorrhea, or the absence of three to six consecutive menstrual cycles, has been found to be present in 69% of dancers and 65% of long distance runners, as compared to 2–5% in the general population (48).

The link between cessation of menses and osteoporosis is estrogen deficiency, which increases bone resorption. Energy deficiency resulting from disordered eating is also likely to independently contribute to altered bone metabolism and reduced bone density (14). Although the incidence of osteoporosis among female athletes is unknown, the consequences of this disorder in young women are potentially tragic. Among one group of over 200 premenopausal female runners, those with amenorrhea had 10% less lumbar bone density than those with normal menses (25). This is of particular concern for adolescent athletes, because roughly 50% of bone mineralization and 15% of adult height are normally established during the teenage years (2). Not surprisingly, amenorrheic premenopausal female athletes have a high rate of stress fractures, with more fractures related to later onset of menarche (48). Moreover, the loss of bone that occurs may be irreversible, and osteoporotic wedge fractures can ruin posture for life.

Preventing and Treating Osteoporosis

Osteoporosis is neither a disease with acute onset nor an inevitable accompaniment of aging, but is the result of a lifetime of habits that are erosive to the skeletal system. Early detection of low bone mineral density is advantageous, because once osteoporotic fractures begin to occur, there has been irreversible loss of trabecular structure (60). Although proper diet, hormone levels, and exercise can work to increase bone mass at any stage in life, evidence suggests that it is easier to prevent osteoporosis than it is to treat it. The single most important factor for preventing or prolonging the onset of osteoporosis is the optimization of peak bone mass during childhood and adolescence (6, 9, 24, 32, 50, 74). Researchers hypothesize that weight-bearing exercise is particularly crucial during the prepubertal years, because the presence of high levels of growth hormone may act with exercise in a synergistic fashion to increase bone density (3, 6, 22, 33, 36). Activities involving osteogenic impact forces, such as jumping, have been shown to be effective in increasing bone mass in children (19).

Weight-bearing physical activity is necessary for maintaining skeletal integrity in both humans and animals. Importantly, studies show that a regular program of weight-bearing exercise, such as walking, can increase bone health and strength even among individuals with osteoporosis.

Because impact loading is particularly osteogenic, jumping in place, with 5–10 sets of 10 jumps done 3–5 times per week, is also recommended for maintenance of bone mass (77). Jumps should be performed with 10–15-second rest intervals between jumps, as this appears to enhance fluid flow within the bone matrix and the related stimulation of osteocytes, potentially doubling the effects of mechanical loading on bone building (21, 52). In practical terms, a very slow childhood game of hopscotch favors bone building over a fast one!

Increased dietary calcium intake exerts a positive influence on bone mass for women with a dietary deficiency, with the amount of calcium absorbed influenced positively by calcitriol (the active form of vitamin D) and negatively by dietary fiber (63). Although adequate dietary calcium is particularly important during the teenage years, unfortunately the median American girl falls below the recommended daily intake of 1200 mg per day by age 11 (13). A modified diet or calcium supplementation can be critical for the development of peak bone mass among adolescent females at a dietary deficiency. The role of vitamin D in enabling absorption of calcium by bone is also important, with over half the women receiving treatment for low bone density in North America having a vitamin D deficiency (26). Clinicians are now recognizing that a predisposition for osteoporosis can begin in childhood and adolescence when a poor diet interferes with bone mass development (7).

Other lifestyle factors also affect bone mineralization. Known risk factors for developing osteoporosis include physical inactivity; weight loss or excessive thinness; tobacco smoking; deficiencies in estrogen, calcium, and vitamin D; and excessive consumption of protein and caffeine (54, 62, 72, 78). A study of female twins, one of whom smoked more heavily than the other, showed that women who smoke one pack of cigarettes a day through adulthood will have a reduction in bone density of 5–10% by the time of menopause, which is sufficient to increase the risk of fracture (27). Although caffeine consumption may negatively affect bone mineral density among postmenopausal women who consume low amounts of dietary calcium, it has been shown *not* to affect bone mineral density among young women (11). Genetic factors also influence bone mass but do not appear to be as important as diet and exercise.

Recent, detailed studies of bone are increasingly showing that subtleties in bone microarchitecture may be more important in determining bone's resistance to fracture than bone mineral density (4, 59). In simple terms, bone quality may be more important in some ways than bone quantity. However, factors affecting bone structure within and around the trabeculae are currently unknown. Until much more is understood about osteoporosis, young women in particular are encouraged to maximize peak bone mass and to minimize its loss by engaging in regular physical activity and avoiding the lifestyle factors that negatively affect bone health.

•*Estrogen and testosterone deficiencies promote the development of osteoporosis.*

COMMON BONE INJURIES

Because of the important mechanical functions performed by bone, bone health is an important part of general health. Bone health can be impaired by injuries and pathologies.

Fractures

A fracture is a disruption in the continuity of a bone. The nature of a fracture depends on the direction, magnitude, loading rate, and duration of the mechanical load sustained, as well as the health and maturity of

fracture
disruption in the continuity of a bone

FIGURE 4-6

Types of fractures. From Shier, Butler, and Lewis. *Hole's Human Anatomy and Physiology,* © 1996. Reprinted by permission of The McGraw-Hill Companies, Inc.

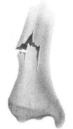

A *greenstick* fracture is incomplete, and the break occurs on the convex surface of the bend in the bone.

A *fissured* fracture involves an incomplete longitudinal break.

A *comminuted* fracture is complete and fragments the bone.

A *transverse* fracture is complete, and the break occurs at a right angle to the axis of the bone.

An *oblique* fracture occurs at an angle other than a right angle to the axis of the bone.

A *spiral* fracture is caused by twisting a bone excessively.

●*Under excessive bending loads, bone tends to fracture on the side loaded in tension.*

the bone at the time of injury. Fractures are classified as simple when the bone ends remain within the surrounding soft tissues and compound when one or both bone ends protrude from the skin. When the loading rate is rapid, a fracture is more likely to be comminuted, containing multiple fragments (Figure 4-6).

Avulsions are fractures caused by tensile loading in which a tendon or ligament pulls a small chip of bone away from the rest of the bone. Explosive throwing and jumping movements may result in avulsion fractures of the medial epicondyle of the humerus and the calcaneus.

Excessive bending and torsional loads can produce spiral fractures of the long bones (Figure 4-6). The simultaneous application of forces from opposite directions at different points along a structure such as a long bone generates a torque known as a *bending moment,* which can cause bending and ultimately fracture of the bone. A bending moment is created on a football player's leg when the foot is anchored to the ground and tacklers apply forces at different points on the leg in opposite directions. When bending is present, the structure is loaded in tension on one side and in compression on the opposite side, as discussed in Chapter 3. Because bone is stronger in resisting compression than in resisting tension, the side of the bone loaded in tension will fracture first.

Torque applied about the long axis of a structure such as a long bone causes torsion, or twisting of the structure. Torsion creates shear stress throughout the structure, as explained in Chapter 3. When a skier's body rotates with respect to one boot and ski during a fall, torsional loads can cause a spiral fracture of the tibia. In such cases, a combined loading pattern of shear and tension produces failure at an oblique orientation to the longitudinal axis of the bone.

Since bone is stronger in resisting compression than in resisting tension and shear, acute compression fractures of bone (in the absence of osteoporosis) are rare. However, under combined loading, a fracture

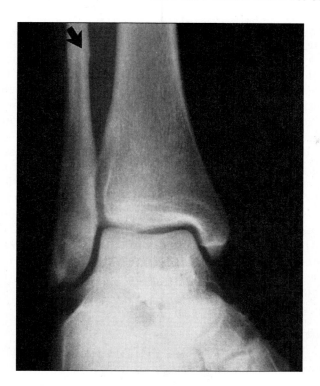

This X-ray shows a stress fracture toward the distal end of the fibula. Courtesy: Lester Cohn, MD.

resulting from a torsional load may also be impacted by the presence of a compressive load. An impacted fracture is one in which the opposite sides of the fracture are compressed together. Fractures that result in depression of bone fragments into the underlying tissues are termed *depressed*.

Since the bones of children contain relatively larger amounts of collagen than do adult bones, they are more flexible and more resistant to fracture under day-to-day loading than are adult bones. Consequently, greenstick fractures, or incomplete fractures, are more common in children than in adults (Figure 4-6). A greenstick fracture is an incomplete fracture caused by bending or torsional loads.

Stress fractures, also known as fatigue fractures, result from low-magnitude forces sustained on a repeated basis. Any increase in the magnitude or frequency of bone loading produces a stress reaction, which may involve microdamage. Bone responds to microdamage by remodeling: First, osteoclasts resorb the damaged tissue; then, osteoblasts deposit new bone at the site. When there is not time for the repair process to complete itself before additional microdamage occurs, the condition can progress to a stress fracture. Stress fractures begin as a small disruption in the continuity of the outer layers of cortical bone but can worsen over time, eventually resulting in complete cortical fracture.

In runners, a group particularly prone to stress fractures, about 50% of fractures occur in the tibia and approximately 20% of fractures are in the metatarsals, with fractures of the femoral neck and pubis also reported (28, 44). Increases in training duration or intensity that do not allow enough time for bone remodeling to occur are the primary culprits. Other factors that predispose runners to stress fractures include muscular fatigue and abrupt changes in either the running surface or the running direction (44).

impacted
pressed together by a compressive load

stress fracture
fracture resulting from repeated loading of relatively low magnitude

stress reaction
progressive bone pathology associated with repeated loading

Epiphyseal Injuries

About 10% of acute skeletal injuries in children and adolescents involve the epiphysis (42). Epiphyseal injuries include injuries to the cartilaginous epiphyseal plate, the articular cartilage, and the apophysis. The apophyses are the sites of tendon attachments to bone, where bone shape is influenced by the tensile loads to which these sites are subjected. The epiphyses of long bones are termed *pressure epiphyses* and the apophyses are called *traction epiphyses,* after the types of physiological loading present. Both acute and repetitive loading can injure the growth plate, potentially resulting in premature closure of the epiphyseal junction and termination of bone growth.

• *Injuries of an epiphyseal plate can terminate bone growth early.*

Another form of epiphyseal injury, osteochondrosis, involves disruption of blood supply to an epiphysis, with associated tissue necrosis and potential deformation of the epiphysis. The cause of the condition is poorly understood. Osteochondrosis occurs most commonly between the ages of 3 and 10 and is more prevalent among boys than among girls (44).

Osteochondrosis of an apophysis, known as *apophysitis,* is often associated with traumatic avulsions. Common sites for apophysitis are the calcaneus and the tibial tubercle at the site of the patellar tendon attachment, where the disorder is referred to respectively as Sever's disease and Osgood-Schlatter's disease.

SUMMARY

Bone is an important and dynamic living tissue. Its mechanical functions are to support and protect other body tissues and to act as a system of rigid levers that can be manipulated by the attached muscles.

Bone's strength and resistance to fracture depend on its material composition and organizational structure. Minerals contribute to a bone's hardness and compressive strength, and collagen provides its flexibility and tensile strength. Cortical bone is stiffer and stronger than trabecular bone, whereas trabecular bone has greater shock-absorbing capabilities.

Bone is an extremely dynamic tissue that is continually being modeled and remodeled in accordance with Wolff's law. Although bones grow in length only until the epiphyseal plates close at adolescence, bones continually change in density, and to some extent in size and shape, through the actions of osteoblasts and osteoclasts.

Osteoporosis, a disorder characterized by excessive loss of bone mineral mass and strength, is extremely prevalent among the elderly. It affects women at an earlier age and more severely than men. It is also present in an alarming frequency among young, eating-disordered, amenorrheic female athletes. Although the cause of osteoporosis remains unknown, the condition can often be improved through hormone therapy, avoidance of negative lifestyle factors, and a regular exercise program.

INTRODUCTORY PROBLEMS

1. Explain why the bones of the human body are stronger in resisting compression than in resisting tension and shear.
2. In the human femur, bone tissue is strongest in resisting compressive force, approximately half as strong in resisting tensile force, and only about one-fifth as strong in resisting shear force. If a tensile force of 8000 N is sufficient to produce a fracture, how much compressive force will produce a fracture? How much shear force will produce a fracture? (Answer: compressive force = 16,000 N; shear force = 3200 N)
3. Explain why bone density is related to an individual's body weight.

4. Rank the following activities according to their effect on increasing bone density: running, backpacking, swimming, cycling, weight lifting, polo, tennis. Write a paragraph providing the rationale for your ranking.
5. Why is bone tissue organized differently (cortical versus trabecular bone)?
6. What kinds of fractures are produced by compression, tension, and shear, respectively?
7. What kinds of fractures are produced only by combined loading? (Identify fracture types with their associated causative loadings.)
8. Approximately 56% of body weight is supported by the fifth lumbar vertebra. How much stress is present on the 22 cm^2 surface area of that vertebra in an erect 756 N man? (Assume that the vertebral surface is horizontal.) (Answer: 19.2 N/cm^2)
9. In Problem 8, how much total stress is present on the fifth lumbar vertebra if the individual holds a 222 N weight bar balanced across his shoulders? (Answer: 29.3 N/cm^2)
10. Why are men less prone than women to vertebral compression fractures?

ADDITIONAL PROBLEMS

1. Hypothesize about the way or ways in which each of the following bones is loaded when a person stands in anatomical position. Be as specific as possible in identifying which parts of the bones are loaded.
   a. Femur
   b. Tibia
   c. Scapula
   d. Humerus
   e. Third lumbar vertebra
2. Outline a six-week exercise program that might be used with a group of osteoporotic elderly persons who are ambulatory.
3. Speculate about what exercises or other strategies may be employed in outer space to prevent the loss of bone mineral density in humans.
4. Hypothesize as to the ability of bone to resist compression, tension, and shear, compared to the same properties in wood, steel, and plastic.
5. How are the bones of birds and fish adapted for their methods of transportation?
6. Why is it important to correct eating disorders?
7. Hypothesize as to why men are less prone to osteoporosis than are women.
8. When an impact force is absorbed by the foot, the soft tissues at the joints act to lessen the amount of force transmitted upward through the skeletal system. If a ground reaction force of 1875 N is reduced 15% by the tissues of the ankle joint and 45% by the tissues of the knee joint, how much force is transmitted to the femur? (Answer: 750 N)
9. How much compression is exerted on the radius at the elbow joint when the biceps brachii, oriented at a 30° angle to the radius, exerts a tensile force of 200 N? (Answer: 173 N)
10. If the anterior and posterior deltoids both insert at an angle of 60° on the humerus and each muscle produces a force of 100 N, how much force is acting perpendicular to the humerus? (Answer: 173.2 N)

LABORATORY EXPERIENCES

1. Using an anatomical model or the *Dynamic Human* CD in conjunction with the material in this chapter, review the bones of the human skeleton. Select a specific bone from each of the four shape categories and explain how each bone's size, shape, and internal structure are suited to its biomechanical function.

Short bone:_____

Relationship of form to function:_____

_____

_____

_____

_____

Flat bone: _____

Relationship of form to function:_____

_____

_____

_____

_____

Irregular bone:_____

Relationship of form to function:_____

_____

_____

_____

_____

Long bone: _____

Relationship of form to function:_____

_____

_____

_____

_____

2. Select three bones on an anatomical model or on the *Dynamic Human* CD and study each bone's shape. What do the bone shapes indicate about the probable locations of muscle tendon attachments and the directions in which the muscles exert force?

Bone 1: _____

Description of shape: _____

_____

Locations of tendon attachments: _____

_____

_____

_____

Directions of muscle forces: _____

_____

_____

_____

Bone 2: _____

Description of shape: _____

_____

Locations of tendon attachments: _____

_____

_____

_____

Directions of muscle forces: _____

_____

_____

_____

Bone 3: _____

Description of shape: _____

_____

Locations of tendon attachments: _____

_____

_____

_____

Directions of muscle forces: _____

_____

_____

_____

3. Review an anatomy text book or the Histology section under The Skeletal System on the *Dynamic Human* CD and compare the microstructure of compact (cortical) and spongy (trabecular) bone. Write a paragraph summarizing how the structure of each bone type contributes to its function.

Summary for compact bone: _____

_____

_____

_____

_____

_____

Summary for spongy bone: _____

_____

_____

_____

_____

_____

4. Using a paper soda straw as a model of a long bone, progressively apply compression to the straw by loading it with weights until it buckles. Using a system of clamps and a pulley, repeat the experiment, progressively loading straws in tension and shear to failure. Record the weight at which each straw failed, and write a paragraph discussing your results and relating them to long bones.

Failure weights for compression:_____ tension:_____ shear:_____

Discussion: _____

_____

_____

_____

_____

_____

_____

5. Visit some of the websites listed at the end of this chapter, or go to the Clinical Concepts section under The Skeletal System on the *Dynamic Human* CD, and locate a picture and description of a surgical repair of a bone injury of interest. Write a paragraph summarizing how the surgical repair was performed.

Website:_____

Bone:_____

Type of surgical repair: _____

Description: _____

_____

_____

_____

_____

_____

_____

_____

_____

_____

REFERENCES

1. Adrian MJ and Cooper JM: *Biomechanics of human movement,* Indianapolis, 1989, Benchmark Press.

2. Bachrach LK: Bone mineralization in childhood and adolescence, *Curr Opin Pediatr* 5:467, 1993.

3. Bass SL: The prepubertal years: a uniquely opportune stage of growth when the skeleton is most responsive to exercise? *Sports Med* 30:73, 2000.

4. Bauer JS and Link TM: Advances in osteoporosis imaging, *Eur J Radiol* 71:440, 2009.

5. Bloomfield SA: Changes in musculoskeletal structure and function with prolonged bed rest, *Med Sci Sports Exerc* 29:197, 1997.

6. Blum M, Harris SS, Must A, Phillips SM, Rand WM, and Dawson-Hughes B: Weight and body mass index at menarche are associated with premenopausal bone mass, *Osteoporos Int* 12:588, 2001.

7. Carrié-Fässler AL and Bonjour JP: Osteoporosis as a pediatric problem, *Pediatr Clin North Am* 42:811, 1995.

8. Cavanagh PR, Licata AA, and Rice AJ: Exercise and pharmacological countermeasures for bone loss during long-duration space flight, *Gravit Space Biol Bull* 18:39, 2005.

9. Cobb KL, Bachrach LK, Greendale G, Marcus R, Neer RM, Nieves J, Sowers MF, Brown BW, Gopalakrishnan G, Luetters C, Tanner HK, Ward B, and Kelsey J: Disordered eating, menstrual irregularity, and bone mineral density in female runners, *Med Sci Sports Exer* 35:711, 2003.

10. Colleran PN, Wilkerson MK, Bloomfield SA, Suva LJ, Turner RT, and Delp MD: Alterations in skeletal perfusion with simulated microgravity: a possible mechanism for bone remodeling, *J Appl Physiol* 89:4046, 2000.

11. Conlisk AJ and Galuska DA: Is caffeine associated with bone mineral density in young adult women?, *Prev Med* 31:562, 2000.

12. Creighton DL, Morgan AL, Boardley D, and Brolinson PG: Weight-bearing exercise and markers of bone turnover in female athletes, *J Appl Physiol* 90:565, 2001.

13. Cromer B and Harel Z: Adolescents: at increased risk for osteoporosis?, *Clin Pediatr* 39:565, 2000.

14. De Souza MI and Williams NI: Beyond hypoestrogenism in amenorrheic athletes: energy deficiency as a contributing factor for bone loss, *Curr Sports Med Rep* 4:38, 2005.

15. Ding M: Age variations in the properties of human tibial trabecular bone and cartilage, *Acta Orthop Scand Suppl* 292:1, 2000.

16. Dowthwaite JN, Kanaley JA, Spadaro JA, Hickman RM, and Scerpella TA: Muscle indices do not fully account for enhanced upper extremity bone mass and strength in gymnasts, *J Musculoskelet Neuronal Interact* 9:2, 2009.

17. Fogelholm GM, Sievanen HT, Kukkonen-Harjula TK, and Pasanen ME: Bone mineral density during reduction, maintenance and regain of body weight in premenopausal, obese women, *Osteoporosis* 12:199, 2001.

18. Frost HM: From Wolff's law to the Utah paradigm: insights about bone physiology and its clinical applications, *Anat Rec* 262:398, 2001.

19. Fuchs RK, Bauer JJ, and Snow CM: Jumping improves hip and lumbar spine bone mass in prepubescent children: a randomized controlled trial, *J Bone Miner Res* 16:148, 2001.

20. Greendale GA, Huang M-H, Wang Y, Finkelstein JS, Danielson ME, and Sternfeld B: Sport and home physical activity are independently associated with bone density, *Med Sci Sports Exer* 35:506, 2003.

21. Gross TS, Poliachik SL, Ausk BJ, Sanford DA, Becker BA, and Srinivasan S: Why rest stimulates bone formation: a hypothesis based on complex adaptive phenomenon, *Exer Sport Sci Rev* 32:9, 2004.

22. Haller E: Eating disorders. A review and update, *West J Med* 157:658, 1992.

23. Hawkey A: Physiological and biomechanical considerations for a human Mars mission, *J Br Interplanet Soc* 58:117, 2005.

24. Heinonen A, Sievanen H, Kannus P, Oja P, Pasanen M, Vuori I: High-impact exercise and bones of growing girls: a 9-month controlled trial, *Osteoporosis Int* 11:1010, 2000.

25. Hetland ML, Haarbo J, and Christiansen C: Running induces menstrual disturbances but bone mass is unaffected, except in amenorrheic women, *Am J Med* 95:53, 1993.

26. Holick MF, Siris ES, Binkley N, Beard MK, Khan A, Katzer JT, Petruschke RA, Chen E, and de Papp AE: Prevalence of vitamin D inadequacy among postmenopausal North American women receiving osteoporosis therapy, *J Clin Endocrinol Metab* 90:3215, 2005.

27. Hopper JL and Seeman E: The bone density of female twins discordant for tobacco use, *New Eng J Med* 330:387, 1994.

28. Hreljac A: Impact and overuse injuries in runners, *Med Sci Sports Exer* 36:845, 2004.

29. Huiskes R, Ruimerman R, van Lenthe GH, and Janssen JD: Effects of mechanical forces on maintenance and adaptation of form in trabecular bone, *Nature* 405:704, 2000.

30. Humphries B, Fenning A, Dugan E, Guinane J, and MacRae K: Whole-body vibration effects on bone mineral density in women with or without resistance training, *Aviat Space Environ Med* 80:1025, 2009.

31. Janz KF, Burns TL, Levy SM, Torner JC, Willing MC, Beck TJ, Gilmore JM, and Marshall TA: Everyday activity predicts bone geometry in children: the Iowa bone development study, *Med Sci Sports Exer* 36:1124, 2004.

32. Kenny AM and Prestwood KM: Osteoporosis: pathogenesis, diagnosis, and treatment in older adults, *Rheum Dis Clin North Am* 26:569, 2000.

33. Khan K, McKay HA, Haapasalo H, Bennell KL, Forwood MR, Kannus P, Wark JD: Does childhood and adolescence provide a unique opportunity for exercise to strengthen the skeleton? *J Sci Med Sport* 3:150, 2000.

34. Kiiski J, Heinonen A, Järvinen TL, Kannus P, and Sievänen H: Transmission of vertical whole body vibration to the human body, *J Bone Miner Res* 23:1318, 2008.

35. Kontulainen S, Kannus P, Haapasalo H, Sievanen H, Pasanen M, Heinonen A, Oja P, and Vuori I: Good maintenance of exercise-induced bone gain with decreased training of female tennis and squash players: a prospective 5-year follow-up study of young and old starters and controls, *J Bone Miner Res* 16:202, 2001.

36. Kohrt WM, Bloomfield SA, Little KD, Nelson ME, and Yingling VR: Physical activity and bone health: ACSM pronouncement, *Med Sci Sports Exer* 36:1985, 2004.

37. Krolner B and Pors Nielsen S: Bone mineral content of the lumbar spine in normal and osteoporotic women: cross-sectional and longitudinal studies, *Clin Sci* 62:329, 1982.

38. Li S, Wagner R, Holm K, Lehotsky J, and Zinaman MJ: Relationship between soft tissue body composition and bone mass in perimenopausal women, *Maturitas* 47:99, 2004.

39. Lips P: Epidemiology and predictors of fractures associated with osteoporosis, *Am J Med* 103:3S, 1997.

40. Loomer PM: The impact of microgravity on bone metabolism in vitro and in vivo, *Crit Rev Oral Biol Med* 12:252, 2001.

41. MacIntyre NJ, Adachi JD, and Webber CE: Gender differences in normal age-dependent patterns of radial bone structure and density: a cross-sectional study using peripheral quantitative computed tomography, *J Clin Densitom* 2:163, 1999.

42. Maffulli N: Intensive training in young athletes: the orthopaedic surgeon's viewpoint, *Sports Med* 9:229, 1990.

43. Martin TJ and Seeman E: Bone remodelling: its local regulation and the emergence of bone fragility, *Best Pract Res Clin Endocrinol Metab* 22:701, 2008.

44. Matheson GO, Clement DB, McKenzie DC, Taunton JE, Lloyd-Smith DR, MacIntyre JG. et al: Stress fractures in athletes: a study of 320 cases, *Am J Sports Med* 15:46, 1987.

45. Merriman H and Jackson K: The effects of whole-body vibration training in aging adults: a systematic review, *J Geriatr Phys Ther* 32:134, 2009.

46. Mosekilde L: Age-related changes in bone mass, structure, and strength—effects of loading, *Z Rheumatol* 59 (Suppl 1):1, 2000.

47. Naganathan V, and Sambrook P: Gender differences in volumetric bone density: a study of opposite-sex twins, *Osteoporos Int* 14:564, 2003.

48. Nattiv A, Loucks AB, Manore MM, Sanborn CF, Sundgot-Borgen J, and Warren MP: American College of Sports Medicine position stand: the female athlete triad, *Med Sci Sports Exerc* 39:1867, 2007.

49. Neil JM and Schweitzer ME: Humeral cortical and trabecular changes in the throwing athlete: a quantitative computed tomography study of male college baseball players, *J Comput Assist Tomogr* 32:492, 2008.

50. Nuti R, Brandi ML, Isaia G, Tarantino U, Silvestri S, and Adami S: New perspectives on the definition and the management of severe osteoporosis: the patient with two or more fragility fractures, *J Endocrinol Invest* 32:783, 2009.

51. Pettersson U, Nordstrom P, and Lorentzon R: A comparison of bone mineral density and muscle strength in young male adults with different exercise level, *Calcif Tissue Int* 64:490, 1999.

52. Ralston SH: Genetic determinants of osteoporosis, *Curr Opin Rheumatol* 17:475, 2005.

53. Rauch F: Vibration therapy, *Dev Med Child Neurol* 51:166, 2009.

54. Ravn P, Cizza G, Bjarnason NH, Thompson D, Daley M, Wasnich RD, McClung M, Hosking D, Yates AJ, and Christiansen C: Low body mass index is an important risk factor for low bone mass and increased bone loss in early postmenopausal women, *J Bone Miner Res* 14:1622, 1999.

55. Recker RR, Davies KM, Hinders SM, et al: Bone gain in young adult women, *JAMA* 268:2403, 1992.

56. Risser WL et al: Bone density in eumenorrheic female college athletes, *Med Sci Sports Exerc* 22:570, 1990.

57. Rittweger J, Beller G, Armbrecht G, Mulder E, Buehring B, Gast U, Dimeo F, Schubert H, de Haan A, Stegeman DF, Schiessl H, and Felsenberg D: Prevention of bone loss during 56 days of strict bed rest by side-alternating resistive vibration exercise, *Bone* 46:137, 2010.

58. Ross JA: Hypertrophy of the little finger, *Brit Med J* 2:987, 1950.

59. Ruppel ME, Miller LM, and Burr DB: The effect of the microscopic and nanoscale structure on bone fragility, *Osteoporos Int* 19:1251, 2008.

60. Salamone LM, Cauley JA, Black DM, Simkin-Silverman L, Lang W, Gregg E, Palermo L, Epstein RS, Kuller LH, and Wing R: Effect of a lifestyle intervention on bone mineral density in premenopausal women: a randomized trial, *Am J Clin Nutr* 70:97, 1999.

61. Smit TH, Huyghe JM, and Burger EH: The mechanical regulation of BMU-coupling by fluid flow, *J Biomech* 34 (Suppl 1):S31, 2001.

62. Sowers MF: Lower peak bone mass and its decline, *Baillieres Best Pract Res Clin Endocrinol Metab* 14:317, 2000.

63. Sowers M: Epidemiology of calcium and vitamin D in bone loss, *J Nutr* 123 (2 Suppl):413, 1993.

64. Spector ER, Smith SM, and Sibonga JD: Skeletal effects of long-duration head-down bed rest, *Aviat Space Environ Med* 80:A23, 2009.

65. Stein MS, Thomas CDL, Feik SA, Wark JD, and Clement JG: Bone size and mechanics at the femoral diaphysis across age and sex, *J Biomech* 31:1101, 1998.

66. Stewart AD and Hannan J: Total and regional bone density in male runners, cyclists, and controls, *Med Sci Sports Exerc* 32:1373, 2000.

67. Tanno M, Horiuchi T, Nakajima I, Maeda S, Igarashi M, and Yamada H: Age-related changes in cortical and trabecular bone mineral status: a quantitative CT study in lumbar vertebrae, *Acta Radiol* 42:15, 2001.

68. Tate MLK: "Whither flows the fluid in bone?" An osteocyte's perspective, *J Biomech* 36:1409, 2003.

69. Thomas-John M, Codd MB, Manne S, Watts NB, and Mongey AB: Risk factors for the development of osteoporosis and osteoporotic fractures among older men, *J Rheumatol* 36:1947, 2009.

70. Turner LW, Bass MA, Ting L, and Brown B: Influence of yard work and weight training on bone mineral density among older US women, *J Women Aging* 14:139, 2002.

71. Turner RT: Physiology of a microgravity environment invited review: what do we know about the effects of spaceflight on bone? *J Appl Physiol* 89:840, 2000.

72. Uusi-Rasi K, Sievanen H, Pasanen M, Oja P, and Vuori I: Maintenance of body weight, physical activity and calcium intake helps preserve bone mass in elderly women, *Osteoporosis Int* 12:373, 2001.
73. van der Linden JC, Day JS, Verhaar JAN, and Weinans H: Altered tissue properties induce changes in cancellous bone architecture in aging and diseases, *J Biomech* 37:367, 2004.
74. Wang Q and Seeman E: Skeletal growth and peak bone strength, *Best Pract Res Clin Endocrinol Metab* 22:687, 2008.
75. Warren MP and Shantha S: The female athlete, *Baillieres Best Pract Res Clin Endocrinol Metab* 14:37, 2000.
76. Wetzsteon RJ, Petit MA, Macdonald HM, Hughes JM, Beck TJ, and McKay HA: Bone structure and volumetric BMD in overweight children: a longitudinal study, *J Bone Miner Res* 23:1946, 2008.
77. Winters-Stone K: *Action plan for osteoporosis,* Champaign, IL: Human Kinetics, p. 45, 2005.
78. Wohl GR, Boyd SK, Judex S, and Zernicke RF: Functional adaptation of bone to exercise and injury, *J Sci Med Sport* 3:313, 2000.
79. Wolff JD: *Das geretz der Transformation der Knochen,* Berlin, 1892, Hirschwald.
80. Yamazaki S, Ichimura S, Iwamoto J, Takeda T, and Toyama Y: Effect of walking exercise on bone metabolism in postmenopausal women with osteopenia/osteoporosis, *J Bone Miner Metab* 22:500, 2004.

ANNOTATED READINGS

Cowin SC: *Bone mechanics handbook* (2nd ed.), Boca Raton, FL, 2001, CRC Press.
Summarizes current understanding of the mechanical properties of bone, as well as clinically related issues, such as bone prostheses, implants, and imaging of bone structure.

Currey JD: *Bones: structure and mechanics,* Princeton, NJ, 2002, Princeton University Press.
Review of bone structure and mechanics, with numerous insights on form following function.

Nattiv A, Loucks AB, Manore MM, Sanborn CF, Sundgot-Borgen J, and Warren MP: American College of Sports Medicine position stand. The female athlete triad, *Med Sci Sports Exerc* 39:1867, 2007.
Pronouncement of the American College of Sports Medicine updating scientific knowledge on the female athlete triad.

Payne MW, Williams DR, and Trudel G: Space flight rehabilitation, *Am J Phys Med Rehabil* 86:583, 2007.
Reviews knowledge related to requirements for rehabilitation of the musculoskeletal system following space flight.

RELATED WEBSITES

American Academy of Orthopaedic Surgeons
http://www.aaos.org/
Includes information for the public, as well as for orthopods, with links for current news, a library, research, and a quiz with flashcard photos of orthopedic conditions.

Clinical Orthopaedics and Related Research
http://www.clinorthop.org/index.html
The online version of Clinical Orthopaedics and Related Research *contains peer-reviewed, original articles on general orthopaedics, specialty topics, the basic sciences, and pathology reports on the latest advances in current research and practice.*

duPont Hospital for Children
http://gait.aidi.udel.edu/res695/homepage/pd_ortho/educate/clincase/clcasehp.htm
Contains numerous clinical case presentations.

Medscape Orthopaedics
http://www.medscape.com/orthopaedicshome
Orthopedic site that includes descriptions of medical conferences, news, a bookstore, specialty pages, and Medline access.
National Institutes of Health Osteoporosis and Related Bone Diseases—National Resource Center
http://www.osteo.org/
Fact sheets, research, newsletters, the Surgeon General's Report, bone links, and a search engine for bone-related resources.
Orthopaedic Research Laboratories
http://orl-inc.com/
Provides links to orthopedic biomechanics publications and presentations..
Osteoporosis and Bone Physiology
http://courses.washington.edu/bonephys/
Information on osteoporosis prevention and treatment, bone density, secondary osteomalacia, cases. Images, FAQs, and advice.
The American Orthopaedic Society for Sports Medicine
http://www.sportsmed.org
Contains resources related to orthopaedic sports medicine education and research, including selected abstracts of presentations from the most recent annual AOSSM meeting.
Wheeless' Textbook of Orthopaedics
http://www.wheelessonline.com/
This comprehensive, online medical textbook has 11,000 pages with more than 5,000 images, with each topic fully searchable by alphabetical, anatomical, and keyword searches. The site is updated daily.
World Ortho
http://www.worldortho.com
Includes links to over 3000 pictures with accompanying text, quizzes, and online books, directed at students, junior doctors, paramedics, and other health professionals.

KEY TERMS

amenorrhea	cessation of menses
anisotropic	exhibiting different mechanical properties in response to loads from different directions
appendicular skeleton	bones composing the body appendages
articular cartilage	protective layer of firm, flexible connective tissue over the articulating ends of long bones
axial skeleton	the skull, vertebrae, sternum, and ribs
bone atrophy	decrease in bone mass resulting from a predominance of osteoclast activity
bone hypertrophy	increase in bone mass resulting from a predominance of osteoblast activity
compressive strength	ability to resist pressing or squeezing force
cortical bone	compact mineralized connective tissue with low porosity that is found in the shafts of long bones
epiphysis	growth center of a bone that produces new bone tissue as part of the normal growth process until it closes during adolescence or early adulthood
flat bones	skeletal structures that are largely flat in shape—for example, the scapula
fracture	disruption in the continuity of a bone
impacted	pressed together by a compressive load
irregular bones	skeletal structures of irregular shapes—for example, the sacrum
lever	a relatively rigid object that may be made to rotate about an axis by the application of force
long bones	skeletal structures consisting of a long shaft with bulbous ends, for example—the femur
osteoblasts	specialized bone cells that build new bone tissue
osteoclasts	specialized bone cells that resorb bone tissue

osteopenia	condition of reduced bone mineral density that predisposes the individual to fractures
osteoporosis	disorder involving decreased bone mass and strength with one or more resulting fractures
periosteum	double-layered membrane covering bone; muscle tendons attach to the outside layer, and the internal layer is a site of osteoblast activity
porous	containing pores or cavities
short bones	small, cubical skeletal structures, including the carpals and tarsals
stiffness	ratio of stress to strain in a loaded material; that is, the stress divided by the relative amount of change in the structure's shape
strain	amount of deformation divided by the original length of the structure or by the original angular orientation of the structure
stress fracture	fracture resulting from repeated loading of relatively low magnitude
stress reaction	progressive bone pathology associated with repeated loading
tensile strength	ability to resist pulling or stretching force
trabecular bone	less compact mineralized connective tissue with high porosity that is found in the ends of long bones and in the vertebrae

The Biomechanics of Human Skeletal Articulations

After completing this chapter, you will be able to:

Categorize joints based on structure and movement capabilities.

Explain the functions of articular cartilage and fibrocartilage.

Describe the material properties of articular connective tissues.

Explain advantages and disadvantages of different approaches to increasing or maintaining joint flexibility.

Describe the biomechanical contributions to common joint injuries and pathologies.

ONLINE LEARNING CENTER RESOURCES

www.mhhe.com/hall6e

Log on to our Online Learning Center (OLC) for access to these additional resources:

- Online Lab Manual
- Flashcards with definitions of chapter key terms
- Chapter objectives
- Chapter lecture PowerPoint presentation
- Self-scoring chapter quiz
- Additional chapter resources
- Web links for study and exploration of chapter-related topics

The mid-radioulnar joint is an example of a syndesmosis, where fibrous tissue binds the bones together. Courtesy of McGraw-Hill Companies, Inc.

The joints of the human body largely govern the directional motion capabilities of body segments. The anatomical structure of a given joint, such as the uninjured knee, varies little from person to person; as do the directions in which the attached body segments, such as the thigh and lower leg, are permitted to move at the joint. However, differences in the relative tightness or laxity of the surrounding soft tissues result in differences in joint ranges of movement. This chapter discusses the biomechanical aspects of joint function, including the concepts of joint stability and joint flexibility, and related implications for injury potential.

JOINT ARCHITECTURE

Anatomists have categorized joints in several ways, based on joint complexity, the number of axes present, joint geometry, or movement capabilities (61). Since this book focuses on human movement, a joint classification system based on motion capabilities is presented.

Immovable Joints

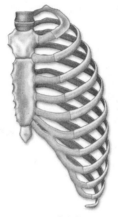

The sternocostal joints are examples of synchondroses, wherein the articulating bones are joined by a thin layer of hyaline cartilage. Courtesy of McGraw-Hill Companies, Inc.

1. *Synarthroses* (immovable) (*syn* = together; *arthron* = joint): These fibrous joints can attenuate force (absorb shock) but permit little or no movement of the articulating bones.
    a. *Sutures:* In these joints, the irregularly grooved articulating bone sheets mate closely and are tightly connected by fibers that are continuous with the periosteum. The fibers begin to ossify in early adulthood and are eventually replaced completely by bone. The only example in the human body is the sutures of the skull.
    b. *Syndesmoses* (*syndesmosis* = held by bands): In these joints, dense fibrous tissue binds the bones together, permitting extremely limited movement. Examples include the coracoacromial, mid-radioulnar, mid-tibiofibular, and inferior tibiofibular joints.

Slightly Movable Joints

2. *Amphiarthroses* (slightly movable) (*amphi* = on both sides): These cartilaginous joints attenuate applied forces and permit more motion of the adjacent bones than synarthrodial joints.
    a. *Synchondroses* (*synchondrosis* = held by cartilage): In these joints, the articulating bones are held together by a thin layer of hyaline cartilage. Examples include the sternocostal joints and the epiphyseal plates (before ossification).
    b. *Symphyses:* In these joints, thin plates of hyaline cartilage separate a disc of fibrocartilage from the bones. Examples include the vertebral joints and the pubic symphysis.

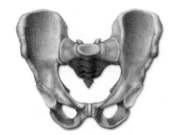

Note the hyaline cartilage disc separating the bones of the pubic symphysis, typical of a symphysis joint. Courtesy of McGraw-Hill Companies, Inc.

Sutures between the occipital and parietal bones of the skull represent synarthroses (immovable joints). From Shier, Butler, and Lewis. *Hole's Human Anatomy and Physiology,* © 1996. Reprinted by permission of The McGraw-Hill Companies, Inc.

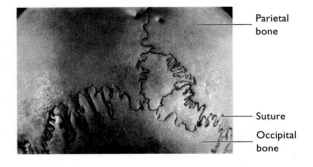

Parietal bone

Suture

Occipital bone

Freely Movable Joints

3. *Diarthroses* or *synovial* (freely movable) (*diarthrosis* = "through joint," indicating only slight limitations to movement capability): At these joints, the articulating bone surfaces are covered with articular cartilage, an articular capsule surrounds the joint, and a synovial membrane lining the interior of the joint capsule secretes a lubricant known as synovial fluid (Figure 5-1). There are many types of synovial joints.

   a. *Gliding* (plane; arthrodial): In these joints, the articulating bone surfaces are nearly flat, and the only movement permitted is nonaxial gliding. Examples include the intermetatarsal, intercarpal, and intertarsal joints, and the facet joints of the vertebrae.

   b. *Hinge* (ginglymus): One articulating bone surface is convex and the other is concave in these joints. Strong collateral ligaments restrict movement to a planar, hingelike motion. Examples include the ulnohumeral and interphalangeal joints.

   c. *Pivot* (screw; trochoid): In these joints, rotation is permitted around one axis. Examples include the atlantoaxial joint and the proximal and distal radioulnar joints.

   d. *Condyloid* (ovoid; ellipsoidal): One articulating bone surface is an ovular convex shape, and the other is a reciprocally shaped concave surface in these joints. Flexion, extension, abduction, adduction, and circumduction are permitted. Examples include the second through fifth metacarpophalangeal joints and the radiocarpal joints.

   e. *Saddle* (sellar): The articulating bone surfaces are both shaped like the seat of a riding saddle in these joints. Movement capability is the same as that of the condyloid joint, but greater range of movement is allowed. An example is the carpometacarpal joint of the thumb.

   f. Ball and socket (spheroidal): In these joints, the surfaces of the articulating bones are reciprocally convex and concave. Rotation in all three planes of movement is permitted. Examples include the hip and shoulder joints.

articular cartilage
protective layer of dense white connective tissue covering the articulating bone surfaces at diarthrodial joints

articular capsule
double-layered membrane that surrounds every synovial joint

synovial fluid
clear, slightly yellow liquid that provides lubrication inside the articular capsule at synovial joints

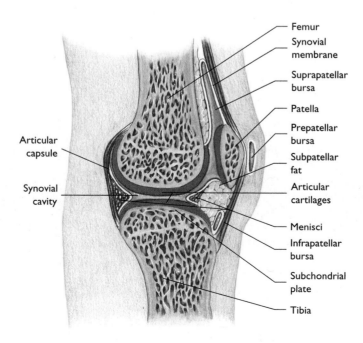

Femur
Synovial membrane
Suprapatellar bursa
Patella
Prepatellar bursa
Subpatellar fat
Articular cartilages
Menisci
Infrapatellar bursa
Subchondral plate
Tibia
Articular capsule
Synovial cavity

FIGURE 5-1

The knee is an example of a synovial joint, with a ligamentous capsule, an articular cavity, and articular cartilage. From Shier, Butler, and Lewis, *Hole's Human Anatomy and Physiology*, © 1996. Reprinted by permission of The McGraw-Hill Companies, Inc.

Synovial joints vary widely in structure and movement capabilities, as shown in Figure 5-2. They are commonly categorized according to the number of axes of rotation present. Joints that allow motion around one, two, and three axes of rotation are referred to respectively as *uniaxial*, *biaxial*, and *triaxial joints*. A few joints where only limited motion is permitted in any direction are termed *nonaxial joints*. Joint motion capabilities are also sometimes described in terms of degrees of freedom (df), or the number of planes in which the joint allows motion. A uniaxial joint has one df, a biaxial joint has two df, and a triaxial joint has three df.

Two synovial structures often associated with diarthrodial joints are *bursae* and *tendon sheaths*. Bursae are small capsules, lined with synovial membranes and filled with synovial fluid, that cushion the structures they separate. Most bursae separate tendons from bone, reducing the friction on the tendons during joint motion. Some bursae, such as the olecranon bursa of the elbow, separate bone from skin. Tendon sheaths are double-layered synovial structures that surround tendons positioned in close association with bones. Many of the long muscle tendons crossing the wrist and finger joints are protected by tendon sheaths.

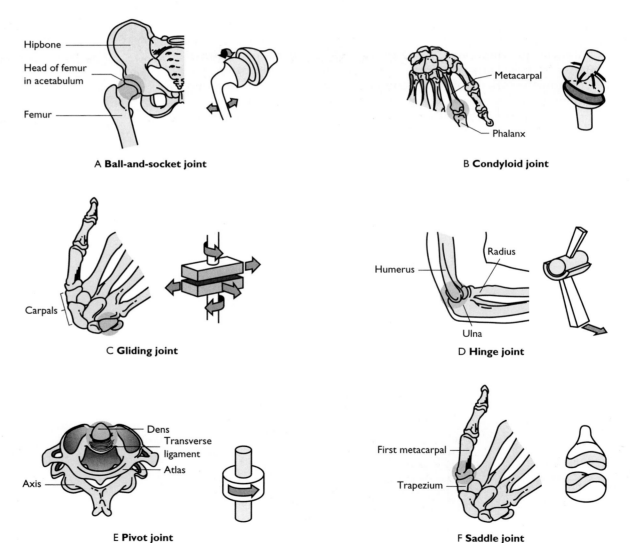

FIGURE 5-2   Examples of the synovial joints of the human body. From Shier, Butler, and Lewis, *Hole's Human Anatomy and Physiology*, © 1996. Reprinted by permission of The McGraw-Hill Companies, Inc.

Articular Cartilage

The joints of a mechanical device must be properly lubricated if the movable parts of the machine are to move freely and not wear against each other. In the human body, a special type of dense, white connective tissue known as *articular cartilage* provides a protective lubrication. A 1- to 5-mm-thick protective layer of this material coats the ends of bones articulating at diarthrodial joints. Articular cartilage serves two important purposes: (a) It spreads loads at the joint over a wide area so that the amount of stress at any contact point between the bones is reduced, and (b) it allows movement of the articulating bones at the joint with minimal friction and wear (4).

Articular cartilage is a soft, porous, and permeable tissue that is hydrated. It consists of specialized cells called *chondrocytes* embedded in a matrix of collagen fibers, proteoglycans, and noncollagenous proteins. The matrix protects the chondrocytes and also signals changes in local pressure to the chondrocytes (6). The chondrocytes maintain and restore cartilage from wear, although this ability diminishes with aging, disease, and injury (68). Chondrocyte densities and matrix structures have been found to vary across joints, as well as within a given joint, depending on the mechanical loading sustained (51).

Under loading at the joint, articular cartilage deforms, exuding synovial fluid. At healthy synovial joints, where the articulating bone ends are covered with articular cartilage, motion of one bone end over the other is typically accompanied by a flow of synovial fluid that is pressed out ahead of the moving contact area and also sucked in behind the contact area (44). At the same time, the permeability of the cartilage is reduced in the area of direct contact, providing a surface on which fluid film can form under the load (44). When joint loading occurs at a low rate, the solid components of the cartilage matrix resist the load. When loading is faster, however, it is the fluid within the matrix that primarily sustains the pressure (35, 49).

Cartilage can reduce the maximum contact stress acting at a joint by 50% or more (71). The lubrication supplied by the articular cartilage is so effective that the friction present at a joint is only approximately 17–33% of the friction of a skate on ice under the same load, and only one-half that of a lubricated bearing (5, 47).

During normal growth, articular cartilage at a joint such as the knee increases in volume as the child's height increases (28). Interestingly, there is no relationship between cartilage accrual at the knee and weight change. Children participating in vigorous sport activities accumulate knee cartilage faster than those who do not, and males tend to gain knee cartilage faster than do females (28).

Unfortunately, once damaged, articular cartilage has little to no ability to heal or regenerate on its own (55). Instead, injuries to this tissue tend to progress, with more and more of the protective coating of the articulating bone ends worn away, resulting in degenerative arthritis. A promising approach for repairing damage to articular cartilage is autologous cartilage regeneration, a procedure through which healthy chondrocytes (cartilage cells) are arthroscopically removed from the patient's joint and then cultured in a laboratory using principles of tissue engineering (8). After a few weeks, the cells have grown into articular cartilage plugs that can be arthroscopically inserted into the damaged area of cartilage. A review of 20 studies revealed that among athletes treated for joint damage with this procedure, 73% sufficiently recovered joint function to return to sports participation (42). Factors influencing an athlete's ability to participate in competitive sports include the athlete's age, duration of injury, level of play, extent of cartilage damage, and repair tissue morphology (42). Research is under way to investigate the potential for a variety of new approaches for treating degenerated cartilage, including the use of mesenchymal stem cells, tissue engineering, and gene transfer technology (53, 66, 69).

Articular Fibrocartilage

articular fibrocartilage
soft-tissue discs or menisci that intervene between articulating bones

●*Intervertebral discs act as cushions between the vertebrae, reducing stress levels by spreading loads.*

At some joints, articular fibrocartilage, in the form of either a fibrocartilaginous disc or partial discs known as menisci, is also present between the articulating bones. The intervertebral discs (Figure 5-3) and the menisci of the knee (Figure 5-4) are examples. Although the function of discs and menisci is not clear, possible roles include the following:

1. Distribution of loads over the joint surfaces
2. Improvement of the fit of the articulating surfaces
3. Limitation of translation or slip of one bone with respect to another
4. Protection of the periphery of the articulation
5. Lubrication
6. Shock absorption

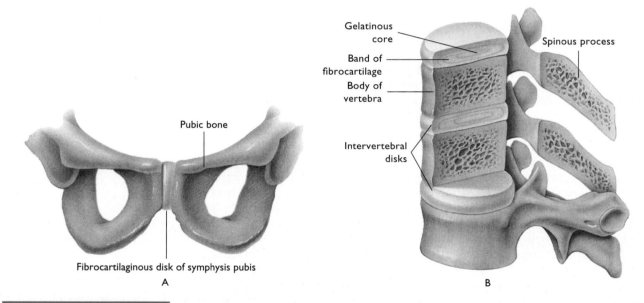

Pubic bone

Fibrocartilaginous disk of symphysis pubis

A

Gelatinous core

Band of fibrocartilage

Body of vertebra

Intervertebral disks

Spinous process

B

| **FIGURE 5-3** | Fibrocartilage is present in (**A**) the symphysis pubis that separates the pubic bones and (**B**) the intervertebral discs between adjacent vertebrae. From Shier, Butler, and Lewis, *Hole's Human Anatomy and Physiology,* © 1996. Reprinted by permission of The McGraw-Hill Companies, Inc. |

FIGURE 5-4

The menisci at the knee joint help to distribute loads, lessening the stress transmitted across the joint.

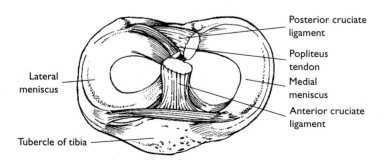

Posterior cruciate ligament

Popliteus tendon

Medial meniscus

Anterior cruciate ligament

Lateral meniscus

Tubercle of tibia

Articular Connective Tissue

Tendons, which connect muscles to bones, and ligaments, which connect bones to other bones, are passive tissues composed primarily of collagen and elastic fibers. Tendons and ligaments do not have the ability to contract like muscle tissue, but they are slightly extensible. These tissues are elastic and will return to their original length after being stretched, unless they are stretched beyond their elastic limits (see Chapter 3). A tendon or ligament stretched beyond its elastic limit during an injury remains stretched and can be restored to its original length only through surgery. The results of modeling studies suggest that tendons routinely undergo healing to repair internal microfailures over the course of the life span in order to remain intact (37).

Tendons and ligaments, like bone, respond to altered habitual mechanical stress by hypertrophying or atrophying. Research has shown that regular exercise over time results in increased size and strength of both tendons (57) and ligaments (36), as well as increased strength of the junctions between tendons or ligaments and bone (65).

Evidence also suggests that the size of a ligament such as the anterior cruciate ligament (ACL) is proportionate to the strength of its antagonists (in this case, the quadriceps muscles) (1). Tendons and ligaments can not only heal following rupturing but in some cases regenerate in their entirety, as evidenced by examples of complete regeneration of the semitendinosus tendon following its surgical removal for repair of anterior cruciate ligament ruptures (14, 16, 48).

●A material stretched beyond its elastic limit remains lengthened beyond its original length after tension is released.

JOINT STABILITY

The stability of an articulation is its ability to resist dislocation. Specifically, it is the ability to resist the displacement of one bone end with respect to another while preventing injury to the ligaments, muscles, and muscle tendons surrounding the joint. Different factors influence joint stability.

joint stability
ability of a joint to resist abnormal displacement of the articulating bones

Shape of the Articulating Bone Surfaces

In many mechanical joints, the articulating parts are exact opposites in shape so that they fit tightly together (Figure 5-5). In the human body, the articulating ends of bones are usually shaped as mating convex and concave surfaces.

Although most joints have reciprocally shaped articulating surfaces, these surfaces are not symmetrical, and there is typically one position of best fit in which the area of contact is maximum. This is known as the close-packed position, and it is in this position that joint stability is usually greatest. Any movement of the bones at the joint away from the

●The articulating bone surfaces at all joints are of approximately matching (reciprocal) shapes.

close-packed position
joint orientation for which the contact between the articulating bone surfaces is maximum

FIGURE 5-5

Mechanical joints are often composed of reciprocally shaped parts.

Ball and socket

Saddle joint

Hinge

loose-packed position
any joint orientation other than the close-packed position

• *The close-packed position occurs for the knee, wrist, and interphalangeal joints at full extension and for the ankle at full dorsiflexion (30).*

• *One factor enhancing the stability of the glenohumeral joint is a posteriorly tilted glenoid fossa and humeral head. Individuals with anteriorly tilted glenoids and humeral heads are predisposed to shoulder dislocation.*

• *Stretching or rupturing of the ligaments at a joint can result in abnormal motion of the articulating bone ends, with subsequent damage to the articular cartilage.*

close-packed position results in a loose-packed position, with reduction of the area of contact.

Some articulating surfaces are shaped so that in both close- and loose-packed positions, there is either a large or a small amount of contact area and consequently more or less stability. For example, the acetabulum provides a relatively deep socket for the head of the femur, and there is always a relatively large amount of contact area between the two bones, which is one reason the hip is a stable joint. At the shoulder, however, the small glenoid fossa has a vertical diameter that is approximately 75% of the vertical diameter of the humeral head and a horizontal diameter that is 60% of the size of the humeral head (46). Therefore, the area of contact between these two bones is relatively small, contributing to the relative instability of the shoulder complex. Slight anatomical variations in shapes and sizes of the articulating bone surfaces at any given joint among individuals are found; therefore, some people have joints that are more or less stable than average.

Arrangement of Ligaments and Muscles

Ligaments, muscles, and muscle tendons affect the relative stability of joints. At joints such as the knee and the shoulder, in which the bone configuration is not particularly stable, the tension in ligaments and muscles contributes significantly to joint stability by helping to hold the articulating bone ends together. If these tissues are weak from disuse or lax from being overstretched, the stability of the joint is reduced. Strong ligaments and muscles often increase joint stability. For example, strengthening of the quadriceps and hamstring groups enhances the stability of the knee (52). The complex array of ligaments and tendons crossing the knee is illustrated in Figure 5-6.

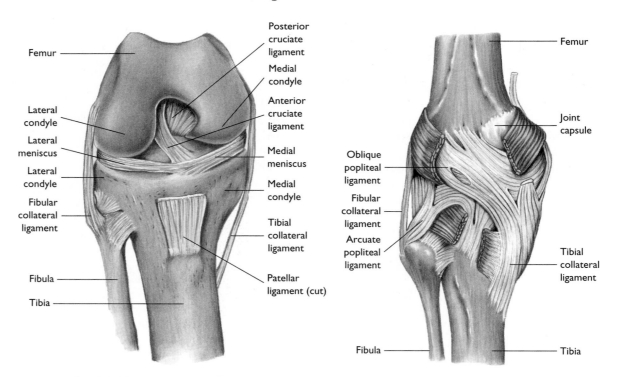

FIGURE 5-6

At the knee joint, stability is derived primarily from the tension in the ligaments and muscles that cross the joint. From Shier, Butler, and Lewis, *Hole's Human Anatomy and Physiology*, © 1996. Reprinted by permission of The McGraw-Hill Companies, Inc.

The angle of attachment of most tendons to bones is arranged so that when the muscle exerts tension, the articulating ends of the bones at the joint crossed are pulled closer together, enhancing joint stability. This situation is usually found when the muscles on opposite sides of a joint produce tension simultaneously. When muscles are fatigued, however, they are less able to contribute to joint stability, and injuries are more likely to occur (13). Rupture of the cruciate ligaments is most likely when the tension in fatigued muscles surrounding the knee is inadequate to protect the cruciate ligaments from being stretched beyond their elastic limits (52).

●*Engaging in athletic participation with fatigued muscles increases the likelihood of injury.*

Other Connective Tissues

White fibrous connective tissue known as *fascia* surrounds muscles and the bundles of muscle fibers within muscles, providing protection and support. A particularly strong, prominent tract of fascia known as the *iliotibial band* crosses the lateral aspect of the knee, contributing to its stability (Figure 5-7). The fascia and the skin on the exterior of the body are other tissues that contribute to joint integrity.

joint flexibility
a term representing the relative ranges of motion allowed at a joint

range of motion
angle through which a joint moves from anatomical position to the extreme limit of segment motion in a particular direction

JOINT FLEXIBILITY

Joint flexibility is a term used to describe the range of motion (ROM) allowed in each of the planes of motion at a joint. *Static flexibility* refers to the ROM present when a body segment is passively moved (by an exercise

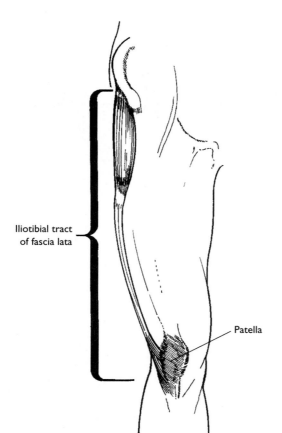

FIGURE 5-7

The iliotibial band is a strong, thickened region of the fascia lata that crosses the knee, contributing to the knee's stability.

Iliotibial tract of fascia lata

Patella

Anterior view

partner or clinician), whereas *dynamic flexibility* refers to the ROM that can be achieved by actively moving a body segment by virtue of muscle contraction. Static flexibility is considered to be the better indicator of the relative tightness or laxity of a joint in terms of implications for injury potential. Dynamic flexibility, however, must be sufficient not to restrict the ROM needed for daily living, work, or sport activities. Research indicates that these two components of flexibility are independent of each other (26).

Although people's general flexibility is often compared, flexibility is actually joint-specific. That is, an extreme amount of flexibility at one joint does not guarantee the same degree of flexibility at all joints.

Measuring Joint Range of Motion

Joint ROM is measured directionally in units of degrees. In anatomical position, all joints are considered to be at zero degrees. The ROM for flexion at the hip is therefore considered to be the size of the angle through which the extended leg moves from zero degrees to the point of maximum flexion (Figure 5-8). The ROM for extension (return to anatomical position) is the same as that for flexion, with movement past anatomical position in the other direction quantified as the ROM for hyperextension. A goniometer used for measuring joint ROM is shown in Figure 5-9.

Factors Influencing Joint Flexibility

Different factors influence joint flexibility. The shapes of the articulating bone surfaces and intervening muscle or fatty tissue may terminate movement at the extreme of a ROM. When the elbow is in extreme hyperextension, for example, contact of the olecranon of the ulna with

FIGURE 5-8

The range of motion for flexion at the hip is typically measured with the individual supine.

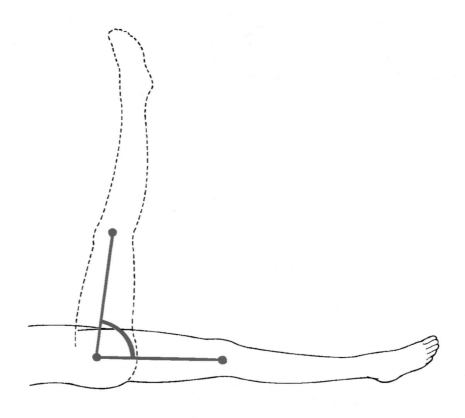

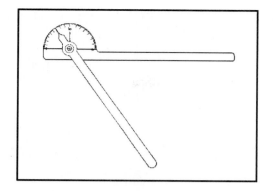

FIGURE 5-9

A goniometer is basically a protractor with two arms. The point where the arms intersect is aligned over the joint center while the arms are aligned with the longitudinal axes of the body segments, to measure the angle present at a joint.

the olecranon fossa of the humerus restricts further motion in that direction. Muscle and/or fat on the anterior aspect of the arm may terminate elbow flexion. Regular participants in bilaterally asymmetrical sports such as tennis are likely to have less range of motion for the dominant arm than for the nondominant arm at the glenohumeral joint of the shoulder (15).

For most individuals, joint flexibility is primarily a function of the relative laxity and/or extensibility of the collagenous tissues and muscles crossing the joint. Tight ligaments and muscles with limited extensibility are the most common inhibitors of a joint's ROM (41). In one study researchers showed that a four-week stretching protocol resulted in increased joint flexibility but with no change in muscle compliance, or extensibility, suggesting that it was the ligaments and tendons that became easier to stretch (75).

Laboratory studies have shown that the extensibility of collagenous tissues increases slightly with temperature elevation (54). Although this finding suggests that "warm-up" exercises should increase joint ROM, this has not been well documented in humans. In a study comparing the effects of static stretching on ankle range of motion, as compared to static stretching preceded by exercise warm-up, superficial heat application, or ultrasound, all protocols produced similar effects (33). More research is needed to identify the specific mechanism responsible for the effects of warm-up on joint ROM.

Flexibility and Injury

Research has shown that the risk of injury is heightened when joint flexibility is extremely low, extremely high, or significantly imbalanced between dominant and nondominant sides of the body (32). Severely limited joint flexibility is undesirable because, if the collagenous tissues and muscles crossing the joint are tight, the likelihood of their tearing or rupturing if the joint is forced beyond its normal ROM increases. Tight ligaments and muscles were found to be related to lower-extremity injury incidence among male, but not female, college athletes, possibly because the female athletes studied were more flexible and less tight at the lower-extremity joints (34). In a study of competitive female gymnasts, those in a highly injury-prone category had less flexibility of the shoulder, elbow, wrist, hip, and knee joints than those in a low injury incidence category (62). Alternatively, an extremely loose, lax joint is lacking in stability and, therefore, prone to displacement-related injuries. Among U.S. Army infantry recruits assessed for hip/low back flexibility with the sit-and-reach test, both the least flexible and the most flexible were over two times as likely to get injured as soldiers in the middle of the flexibility range.

●*A joint with an unusually large range of motion is termed* hypermobile.

Gymnastics is a sport requiring a large amount of flexibility at the major joints of the body. Photo © 2009 Jupiterimages Corporation.

Soldiers who participated in a stretching program for the hamstrings, however, sustained 12.4% fewer lower extremity overuse injuries than those who did not participate (21). Female college athletes with a hip extension flexibility *imbalance* of 15% or more were 2.6 times more likely to suffer lower-extremity injuries (31).

The desirable amount of joint flexibility is largely dependent on the activities in which an individual wishes to engage. Gymnasts and dancers obviously require greater joint flexibility than do nonathletes. However, these athletes also require strong muscles, tendons, and ligaments to perform well and avoid injury. Although large, bulky muscles may inhibit joint ROM, an extremely strong, stable joint can also enable large ROMs.

Athletes and recreational runners commonly stretch before engaging in activity for purposes of reducing the likelihood of injury. There is some evidence that preparticipation stretching reduces the incidence of muscle strains, and recent research shows that increased joint flexibility translates to a lower incidence of eccentric exercise-induced muscle damage (7, 40). Stretching has no effect, however, on overuse-type injuries (40).

Although people usually become less flexible as they age, this phenomenon appears to be primarily related to decreased levels of physical activity rather than to changes inherent in the aging process. No changes in flexibility have been found to be associated with growth during adolescence (17). Regardless of the age of the individual, however, if the collagenous tissues crossing a joint are not stretched, they will shorten. Conversely, when these tissues are regularly stretched, they lengthen and flexibility is increased. Among women, significant, positive relationships have been found between weekly hours of participation in a sport and knee ROM, with active knee extension ROM increasing among swimmers and competitive gymnasts, and active knee flexion ROM increasing among basketball players (20). The results of several studies indicate that flexibility can be significantly increased among elderly individuals who participate in a program of regular stretching and exercise (19, 45).

TECHNIQUES FOR INCREASING JOINT FLEXIBILITY

Increasing joint flexibility is often an important component of thera-peutic and rehabilitative programs and programs designed to train athletes for a particular sport. Increasing or maintaining flexibility in-volves stretching the tissues that limit the ROM at a joint. Several ap-proaches for stretching these tissues can be used, with some being more effective than others because of differential neuromuscular responses elicited.

Neuromuscular Response to Stretch

Sensory receptors known as Golgi tendon organs (GTOs) are located in the muscle–tendon junctions and in the tendons at both ends of muscles (Figure 5-10). Approximately 10–15 muscle fibers are connected in di-rect line, or in series, with each GTO. These receptors are stimulated by tension in the muscle–tendon unit. Although both tension produced by muscle contraction and tension produced by passive muscle stretch can stimulate GTOs, the threshold for stimulation by passive stretch is much higher. Whereas the muscle force arising from passive stretch must reach approximately 2 N, the activation of a single muscle fiber with a force production of 30–90 μN is sufficient to stimulate a GTO (3). The GTOs respond through their neural connections by inhibiting tension develop-ment in the activated muscle (promoting muscle relaxation) and by initi-ating tension development in the antagonist muscles.

Other sensory receptors are interspersed throughout the fibers of muscles. These receptors, which are oriented parallel to the fibers, are known as muscle spindles because of their shape (Figure 5-11). Each muscle spindle is composed of approximately 3–10 small muscle fibers, termed *intrafusal fibers,* that are encased in a sheath of connective tissue.

Muscle spindles respond to both the amount of muscle lengthening (static response) and the rate of muscle lengthening (dynamic response). Intrafusal fibers known as nuclear chain fibers are primarily responsible for the static component, and intrafusal fibers known as nuclear bag fi-bers are responsible for the dynamic component. These two types of in-trafusal fiber have been shown to function independently, but because the dynamic response is much stronger than the static response, a slow

Golgi tendon organ
sensory receptor that inhibits tension development in a muscle and initiates tension development in antagonist muscles

muscle spindle
sensory receptor that provokes reflex contraction in a stretched muscle and inhibits tension development in antagonist muscles

FIGURE 5-10

Schematic representation of a Golgi tendon organ. From Shier, Butler, and Lewis, *Hole's Human Anatomy and Physiology,* © 1996. Reprinted by permission of The McGraw-Hill Companies, Inc.

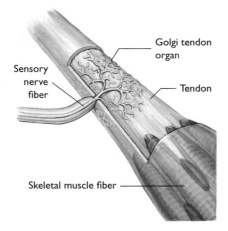

Golgi tendon organ

Sensory nerve fiber

Tendon

Skeletal muscle fiber

Schematic representation of a muscle spindle. From Shier, Butler, and Lewis, *Hole's Human Anatomy and Physiology,* © 1996. Reprinted by permission of The McGraw-Hill Companies, Inc.

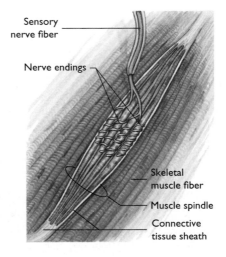

stretch reflex
monosynaptic reflex initiated by stretching of muscle spindles and resulting in immediate development of muscle tension

reciprocal inhibition
inhibition of tension development in the antagonist muscles resulting from activation of muscle spindles

rate of stretching does not activate the muscle spindle response until the muscle is significantly stretched (9). Some muscles receive greater muscle spindle response than others. For example, the soleus receives more muscle spindle feedback than the gastrocnemius during both rest and muscle activation (67).

The spindle response includes activation of the stretch reflex and inhibition of tension development in the antagonist muscle group, a process known as reciprocal inhibition. The stretch reflex, also known as the *myotatic reflex,* is provoked by the activation of the spindles in a stretched muscle. This rapid response involves neural transmission across a single synapse, with afferent nerves carrying stimuli from the spindles to the spinal cord and efferent nerves, returning an excitatory signal directly from the spinal cord to the muscle, resulting in tension development in the muscle. The knee-jerk test, a common neurological test of motor function, is an example of muscle spindle function producing a quick, brief contraction in a stretched muscle. A tap on the patellar tendon initiates the stretch reflex, resulting in the jerk caused by the immediate development of tension in the quadriceps group (Figure 5-12).

The myotatic (stretch) reflex is initiated by stretching of the muscle spindles. From Shier, Butler, and Lewis, *Hole's Human Anatomy and Physiology,* © 1996. Reprinted by permission of The McGraw-Hill Companies, Inc.

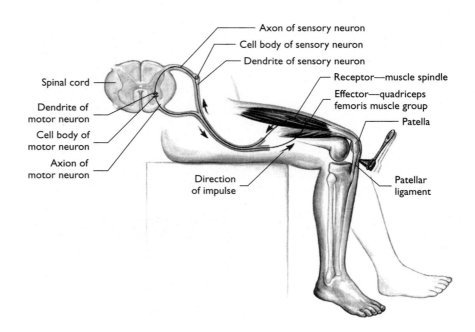

CHARACTERISTIC	GOLGI TENDON ORGANS	MUSCLE SPINDLES
Location	Within tendons near the muscle–tendon junction in series with muscle fibers	Interspersed among muscle fibers in parallel with the fibers
Stimulus	Increase in muscle tension	Increase in muscle length
Response	1. Inhibits tension development in stretched muscle, 2. Initiates tension development in antagonist muscles	1. Initiates rapid contraction of stretched muscle, 2. Inhibits tension development in antagonist muscles
Overall effect	Promotes relaxation in muscle developing tension	Inhibits stretch in muscle being stretched

TABLE 5-1

Golgi Tendon Organs (GTOs) and Muscle Spindles: How Do They Compare?

Because muscle spindle activation produces tension development in stretching muscle, whereas GTO activation promotes relaxation of muscle developing tension, the general goals of any procedure for stretching muscle are minimizing the spindle effect and maximizing the GTO effect. A summary comparison of GTOs and muscle spindles is presented in Table 5-1.

Active and Passive Stretching

Stretching can be done either actively or passively. Active stretching is produced by contraction of the antagonist muscles (those on the side of the joint opposite the muscles, tendons, and ligaments to be stretched). Thus, to actively stretch the hamstrings (the primary knee flexors), the quadriceps (primary knee extensors) should be contracted. Passive stretching involves the use of gravitational force, force applied by another body segment, or force applied by another person, to move a body segment to the end of the ROM. Active stretching provides the advantage of exercising the muscle groups used to develop force. With passive stretching, movement can be carried farther beyond the existing ROM than with active stretching, but with the concomitant disadvantage of increased injury potential.

active stretching
stretching of muscles, tendons, and ligaments produced by active development of tension in the antagonist muscles

passive stretching
stretching of muscles, tendons, and ligaments produced by a stretching force other than tension in the antagonist muscles

Ballistic and Static Stretching

Ballistic stretching, or performance of bouncing stretches, makes use of the momentum of body segments to repeatedly extend joint position to or beyond the extremes of the ROM. Because a ballistic stretch activates the stretch reflex and results in the immediate development of tension in the muscle being stretched, microtearing of the stretched muscle tissue may occur. Because the extent of the stretch is not controlled, the potential for injury to all of the stretched tissues is heightened.

With static stretching, the movement is slow, and when the desired joint position is reached, it is maintained statically, usually for about 30–60 seconds (s). There seems to be general agreement that for optimal effect, the static stretch of each muscle group should be sequentially repeated three to five times (46). Other research demonstrates that it is the total stretch time during each day, rather than the stretching protocol, that determines the effect on tissue extensibility (10).

Although static stretching has been shown to be effective for increasing joint flexibility, there is also overwhelming evidence that a single, 30-s bout of static stretching has a noticeably detrimental effect on muscle strength, with additional stretching further decreasing

ballistic stretching
a series of quick, bouncing-type stretches

●*Ballistic, bouncing types of stretches can be dangerous because they tend to promote contraction of the muscles being stretched, and the momentum generated may carry the body segments far enough beyond the normal ROM to tear or rupture collagenous tissues.*

static stretching
maintaining a slow, controlled, sustained stretch over time, usually about 30 seconds

Active, static stretching involves holding a position at the extreme of the range of motion. Photo © Lars A. Niki.

strength (40, 74). Following static stretching this decrease in muscle strength has also been shown to translate to a significant decrement in performance in both 60- and 100-m sprints, as well as in endurance running events (30, 73). Although some coaches seem to believe that performing concentric contraction exercises after stretching will ameliorate the negative effects of stretching on muscular strength, research shows this to be false, even when the exercises involve maximal contractions (29, 70). Studies comparing static and ballistic stretching have shown that static stretching is more effective in increasing joint range of motion, both after a single bout of stretching and after a four-week stretching protocol (2, 11). However, whereas static stretching produces a transient decrease in muscle strength, there is no such effect with ballistic stretching (2).

Dynamic stretching involves motion of the body as in ballistic stretching, but unlike ballistic stretching, the motion is controlled and not a bouncing-type movement. Recent research demonstrates that following a bout of dynamic stretching there is a beneficial effect for activities requiring muscular power (12, 18, 38, 56). The current literature suggests that prior to athletic competition a warm-up including dynamic stretching may be desirable, with static stretching being most beneficial following a performance to maintain or increase joint range of motion. Both forms of stretching can induce soreness in muscles that are not habitually stretched (60).

Proprioceptive Neuromuscular Facilitation

**proprioceptive neuro-
muscular facilitation**
*a group of stretching procedures
involving alternating contraction
and relaxation of the muscles being
stretched*

The most effective stretching procedures are known collectively as proprioceptive neuromuscular facilitation (PNF). PNF techniques were originally used by physical therapists for treating patients with neuromuscular paralysis. All PNF procedures involve some pattern of alternating contraction and relaxation of agonist and antagonist muscles designed to take advantage of the GTO response. All PNF techniques require a partner or clinician. Stretching the hamstrings from a supine position provides a good illustration for several of the popular PNF approaches (see Figure 5-12).

The contract-relax-antagonist-contract technique (also referred to as slow-reversal-hold-relax), involves passive static stretch of the hamstrings by a partner, followed by active contraction of the hamstrings against the partner's resistance. Next, the hamstrings are relaxed and

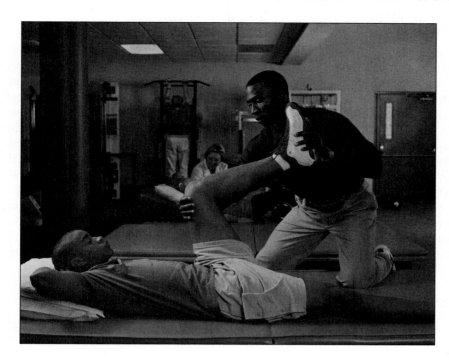

Passive stretching can be accomplished with the assistance of a partner. Photo courtesy Royalty-Free/CORBIS.

the quadriceps are contracted as the partner pushes the leg into increasing flexion at the hip. There is then a phase of complete relaxation, with the leg held in the new position of increased hip flexion. Each phase of this process is typically maintained for a duration of 5–10 s, and the entire sequence is carried out at least four times.

The contract-relax and hold-relax procedures begin as in the slow-reversal-hold method, with a partner applying passive stretch to the hamstrings, followed by active contraction of the hamstrings against the partner's resistance. With the contract-relax approach, the contraction of the hamstrings is isotonic, resulting in slow movement of the leg in the direction of hip extension. In the hold-relax method, the contraction of the hamstrings is isometric against the partner's unmoving resistance. Following contraction, both methods involve relaxation of the hamstrings and quadriceps while the hamstrings are passively stretched. Again, the duration of each phase is usually 5–10 s, and the entire sequence is repeated several times.

The agonist-contract-relax method is another PNF variation, with 5–20 s sequential phases. This procedure begins with active, maximal contraction of the quadriceps to extend the knee, followed by relaxation as the partner manually supports the leg in the position actively attained.

Studies show that PNF techniques can significantly increase joint ROM transiently after a single stretching session and with a more long-lasting effect when three bouts of PNF stretching are performed three times per week (24). Researchers have found the optimal contraction intensity for individuals using PNF techniques to be approximately 65% of maximum voluntary isometric contraction (58).

COMMON JOINT INJURIES AND PATHOLOGIES

The joints of the human body support weight, are loaded by muscle forces, and at the same time provide range of movement for the body segments. They are consequently subject to both acute and overuse injuries, as well as to infection and degenerative conditions.

Sprains

Sprains are injuries caused by abnormal displacement or twisting of the articulating bones that results in stretching or tearing of ligaments, tendons, and connective tissues crossing a joint. Sprains can occur at any joint, but are most common at the ankle. Lateral ankle sprains are particularly common, because the ankle is a major weight-bearing joint and because there is less ligamentous support on the lateral than on the medial side of the ankle. Sprains can be classified as first, second, and third degree, depending on the severity of the injury. First-degree sprains are the mildest, with symptoms of tenderness and slight swelling, but little loss of joint ROM. With second-degree sprains, more damage to the tissues is present, and there is usually swelling, bruising, localized tenderness, moderate pain, and some restriction of joint ROM. Third-degree sprains involve partial to complete tearing of the ligaments, accompanied by swelling, pain, and typically joint instability. The traditional treatment for sprains is rest, ice, compression, and elevation.

Dislocations

Displacement of the articulating bones at a joint is termed *dislocation.* These injuries usually result from falls or other mishaps involving a large magnitude of force. Common sites for dislocations include the shoulders, fingers, knees, elbows, and jaw. Symptoms include visible joint deformity, intense pain, swelling, numbness or tingling, and some loss of joint movement capability. A dislocated joint may result in damage to the surrounding ligaments, nerves, and blood vessels. It is important to reduce (or properly relocate) a dislocated joint as soon as possible both to alleviate the pain and to ensure that the blood supply to the joint is not impeded. Reduction of a dislocated joint should be performed by a trained medical professional.

Bursitis

The bursae are sacs filled with fluid that function to cushion points where muscles or tendons slide over bone. Under normal conditions, the bursae create a smooth, nearly frictionless gliding surface. With bursitis, or inflammation of a bursa, movement around the affected area becomes painful, with more movement increasing the inflammation and aggravating the problem. Bursitis can be caused by overuse-type, repetitive, minor impacts on the area, or from acute injuries, with subsequent inflammation of the surrounding bursae. The condition is treated with rest, ice, and anti-inflammatory medications. For example, runners who increase training mileage too abruptly may experience inflammation of the bursa between the Achilles tendon and the calcaneous. Pain and possibly some swelling are symptoms of bursitis.

Arthritis

Arthritis is a pathology involving joint inflammation accompanied by pain and swelling. It is extremely common with aging, with over 100 different types of arthritis identified.

Rheumatoid Arthritis

The most debilitating and painful form of arthritis is rheumatoid arthritis, an autoimmune disorder that involves the body's immune system attacking healthy tissues. It is more common in adults, but there is also a juvenile rheumatoid arthritis. Characteristics include inflammation and thickening of the synovial membranes and breakdown of the articular cartilage, resulting in limitation of motion and eventually ossification or fusing of the articulating bones. Other symptoms include anemia, fatigue, muscular atrophy, osteoporosis, and other systemic changes.

Osteoarthritis

Osteoarthritis, or degenerative joint disease, is the most common form of arthritis. It is increasingly believed to be an entire family of related disorders that result in progressive degradation of the biomechanical properties of articular cartilage (50). In the early stages of the disorder, the joint cartilage loses its smooth, glistening appearance and becomes rough and irregular. Eventually, the cartilage completely wears away, leaving the articulating bone surfaces bare. Thickening of the subchondral bone and the formation of osteophytes, or bone spurs, are accompanying features (59). Pain, swelling, ROM restriction, and stiffness are all symptoms, with the pain typically relieved by rest, and joint stiffness improved by activity.

The cause of osteoarthritis is usually unknown. Although articular cartilage appears to adapt to changes in habitual loading patterns, efforts to associate the incidence of osteoarthritis with lifestyle factors have produced conflicting results (22, 23, 25, 63). Whereas occupations requiring heavy lifting, farming, and participation in elite sports have been associated with higher incidences of hip osteoarthritis, no relationship has been found between levels of regular physical activity throughout life and the incidence of knee osteoarthritis (25, 63). It has been shown, however, that malalignment of the hip-knee-ankle increases the progression of osteoarthritis at the knee, with varus and valgus alignments respectively increasing loading and osteoarthritis progression on the medial and lateral aspects of the knee (see Chapter 8) (57).

Because articular cartilage is avascular in adults, it relies on cyclic mechanical loading for fluid exchange to deliver nutrients and remove waste products. Consequently, too little cyclic mechanical stress at synovial joints results in deterioration of the cartilage. Research suggests that some degenerative joint disease may actually stem from remodeling and related vascular insufficiency in the underlying subchondral bone, a pattern also associated with disuse (27, 39, 43). Current thinking is that both too little mechanical stress and excessive mechanical stress can promote the development of osteoarthritis, with an intermediate zone of regular cyclic loading that optimizes the health of articular cartilage (72).

SUMMARY

The anatomical configurations of the joints of the human body govern the directional movement capabilities of the articulating body segments. From the perspective of movements permitted, there are three major categories of joints: synarthroses (immovable joints), amphiarthroses (slightly movable joints), and diarthroses (freely movable joints). Each major category is further subdivided into classes of joints with common anatomical characteristics.

The ends of bones articulating at diarthrodial joints are covered with articular cartilage, which reduces contact stress and regulates joint lubrication. Fibrocartilaginous discs or menisci present at some joints also may contribute to these functions.

Tendons and ligaments are strong collagenous tissues that are slightly extensible and elastic. These tissues are similar to muscle and bone in that they adapt to levels of increased or decreased mechanical stress by hypertrophying or atrophying.

Joint stability is the ability of the joint to resist displacement of the articulating bones. The major factors influencing joint stability are the size and shape of the articulating bone surfaces, and the arrangement and strength of the surrounding muscles, tendons, and ligaments.

Joint flexibility is primarily a function of the relative tightness of the muscles and ligaments that span the joint. If these tissues are not stretched, they tend to shorten. Approaches for increasing flexibility include active versus passive stretching, and static versus dynamic stretching. PNF is a particularly effective procedure for stretching muscles and ligaments.

INTRODUCTORY PROBLEMS

(Reference may be made to Chapters 7–9 for additional information on specific joints.)

1. Construct a table that identifies joint type and the plane or planes of allowed movement for the shoulder (glenohumeral joint), elbow, wrist, hip, knee, and ankle.
2. Describe the directions and approximate ranges of movement that occur at the joints of the human body during each of the following movements:
   a. Walking
   b. Running
   c. Performing a jumping jack
   d. Rising from a seated position
3. What factors contribute to joint stability?
4. Explain why athletes' joints are often taped before the athletes participate in an activity. What are some possible advantages and disadvantages of taping?
5. What factors contribute to flexibility?
6. What degree of joint flexibility is desirable?
7. How is flexibility related to the likelihood of injury?
8. Discuss the relationship between joint stability and joint flexibility.
9. Explain why grip strength diminishes as the wrist is hyperextended.
10. Why is ballistic stretching contraindicated?

ADDITIONAL PROBLEMS

1. Construct a table that identifies joint type and the plane or planes of movement for the atlanto-occipital joint, the L5-S1 vertebral joint, the metacarpophalangeal joints, the interphalangeal joints, the carpometacarpal joint of the thumb, the radioulnar joint, and the talocrural joint.
2. Identify the position (for example, full extension, 90° of flexion) for which each of the following joints is close packed:
   a. Shoulder
   b. Elbow
   c. Knee
   d. Ankle
3. How is articular cartilage similar to and different from ordinary sponge? (You may wish to consult the Annotated Readings.)
4. Comparatively discuss the properties of muscle, tendon, and ligament. (You may wish to consult the Annotated Readings.)

5. Discuss the relative importance of joint stability and joint mobility for athletes participating in each of the following sports:
   a. Gymnastics
   b. Football
   c. Swimming
6. What specific exercises would you recommend for increasing the stability of each of the following joints?
   a. Shoulder
   b. Knee
   c. Ankle
   Explain the rationale for your recommendations.
7. What specific exercises would you recommend for increasing the flexibility of each of the following joints?
   a. Hip
   b. Shoulder
   c. Ankle
   Explain the rationale for your recommendations.
8. In which sports are athletes more likely to incur injuries that are related to insufficient joint stability? Explain your answer.
9. In which sports are athletes likely to incur injuries related to insufficient joint flexibility? Explain your answer.
10. What exercises would you recommend for senior citizens interested in maintaining an appropriate level of joint flexibility?

LABORATORY EXPERIENCES

1. Using a skeleton, an anatomical model, or the *Dynamic Human* CD, locate and provide a brief description for an example of each type of joint.

   a. Synarthroses (immovable joints)

Suture: _____

Description: _____

_____

Syndesmosis: _____

Description: _____

_____

   b. Amphiarthroses (slightly movable joints)

Synchondrosis: _____

Description: _____

_____

Symphysis: _____

Description: _____

_____

   c. Diarthroses (Freely movable joints)

Gliding: _____

Description: _____

_____

Hinge: _____

Description: _____

_____

Pivot: _____

Description: _____

_____

Condyloid: _____

Description: _____

Saddle: _____

Description: _____

_____

Ball and socket: _____

Description: _____

_____

2. Using online resources or the *Dynamic Human* CD, review the histology of fibrocartilage and hyaline cartilage. List the locations in the body where each of these are found.

Fibrocartilage: _____

_____

Hyaline cartilage: _____

_____

3. With a partner, use a goniometer to measure the range of motion for hip flexion with the leg fully extended before and after a 30-second *active* static hamstring stretch. Explain your results.

ROM before stretch: _____ after stretch: _____

Explanation: _____

_____

_____

4. With a partner, use a goniometer to measure the range of motion for hip flexion with the leg fully extended before and after a 30-second *passive* static hamstring stretch. Explain your results.

ROM before stretch: _____ after stretch: _____

Explanation: _____

_____

_____

5. With a partner, use a goniometer to measure the range of motion for hip flexion, with the leg fully extended before and after stretching the hamstrings with one of the PNF techniques described in the chapter. Explain your results.

ROM before stretch: _____ after stretch: _____

Explanation: _____

_____

_____

REFERENCES

1. Anderson AF, Dome DC, Gautam S, Awh MH, and Rennirt GW: Correlation of anthropometric measurements, strength, anterior cruciate ligament size, and intercondylar notch characteristics to sex differences in anterior cruciate ligament tear rates, *Am J Sports Med* 29:58, 2001.

2. Bacurau RF, Monteiro GA, Ugrinowitsch C, Tricoli V, Cabral LF, and Aoki MS: Acute effect of a ballistic and a static stretching exercise bout on flexibility and maximal strength, *J Strength Cond Res* 23:304, 2009.

3. Binder MD, Krion JS, Moore GP, and Stuart DG: The response of Golgi tendon organs to single motor unit contractions, *J Physiol* (London) 271:337, 1977.

4. Boschetti F, Pennati G, Gervaso F, Peretti GM, and Dubini G: Biomechanical properties of human articular cartilage under compressive loads, *Biorheology* 41:159, 2004.

5. Brand RA: Joint lubrication. In Albright JA and Brand RA, eds: *The scientific basis of orthopedics,* New York, 1979, Appleton-Century-Crofts.

6. Buckwalter JA, Mankin HJ, and Grodzinsky AJ: Articular cartilage and osteoarthritis, *Instr Course Lect* 54:465, 2005.

7. Chen CH, Nosaka K, Chen HL, Lin MJ, Tseng KW, and Chen TC: Effects of flexibility training on eccentric exercise-induced muscle damage, Aug 2. [Epub ahead of print] 2010.

8. Chiang H and Jiang CC: Repair of articular cartilage defects: review and perspectives, *J Formos Med Assoc,* 108:87, 2009.

9. Chou SW, Abraham LD, Huang IS, Pei YC, Lai CH, and Wong AM: Starting position and stretching velocity effects on the reflex threshold angle of stretch reflex in the soleus muscle of normal and spastic subjects, *J Formos Med Assoc* 104:493, 2005.

10. Cipriani D, Abel B, and Pirrwitz D: A comparison of two stretching protocols on hip range of motion: implications for total daily stretch duration, *J Strength Cond Res* 17:274, 2003.

11. Covert CA, Alexander MP, Petronis JJ, and Davis DS: Comparison of ballistic and static stretching on hamstring muscle length using an equal stretching dose, Apr 1. [Epub ahead of print], 2010.

12. Curry BS, Chengkalath D, Crouch GJ, Romance M, and Manns PJ: Acute effects of dynamic stretching, static stretching, and light aerobic activity on muscular performance in women, *J Strength Cond Res* 23:1811, 2009.

13. Dugan SA and Frontera WR: Muscle fatigue and muscle injury, *Phys Med Rehabil Clin N Am* 11:385, 2000.

14. Durselen L, Hehl G, Simnacher M, Kinzl L, and Claes L: Augmentation of a ruptured posterior cruciate ligament provides normal knee joint stability during ligament healing, *Clin Biomech* 16:222, 2001.

15. Ellenbecker TS, Roetert EP, Piorkowski PA, and Schulz DA: Glenohumeral joint internal and external range of motion in elite junior tennis players, *J Orthop Sports Phys Ther* 24:336, 1996.

16. Eriksson K, Kindblom LG, Hamberg P, Larsson H, and Wredmark T: The semitendinosus tendon regenerates after resection: a morphologic and MRI analysis in 6 patients after resection for anterior cruciate ligament reconstruction, *Acta Orthop Scand* 72:379, 2001.

17. Feldman D, Shrier I, Rossignol M, and Abenhaim L: Adolescent growth is not associated with changes in flexibility, *Clin J Sport Med* 24, 1999.

18. Gelen E: Acute effects of different warm-up methods on sprint, slalom dribbling, and penalty kick performance in soccer players, *J Strength Cond Res* 24(4):950, 2010.

19. Girouard CK and Hurley BF: Does strength training inhibit gains in range of motion from flexibility training in older adults?, *Med Sci Sports Exerc* 27:1444, 1995.

20. Hahn T, Foldspang A, Vestergaard E, and Ingemann-Hansen T: Active knee joint flexibility and sports activity, *Scand J Med Sci Sports* 9:74, 1999.

21. Hartig DE and Henderson JM: Increasing hamstring flexibility decreases lower extremity overuse injuries in military basic trainees, *Am J Sports Med* 27:173, 1999.

22. Herzog W, Diet S, Suter E, Mayzus P, Leonard TR, Müller C, Wu JZ, and Epstein M: Material and functional properties of articular cartilage and patellofemoral contact mechanics in an experimental model of osteoarthritis, *J Biomech* 31:1137, 1998.

23. Herzog W, Wu JZ, and Clark A: Experimental and theoretical investigations in osteoarthritis research, *J Biomech* 34:S29, 2001.

24. Higgs F, Winter SL: The effect of a four-week proprioceptive neuromuscular facilitation stretching program on isokinetic torque production, *J Strength Cond Res* 23:1442, 2009.

25. Hoaglund FT and Steinbach LS: Primary osteoarthritis of the hip: etiology and epidemiology, *J Am Acad Orthop Surg* 9:320, 2001.

26. Hunter DG and Spriggs J: Investigation into the relationship between the passive flexibility and active stiffness of the ankle plantar-flexor muscles, *Clin Biomech* 15:600, 2000.

27. Imhof H, Breitenseher M, Kainberger F, and Trattnig S: Degenerative joint disease: cartilage or vascular disease?, *Skeletal Radiol* 26:398, 1997.

28. Jones G, Ding C, Glisson M, Hynes K, Ma D, and Cicuttini F: Knee articular cartilage development in children: a longitudinal study of the effect of sex, growth, body composition, and physical activity, *Pediatr Res* 54:230, 2003.

29. Kay AD and Blazevich AJ: Concentric muscle contractions before static stretching minimize, but do not remove, stretch-induced force deficits, *J Appl Physiol,* 108:637 Epub Jan 14, 2010.

30. Kistler BM, Walsh MS, Horn TS, and Cox RH: The acute effects of static stretching on the sprint performance of collegiate men in the 60- and 100-m dash after a dynamic warm-up, *J Strength Cond Res* [Epub ahead of print] 2010.

31. Knapik JJ, Jones BH, Bauman CL, and Harris JM: Strength, flexibility and athletic injuries, *Sports Med* 14:277, 1992.

32. Knapik JJ, Bauman CL, Jones BH, Harris JM, Vaughan L.: Preseason strength and flexibility imbalances associated with athletic injuries in female college athletes, *Am J Sports Med* 19:76, 1991.

33. Knight CA, Rutledge CR, Cox ME, Acosta M, and Hall SJ: Effect of superficial heat, deep heat, and active exercise warm-up on the extensibility of the plantar flexors, *Phys Ther* 81:1206, 2001.

34. Krivickas LS and Feinberg JH: Lower extremity injuries in college athletes: relation between ligamentous laxity and lower extremity muscle tightness, *Arch Phys Med Rehabil* 77:1139, 1996.

35. Li LP, Buschmann MD, and Shirazi-Adl A: Strain-rate dependent stiffness of articular cartilage in unconfined compression, *J Biomech Eng* 125:161, 2003.

36. Loitz BJ and Frank CB: Biology and mechanics of ligament and ligament healing, *Exerc Sport Sci Rev* 21:33, 1993.

37. Maffulli N, Ewen SW, Waterson SW, Reaper J, and Barrass V: Tenocytes from ruptured tendinpathic achilles tendons produce greater quantities of type III collagen than tenocytes from normal achilles tendons, *Am J Sports Med* 28:499, 2000.

38. Manoel ME, Harris-Love MO, Danoff JV, and Miller TA: Acute effects of static, dynamic, and proprioceptive neuromuscular facilitation stretching on muscle power in women, *J Strength Cond Res* 22:1528, 2008.

39. Matsui H, Shimizu M, and Tsuji H: Cartilage and subchondral bone interaction in osteoarthrosis of human knee joint: a histological and histomorphometric study, *Microsc Res Tech* 37:333, 1997.

40. McHugh MP and Cosgrave CH: To stretch or not to stretch: the role of stretching in injury prevention and performance, *Scand J Med Sci Sports* 20:169, 2009.

41. McHugh MP, Kremenic IJ, Fox MB, and Gleim GW: The role of mechanical and neural restraints to joint range of motion during passive stretch, *Med Sci Sports Exerc* 30:928, 1998.

42. Mithoefer K, Hambly K, Della Villa S, Silvers H, and Mandelbaum BR: Return to sports participation after articular cartilage repair in the knee: scientific evidence, *Am J Sports Med* 37 (Suppl 1):167S, 2009.

43. Mithoefer K, Minas T, Peterson L, Yeon H, and Micheli LJ: Functional outcome of knee articular cartilage repair in adolescent athletes, *Am J Sports Med* 33:1147, 2005.

44. Mow VC and Wang CC: Some bioengineering considerations for tissue engineering of articular cartilage, *Clin Orthop* 367:S204, 1999.

45. Munns K: Effects of exercise on the range of joint motion in elderly subjects. In Smith EL and Serfass RC, eds: *Exercise and aging: The scientific basis,* Hillside, NJ, 1981, Enslow Publishers.

46. Nordin M and Frankel VH: *Basic biomechanics of the skeletal system* (3rd ed.), Baltimore, 2001, Lippincott Williams & Wilkins.

47. Panjabi MM and White AA: *Biomechanics in the musculoskeletal system,* New York, 2001, Churchill Livingstone.

48. Papandrea P, Vulpiani MC, Ferretti A, and Conteduca F: Regeneration of the semitendinosus tendon harvested for anterior cruciate ligament reconstruction, *Am J Sports Med* 28:556, 2000.

49. Park S, Krishnan R, Nicoll SB, and Ateshian GA: Cartilage interstitial fluid load support in unconfined compression, *J Biomech* 36:1785, 2003.

50. Pearle AD, Warren RF, and Rodeo SA: Basic science of articular cartilage and osteoarthritis, *Clin Sports Med* 24:1, 2005.

51. Quinn TM, Hunziker EB, and Hauselmann HJ: Variation of cell and matrix morphologies in articular cartilage among locations in the adult human knee, *Osteoarthritis Cartilage* 13:672, 2005.

52. Radin EL: Role of muscles in protecting athletes from injury, *Acta Med Scand Suppl* 711:143, 1986.

53. Redman SN, Oldfield SF, and Archer CW: Current strategies for articular cartilage repair, *Eur Cell Mater* 14:23, 2005.

54. Safran MR, Garrett Jr WE, Seaber AV, Glisson RR, and Ribbeck BM: The role of warm-up in muscular injury prevention, *Am J Sports Med* 16:123, 1988.

55. Safran MR and Seiber K: The evidence for surgical repair of articular cartilage in the knee, *J Am Acad Orthop Surg* 18:259, 2010.

56. Sekir U, Arabaci R, Akova B, and Kadagan SM: Acute effects of static and dynamic stretching on leg flexor and extensor isokinetic strength in elite women athletes, *Scand J Med Sci Sports* 20:268, 2010.

57. Sharma L, Song J, Felson DT, Cahue S, Shamiyeh E, and Dunlop DD: The role of knee alignment in disease progression and functional decline in knee osteoarthritis, *JAMA* 286:188, 2001.

58. Sheard PW and Paine TJ: Optimal contraction intensity during proprioceptive neuromuscular facilitation for maximal increase of range of motion, *J Strength Cond Res* 24:416, 2010.

59. Silver, FH, Bradica G, and Tria A: Do changes in the mechanical properties of articular cartilage promote catabolic destruction of cartilage and osteoarthritis? *Matrix Biol* 23:467, 2004.

60. Smith LL, Brunetz MH, Chenier TC, McCammon MR, Houmard JA, Franklin ME, Israel RG.: The effects of static and ballistic stretching on delayed onset muscle soreness and creatine kinase, *Res Q Exerc Sport* 64:103, 1993.

61. Standring S (ed.): *Gray's anatomy ed 40,* UK, 2008, Elsevier Limited.

62. Steele VA and White JA: Injury prediction in female gymnasts, *Br J Sports Med* 20:31, 1986.

63. Sutton AJ, Muir KR, Mockett S, and Fentem P: A case-controlled study to investigate the relation between low and moderate levels of physical activity and osteoarthritis of the knee using data collected as part of the Allied Dunbar National Fitness Survey, *Ann Rheum Dis* 60:756, 2001.

64. Tipton CM, James SL, Mergner W, Tcheng TK.: Influence of exercise on strength of medial collateral ligaments of dogs, *Am J Physiol* 218:894, 1970.

65. Tipton CM, Matthes RD, Maynard JA, Carey RA.: The influence of physical activity on ligaments and tendons, *Med Sci Sports Exerc* 7:165, 1975.

66. Trippel SB, Ghivizzani SC, and Nixon AJ: Gene-based approaches for the repair of articular cartilage, *Gene Ther* 11:351, 2004.

67. Tucker KJ and Turker KS: Muscle spindle feedback differs between the soleus and gastrocnemius in humans, *Somatosens Mot Res* 21:189, 2004.

68. Ulrich-Vinther M, Maloney MD, Schwarz EM, Rosier R, and O'Keefe RJ: Articular cartilage biology, *J Am Acad Orthop Surg* 11:421, 2003.

69. Vangsness CT Jr, Kurzweil PR, and Lieberman JR: Restoring articular cartilage in the knee, *Am J Orthop* 33:29, 2004.

70. Viale F, Nana-Ibrahim S, and Martin RJ: Effect of active recovery on acute strength deficits induced by passive stretching, *J Strength Cond Res* 21:1233, 2007.
71. Weightman BO and Kempson GE: Load carriage. In Freeman MAR, ed: *Adult articular cartilage,* London, 1979, Pitman.
72. Whiting WC and Zernicke RF: *Biomechanics of musculoskeletal injury (2nd ed),* Champaign, IL, Human Kinetics, 2008.
73. Wilson JM, Hornbuckle LM, Kim JS, Ugrinowitch C, Lee SR, Zoundos MC, Sommer B, and Panton LB: Effects of static stretching on energy cost and running endurance performance, *J Strength Cond Res* Nov 13. [Epub ahead of print] 2009.
74. Winchester JB, Nelson AG, and Kokkonen J: A single 30-s stretch is sufficient to inhibit maximal voluntary strength, *Res Q Exerc Sport* 80:257, 2009.
75. Ylinen J, Kankainen T, Kautiainen H, Rezasoltani A, Kuukkanen T, and Häkkinen A: Effect of stretching on hamstring muscle compliance, *J Rehabil Med* 41:80, 2009.

ANNOTATED READINGS

Levangie PK and Norkin CC: *Joint structure and function: A comprehensive analysis* (4th ed.), Philadelphia, 2005, F.A. Davis.
Comprehensive presentation of joint structure and muscle actions with reference to normal and pathologic functions.

Messier, SP: Arthritic diseases and conditions. In Kaminsky LA: *ACSM's resource manual for guidelines for exercise testing and prescription* (5th ed.), Philadelphia, 2006, Lippincott Williams & Wilkins.
Includes a comprehensive, research-based overview of current knowledge about osteoporosis, including clinical factors, etiology, nonpharmacological and pharmacological treatments, and exercise prescription.

Mow VC and Hung CT: Biomechanics of articular cartilage. In Nordin M and Frankel VH: *Basic biomechanics of the skeletal system* (3rd ed.), Baltimore, 2001, Lippincott Williams & Wilkins.
Provides in-depth information from the research literature on the structure and function of joint cartilage. Extensive reference list included.

Robinson P (ed.): *Essential radiology for sports medicine,* New York, 2010, Springer.
Chapters on all major joints and other anatomical areas present detailed anatomical descriptions, common injury mechanisms, and radiographs illustrating these.

RELATED WEBSITES

The Center for Orthopaedics and Sports Medicine
http://www.arthroscopy.com/sports.htm
Includes information and color graphics on the anatomy and function of the upper extremity, foot and ankle, and knee, as well as description of knee surgery techniques and articular surface grafting.

Rothman Institute
http://www.rothmaninstitute.com/
Includes information on common sports injuries to the knee, shoulder, and elbow; arthritis of the hip and knee; total joint replacements; spinal anatomy and spine abnormalities and pathologies; and the foot and ankle.

University of Washington Orthopaedic Physicians
http://www.orthop.washington.edu
Provides radiographs and information on common injuries and pathological conditions for the neck, back/spine, hand/wrist, hip, knee, and ankle/foot.

Wheeless Textbook of Orthopaedics Online
http://www.wheelessonline.com/
Provides links to comprehensive medical content on joints, arthritis, and arthroscopy.

KEY TERMS

active stretching	stretching of muscles, tendons, and ligaments produced by active development of tension in the antagonist muscles
articular capsule	double-layered membrane that surrounds every synovial joint
articular cartilage	protective layer of dense white connective tissue covering the articulating bone surfaces at diarthrodial joints
articular fibrocartilage	soft tissue discs or menisci that intervene between articulating bones
ballistic stretching	a series of quick, bouncing-type stretches
close-packed position	joint orientation for which the contact between the articulating bone surfaces is maximum
Golgi tendon organ	sensory receptor that inhibits tension development in a muscle and initiates tension development in antagonist muscles
joint flexibility	a term representing the relative ranges of motion allowed at a joint
joint stability	ability of a joint to resist abnormal displacement of the articulating bones
loose-packed position	any joint orientation other than the close-packed position
muscle spindle	sensory receptor that provokes reflex contraction in a stretched muscle and inhibits tension development in antagonist muscles
passive stretching	stretching of muscles, tendons, and ligaments produced by a stretching force other than tension in the antagonist muscles
proprioceptive neuromuscular facilitation	a group of stretching procedures involving alternating contraction and relaxation of the muscles being stretched
range of motion	angle through which a joint moves from anatomical position to the extreme limit of segment motion in a particular direction
reciprocal inhibition	inhibition of tension development in the antagonist muscles resulting from activation of muscle spindles
static stretching	maintaining a slow, controlled, sustained stretch over time, usually about 30 seconds
stretch reflex	monosynaptic reflex initiated by stretching of muscle spindles and resulting in immediate development of muscle tension
synovial fluid	clear, slightly yellow liquid that provides lubrication inside the articular capsule at synovial joints

6

The Biomechanics of Human Skeletal Muscle

After completing this chapter, you will be able to:

Identify the basic behavioral properties of the musculotendinous unit.

Explain the relationships of fiber types and fiber architecture to muscle function.

Explain how skeletal muscles function to produce coordinated movement of the human body.

Discuss the effects of the force–velocity and length–tension relationships and electromechanical delay on muscle function.

Discuss the concepts of strength, power, and endurance from a biomechanical perspective.

ONLINE LEARNING CENTER RESOURCES

www.mhhe.com/hall6e

Log on to our Online Learning Center (OLC) for access to these additional resources:

- Online Lab Manual
- Flashcards with definitions of chapter key terms
- Chapter objectives
- Chapter lecture PowerPoint presentation
- Self-scoring chapter quiz
- Additional chapter resources
- Web links for study and exploration of chapter-related topics

What enables some athletes to excel at endurance events such as the marathon and others to dominate in power events such as the shot put or sprinting? What characteristics of the neuromuscular system contribute to quickness of movement? What exercises tend to cause muscular soreness? From a biomechanical perspective, what is muscular strength?

Muscle is the only tissue capable of actively developing tension. This characteristic enables skeletal, or striated, muscle to perform the important functions of maintaining upright body posture, moving the body limbs, and absorbing shock. Because muscle can only perform these functions when appropriately stimulated, the human nervous system and the muscular system are often referred to collectively as the neuromuscular system. This chapter discusses the behavioral properties of muscle tissue, the functional organization of skeletal muscle, and the biomechanical aspects of muscle function.

BEHAVIORAL PROPERTIES OF THE MUSCULOTENDINOUS UNIT

● *The characteristic behavioral properties of muscle are extensibility, elasticity, irritability, and the ability to develop tension.*

The four behavioral properties of muscle tissue are extensibility, elasticity, irritability, and the ability to develop tension. These properties are common to all muscle, including the cardiac, smooth, and skeletal muscle of human beings, as well as the muscles of other mammals, reptiles, amphibians, birds, and insects.

Extensibility and Elasticity

The properties of extensibility and elasticity are common to many biological tissues. As shown in Figure 6-1, extensibility is the ability to be stretched or to increase in length, and elasticity is the ability to return to normal length after a stretch. Muscle's elasticity returns it to normal resting length following a stretch and provides for the smooth transmission of tension from muscle to bone.

parallel elastic component
passive elastic property of muscle derived from the muscle membranes

The elastic behavior of muscle has been described as consisting of two major components (32, 57). The parallel elastic component (PEC), provided by the muscle membranes, supplies resistance when a muscle is

FIGURE 6-1

The characteristic properties of muscle tissue enable it to extend, recoil, and contract.

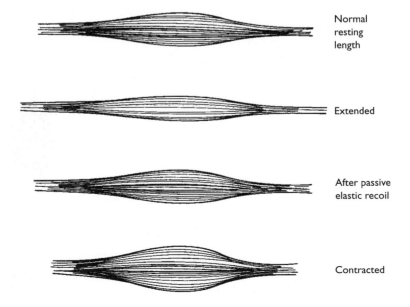

Normal resting length

Extended

After passive elastic recoil

Contracted

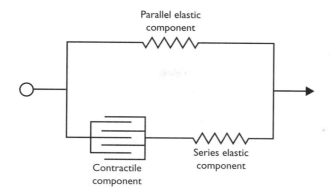

Parallel elastic component

Contractile component

Series elastic component

FIGURE 6-2

From a mechanical perspective, the musculotendinous unit behaves as a contractile component (the muscle fibers) in parallel with one elastic component (the muscle membranes) and in series with another elastic component (the tendons).

passively stretched. The series elastic component (SEC), residing in the tendons, acts as a spring to store elastic energy when a tensed muscle is stretched. These components of muscle elasticity are so named because the membranes and tendons are respectively parallel to and in series (or in line) with the muscle fibers, which provide the contractile component (Figure 6-2). The elasticity of human skeletal muscle is believed to be due primarily to the SEC. Modeling studies show that the height of a jump increases when a countermovement (knee flexion) immediately precedes it due to increased elasticity of the SEC in the lower-extremity muscles (59). Other research supporting the increase in muscle force following stretch has shown that part of the force enhancement also comes from the PEC (66).

Both the SEC and the PEC have a viscous property that enables muscle to stretch and recoil in a time-dependent fashion. When a static stretch of a muscle group such as the hamstrings is maintained over time, the muscle progressively lengthens, increasing joint range of motion. Likewise, after a muscle group has been stretched, it does not recoil to resting length immediately, but shortens gradually over time. This viscoelastic response is independent of gender.

series elastic component
passive elastic property of muscle derived from the tendons

contractile component
muscle property enabling tension development by stimulated muscle fibers

•*Muscle's viscoelastic property enables it to progressively lengthen over time when stretched.*

viscoelastic
having the ability to stretch or shorten over time

Irritability and the Ability to Develop Tension

Another of muscle's characteristic properties, irritability, is the ability to respond to a stimulus. Stimuli affecting muscles are either electrochemical, such as an action potential from the attaching nerve, or mechanical, such as an external blow to a portion of a muscle. When activated by a stimulus, muscle responds by developing tension.

The ability to develop tension is the one behavioral characteristic unique to muscle tissue. Historically, the development of tension by muscle has been referred to as *contraction,* or the contractile component of muscle function. Contractility is the ability to shorten in length. However, as discussed in a later section, tension in a muscle may not result in the muscle's shortening.

STRUCTURAL ORGANIZATION OF SKELETAL MUSCLE

There are approximately 434 muscles in the human body, making up 40–45% of the body weight of most adults. Muscles are distributed in pairs on the right and left sides of the body. About 75 muscle pairs are responsible for body movements and posture, with the remainder involved in activities such as eye control and swallowing. When tension is developed in a muscle, biomechanical considerations such as the magnitude of the force generated, the speed with which the force is developed, and the length of time that the force may be maintained are affected by the particular anatomical and physiological characteristics of the muscle.

Muscle Fibers

A single muscle cell is termed a *muscle fiber* because of its threadlike shape. The membrane surrounding the muscle fiber is sometimes called the *sarcolemma,* and the specialized cytoplasm is termed *sarcoplasm*. The sarcoplasm of each fiber contains a number of nuclei and mitochondria, as well as numerous threadlike myofibrils that are aligned parallel to one another. The myofibrils contain two types of protein filaments whose arrangement produces the characteristic striated pattern after which skeletal, or striated, muscle is named.

Observations through the microscope of the changes in the visible bands and lines in skeletal muscle during muscle contraction have prompted the naming of these structures for purposes of reference (Figure 6-3). The sarcomere, compartmentalized between two Z lines, is the basic structural unit of the muscle fiber (Figure 6-4). Each sarcomere is bisected by an M line. The A bands contain thick, rough myosin filaments, each of which is surrounded by six thin, smooth actin filaments. The I bands contain only thin actin filaments. In both bands, the protein filaments are held in place by attachment to Z lines, which adhere to the sarcolemma. In the center of the A bands are the H zones, which contain only the thick myosin filaments. (See Table 6-1 for the origins of the names of these bands.)

During muscle contraction, the thin actin filaments from either end of the sarcomere slide toward each other. As viewed through a microscope, the Z lines move toward the A bands, which maintain their original size, while the I bands narrow and the H zone disappears. Projections from the myosin filaments called cross-bridges form physical linkages with the actin filaments during muscle contraction, with the number of linkages proportional to both force production and energy expenditure.

FIGURE 6-3

The sarcoplasm of a muscle fiber contains parallel, threadlike myofibrils, each composed of myosin and actin filaments. From Shier, Butler, and Lewis, *Hole's Human Anatomy and Physiology,* © 1996. Reprinted by permission of The McGraw-Hill Companies, Inc.

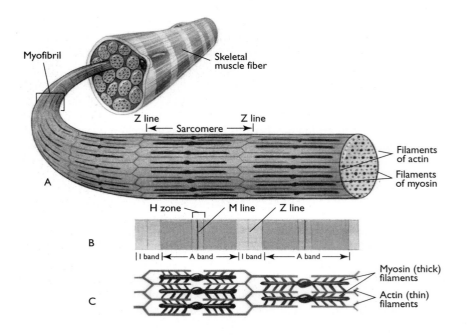

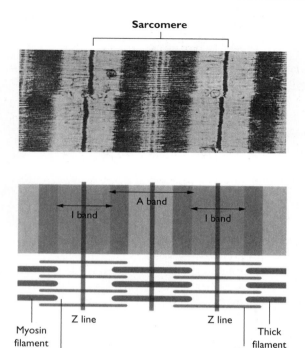

Sarcomere

FIGURE 6-4

The sarcomere is composed of alternating dark and light bands that give muscle its striated appearance. From Shier, Butler, and Lewis, *Hole's Human Anatomy and Physiology*, © 1996. Reprinted by permission of The McGraw-Hill Companies, Inc.

TABLE 6-1

How the Structures within the Sarcomeres Got Their Names

STRUCTURE	HISTORICAL DERIVATION OF NAME
A bands	Polarized light is *anisotropic* as it passes through this region.
I bands	Polarized light is *isotropic* as it passes through this region.
Z lines	The German word *Zwischenscheibe* means "intermediate disc."
H zones	Those were discovered by Hensen.
M band	The German word *Mittelscheibe* means "intermediate band."

A network of membranous channels known as the *sarcoplasmic reticulum* is associated with each fiber externally (Figure 6-5). Internally, the fibers are transected by tiny tunnels called *transverse tubules* that pass completely through the fiber and open only externally. The sarcoplasmic reticulum and transverse tubules provide the channels for transport of the electrochemical mediators of muscle activation.

Several layers of connective tissue provide the superstructure for muscle fiber organization (Figure 6-6). Each fiber membrane, or sarcolemma, is surrounded by a thin connective tissue called the *endomysium*. Fibers are bundled into fascicles by connective tissue sheaths referred to as the *perimysium*. Groups of fascicles forming the whole muscles are then surrounded by the epimysium, which is continuous with the muscle tendons.

Considerable variation in the length and diameter of muscle fibers within muscles is seen in adults. Some fibers may run the entire length of a muscle, whereas others are much shorter. Skeletal muscle fibers grow in length and diameter from birth to adulthood. Fiber diameter can also be increased by resistance training with few repetitions of large loads in adults of all ages.

FIGURE 6-5

The sarcoplasmic reticulum and transverse tubules provide channels for movement of electrolytes. From Shier, Butler, and Lewis, *Hole's Human Anatomy and Physiology*, © 1996. Reprinted by permission of The McGraw-Hill Companies, Inc.

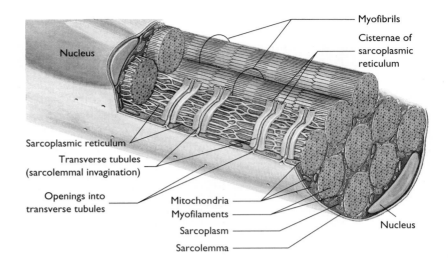

FIGURE 6-6

Muscle is compartmentalized by a series of connective tissue membranes. From Fox, *Human Physiology,* © 1999. Reprinted by permission of The McGraw-Hill Companies, Inc.

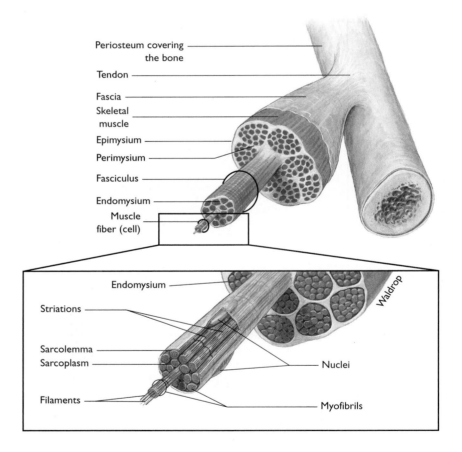

In animals such as amphibians, the number of muscle fibers present also increases with the age and size of the organism. However, this does not appear to occur in human beings. The number of muscle fibers present in humans is genetically determined and varies from person to person. The same number of fibers present at birth is apparently maintained throughout life, except for the occasional loss from injury. The increase in muscle size after resistance training is generally believed to represent an increase in fiber diameters rather than in the number of fibers (50).

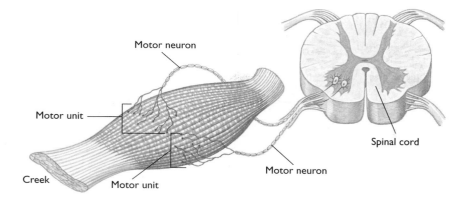

Motor neuron

Motor unit

Motor unit

Creek

Motor neuron

Spinal cord

FIGURE 6-7

A motor unit consists of a single neuron and all muscle fibers innervated by that neuron. From Fox, *Human Physiology,* © 1999. Reprinted by permission of The McGraw-Hill Companies, Inc.

Motor Units

Muscle fibers are organized into functional groups of different sizes. Composed of a single motor neuron and all fibers innervated by it, these groups are known as motor units (Figure 6-7). The axon of each motor neuron subdivides many times so that each individual fiber is supplied with a motor end plate (Figure 6-8). Typically, there is only one end plate per fiber, although multiple innervation of fibers has been reported in vertebrates other than humans (27). The fibers of a motor unit may be spread over a several-centimeter area and be interspersed with the fibers of other motor units. Motor units are typically confined to a single muscle and are localized within that muscle. A single mammalian motor unit may contain from less than 100 to nearly 2000 fibers, depending on the type of movements the muscle executes (9). Movements that are precisely controlled, such as those of the eyes or fingers, are produced by motor units with small numbers of fibers. Gross, forceful movements, such as those produced by the gastrocnemius, are usually the result of the activity of large motor units.

Most skeletal motor units in mammals are composed of *twitch-type* cells that respond to a single stimulus by developing tension in a twitch-like fashion. The tension in a twitch fiber following the stimulus of a single nerve impulse rises to a peak value in less than 100 msec and then immediately declines.

In the human body, however, motor units are generally activated by a volley of nerve impulses. When rapid, successive impulses activate a fiber

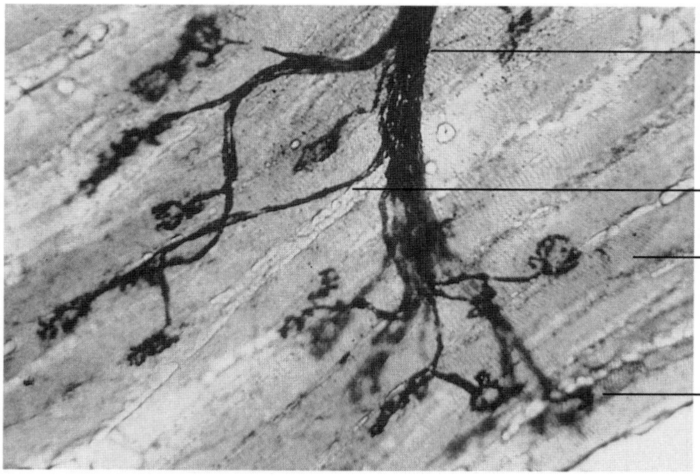

motor unit
a single motor neuron and all fibers it innervates

Motor nerve

Motor neuron axon

Muscle fiber

Motor end plate

FIGURE 6-8

Each muscle fiber in a motor unit receives a motor end plate from the motor neuron. From Fox, *Human Physiology,* © 1999. Reprinted by permission of The McGraw-Hill Companies, Inc.

Tension developed in a muscle fiber (**A**) in response to a single stimulus, (**B**) in response to repetitive stimulation, and (**C**) in response to high-frequency stimulation (tetanus).

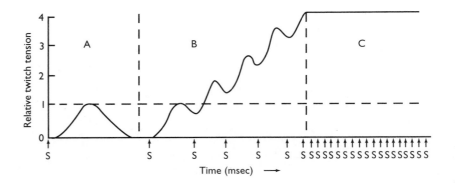

summation
building in an additive fashion

tetanus
state of muscle producing sustained maximal tension resulting from repetitive stimulation

already in tension, summation occurs and tension is progressively elevated until a maximum value for that fiber is reached (Figure 6-9). A fiber repetitively activated so that its maximum tension level is maintained for a time is in tetanus. The tension present during tetanus may be as much as four times peak tension during a single twitch (73). As tetanus is prolonged, fatigue causes a gradual decline in the level of tension produced.

Not all human skeletal motor units are of the twitch type. Motor units of the *tonic type* are found in the oculomotor apparatus. These motor units require more than a single stimulus before the initial development of tension.

Fiber Types

Skeletal muscle fibers exhibit many different structural, histochemical, and behavioral characteristics. Because these differences have direct implications for muscle function, they are of particular interest to many scientists. The fibers of some motor units contract to reach maximum tension more quickly than do others after being stimulated. Based on this distinguishing characteristic, fibers may be divided into the umbrella categories of fast twitch (**FT**) and slow twitch (**ST**). It takes FT fibers only about one-seventh the time required by ST fibers to reach peak tension (Figure 6-10) (18). This difference in time to peak tension is attributed to higher concentrations of myosin-ATPase in FT fibers. The FT fibers are also larger in diameter than ST fibers. Because of these and other differences, FT fibers usually fatigue more quickly than do ST fibers.

fast-twitch fiber
a fiber that reaches peak tension relatively quickly

slow-twitch fiber
a fiber that reaches peak tension relatively slowly

Fast-twitch fibers both reach peak tension and relax more quickly than slow-twitch fibers. Note that the twitch tension levels shown are relative to peak tension and not absolute, since FT fibers tend to reach higher peak tensions than ST fibers.

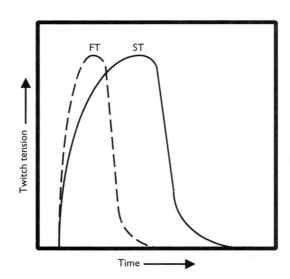

TABLE 6-2

**Skeletal Muscle Fiber
Characteristics**

CHARACTERISTIC	TYPE I SLOW-TWITCH OXIDATIVE (SO)	TYPE IIA FAST-TWITCH OXIDATIVE GLYCOLYTIC (FOG)	TYPE IIB FAST-TWITCH GLYCOLYTIC (FG)
Contraction speed	slow	fast	fast
Fatigue rate	slow	intermediate	fast
Diameter	small	intermediate	large
ATPase concentration	low	high	high
Mitochondrial concentration	high	high	low
Glycolytic enzyme concentration	low	intermediate	high

Although intact FT and ST muscles generate approximately the same amount of peak isometric force per cross-sectional area of muscle, individuals with a high percentage of FT fibers are able to generate higher magnitudes of torque and power during movement than are those with more ST fibers (18).

FT fibers are divided into two categories based on histochemical properties. The first type of FT fiber shares the resistance to fatigue that characterizes ST fibers. The second type of FT fiber is larger in diameter, contains fewer mitochondria, and fatigues more rapidly than the first type (Table 6-2).

Researchers have categorized the three types of muscle fibers using several different schemes. In one, ST fibers are referred to as *Type I,* and the FT fibers as *Type IIa* and *Type IIb.* Another system terms the ST fibers as *slow-twitch oxidative* (SO), with FT fibers divided into *fast-twitch oxidative glycolytic* (FOG) and *fast-twitch glycolytic* (FG) fibers. Yet another scheme includes ST fibers and *fast-twitch fatigue resistant* (FFR) and *fast-twitch fast-fatigue* (FF) fibers. These categorizations are *not* interchangeable, as they are based on different fiber properties. While three categories of muscle fiber are useful for describing gross functional differences, it is important to recognize that there is a continuum of fiber characteristics (90).

Although all fibers in a motor unit are the same type, most skeletal muscles contain both FT and ST fibers, with the relative amounts varying from muscle to muscle and from individual to individual. For example, the soleus, which is generally used only for postural adjustments, contains primarily ST fibers. In contrast, the overlying gastrocnemius may contain more FT than ST fibers. Muscle fiber composition is the same across genders in the normal population, although men tend to have larger fibers than do women (79).

FT fibers are important contributors to a performer's success in events requiring fast, powerful muscular contraction, such as sprinting and jumping. Endurance events such as distance running, cycling, and swimming require effective functioning of the more fatigue-resistant ST fibers. Using muscle biopsies, researchers have shown that highly successful athletes in events requiring strength and power tend to have unusually high proportions of FT fibers (6), and that elite endurance athletes usually have abnormally high proportions of ST fibers (51). As might be

●*A high percentage of FT fibers is advantageous for generating fast movements, and a high percentage of ST fibers is beneficial for activities requiring endurance.*

Elite sprint cyclists are likely to have muscles composed of a high percentage of FT fibers.

expected, in an event such as cycling, the most energetically optimal cycling cadence has been shown to be related to lower-extremity fiber type composition, with a faster pedaling frequency being better for athletes with a higher percentage of FT fibers (82).

As these findings suggest, exercise training over time can result in changes in fiber types within an individual. Today it is accepted that FT fibers can be converted to ST fibers with endurance training and that within the FT fibers conversions from Type IIb to Type IIa fibers can occur with heavy resistance (strength) training, endurance training, and concentric and eccentric isokinetic training (3, 12, 36).

Individuals genetically endowed with a high percentage of FT fibers may gravitate to sports requiring strength, and those with a high percentage of ST fibers may choose endurance sports. However, the fiber type distributions of both elite strength-trained and elite endurance-trained athletes fall within the range of fiber type compositions found in untrained individuals (21). Within the general population, a bell-shaped distribution of FT versus ST muscle composition exists, with most people having an approximate balance of FT and ST fibers, and relatively small percentages having a much greater number of FT or ST fibers.

Two factors known to affect muscle fiber type composition are age and obesity. There is a progressive, age-related reduction in the number of motor units and muscle fibers and in the size of Type II fibers that is not related to gender or to training (71). A longitudinal study of 28 male distance runners showed a significantly increased proportion of Type I fibers over a 20-year period, presumably due to selective loss of Type II fibers (80). However, there is good evidence that regular, lifelong, high-intensity exercise can reduce the loss of motor units typically associated with aging (64). These age-related changes may vary with the muscle, however, as the numbers of Type I and Type II fibers have been found not to change with age in the biceps brachii (47). Infants and young children, on the other hand, also have significantly smaller proportions of Type IIb fibers than do adults, and significantly lower proportions of Type IIb fibers are found in obese than in nonobese adults (54).

Exciting new evidence underscores the role of genetic expression on fiber type and suggests that skeletal muscle adapts to altered functional demands with changes in the genetic phenotype of individual fibers (89). Myogenic stem cells called *satellite cells* are normally inactive but can be stimulated by a change in habitual muscle activity to proliferate and form new muscle fibers (7). It has been hypothesized that muscle regeneration following exercise may provide a stimulus for satellite cell involvement in remodeling muscle by altering genetic expression in terms of muscle fiber appearance and function within the muscle (89).

Fiber Architecture

Another variable influencing muscle function is the arrangement of fibers within a muscle. The orientations of fibers within a muscle and the arrangements by which fibers attach to muscle tendons vary considerably among the muscles of the human body. These structural considerations affect the strength of muscular contraction and the range of motion through which a muscle group can move a body segment.

The two umbrella categories of muscle fiber arrangement are termed parallel and pennate. Although numerous subcategories of parallel and pennate fiber arrangements have been proposed, the distinction between these two broad categories is sufficient for discussing biomechanical features.

parallel fiber arrangement
pattern of fibers within a muscle in which the fibers are roughly parallel to the longitudinal axis of the muscle

pennate fiber arrangement
pattern of fibers within a muscle with short fibers attaching to one or more tendons

FIGURE 6-11

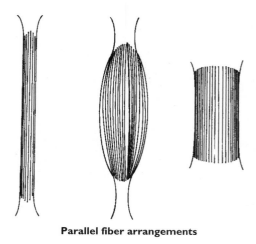

Parallel fiber arrangements

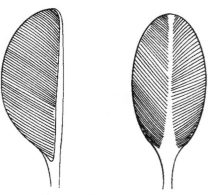

Pennate fiber arrangements

In a parallel fiber arrangement, the fibers are oriented largely in parallel with the longitudinal axis of the muscle (Figure 6-11). The sartorius, rectus abdominis, and biceps brachii have parallel fiber orientations. In most parallel-fibered muscles, there are fibers that do not extend the entire length of the muscle, but terminate somewhere in the muscle belly. Such fibers have structural specializations that provide interconnections with neighboring fibers at many points along the fiber's surface to enable delivery of tension when the fiber is stimulated (75).

A pennate fiber arrangement is one in which the fibers lie at an angle to the muscle's longitudinal axis. Each fiber in a pennate muscle attaches to one or more tendons, some of which extend the entire length of the muscle. The fibers of a muscle may exhibit more than one angle of pennation (angle of attachment) to a tendon. The tibialis posterior, rectus femoris, and deltoid muscles have pennate fiber arrangements.

When tension is developed in a parallel-fibered muscle, any shortening of the muscle is primarily the result of the shortening of its fibers. When the fibers of a pennate muscle shorten, they rotate about their tendon attachment or attachments, progressively increasing the angle of pennation (74) (Figure 6-12). As demonstrated in Sample Problem 6.1, the greater the angle of pennation, the smaller the amount of effective force actually transmitted to the tendon or tendons to move the attached bones. Once the angle of pennation exceeds 60°, the amount of effective force transferred to the tendon is less than one-half of the force actually produced by the muscle fibers. Sprinters have been found to have leg muscle

FIGURE 6-12

The angle of pennation
increases as tension
progressively increases in the
muscle fibers.

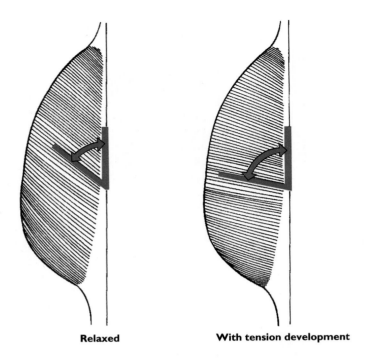

Relaxed   **With tension development**

SAMPLE PROBLEM 6.1

How much force is exerted by the tendon of a pennate muscle when the tension in the fibers is 100 N, given the following angles of pennation?

1. 40°
2. 60°
3. 80°

Known

$$F_{fibers} = 100 \text{ N}$$

angle of pennation = 40°, 60°, 80°

Solution
Wanted: F_{tendon}

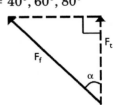

The relationship between the tension in the fibers and the tension in the tendon is

$$F_{tendon} = F_{fibers} \cos \alpha$$

1. For $\alpha = 40°$, $F_{tendon} = (100 \text{ N}) (\cos 40)$

$$F_{tendon} = 76.6 \text{ N}$$

2. For $\alpha = 60°$, $F_{tendon} = (100 \text{ N}) (\cos 60)$

$$F_{tendon} = 50 \text{ N}$$

3. For $\alpha = 80°$, $F_{tendon} = (100 \text{ N}) (\cos 80)$

$$F_{tendon} = 17.4 \text{ N}$$

pennation angles less than those in distance runners, with the smaller pennation angle favoring greater shortening velocity for faster running speeds (77).

Although pennation reduces the effective force generated at a given level of fiber tension, this arrangement allows the packing of more fibers than can be packed into a longitudinal muscle occupying equal space. Because pennate muscles contain more fibers per unit of muscle volume, they can generate more force than parallel-fibered muscles of the same size. Interestingly, when muscle hypertrophies, there is a concomitant increase in the angulation of the constituent fibers, and even in the absence of hypertrophy, thicker muscles have larger pennation angles (45).

• *Pennate fiber arrangement promotes muscle force production, and parallel fiber arrangement facilitates muscle shortening.*

The parallel fiber arrangement, on the other hand, enables greater shortening of the entire muscle than is possible with a pennate arrangement. Parallel-fibered muscles can move body segments through larger ranges of motion than can comparably sized pennate-fibered muscles. Increasing research findings point to differences in regional structural organization and regional functional differences within a given muscle (23).

SKELETAL MUSCLE FUNCTION

When an activated muscle develops tension, the amount of tension present is constant throughout the length of the muscle, as well as in the tendons, and at the sites of the musculotendinous attachments to bone. The tensile force developed by the muscle pulls on the attached bones and creates torque at the joints crossed by the muscle. As discussed in Chapter 3, the magnitude of the torque generated is the product of the muscle force and the force's moment arm (Figure 6-13). In keeping with the laws of vector addition, the net torque present at a joint determines the direction of any resulting movement. The weight of the attached body segment, external forces acting on the body, and tension in any muscle crossing a joint can all generate torques at that joint (Figure 6-14).

• *The net torque at a joint is the vector sum of the muscle torque and the resistive torque.*

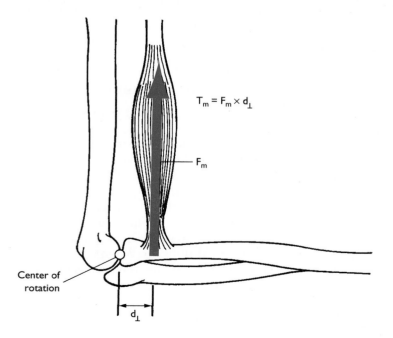

$$T_m = F_m \times d_\perp$$

F_m

Center of rotation

$d_\perp$

FIGURE 6-13

Torque (T_m) produced by a muscle at the joint center of rotation is the product of muscle force (F_m) and muscle moment arm ($d_\perp$).

FIGURE 6-14

The torque exerted by the biceps brachii (F_b) must counteract the torques created by the force developed in the triceps brachii (F_t), the weight of the forearm and hand (wt_f), and the weight of the shot held in the hand (wt_s).

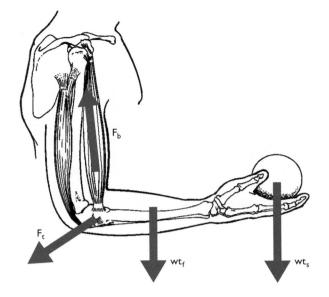

• *Slow-twitch motor units always produce tension first, whether the resulting movement is slow or fast.*

Recruitment of Motor Units

The central nervous system exerts an elaborate system of control that enables matching of the speed and magnitude of muscle contraction to the requirements of the movement so that smooth, delicate, and precise movements can be executed. The neurons that innervate ST motor units generally have low thresholds and are relatively easy to activate, whereas FT motor units are supplied by nerves more difficult to activate. Consequently, the ST fibers are the first to be activated, even when the resulting limb movement is rapid (13).

As the force requirement, speed requirement, or duration of the activity increases, motor units with higher thresholds are progressively activated, with Type IIa, or FOG, fibers added before the Type IIb, or FG, fibers. Within each fiber type, a continuum of ease of activation exists, and the central nervous system may selectively activate more or fewer motor units.

During low-intensity exercise, the central nervous system may recruit ST fibers almost exclusively. As activity continues and fatigue sets in, Type IIa and then Type IIb motor units are activated until all motor units are involved (22).

Change in Muscle Length with Tension Development

When muscular tension produces a torque larger than the resistive torque at a joint, the muscle shortens, causing a change in the angle at the joint. When a muscle shortens, the contraction is concentric, and the resulting joint movement is in the same direction as the net torque generated by the muscles. A single muscle fiber is capable of shortening to approximately one-half of its normal resting length.

Muscles can also develop tension without shortening. If the opposing torque at the joint crossed by the muscle is equal to the torque produced by the muscle (with zero net torque present), muscle length remains unchanged, and no movement occurs at the joint. When muscular tension develops but no change in muscle length occurs, the contraction is isometric. Because the development of tension increases the diameter of the muscle, body builders develop isometric tension to display their muscles when competing. Developing isometric tension simultaneously in muscles on

concentric
describing a contraction involving shortening of a muscle

isometric
describing a contraction involving no change in muscle length

opposite sides of a limb, such as in the triceps brachii and the biceps brachii, enlarges the cross-sectional area of the tensed muscles, although no movement occurs at either the shoulder or the elbow joints.

When opposing joint torque exceeds that produced by tension in a muscle, the muscle lengthens. When a muscle lengthens as it is being stimulated to develop tension, the contraction is eccentric, and the direction of joint motion is opposite that of the net muscle torque. Eccentric tension occurs in the elbow flexors during the elbow extension or weight-lowering phase of a curl exercise. The eccentric tension acts as a braking mechanism to control movement speed. Without the presence of eccentric tension in the muscles, the forearm, hand, and weight would drop in an uncontrolled way because of the force of gravity. Research indicates that enhanced ability to develop tension under concentric, isometric, and eccentric conditions is best achieved by training in the same respective exercise mode (88).

eccentric
describing a contraction involving lengthening of a muscle

Roles Assumed by Muscles

An activated muscle can do only one thing: Develop tension. Because one muscle rarely acts in isolation, however, we sometimes speak in terms of the function or role that a given muscle is carrying out when it acts in concert with other muscles crossing the same joint (65).

When a muscle contracts and causes movement of a body segment at a joint, it is acting as an agonist, or mover. Because several different muscles often contribute to a movement, the distinction between primary and assistant agonists is sometimes also made. For example, during the elbow flexion phase of a forearm curl, the brachialis and the biceps brachii act as the primary agonists, with the brachioradialis, extensor carpi radialis longus, and pronator teres serving as assistant agonists. All one-joint muscles functioning as agonists either develop tension simultaneously or are quiescent (2).

Muscles with actions opposite those of the agonists can act as antagonists, or opposers, by developing eccentric tension at the same time that the agonists are causing movement. Agonists and antagonists are typically positioned on opposite sides of a joint. During elbow flexion, when the brachialis and the biceps brachii are primary agonists, the triceps could act as antagonists by developing resistive tension. Conversely, during elbow extension, when the triceps are the agonists, the brachialis and biceps brachii could perform as antagonists. Although skillful movement is not characterized by continuous tension in antagonist muscles, antagonists often provide controlling or braking actions, particularly at the end of fast, forceful movements. Whereas agonists are particularly active during acceleration of a body segment, antagonists are primarily active during deceleration, or negative acceleration (41). When a person runs down a hill, for example, the quadriceps function eccentrically as antagonists to control the amount of knee flexion occurring. Co-contraction of agonist and antagonist muscles also enhances stability at the joint the muscles cross (19). Simultaneous tension development in the quadriceps and hamstrings helps stabilize the knee against potentially injurious rotational forces.

Another role assumed by muscles involves stabilizing a portion of the body against a particular force. The force may be internal, from tension in other muscles, or external, as provided by the weight of an object being lifted. The rhomboids act as stabilizers by developing tension to stabilize the scapulae against the pull of the tow rope during waterskiing.

A fourth role assumed by muscles is that of neutralizer. Neutralizers prevent unwanted accessory actions that normally occur when agonists develop concentric tension. For example, if a muscle causes both flexion and abduction at a joint but only flexion is desired, the action of a neutralizer

agonist
role played by a muscle acting to cause a movement

antagonist
role played by a muscle acting to slow or stop a movement

stabilizer
role played by a muscle acting to stabilize a body part against some other force

neutralizer
role played by a muscle acting to eliminate an unwanted action produced by an agonist

Body builders commonly develop isometric tension in their muscles to display muscle size and definition.

During the elbow flexion phase of a forearm curl, the brachialis and the biceps brachii act as the primary agonists, with the brachioradialis, flexor carpi radialis, and pronator teres serving as assistant agonists.

causing adduction can eliminate the unwanted abduction. When the biceps brachii develops concentric tension, it produces both flexion at the elbow and supination of the forearm. If only elbow flexion is desired, the pronator teres act as a neutralizer to counteract the supination of the forearm.

Performance of human movements typically involves the cooperative actions of many muscle groups acting sequentially and in concert. For example, even the simple task of lifting a glass of water from a table requires several different muscle groups to function in different ways. Stabilizing roles are performed by the scapular muscles and both flexor and extensor muscles of the wrist. The agonist function is performed by the flexor muscles of the fingers, elbow, and shoulder. Because the major shoulder flexors, the anterior deltoid and pectoralis major, also produce horizontal adduction, horizontal abductors such as the middle deltoid and supraspinatus act as neutralizers. Movement speed during the motion may also be partially controlled by antagonist activity in the elbow extensors. When the glass of water is returned to the table, gravity serves as the prime mover, with antagonist activity in the elbow and shoulder flexors controlling movement speed.

Two-Joint and Multijoint Muscles

Many muscles in the human body cross two or more joints. Examples are the biceps brachii, the long head of the triceps brachii, the hamstrings, the rectus femoris, and a number of muscles crossing the wrist and all finger joints. Since the amount of tension present in any muscle is essentially constant throughout its length, as well as at the sites of its tendinous attachments to bone, these muscles affect motion at both or all of the joints crossed simultaneously. The effectiveness of a two-joint or multijoint muscle in causing movement at any joint crossed depends on the location and orientation of the muscle's attachment relative to the joint, the tightness or laxity present in the musculotendinous unit, and the actions of other muscles that cross the joint. Whereas one-joint muscles produce force directed primarily in line with a body segment, two-joint muscles can produce force with a significant transverse component (34). During power-based activities such as jumping and sprinting, the biarticular muscles crossing the hip and knee have been shown to be particularly effective in converting body segment rotations into the desired translational motion of the total-body center of gravity (43).

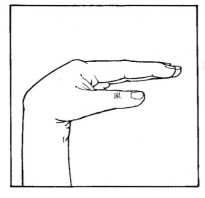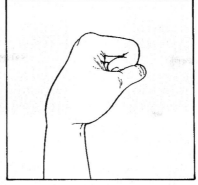

FIGURE 6-15

When the wrist is fully flexed, the finger flexors (that cross the wrist) are placed on slack and cannot develop sufficient tension to form a fist until the wrist is extended to a more neutral position. The inability to develop tension in a two- or multijoint muscle is referred to as *active insufficiency*.

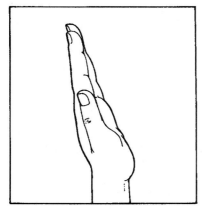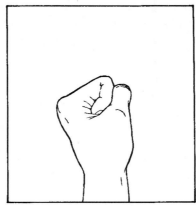

FIGURE 6-16

When the finger flexors are maximally stretched with both the wrist and the fingers in full extension, the range of motion for wrist extension is restricted. Flexion of the fingers enables further extension at the wrist. Restriction of range of motion at a joint because of tightness in a two- or multijoint muscle is referred to as *passive insufficiency*.

There are also, however, two disadvantages associated with the function of two-joint and multijoint muscles. They are incapable of shortening to the extent required to produce a full range of motion at all joints crossed simultaneously, a limitation that is termed active insufficiency. For example, the finger flexors cannot produce as tight a fist when the wrist is in flexion as when it is in a neutral position (Figure 6-15). Some two-joint muscles are not able to produce force at all when the positions of both joints crossed place the muscles in a severely slackened state (31). A second problem is that for most people, two-joint and multijoint muscles cannot stretch to the extent required for full range of motion in the opposite direction at all joints crossed. This problem is referred to as passive insufficiency. For example, a larger range of hyperextension is possible at the wrist when the fingers are not fully extended (Figure 6-16). Likewise, a larger range of ankle dorsiflexion can be accomplished when the knee is in flexion due to the change in the tightness of the gastrocnemius.

● Two-joint muscles can fail to produce force when slack (active insufficiency) and can restrict range of motion when fully stretched (passive insufficiency).

active insufficiency
limited ability of a two-joint muscle to produce force when joint position places the muscle on slack

passive insufficiency
inability of a two-joint muscle to stretch to the extent required to allow full range of motion at all joints crossed

FACTORS AFFECTING MUSCULAR FORCE GENERATION

The magnitude of the force generated by muscle is also related to the velocity of muscle shortening, the length of the muscle when it is stimulated, and the period of time since the muscle received a stimulus. Because these factors are significant determiners of muscle force, they have been extensively studied.

Force–Velocity Relationship

The maximal force that a muscle can develop is governed by the velocity of the muscle's shortening or lengthening, with the relationship respectively shown in the concentric and eccentric zones of the graph in Figure 6-17. This force–velocity relationship was first described for concentric tension development in muscle by Hill in 1938 (32). Because the relationship holds true only for maximally activated muscle, it does not apply to muscle actions during most daily activities.

Accordingly, the force–velocity relationship does *not* imply that it is impossible to move a heavy resistance at a fast speed. The stronger the muscle is, the greater the magnitude of maximum isometric tension (shown in the center of Figure 6-17). This is the maximum amount of force that a muscle can generate before actually lengthening as the resistance is increased. However, the general shape of the force–velocity curve remains the same, regardless of the magnitude of maximum isometric tension.

The force–velocity relationship also does *not* imply that it is impossible to move a light load at a slow speed. Most activities of daily living require slow, controlled movements of submaximal loads. With submaximal loads, the velocity of muscle shortening is subject to volitional control. Only the number of motor units required are activated. For example, a pencil can be picked up from a desktop quickly or slowly, depending on the controlled pattern of motor unit recruitment in the muscle groups involved.

The force–velocity relationship has been tested for skeletal, smooth, and cardiac muscle in humans, as well as for muscle tissues from other species (33). The general pattern holds true for all types of muscle, even the tiny muscles responsible for the rapid fluttering of insect wings. Maximum values of force at zero velocity and maximum values of velocity at a minimal load vary with the size and type of muscle. Although the physiological basis for the force–velocity relationship is not completely understood, the shape of the concentric portion of the curve corresponds to the rate of energy production in a muscle.

The force–velocity relationship for muscle loaded beyond the isometric maximum is shown in the top half of Figure 6-17 (44). Under eccentric conditions, the maximal force a muscle can produce exceeds the isometric maximum by a factor of 1.5–2.0 (30). Achievement of such a high force level, however, appears to require electrical stimulation of the motor neuron (86). Maximal eccentric forces produced volitionally are similar to the isometric maximum (86). It is likely that this is true because the nervous system provides inhibition through reflex pathways to protect against

•*The stronger a muscle, the greater the magnitude of its isometric maximum on the force–velocity curve.*

The force–velocity relationship for muscle tissue. When the resistance (force) is negligible, muscle contracts with maximal velocity. As the load progressively increases, concentric contraction velocity slows to zero at isometric maximum. As the load increases further, the muscle lengthens eccentrically.

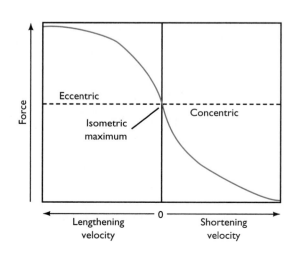

injury to muscles and tendons (86). Elevated force production under eccentric conditions with volitional muscle activation is not a function of greater neural activation of the muscle, but appears to represent the contribution of the elastic components of muscle (46, 49).

Eccentric strength training involves the use of resistances that are greater than the athlete's maximum isometric force generation capability. As soon as the load is assumed, the muscle begins to lengthen. Research shows this type of training to be more effective than concentric training in increasing muscle size and strength (36). As compared with concentric and isometric training, however, eccentric training is also associated with delayed onset muscle soreness (40).

Length–Tension Relationship

The amount of maximum isometric tension a muscle is capable of producing is partly dependent on the muscle's length. In single muscle fibers, isolated muscle preparations, and in vivo human muscles, force generation is at its peak when the muscle is slightly stretched (66). Conversely, muscle tension development capability is less following muscle shortening (70). Both the duration of muscle stretch or shortening and the time since stretch or shortening affect force generation capability (29, 67).

Within the human body, force generation capability increases when the muscle is slightly stretched. Parallel-fibered muscles produce maximum tensions at just over resting length, and pennate-fibered muscles generate maximum tensions at between 120% and 130% of resting length (25). This phenomenon is due to the contribution of the elastic components of muscle (primarily the SEC), which add to the tension present in the muscle when the muscle is stretched. Figure 6-18 shows the pattern of maximum tension development as a function of muscle length, with the active contribution of the contractile component and the passive contribution of the SEC and PEC indicated. Research indicates that following eccentric exercise there may be a slight, transient increase in muscle length that impairs force development when joint angle does not place the muscle in sufficient stretch (73).

Stretch-Shortening Cycle

When an actively tensed muscle is stretched just prior to contraction, the resulting contraction is more forceful than in the absence of the

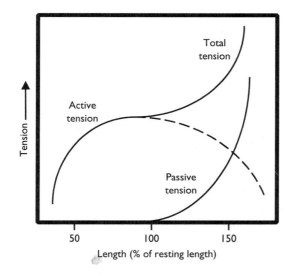

FIGURE 6-18

The total tension present in a stretched muscle is the sum of the active tension provided by the muscle fibers and the passive tension provided by the tendons and muscle membranes.

stretch-shortening cycle
eccentric contraction followed immediately by concentric contraction

Baseball pitchers initiate a forceful stretch of the shoulder flexors and horizontal adductors immediately before throwing the ball. The stretch reflex then contributes to forceful tension development in these muscles. Photo courtesy of Donald Miralle/Getty Images.

prestretch. This pattern of eccentric contraction followed immediately by concentric contraction is known as the stretch-shortening cycle (SSC). A muscle can perform substantially more work when it is actively stretched prior to shortening than when it simply contracts. In an experiment involving forceful dorsiflexion followed by plantar flexion at slow and fast frequencies, the SSC contributed an estimated 20.2% and 42.5%, respectively, to the positive work done (56). The metabolic cost of performing a given amount of mechanical work is also less when the SSC is invoked than the cost without it.

The mechanisms responsible for the SSC are not fully understood. However, one contributor in at least some cases is likely to be the SEC, with the elastic recoil effect of the actively stretched muscle enhancing force production. It has been estimated that during running at a slow speed, the triceps surae complex stores 45 J of elastic energy in the first half of stance, with 60 J produced during the second half (35). Eccentric training enhances the ability of the musculotendinous unit to store and return more elastic energy (68). Another potential contributor to the SSC is activation of the stretch reflex provoked by the forced lengthening of the muscle. Muscle spindle activity has been shown to provide a brief but substantial facilitation of neural drive during volitional contraction following prestretch (81). Force potentiation has also been shown to be significantly diminished following fatiguing exercise involving the SSC and following cooling-induced suppression of muscle spindle activity (48, 62).

Regardless of its cause, the SSC contributes to effective development of concentric muscular force in many sport activities. Quarterbacks and pitchers typically initiate a forceful stretch of the shoulder flexors and horizontal adductors immediately before throwing the ball. The same action occurs in muscle groups of the trunk and shoulders at the peak of the backswing of a golf club and a baseball bat. Competitive weight lifters use quick knee flexion during the transition phase of the snatch to invoke the SSC and enhance performance (24). The SSC also promotes storage and use of elastic energy during running, particularly with the alternating eccentric and concentric tension present in the gastrocnemius (42). Researchers have found that the muscles, tendons, and ligaments in the lower extremity behave very much like a spring during running, with higher stride frequencies associated with increased spring stiffness (17).

Electromechanical Delay

electromechanical delay
time between the arrival of neural stimulus and tension development by the muscle

When a muscle is stimulated, a brief period elapses before the muscle begins to develop tension (Figure 6-19). Referred to as electromechanical delay (EMD), this time is believed to be needed for the contractile component of the muscle to stretch the SEC. During this time, muscle laxity is eliminated. Once the SEC is sufficiently stretched, tension development proceeds.

The length of EMD varies considerably among human muscles, with values of 20–100 msec reported (50). Researchers have found shorter EMDs produced by muscles with high percentages of FT fibers as compared to muscles with high percentages of ST fibers (60). Development of higher contraction forces is also associated with shorter EMDs (85). Factors such as muscle length, contraction type, contraction velocity, and fatigue, however, do not appear to affect EMD (85). EMD is longer under the following conditions: immediately following passive stretching, several days after eccentric exercise resulting in muscle damage, after a period of endurance training, and when contraction is initiated from a resting state as compared to an activated state (11, 26, 38, 84). EMD in children is also significantly longer than in adults (16).

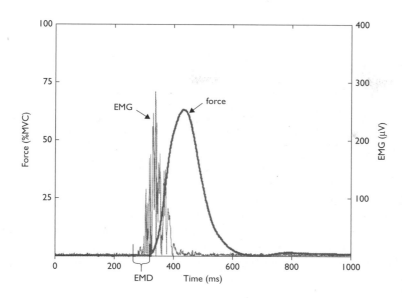

FIGURE 6-19

Myoelectric activity (EMG) in the vastus lateralis during isometric knee extension superimposed on a trace of the force output from the leg. Notice that the burst of EMG activity clearly precedes the onset of force production, demonstrating electromechanical delay (EMD). Graph courtesy of Dr. Chris Knight, University of Delaware.

The time required for a muscle to develop maximum isometric tension may be a full second following EMD (36). Shorter maximum force development times are associated with a high percentage of FT fibers in the muscle and with a trained state (83).

MUSCULAR STRENGTH, POWER, AND ENDURANCE

In practical evaluations of muscular function, the force-generating characteristics of muscle are discussed within the concepts of muscular strength, power, and endurance. These characteristics of muscle function have significant implications for success in different forms of strenuous physical activity, such as splitting wood, throwing a javelin, or hiking up a mountain trail. Among senior citizens and individuals with neuromuscular disorders or injuries, maintaining adequate muscular strength and endurance is essential for carrying out daily activities and avoiding injury.

Muscular Strength

When scientists excise a muscle from an experimental animal and electrically stimulate it in a laboratory, they can directly measure the force generated by the muscle. It is largely from controlled experimental work of this kind that our understandings of the force–velocity and length–tension relationships for muscle tissue are derived.

In the human body, however, it is not convenient to directly assess the force produced by a given muscle. The most direct assessment of "muscular strength" commonly practiced is a measurement of the maximum torque generated by an entire muscle group at a joint. Muscular strength, then, is measured as a function of the collective force-generating capability of a given functional muscle group. More specifically, muscular strength is the ability of a given muscle group to generate torque at a particular joint.

As discussed in Chapter 3, torque is the product of force and the force's moment arm, or the perpendicular distance at which the force acts from an axis of rotation. Resolving a muscle force into two orthogonal components, perpendicular and parallel to the attached bone, provides a clear picture of the muscle's torque-producing effect (Figure 6-20). Because the component of muscle force directed perpendicular to the attached bone

FIGURE 6-20

The component of muscular force that produces torque at the joint crossed (F_t) is directed perpendicular to the attached bone.

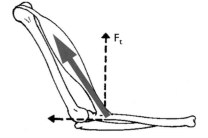 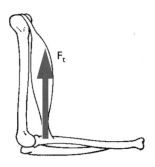

● *Muscular strength is most commonly measured as the amount of torque a muscle group can generate at a joint.*

produces torque, or a rotary effect, this component is termed the *rotary component* of muscle force. The size of the rotary component is maximum when the muscle is oriented at 90° to the bone, with change in angle of orientation in either direction progressively diminishing it. Isokinetic resistance machines are designed to match the size of the rotary component of muscle force throughout the joint range of motion. Sample Problem 6.2 demonstrates how the torque generated by a given muscle force changes as the angle of the muscle's attachment to the bone changes.

The component of muscle force acting parallel to the attached bone does not produce torque, since it is directed through the joint center and therefore has a moment arm of zero (Figure 6-21). This component, however, can provide either a stabilizing influence or a dislocating influence, depending on whether it is directed toward or away from the joint center. Actual dislocation of a joint rarely occurs from the tension developed by a muscle, but if a dislocating component of muscle force is present,

SAMPLE PROBLEM 6.2

How much torque is produced at the elbow by the biceps brachii inserting at an angle of 60° on the radius when the tension in the muscle is 400 N? (Assume that the muscle attachment to the radius is 3 cm from the center of rotation at the elbow joint.)

Known

$$F_m = 400 \text{ N}$$
$$\alpha = 60°$$
$$d_\perp = 0.03 \text{ m}$$

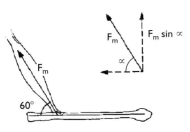

Solution

Wanted: T_m

Only the component of muscle force perpendicular to the bone generates torque at the joint. From the diagram, the perpendicular component of muscle force is

$$F_p = F_m \sin \alpha$$
$$F_p = (400 \text{ N}) (\sin 60)$$
$$= 346.4 \text{ N}$$

$$T_m = F_p d_\perp$$
$$= (346.4 \text{ N})(0.03 \text{ m})$$

$$T_m = 10.4 \text{ N-m}$$

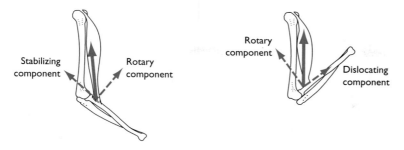

FIGURE 6-21

Contraction of the biceps brachii produces a component of force at the elbow that may tend to be stabilizing or dislocating, depending on the angle present at the elbow when contraction occurs.

a tendency for dislocation occurs. If the elbow is at an acute angle in greater than 90° of flexion, for example, tension produced by the biceps tends to pull the radius away from its articulation with the humerus, thereby lessening the stability of the elbow in that particular position.

Therefore, muscular strength is derived both from the amount of tension the muscles can generate and from the moment arms of the contributing muscles with respect to the joint center. Both sources are affected by several factors.

The tension-generating capability of a muscle is related to its cross-sectional area and its training state. The force generation capability per cross-sectional area of muscle is approximately 90 N/cm² (61), as illustrated in Sample Problem 6.3. With both concentric and eccentric strength

SAMPLE PROBLEM 6.3

How much tension may be developed in muscles with the following cross-sectional areas?

1. 4 cm²
2. 10 cm²
3. 12 cm²

Known

muscle cross-sectional areas = 4 cm², 10 cm², and 12 cm²

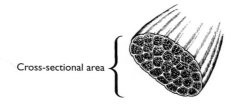

Cross-sectional area

Solution

Wanted: tension development capability

The tension-generating capability of muscle tissue is 90 N/cm². The force produced by a muscle is the product of 90 N/cm² and the muscle's cross-sectional area. So,

1. F = (90 N/cm²) (4 cm²)

   F = 360 N

2. F = (90 N/cm²) (10 cm²)

   F = 900 N

3. F = (90 N/cm²) (12 cm²)

   F = 1080 N

FIGURE 6-22

The relationships among concentric tension, shortening velocity, and power for muscle.

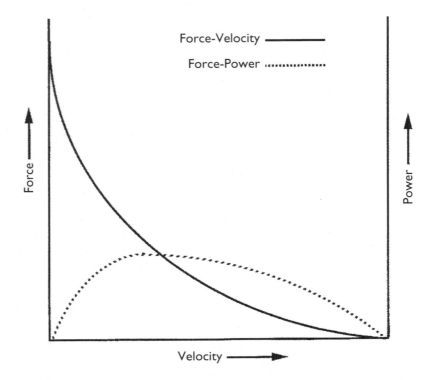

training, gains in strength over approximately the first 12 weeks appear to be related more to improved innervation of the trained muscle than to the increase in its cross-sectional area (53). This notion is further strengthened by the finding that unilateral strength training also produces strength gains in the untrained contralateral limb (37). The neural adaptations that occur with resistance training may include increased neuronal firing rates, increased motoneuron excitability and decreased presynaptic inhibition, lessening of inhibitory neural pathways, and increased levels of motor output from the central nervous system (1). Recent research findings suggest that muscle hypertrophy in response to resistance exercise is at least partially regulated by each individual's genetic composition (8).

A muscle's moment arm is affected by two equally important factors: (a) the distance between the muscle's anatomical attachment to bone and the axis of rotation at the joint center, and (b) the angle of the muscle's attachment to bone, which is typically a function of relative joint angle. The greatest amount of torque is produced by maximum tension in a muscle that is oriented at a 90° angle to the bone, and anatomically attached as far from the joint center as possible.

Muscular Power

● *Explosive movements require muscular power.*

Mechanical power (discussed in Chapter 12) is the product of force and velocity. Muscular power is therefore the product of muscular force and the velocity of muscle shortening. Maximum power occurs at approximately one-third of maximum velocity (32) and at approximately one-third of maximum concentric force (50) (Figure 6-22). Research indicates that training designed to increase muscular power over a range of resistance occurs most effectively with loads of one-third of one maximum repetition (58).

Because neither muscular force nor the speed of muscle shortening can be directly measured in an intact human being, muscular power is more generally defined as the rate of torque production at a joint, or the product of the net torque and the angular velocity at the joint. Accordingly, muscular power is affected by both muscular strength and movement speed.

Muscular power is an important contributor to activities requiring both strength and speed. The strongest shot-putter on a team is not necessarily the best shot-putter, because the ability to accelerate the shot is a critical component of success in the event. Athletic endeavors that require explosive movements, such as Olympic weight lifting, throwing, jumping, and sprinting, are based on the ability to generate muscular power.

Since FT fibers develop tension more rapidly than do ST fibers, a large percentage of FT fibers in a muscle is an asset for an individual training for a muscular power–based event. Individuals with a predominance of FT fibers generate more power at a given load than do individuals with a high percentage of ST compositions. Those with primarily FT compositions also develop their maximum power at faster velocities of muscle shortening (78). The ratio for mean peak power production by Type IIb, Type IIa, and Type I fibers in human skeletal muscle is 10:5:1 (18).

Sprinting requires muscular power, particularly in the hamstrings and the gastrocnemius. Photo courtesy of Digital Vision/Getty Images.

Muscular Endurance

Muscular endurance is the ability of the muscle to exert tension over time. The tension may be constant, as when a gymnast performs an iron cross, or may vary cyclically, as during rowing, running, and cycling. The longer the time tension is exerted, the greater the endurance. Although maximum muscular strength and maximum muscular power are relatively specific concepts, muscular endurance is less well understood because the force and speed requirements of the activity dramatically affect the length of time it can be maintained.

Training for muscular endurance typically involves large numbers of repetitions against relatively light resistance. This type of training does not increase muscle fiber diameter.

Muscle Fatigue

Muscle fatigue has been defined as an exercise-induced reduction in the maximal force capacity of muscle (20). Fatigability is also the opposite of endurance. The more rapidly a muscle fatigues, the less endurance it has. A complex array of factors affects the rate at which a muscle fatigues, including the type and intensity of exercise, the specific muscle groups involved, and the physical environment in which the activity occurs (39). Moreover, within a given muscle, fiber type composition and the pattern of motor unit activation play a role in determining the rate at which a muscle fatigues. However, this is an evolving area of understanding, with a considerable amount of related research in progress (10).

Characteristics of muscle fatigue include reduction in muscle force production capability and shortening velocity, as well as prolonged relaxation of motor units between recruitment (4). High-intensity muscle activity over time also results in prolonged twitch duration and a prolonged sarcolemma action potential of reduced amplitude (18).

A muscle fiber reaches absolute fatigue once it is unable to develop tension when stimulated by its motor axon. Fatigue may also occur in the motor neuron itself, rendering it unable to generate an action potential. FG fibers fatigue more rapidly than FOG fibers, and SO fibers are the most resistant to fatigue. Research has shown that the proportion of ST fibers in the vastus lateralis is directly related to the length of time that a level of 50% of maximum isometric tension can be maintained (52).

The specific causes of muscle fatigue are not well understood. However, a growing body of evidence indicates that reduction in the rate of

FIGURE 6-23

When muscle temperature is slightly elevated, the force–velocity curve is shifted. This is one benefit of warming up before an athletic endeavor.

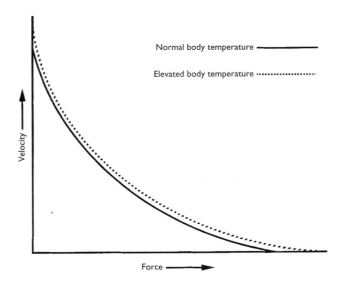

Normal body temperature ——————

Elevated body temperature ·····················

Velocity

Force

intracellular calcium release and uptake by the sarcoplasmic reticulum is involved (87). As many as three different mechanisms of reduced calcium release have been identified but are incompletely understood (4). Some experimental evidence suggests that sliding of the actin and myosin filaments during repeated muscle contraction reduces the affinity for calcium at uptake sites on the thin actin filaments (15). A variety of other factors have also been implicated in the development of fatigue, including increases in muscle acidity and intracellular potassium (5) and decreases in muscle energy supplies and intracellular oxygen (63).

Effect of Muscle Temperature

As body temperature elevates, the speeds of nerve and muscle functions increase. This causes a shift in the force–velocity curve, with a higher value of maximum isometric tension and a higher maximum velocity of shortening possible at any given load (Figure 6-23). At an elevated temperature, the activation of fewer motor units is needed to sustain a given load (69). The metabolic processes supplying oxygen and removing waste products for the working muscle also quicken with higher body temperatures. These benefits result in increased muscular strength, power, and endurance, and provide the rationale for warming up before an athletic endeavor. Notably, these benefits are independent of any change in the elasticity of the musculotendinous units, as research has demonstrated that the mechanical properties of muscle and tendon are not altered with either heating or cooling over the physiological range (55).

Muscle function is most efficient at 38.5°C (101°F) (6). Elevation of body temperature beyond this point may occur during strenuous exercise under conditions of high ambient temperature and/or humidity and can be extremely dangerous, possibly resulting in heat exhaustion or heat stroke. Organizers of long-distance events involving running or cycling should be particularly cognizant of the potential hazards associated with competition in such environments.

COMMON MUSCLE INJURIES

Muscle injuries are common, with most being relatively minor. Fortunately, healthy skeletal muscle has considerable ability to self-repair.

Strains

Muscular strains result from overstretching of muscle tissue. Most typically, an active muscle is overloaded, with the magnitude of the injury related to the size of the overload and the rate of overloading. Strains can be mild, moderate, or severe. Mild strains involve minimal structural damage and are characterized by a feeling of tightness or tension in the muscle. Second-degree strains involve a partial tear in the muscle tissue, with symptoms of pain, weakness, and some loss of function. With third-degree sprains, there is severe tearing of the muscle, functional loss, and accompanying hemorrhage and swelling. The hamstrings are the most frequently strained muscles in the human body. Hamstring strains are particularly problematic for athletes because they are slow to heal and have nearly a one-third incidence of recurrence within the first year following a return to sport participation (28).

Contusions

Contusions, or muscle bruises, are caused by compressive forces sustained during impacts. They consist of hematomas within the muscle tissue. A serious muscle contusion, or a contusion that is repeatedly impacted, can lead to the development of a much more serious condition known as *myositis ossificans*. Myositis ossificans consists of the presence of a calcified mass within the muscle. Apparently, the fibroblasts recruited during the healing process begin to differentiate into osteoblasts, with calcification becoming visible on a radiograph after three or four weeks (72). After six or seven weeks, resorption of the calcified mass usually begins, although sometimes a bony lesion in the muscle remains.

Cramps

The etiology of muscle cramps is not well understood, with possible causative factors including electrolyte imbalances, deficiencies in calcium and magnesium, and dehydration. Cramps can also occur secondary to direct impacts. Cramps may involve moderate to severe muscle spasms, with proportional levels of accompanying pain.

Delayed-Onset Muscle Soreness

Muscle soreness often occurs after some period of time following unaccustomed exercise. Delayed-onset muscle soreness (DOMS) arises 24–72 hours after participation in a long or strenuous bout of exercise and is characterized by pain, swelling, and the same kinds of histological changes that accompany acute inflammation (76). Microtearing of the muscle tissue is involved, with symptoms of pain, stiffness, and restricted range of motion. Researchers have hypothesized that the increase in joint stiffness may serve as a protective mechanism that helps prevent added damage and pain (14).

Compartment Syndrome

Hemorrhage or edema within a muscle compartment can result from injury or excessive muscular exertion. Pressure increases within the compartment, and severe damage to the neural and vascular structures within the compartment follows in the absence of pressure release. Swelling, discoloration, diminished distal pulse, loss of sensation, and loss of motor function are all progressively apparent symptoms.

SUMMARY

Muscle is elastic and extensible and responds to stimulation. Most importantly, however, it is the only biological tissue capable of developing tension.

The functional unit of the neuromuscular system is the motor unit, consisting of a single motor neuron and all the fibers it innervates. The fibers of a given motor unit are either slow twitch, fast-twitch fatigue-resistant, or fast-twitch fast-fatigue. Both ST and FT fibers are typically found in all human muscles, although the proportional fiber composition varies. The number and distribution of fibers within muscles appear to be genetically determined and related to age. Within human skeletal muscles, fiber arrangements are parallel or pennate. Pennate fiber arrangements promote force production, whereas parallel fiber arrangement enables greater shortening of the muscle.

Muscle responds to stimulation by developing tension. Depending on what other forces act, however, the resulting action can be concentric, eccentric, or isometric, for muscle shortening, lengthening, or remaining unchanged in length. The central nervous system directs the recruitment of motor units such that the speed and magnitude of muscle tension development are well matched to the requirements of the activity.

There are well-defined relationships between muscle force output and the velocity of muscle shortening, the length of the muscle at the time of stimulation, and the time since the onset of the stimulus. Because of the added contribution of the elastic components of muscle and neural facilitation, force production is enhanced when a muscle is actively prestretched.

Muscle performance is typically described in terms of muscular strength, power, and endurance. From a biomechanical perspective, strength is the ability of a muscle group to generate torque at a joint, power is the rate of torque production at a joint, and endurance is resistance to fatigue.

INTRODUCTORY PROBLEMS

1. List three examples of activities requiring concentric muscle action and three examples of activities requiring eccentric muscle action, and identify the specific muscles or muscle groups involved.
2. List five movement skills for which a high percentage of fast-twitch muscle fibers is an asset and five movement skills for which a high percentage of slow-twitch fibers is an asset. Provide brief statements of rationale for each of your lists.
3. Hypothesize about the pattern of recruitment of motor units in the major muscle group or groups involved during each of the following activities:
   a. Walking up a flight of stairs
   b. Sprinting up a flight of stairs
   c. Throwing a ball
   d. Cycling in a 100 km race
   e. Threading a needle
4. Identify three muscles that have parallel fiber arrangements, and explain the ways in which the muscles' functions are enhanced by this arrangement.
5. Answer Problem 4 for pennate fiber arrangement.
6. How is the force–velocity curve affected by muscular strength training?
7. Write a paragraph describing the biomechanical factors determining muscular strength.
8. List five activities in which the production of muscular force is enhanced by the series elastic component and the stretch reflex.

9. Muscle can generate approximately 90 N of force per square centimeter of cross-sectional area. If a biceps brachii has a cross-sectional area of 10 cm², how much force can it exert? (Answer: 900 N)

10. Using the same force/cross-sectional area estimate as in Problem 9, and estimating the cross-sectional area of your own biceps brachii, how much force should the muscle be able to produce?

ADDITIONAL PROBLEMS

1. Identify the direction of motion (flexion, extension, etc.) at the hip, knee, and ankle, and the source of the force(s) causing motion at each joint for each of the following activities:
   a. Sitting down in a chair
   b. Taking a step up on a flight of stairs
   c. Kicking a ball

2. Considering both the force–length relationship and the rotary component of muscle force, sketch what you would hypothesize to be the shape of a force versus joint angle curve for the elbow flexors. Write a brief rationale in support of the shape of your graph.

3. Certain animals, such as kangaroos and cats, are well known for their jumping abilities. What would you hypothesize about the biomechanical characteristics of their muscles?

4. Identify the functional roles played by the muscle groups that contribute to each of the following activities:
   a. Carrying a suitcase
   b. Throwing a ball
   c. Rising from a seated position

5. If the fibers of a pennate muscle are oriented at a 45° angle to a central tendon, how much tension is produced in the tendon when the muscle fibers contract with 150 N of force? (Answer: 106 N)

6. How much force must be produced by the fibers of a pennate muscle aligned at a 60° angle to a central tendon to create a tensile force of 200 N in the tendon? (Answer: 400 N)

7. What must be the effective minimal cross-sectional areas of the muscles in Problems 5 and 6, given an estimated 90 N of force-producing capacity per square centimeter of muscle cross-sectional area? (Answer: 1.2 cm²; 4.4 cm²)

8. If the biceps brachii, attaching to the radius 2.5 cm from the elbow joint, produces 250 N of tension perpendicular to the bone, and the triceps brachii, attaching 3 cm away from the elbow joint, exerts 200 N of tension perpendicular to the bone, how much net torque is present at the joint? Will there be flexion, extension, or no movement at the joint? (Answer: 0.25 N-m; flexion)

9. Calculate the amount of torque generated at a joint when a muscle attaching to a bone 3 cm from the joint center exerts 100 N of tension at the following angles of attachment:
   a. 30°
   b. 60°
   c. 90°
   d. 120°
   e. 150°
   (Answers: a. 1.5 N-m; b. 2.6 N-m; c. 3 N-m; d. 2.6 N-m; e. 1.5 N-m)

10. Write a quantitative problem of your own involving the following variables: muscle tension, angle of muscle attachment to bone, distance of the attachment from the joint center, and torque at the joint. Provide a solution for your problem.

LABORATORY EXPERIENCES

1. With a partner, use a goniometer to measure the ankle range of motion for dorsiflexion and plantar flexion both when the knee is fully extended and when it is comfortably flexed. Explain your results.

Dorsiflexion ROM with full knee extension: _____ with knee flexion: _____

Plantar flexion ROM with full knee extension: _____ with knee flexion: _____

Explanation: _____

_____

_____

_____

2. Using a series of dumbbells, determine your maximum weight for the forearm curl exercise when the elbow is at angles of 5°, 90°, and 140°. Explain your findings.

Max at 5°: _____ at 90°: _____ at 140°: _____

Explanation: _____

_____

_____

_____

3. Using electromyography apparatus with surface electrodes positioned over the biceps brachii, perform a forearm curl exercise with light and heavy weights. Explain the changes evident in the electromyogram.

Comparison of traces: _____

_____

_____

_____

Explanation: _____

_____

_____

4. Using electromyography apparatus with surface electrodes positioned over the pectoralis major and triceps brachii, perform bench presses with wide, medium, and narrow grip widths on the bar. Explain the differences in muscle contributions evident.

Comparison of traces: _____

_____

_____

_____

Explanation: _____

_____

_____

_____

5. Using electromyography apparatus with surface electrodes positioned over the biceps brachii, perform a forearm curl exercise to fatigue. What changes are evident in the electromyogram with fatigue? Explain your results.

Comparison of pre- and postfatigue traces: _____

_____

_____

_____

Explanation: _____

_____

_____

_____

REFERENCES

1. Aagaard P: Training-induced changes in neural function, *Exerc Sport Sci Rev* 31:61, 2003.
2. Ait-Haddou R, Binding P, and Herzog W: Theoretical considerations on co-contraction of sets of agonistic and antagonistic muscles, *J Biomech* 33:1105, 2000.
3. Allemeier CA, Fry AC, Johnson P, Hikida RS, Hagerman FC, and Staron RS: Effects of sprint cycle training on human skeletal muscle, *J Appl Physiol* 77:2385, 1994.
4. Allen DG, Lännergren J, and Westerblad H: Muscle cell function during prolonged activity: cellular mechanisms of fatigue, *Exp Physiol* 80:497, 1995.
5. Bangsbo J, Madsen K, Kiens B, and Richter EA: Effect of muscle acidity on muscle metabolism and fatigue during intense exercise in man, *J Physiol* 495:587, 1996.
6. Bar-Or O et al: Anaerobic capacity and muscle fiber type distribution in man, *Int J Sports Med* 10:82, 1980.
7. Bischoff R: The satellite cell and muscle regeneration. In Engel AG and Franzini-Armstrong C, eds: *Myology,* New York, McGraw-Hill, 1994, pp. 97–118.
8. Bolster DR, Kimball SR, and Jefferson LS: Translational control mechanisms modulate skeletal muscle gene expression during hypertrophy, *Exerc Sport Sci Rev* 31:111, 2003.
9. Buchthal F and Schalbruch H: Motor unit of mammalian muscle, *Physiol Rev* 60:90, 1980.
10. Cairns SP, Knicker AJ, Thompson MW, and Sjøgaard G: Evaluation of models used to study neuromuscular fatigue, *Exerc Sport Sci Rev* 33:9, 2005.
11. Costa PB, Ryan ED, Herda TJ, Walter AA, Hoge KM, and Cramer JT: Acute effects of passive stretching on the electromechanical delay and evoked twitch properties, *Eur J Appl Physiol,* 108:301, 2010.
12. Delecluse C: Influence of strength training on sprint running performance. Current findings and implications for training, *Sports Med* 24:147, 1997.
13. Desmedt JE and Godaux E: Fast motor units are not preferentially activated in rapid voluntary contractions in man, *Nature* 267:717, 1977.
14. Dutto DJ and Braun WA: DOMS-Associated changes in ankle and knee joint dynamics during running, *Med Sci Sports Exerc* 36:560, 2004.
15. Edman KA: Fatigue vs. shortening-induced deactivation in striated muscle, *Acta Physiol Scand* 156:183, 1996.
16. Falk B, Usselman C, Dotan R, Brunton L, Klentrou P, Shaw J, and Gabriel D: Child-adult differences in muscle strength and activation pattern during isometric elbow flexion and extension, *Appl Physiol Nutr Metab,* 34:609, 2009.
17. Farley CT and González O: Leg stiffness and stride frequency in human running, *J Biomech* 29:181, 1996.
18. Fitts RH: Muscle fatigue: the cellular aspects, *Am J Sports Med* 24:S9, 1996.
19. Forster E, Simon U, Augat P, and Claes L: Extension of a state-of-the-art optimization criterion to predict co-contraction, *J Biomech* 37:5577, 2004.
20. Gandevia SC: Spinal and supraspinal factors in human muscle fatigue, *Physiol Rev* 81:1725, 2001.
21. Gollnick PD: Muscle characteristics as a foundation of biomechanics. In Matsui H and Kobayashi K, eds: *Biomechanics VIII-A,* Champaign, IL, 1983, Human Kinetics Publishers.
22. Gollnick PD, Piehl K, and Saltin B: Selective glycogen depletion pattern in muscle fibres after exercise of varying intensity and at varying pedaling rates, *J Physiol* 241:45, 1974.
23. Gordon T and Pattullo MC: Plasticity of muscle fiber and motor unit types, *Exerc Sports Sci Rev,* 21:331, 1993.
24. Gourgoulis V, Aggelousis N, Mavromatis G, and Garas A: Three-dimensional kinematic analysis of the snatch of elite Greek weightlifters, *J Sports Sci* 18:643, 2000.
25. Gowitzke BA and Milner M: *Understanding the scientific bases of human movement* (2nd ed.), Baltimore, 1980, Williams & Wilkins.

26. Grosset JF, Piscione J, Lambertz D, and Pérot C: Paired changes in electrome-chanical delay and musculo-tendinous stiffness after endurance or plyometric training, *Eur J Appl Physiol,* 105:131, 2009.

27. Guthe K: Reptilian muscle: fine structure and physiological parameters. In Gans C and Parsons TS, eds: *Biology of the reptilia,* vol. 11, London, 1981, Academic Press.

28. Heiderscheit BC, Sherry MA, Silder A, Chumanov ES, and Thelen DG: Hamstring strain injuries: recommendations for diagnosis, rehabilitation, and injury prevention, *J Orthop Sports Phys Ther,* 40:67, 2010.

29. Herzog W and Leonard TR: The role of passive structures in force en-hancement of skeletal muscles following active stretch, *J Biomech* 38:409, 2005.

30. Herzog W: Force production in human skeletal muscle. In Nigg BM, MacIntosh BR, and Mester J, eds: *Biomechanics and biology of movement,* Champaign, IL, 2000, Human Kinetics, pp. 269–281.

31. Herzog W, Abrahamse SK, and ter Keurs HE, Theoretical determination of force-length relations of intact human skeletal muscles using the cross-bridge model. *Pflugers Arch,* 416:113, 1990.

32. Hill AV: *First and last experiments in muscle mechanics,* Cambridge, MA, 1970, Cambridge University Press.

33. Hill AV: The heat of shortening and the dynamic constants of muscle, *Proc R Soc Lond* B126:136, 1938.

34. Hof AL: The force resulting from the action of mono- and biarticular muscles in a limb, *J Biomech* 34:1085, 2001.

35. Hof AL: In vivo measurement of the series elasticity release curve of human triceps surae muscle, *J Biomech* 31:793, 1998.

36. Hortobágyi T, Hill JP, Houmard JA, Fraser DD, Labert NJ, and Israel RG: Adaptive responses to muscle lengthening and shortening in humans, *J Appl Physiol* 80:765, 1996.

37. Housh DJ and Housh TJ: The effects of unilateral velocity-specific concentric strength training, *J Orthop Sports Phys Ther* 17:252, 1993.

38. Howatson G: The impact of damaging exercise on electromechanical delay in biceps brachii, *J Electromyogr Kinesiol,* 20:477, 2010.

39. Hunter SK, Duchateau J, and Enoka RM: Muscle fatigue and the mechanisms of task failure, *Exerc Sport Sci Rev* 32:44, 2004.

40. Iguchi M and Shields RK: Quadriceps low-frequency fatigue and muscle pain are contraction-type-dependent, *Muscle Nerve* 42:230, 2010.

41. Jacobs R, Bobbert MF, and van Ingen Schenau GJ: Mechanical output from individual muscles during explosive leg extensions: the role of biarticular muscles, *J Biomech* 29:513, 1996.

42. Jacobs R et al: Function of mono- and biarticular muscles in running, *Med Sci Sports Exerc,* 25:1163, 1993.

43. Jaric S, Radovanovic S, Milanovic S, Ljubisavljevic M, and Anastasijevic R: A comparison of the effects of agonist and antagonist muscle fatigue on perfor-mance of rapid movements, *Eur J Appl Physiol* 76:41, 1997.

44. Katz B: The relation between force and speed in muscular contraction, *J Physiol* (Lond) 96:45, 1939.

45. Kawakami Y, Ichinose Y, Kubo K, Ito M, Imai M, and Fukunaga T: Archi-tecture of contracting human muscles and its functional significance, *J Appl Biomech* 16:88, 2000.

46. Kellis E and Baltzpoulos V: Muscle activation differences between eccentric and concentric isokinetic exercise, *Med Sci Sports Exerc,* 30:1616, 1998.

47. Komi PV: Physiological and biomechanical correlates of muscle function: effects of muscle structure and stretch-shortening cycle on force and speed, *Exerc Sport Sci Rev* 12:81, 1984.

48. Klein CS, Marsh GD, Petrella RJ, and Rice CL: Muscle fiber number in the biceps brachii muscle of young and old men, *Muscle Nerve* 28:62, 2003.

49. Komi PV: Stretch-shortening cycle: a powerful model to study normal and fatigued muscle, *J Biomech* 33:1197, 2000.

50. Komi PV, Linnamo V, Silventoinen P, and Sillanpää M: Force and EMG power spectrum during eccentric and concentric actions, *Med Sci Sports Exerc,* 32:1757, 2000.

51. Komi PV et al: Anaerobic performance capacity in athletes, *Acta Physiol Scand* 100:107, 1977.

52. Komi PV et al: Effects of heavy resistance and explosive-type strength training methods on mechanical, functional, and metabolic aspects of performance. In Komi PV, ed: *Exercise and sport biology,* Champaign, IL, 1982, Human Kinetics Publishers.

53. Kraemer WJ, Fleck SJ, and Evans WJ: Strength and power training: physiological mechanisms of adaptation, *Exerc Sport Sci Rev* 24:363, 1996.

54. Kriketos AD, Baur LA, OConnor J, Carey D, King S, Caterson ID, and Storlien LH: Muscle fibre type composition in infant and adult populations and relationships with obesity, *Int J Obes Relat Metab Disord* 21:796, 1997.

55. Kubo K, Kanehisha H, and Fukunaga T: Effects of cold and hot water immersion on the mechanical properties of human muscle and tendon in vivo, *Clin Biomech* 20:291, 2005.

56. Kubo K, Kanehisa H, Takeshita D, Kawakami Y, Fukashiro S, and Fukunaga T: In vivo dynamics of human medial gastrocnemius muscle-tendon complex during stretch-shortening cycle exercise, *Acta Physiol Scand* 170:127, 2000.

57. Levin A and Wyman J: The viscous elastic properties of muscle, *Proc R Soc Lond* B101:218, 1927.

58. Moss BM, Refsnes PE, Abildgaard A, Nicolaysen K, and Jensen J: Effects of maximal effort strength training with different loads on dynamic strength, cross-sectional area, load-power and load-velocity relationships, *Eur J Appl Physiol* 75:193, 1997.

59. Nagano A, Komura T, and Fukashiro S: Effects of series elasticity of the muscle tendon complex on an explosive activity performance with a counter movement, *J Appl Biomech* 20:85, 2004.

60. Nilsson J, Tesch P, and Thorstensson A: Fatigue and EMG of repeated fast and voluntary contractions in man, *Acta Physiol Scand* 101:194, 1977.

61. Norman RW: *The use of electromyography in the calculation of dynamic joint torque,* doctoral dissertation, University Park, PA, 1977, Pennsylvania State University.

62. Oksa J, Rintamaki H, Rissanen S, Rytky S, Tolonen U, and Komi PV: Stretch- and H-reflexes of the lower leg during whole body cooling and local warming, *Aviat Space Environ Med* 71:156, 2000.

63. Pitcher JB and Miles TS: Influence of muscle blood flow on fatigue during intermittent human hand-grip exercise and recovery, *Clin Exp Pharmacol Physiol* 24:471, 1997.

64. Power GA, Dalton BH, Behm DG, Vandervoort AA, Doherty TJ, and Rice CL: Motor unit number estimates in master runners: use it or lose it? *Med Sci Sports Exerc,* 42:1644, 2010.

65. Rasch PJ: *Kinesiology and applied anatomy* (7th ed), Philadelphia, 1989, Lea & Febiger.

66. Rassier DE, Herzog W, Wakeling J, and Syme DA: Stretch-induced, steady-state force enhancement in single skeletal muscle fibers exceeds the isometric force at optimum fiber length, *J Biomech* 36:1309, 2003.

67. Rassier De and Herzog W: Effects of shortening on stretch-induced force enhancement in single skeletal muscle fibers, *J Biomech* 37:1305, 2004.

68. Reich TE, Lindstedt SL, LaStayo PC, and Pierotti DJ: Is the spring quality of muscle plastic? *Am J Physiol* 278:R1661, 2000.

69. Rosenbaum D and Hennig EM: The influence of stretching and warm-up exercises on Achilles tendon reflex activity, *J Sports Sci* 13:481, 1995.

70. Saltin B, Henriksson J, Nygaard E, and Andersen P: Fiber types and metabolic potentials of skeletal muscles in sedentary man and endurance runners, *Ann NY Acad Sci* 301:3, 1977.

71. Saini A, Faulkner S, Al-Shanti N, and Stewart C: Powerful signals for weak muscles, *Ageing Res Rev* 8:251-67, 2009.

72. Sanders B and Nemeth WC: Hip and thigh injuries. In Zachazewski JE, Magee DJ, and Quillen WS, eds: *Athletic injuries and rehabilitation,* Philadelphia, WB Saunders, 1996.

73. Saxton JM and Donnelly AE: Length-specific impairment of skeletal muscle contractile function after eccentric muscle actions in man, *Clin Sci (Colch)* 90:119, 1996.

74. Scott SH and Winter DA: A comparison of three muscle pennation assumptions and their effect on isometric and isotonic force, *J Biomech,* 24:163, 1991.

75. Sheard PW: Tension delivery from short fibers in long muscles, *Exerc and Sport Sci Rev* 28:51, 2000.

76. Smith LL: Acute inflammation: the underlying mechanism in delayed onset muscle soreness? *Med Sci Sports Exerc* 23:542, 1991.

77. Takashi A, Kumagai K, and Brechue WF: Fascicle length of leg muscles is greater in sprinters than distance runners, *Med Sci Sports Exerc* 32:1125, 2000.

78. Tihanyi J, Apor P, and Fekete GY: Force-velocity-power characteristics and fiber composition in human knee extensor muscles, *Eur J Appl Physiol* 48:331, 1982.

79. Toft I, Lindal S, Bonaa Kh, and Jenssen T: Quantitative measurement of muscle fiber composition in a normal population, *Muscle Nerve* 28:101, 2003.

80. Trappe SW, Costill DL, Fink WJ, and Pearson DR: Skeletal muscle characteristics among distance runners: a 20-yr follow-up study, *J Appl Physiol* 78:823, 1995.

81. Trimble MH, Kukulka CG, and Thomas RS: Reflex facilitation during the stretch-shortening cycle, *J Electromyogr Kinesiol* 10:179, 2000.

82. Umberger BR, Gerritsen KG, and Martin PE: Muscle fiber type effects on energetically optimal cadences in cycling, *J Biomech* 38, 2005. [Epub ahead of print]

83. Viitasalo JT and Komi PV: Interrelationships between electromyographic, mechanical, muscle structure and reflex time measurements in man, *Acta Physiol Scand* 111:97, 1981.

84. Vint P, McLean S, and Harron GM: Electromechanical delay in isometric actions initiated from nonresting levels, *Med Sci Sports Exerc* 33:978, 2001.

85. Vos EJ, Harlaar J, and Van Ingen Schenau GJ: Electromechanical delay during knee extensor contractions, *Med Sci Sports Exerc,* 23:1187, 1991.

86. Westing SH, Seger JY, and Thorstensson A: Effects of electrical stimulation on eccentric and concentric torque-velocity relationships during knee extension in man, *Acta Physiol Scand,* 140:17, 1990.

87. Williams JH and Klug GA: Calcium exchange hypothesis of skeletal muscle fatigue: a brief review, *Muscle Nerve* 18:421, 1995.

88. Wilson GJ, Murphy AJ, and Giorgi A: Weight and plyometric training: effects on eccentric and concentric force production, *Can J Appl Physiol* 21:301, 1996.

89. Yan Z: Skeletal muscle adaptation and cell cycle regulation, *Exerc Sport Sci Rev* 28:24, 2000.

90. Zierath JR and Hawley JA: Skeletal muscle fiber type: Influence on contractile and metabolic properties, *PLoS Biol* 2:e348, 2004.

ANNOTATED READINGS

Komi PV: *Neuromuscular aspects of sports performance,* Oxford, 2010, Blackwell.
Describes neuromuscular function as related to optimal performance in different types of sports.

Miller MG, Cheatham CC, and Patel ND: Resistance training for adolescents, *Pediatr Clin North Am,* 57:671, 2010.
Reviews the guidelines for resistance training for health-related fitness for adolescents.

Raj IS, Bird SR, and Shield AJ: Aging and the force-velocity relationship of muscles, *Exp Gerontol,* 45:81, 2010.
Review paper discussing changes in muscle with normal aging, including loss of muscle mass, changes in muscle architecture, alteration of the force–velocity relationship, and different training approaches for ameliorating these changes.

Reich TE, Lindstedt SL, LaStayo PC, and Pierotti DJ: Is the spring quality of muscle plastic? *Am J Physiol* 278:R1661, 2000.
Research paper accompanied by an excellent discussion about the spring-like properties of muscle and the contribution of elastic energy storage and recovery in the stretch-shortening cycle.

RELATED WEBSITES

Duke University Presents Wheeless' Textbook of Orthopaedics
http://www.wheelessonline.com/
A comprehensive review of orthopaedics, with a search engine to locate topics related to fractures, joints, muscles, nerves, etc.
E-Tech: An Orthopaedics & Biomechanics Resource
http://dspace.dial.pipex.com/town/square/fk14/index.htm
Includes links to additional pages on a number of orthopaedic and biomechanic topics, including linear viscoelasticity.
Guided Tour of the Visible Human
http://www.madsci.org/~lynn/VH/
The Visible Human Project generated over 18,000 digitized sections of the human body. This tour introduces key concepts in human anatomy with images and animations from the dataset.
Martindale's The "Virtual" Medical Center: Muscles
http://www.martindalecenter.com/Medical1_1_MhU.html#Mus
Contains numerous images, movies, and course links for human anatomy.
Myology Section from Gray's Anatomy
http://www.bartleby.com/107/102.html
A description of muscle mechanics from this classic anatomy text.
Nicholas Institute of Sports Medicine and Athletic Trauma
http://www.nismat.org/
Provides links to pages on skeletal muscle anatomy and physiology.

KEY TERMS

active insufficiency	limited ability of a two-joint muscle to produce force when joint position places the muscle on slack
agonist	role played by a muscle acting to cause a movement
antagonist	role played by a muscle acting to slow or stop a movement
concentric	describing a contraction involving shortening of a muscle
contractile component	muscle property enabling tension development by stimulated muscle fibers
eccentric	describing a contraction involving lengthening of a muscle
electromechanical delay	time between the arrival of a neural stimulus and tension development by the muscle
fast-twitch fiber	a fiber that reaches peak tension relatively quickly
isometric	describing a contraction involving no change in muscle length
motor unit	a single motor neuron and all fibers it innervates
neutralizer	role played by a muscle acting to eliminate an unwanted action produced by an agonist
parallel elastic component	passive elastic property of muscle derived from the muscle membranes
parallel fiber arrangement	pattern of fibers within a muscle in which the fibers are roughly parallel to the longitudinal axis of the muscle
passive insufficiency	inability of a two-joint muscle to stretch to the extent required to allow full range of motion at all joints crossed
pennate fiber arrangement	pattern of fibers within a muscle with short fibers attaching to one or more tendons
series elastic component	passive elastic property of muscle derived from the tendons
slow-twitch fiber	a fiber that reaches peak tension relatively slowly
stabilizer	role played by a muscle acting to stabilize a body part against some other force
stretch-shortening cycle	eccentric contraction followed immediately by concentric contraction
summation	building in an additive fashion
tetanus	state of muscle producing sustained maximal tension resulting from repetitive stimulation
viscoelastic	having the ability to stretch or shorten over time

7

The Biomechanics of the Human Upper Extremity

After completing this chapter, you will be able to:

Explain how anatomical structure affects movement capabilities of upper-extremity articulations.

Identify factors influencing the relative mobility and stability of upper-extremity articulations.

Identify muscles that are active during specific upper-extremity movements.

Describe the biomechanical contributions to common injuries of the upper extremity.

ONLINE LEARNING CENTER RESOURCES

www.mhhe.com/hall6e

Log on to our Online Learning Center (OLC) for access to these additional resources:

- Online Lab Manual
- Flashcards with definitions of chapter key terms
- Chapter objectives
- Chapter lecture PowerPoint presentation
- Self-scoring chapter quiz
- Additional chapter resources
- Web links for study and exploration of chapter-related topics

Pitching a ball requires the coordination of the muscles of the entire upper extremity.

●*The glenohumeral joint is considered to be the shoulder joint.*

sternoclavicular joint
modified ball-and-socket joint between the proximal clavicle and the manubrium of the sternum

●*The clavicles and the scapulae make up the shoulder girdle.*

●*Most of the motion of the shoulder girdle takes place at the sternoclavicular joints.*

acromioclavicular joint
irregular joint between the acromion process of the scapula and the distal clavicle

Front view of the shoulder.
From Shier, Butler, and Lewis, *Hole's Human Anatomy and Physiology,* © 1996. Reprinted by permission of The McGraw-Hill Companies, Inc.

The capabilities of the upper extremity are varied and impressive. With the same basic anatomical structure of the arm, forearm, hand, and fingers, major league baseball pitchers hurl fastballs at 40 m/s, swimmers cross the English Channel, gymnasts perform the iron cross, travelers carry briefcases, seamstresses thread needles, and students type on computer keyboards. This chapter reviews the anatomical structures enabling these different types of movement and examines the ways in which the muscles cooperate to achieve the diversity of movement of which the upper extremity is capable.

STRUCTURE OF THE SHOULDER

The shoulder is the most complex joint in the human body, largely because it includes five separate articulations: the glenohumeral joint, the sternoclavicular joint, the acromioclavicular joint, the coracoclavicular joint, and the scapulothoracic joint. The glenohumeral joint is the articulation between the head of the humerus and the glenoid fossa of the scapula, which is the ball-and-socket joint typically considered to be *the* major shoulder joint. The sternoclavicular and acromioclavicular joints provide mobility for the clavicle and the scapula—the bones of the shoulder girdle.

Sternoclavicular Joint

The proximal end of the clavicle articulates with the clavicular notch of the manubrium of the sternum and with the cartilage of the first rib to form the sternoclavicular joint. This joint provides the major axis of rotation for movements of the clavicle and scapula (Figure 7-1). The sternoclavicular (SC) joint is a modified ball and socket, with frontal and transverse plane motion freely permitted and some forward and backward sagittal plane rotation allowed. A fibrocartilaginous articular disc improves the fit of the articulating bone surfaces and serves as a shock absorber. Rotation occurs at the SC joint during motions such as shrugging the shoulders, elevating the arms above the head, and swimming. The close-packed position for the SC joint occurs with maximal shoulder elevation.

Acromioclavicular Joint

The articulation of the acromion process of the scapula with the distal end of the clavicle is known as the acromioclavicular joint. It is classified as an

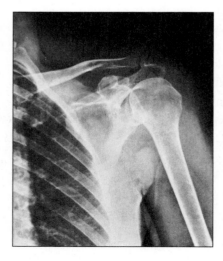

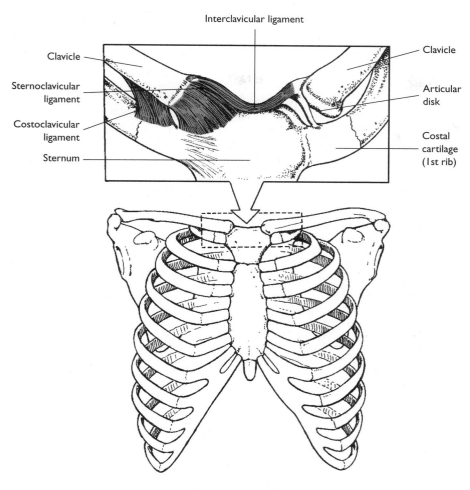

FIGURE 7-1

The sternoclavicular joint.

irregular diarthrodial joint, although the joint's structure allows limited motion in all three planes. There is a significant amount of anatomical variation in the acromioclavicular (AC) joint from individual to individual, with as many as five different morphological types identified (60). Rotation occurs at the AC joint during arm elevation. The close-packed position of the AC joint occurs when the humerus is abducted to 90°.

Coracoclavicular Joint

The coracoclavicular joint is a syndesmosis, formed where the coracoid process of the scapula and the inferior surface of the clavicle are bound together by the coracoclavicular ligament. This joint permits little movement. The coracoclavicular and acromioclavicular joints are shown in Figure 7-2.

coracoclavicular joint
syndesmosis with the coracoid process of the scapula bound to the inferior clavicle by the coracoclavicular ligament

Glenohumeral Joint

The glenohumeral joint is the most freely moving joint in the human body, enabling flexion, extension, hyperextension, abduction, adduction, horizontal abduction and adduction, and medial and lateral rotation of the humerus (Figure 7-3). The almost hemispherical head of the humerus has three to four times the amount of surface area as the shallow glenoid fossa of the scapula with which it articulates. The glenoid fossa is also less curved than the surface of the humeral head, enabling the humerus to move linearly across the surface of the glenoid fossa in addition to its extensive rotational capability (61). There are anatomical variations in the shape of the glenoid fossa from person to person, with an oval or

glenohumeral joint
ball-and-socket joint in which the head of the humerus articulates with the glenoid fossa of the scapula

FIGURE 7-2

The acromioclavicular and coracoclavicular joints.

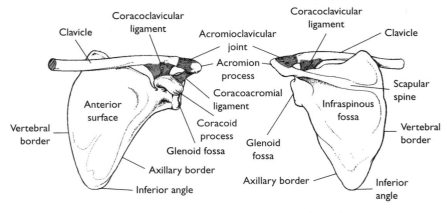

Anterior view   **Posterior view**

FIGURE 7-3

The glenohumeral joint.

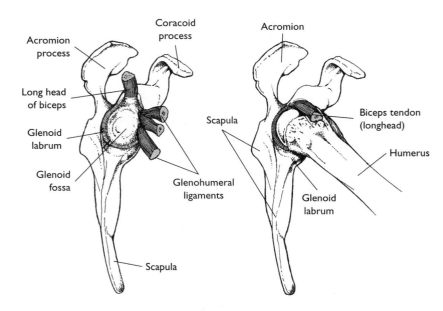

Lateral view

●*The extreme mobility of the glenohumeral joint is achieved at the expense of joint stability.*

glenoid labrum
rim of soft tissue located on the periphery of the glenoid fossa that adds stability to the glenohumeral joint

rotator cuff
band of tendons of the subscapularis, supraspinatus, infraspinatus, and teres minor, which attach to the humeral head

egg-shaped cavity in about 45% of the population and a pear-shaped cavity in the remaining 55% (63). With passive rotation of the arm, large translations of the humeral head on the glenoid fossa are present at the extremes of the range of motion (38). The muscle forces during active rotation tend to limit ranges of motion at the shoulder, thereby limiting the humeral translation that occurs (38).

The glenoid fossa is encircled by the glenoid labrum, a lip composed of part of the joint capsule, the tendon of the long head of the biceps brachii, and the glenohumeral ligaments. This rim of dense collagenous tissue is triangular in cross-section and is attached to the periphery of the fossa. The labrum deepens the fossa and adds stability to the joint. The capsule surrounding the glenohumeral joint is shown in Figure 7-4. Several ligaments merge with the glenohumeral joint capsule, including the superior, middle, and inferior glenohumeral ligaments on the anterior side of the joint and the coracohumeral ligament on the superior side.

The tendons of four muscles also join the joint capsule. These are known as the rotator cuff muscles because they contribute to rotation of the humerus and because their tendons form a collagenous cuff around the glenohumeral joint. These include supraspinatus, infraspinatus, teres

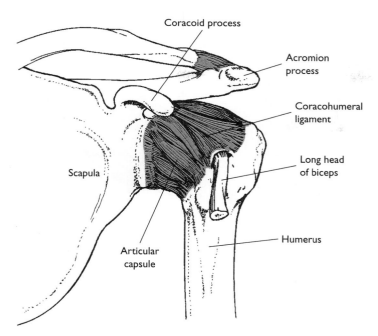

Coracoid process

Acromion process

Coracohumeral ligament

Long head of biceps

Scapula

Humerus

Articular capsule

FIGURE 7-4

The capsule surrounding the glenohumeral joint contributes to joint stability.

minor, and subscapularis, and are also sometimes referred to as the *SITS* muscles after the first letter of the muscles' names. Supraspinatus, infraspinatus, and teres minor participate in lateral rotation, and subscapularis contributes to medial rotation. The muscles of the lateral rotator group exchange muscle bundles with one another, which increases their ability to quickly develop tension and functional power (28). The rotator cuff surrounds the shoulder on the posterior, superior, and anterior sides. Tension in the rotator cuff muscles pulls the head of the humerus toward the glenoid fossa, contributing significantly to the joint's minimal stability. It has been shown that the rotator cuff muscles and the biceps are activated to provide shoulder stability prior to motion of the humerus (13). Negative pressure within the capsule of the glenohumeral joint also helps to stabilize the joint (31). The joint is most stable in its close-packed position, when the humerus is abducted and laterally rotated.

Scapulothoracic Joint

Because the scapula can move in both sagittal and frontal planes with respect to the trunk, the region between the anterior scapula and the thoracic wall is sometimes referred to as the *scapulothoracic joint*. The muscles attaching to the scapula perform two functions. First, they can contract to stabilize the shoulder region. For example, when a suitcase is lifted from the floor, the levator scapula, trapezius, and rhomboids develop tension to support the scapula, and in turn the entire shoulder, through the acromioclavicular joint. Second, the scapular muscles can facilitate movements of the upper extremity through appropriate positioning of the glenohumeral joint. During an overhand throw, for example, the rhomboids contract to move the entire shoulder posteriorly as the humerus is horizontally abducted and externally rotated during the preparatory phase. As the arm and hand then move forward to execute the throw, tension in the rhomboids is released to permit forward movement of the glenohumeral joint.

Bursae

Several small, fibrous sacs that secrete synovial fluid internally in a fashion similar to that of a joint capsule are located in the shoulder region.

FIGURE 7-5

The four muscles of the rotator cuff.

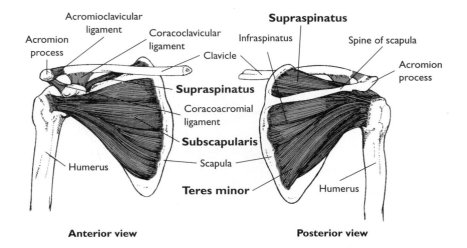

Anterior view   Posterior view

bursae
sacs secreting synovial fluid internally that lessen friction between soft tissues around joints

These sacs, known as bursae, cushion and reduce friction between layers of collagenous tissues. The shoulder is surrounded by several bursae, including the subscapularis, subcoracoid, and subacromial.

The subscapularis and subcoracoid bursae are responsible for managing friction of the superficial fibers of the subscapularis muscle against the neck of the scapula, the head of the humerus, and the coracoid process. In 28% of studied cases, these two bursae physically merge into a single wide bursa (11). Given that the subscapularis undergoes significant changes in orientation during movements of the arm at the glenohumeral joint, especially where the upper portion of the muscle coils around the coracoid process, the role of these bursae is important.

The subacromial bursa lies in the subacromial space, between the acromion process of the scapula and the coracoacromial ligament (above) and the glenohumeral joint (below). This bursa cushions the rotator cuff muscles, particularly the supraspinatus, from the overlying bony acromion (Figure 7-5). The subacromial bursa may become irritated when repeatedly compressed during overhead arm action.

MOVEMENTS OF THE SHOULDER COMPLEX

Although some amount of glenohumeral motion may occur while the other shoulder articulations remain stabilized, movement of the humerus more commonly involves some movement at all three shoulder joints (Figure 7-6). Elevation of the humerus in all planes is accompanied by approximately 55° of lateral rotation (71). As the arm is elevated in both abduction and flexion, rotation of the scapula accounts for part of the total humeral range of motion. Although the absolute positions of the humerus and scapula vary due to anatomical variations among individuals, a general pattern persists (27). During about the first 30° of humeral elevation, the contribution of the scapula is only about one-fifth that of the glenohumeral joint (61). As elevation proceeds beyond 30°, the scapula rotates approximately 1° for every 2° of movement of the humerus (18, 33, 67). This important coordination of scapular and humeral movements, known as scapulohumeral rhythm, enables a much greater range of motion at the shoulder than if the scapula were fixed. During the first 90° of arm elevation (in sagittal, frontal, or diagonal planes), the clavicle is also elevated through approximately 35–45° of motion at the sternoclavicular joint (61). Rotation at the acromioclavicular joint occurs during the first 30° of humeral elevation and again as the arm is moved from 135° to maximum

scapulohumeral rhythm
a regular pattern of scapular rotation that accompanies and facilitates humeral abduction

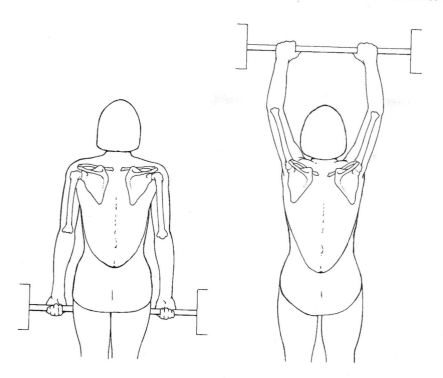

FIGURE 7-6

Elevation of the arm is accompanied by rotation of the clavicle and scapula.

elevation (40). Positioning of the humerus is further facilitated by motions of the spine. When the hands support an external load, the orientation of the scapula and the scapulohumeral rhythm are altered, with muscular stabilization of the scapula reducing scapulothoracic motion as dynamic scapular stabilization provides a platform for upper extremity movements (41). Generally, scapulohumeral relationships are more fixed when the arm is loaded and engaged in purposeful movement as compared to when the arm is moving in an unloaded condition (12).

The movement patterns of the scapula are also different in children and in the elderly. As compared to adults, children receive a greater contribution from the scapulothoracic joint during humeral elevation (15). With aging, there is a lessening of scapular rotation, as well as posterior tilt, with glenohumeral abduction (21). Abnormal motion of the scapula may contribute to a variety of shoulder pathologies, including shoulder impingement, rotator cuff tears, glenohumeral instability, and stiff shoulders (48).

Muscles of the Scapula

The muscles that attach to the scapula are the levator scapula, rhomboids, serratus anterior, pectoralis minor, and subclavius, and the four parts of the trapezius. Figures 7-7 and 7-8 show the directions in which these muscles exert force on the scapula when contracting. Scapular muscles have two general functions. First, they stabilize the scapula so that it forms a rigid base for muscles of the shoulder during the development of tension. For example, when a person carries a briefcase, the levator scapula, trapezius, and rhomboids stabilize the shoulder against the added weight. Second, scapular muscles facilitate movements of the upper extremity by positioning the glenohumeral joint appropriately. For example, during an overhand throw, the rhomboids contract to move the entire shoulder posteriorly as the arm and hand move posteriorly during the preparatory phase. As the arm and hand move anteriorly to deliver the throw, tension in the rhomboids subsides to permit forward movement of the shoulder, facilitating outward rotation of the humerus.

● *The scapular muscles perform two functions: (a) stabilizing the scapula when the shoulder complex is loaded, and (b) moving and positioning the scapula to facilitate movement at the glenohumeral joint.*

FIGURE 7-7

Actions of the scapular muscles.

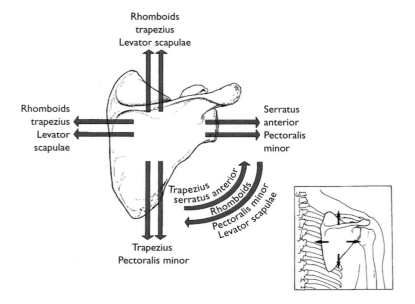

FIGURE 7-8

The muscles of the scapula.

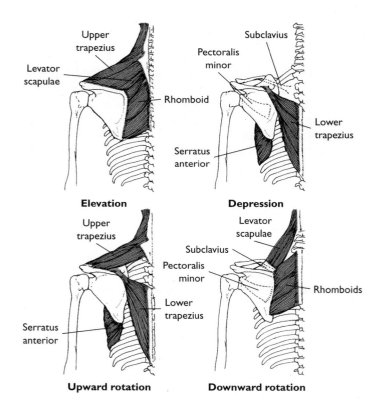

Muscles of the Glenohumeral Joint

Many muscles cross the glenohumeral joint. Because of their attachment sites and lines of pull, some muscles contribute to more than one action of the humerus. A further complication is that the action produced by the development of tension in a muscle may change with the orientation of the humerus because of the shoulder's large range of motion. With the basic instability of the structure of the glenohumeral joint, a significant portion of the joint's stability is derived from tension in the muscles and

•The development of tension in one shoulder muscle must frequently be accompanied by the development of tension in an antagonist to prevent dislocation of the humeral head.

tendons crossing the joint. However, when one of these muscles develops tension, tension development in an antagonist may be required to prevent dislocation of the joint. A review of the muscles of the shoulder is presented in Table 7-1.

TABLE 7-1

Muscles of the Shoulder

MUSCLE	PROXIMAL ATTACHMENT	DISTAL ATTACHMENT	PRIMARY ACTION(S)	INNERVATION
Deltoid	Outer third of clavicle, top of acromion, scapular spine	Deltoid tuberosity of humerus		Axillary (C_5, C_6)
(Anterior)			Flexion, horizontal adduction, medial rotation	
(Middle)			Abduction, horizontal abduction	
(Posterior)			Extension, horizontal abduction, lateral rotation	
Pectoralis major		Lateral aspect of humerus just below head		
(Clavicular)	Medial two-thirds of clavicle		Flexion, horizontal adduction, medial rotation	Lateral pectoral (C_5–T_1)
(Sternal)	Anterior sternum and cartilage of first six ribs		Extension, adduction, horizontal adduction, medial rotation	Medial pectoral (C_5–T_1)
Supraspinatus	Supraspinous fossa	Greater tuberosity of humerus	Abduction, assists with lateral rotation	Suprascapular (C_5, C_6)
Coracobrachialis	Coracoid process of scapula	Medial anterior humerus	Flexion, adduction, horizontal adduction	Musculocutaneous (C_5–C_7)
Latissimus dorsi	Lower six thoracic and all lumbar vertebrae, posterior sacrum, iliac crest, lower three ribs	Anterior humerus	Extension, adduction, medial rotation, horizontal abduction	Thoracodorsal (C_6–C_8)
Teres major	Lower, lateral, dorsal scapula	Anterior humerus	Extension, adduction, medial rotation	Subscapular (C_5, C_6)
Infraspinatus	Infraspinous fossa	Greater tubercle of humerus	Lateral rotation, horizontal abduction	Subscapular (C_5, C_6)
Teres minor	Posterior, lateral border of scapula	Greater tubercle, adjacent shaft of humerus	Lateral rotation, horizontal abduction	Axillary (C_5, C_6)
Subscapularis	Entire anterior surface of scapula	Lesser tubercle of humerus	Medial rotation	Subscapular (C_5, C_6)
Biceps brachii		Radial tuberosity		Musculocutaneous (C_5–C_7)
(Long head)	Upper rim of glenoid fossa		Assists with abduction	
(Short head)	Coracoid process of scapula		Assists with flexion, adduction, medial rotation, horizontal adduction	
Triceps brachii (Long head)	Just inferior to glenoid fossa	Olecranon process of ulna	Assists with extension, adduction	Radial (C_5–T_1)

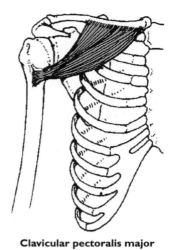

Clavicular pectoralis major

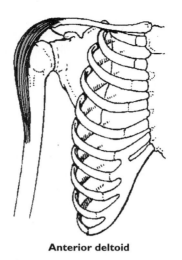

Anterior deltoid

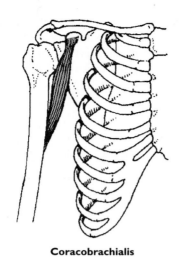

Coracobrachialis

FIGURE 7-9   The major flexor muscles of the shoulder.

Flexion at the Glenohumeral Joint

The muscles crossing the glenohumeral joint anteriorly participate in flexion at the shoulder (Figure 7-9). The prime flexors are the anterior deltoid and the clavicular portion of the pectoralis major. The small coracobrachialis assists with flexion, as does the short head of the biceps brachii. Although the long head of the biceps also crosses the shoulder, it is not active in isolated shoulder motion when the elbow and forearm do not move (43).

Extension at the Glenohumeral Joint

When shoulder extension is unresisted, gravitational force is the primary mover, with eccentric contraction of the flexor muscles controlling or braking the movement. When resistance is present, contraction of the muscles posterior to the glenohumeral joint, particularly the sternocostal pectoralis, latissimus dorsi, and teres major, extend the humerus. The posterior deltoid assists in extension, especially when the humerus is externally rotated. The long head of the triceps brachii also assists, and because the muscle crosses the elbow, its contribution is slightly more effective when the elbow is in flexion. The shoulder extensors are illustrated in Figure 7-10.

Abduction at the Glenohumeral Joint

The middle deltoid and supraspinatus are the major abductors of the humerus. Both muscles cross the shoulder superior to the glenohumeral joint (Figure 7-11). The supraspinatus, which is active through approximately the first 110° of motion, initiates abduction. During the contribution of the middle deltoid (occurring from approximately 90° to 180° of abduction), the infraspinatus, subscapularis, and teres minor neutralize the superiorly dislocating component of force produced by the middle deltoid.

Adduction at the Glenohumeral Joint

As with extension at the shoulder, adduction in the absence of resistance results from gravitational force, with the abductors controlling the speed of motion. With resistance added, the primary adductors are

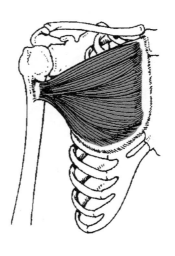

Sternal pectoralis major

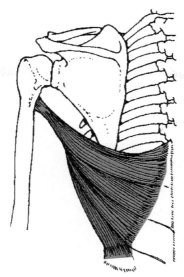

Latissimus dorsi

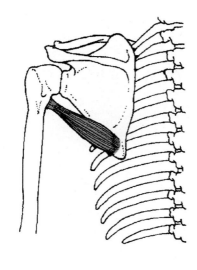

Teres major

FIGURE 7-10   The major extensor muscles of the shoulder.

the latissimus dorsi, teres major, and sternocostal pectoralis, which are located on the inferior side of the joint (Figure 7-12). The short head of the biceps and the long head of the triceps contribute minor assistance, and when the arm is elevated above 90°, the coracobrachialis and subscapularis also assist.

Medial and Lateral Rotation of the Humerus

Medial, or inward, rotation of the humerus results primarily from the action of the subscapularis and teres major, both attaching to the anterior side of the humerus, with the subscapularis having the greatest mechanical advantage for medial rotation (42). Both portions of the pectoralis major, the anterior deltoid, the latissimus dorsi, and the short head of the biceps brachii assist, with the pectoralis major being the primary assistant (8). Muscles attaching to the posterior aspect of the humerus,

FIGURE 7-11

The major abductor muscles of the shoulder.

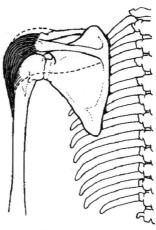

Middle deltoid

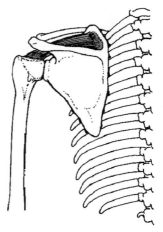

Supraspinatus

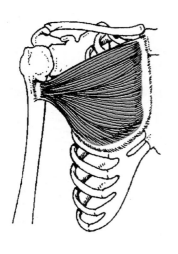

Sternal pectoralis major

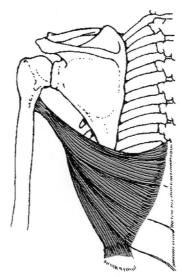

Latissimus dorsi

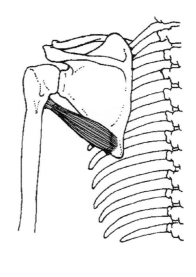

Teres major

FIGURE 7-12   The major adductor muscles of the shoulder.

particularly infraspinatus and teres minor, produce lateral, or outward, rotation, with some assistance from the posterior deltoid.

Horizontal Adduction and Abduction at the Glenohumeral Joint

The muscles anterior to the joint, including both heads of the pectoralis major, the anterior deltoid, and the coracobrachialis, produce horizontal adduction, with the short head of the biceps brachii assisting. Muscles posterior to the joint axis affect horizontal abduction. The major horizontal abductors are the middle and posterior portions of the deltoid, infraspinatus, and teres minor, with assistance provided by the teres major and the latissimus dorsi. The major horizontal adductors and abductors are shown in Figures 7-13 and 7-14.

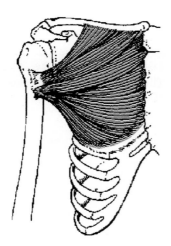

Pectoralis major

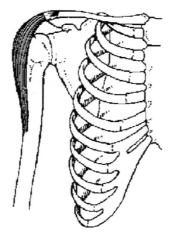

Anterior deltoid

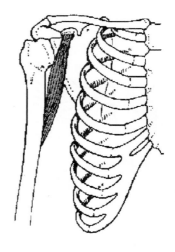

Coracobrachialis

FIGURE 7-13   The major horizontal adductors of the shoulder.

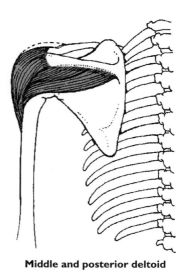

Middle and posterior deltoid

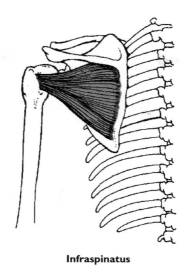

Infraspinatus

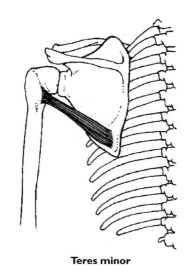

Teres minor

FIGURE 7-14   The major horizontal abductors of the shoulder.

LOADS ON THE SHOULDER

Because the articulations of the shoulder girdle are interconnected, they function to some extent as a unit in bearing loads and absorbing shock. However, because the glenohumeral joint provides direct mechanical support for the arm, it sustains much greater loads than the other shoulder joints.

As indicated in Chapter 3, when analyzing the effect of body position, we may assume that body weight acts at the body's center of gravity. Likewise, when analyzing the effect of the positions of body segments on a joint such as the shoulder, we assume that the weight of each body segment acts at the segmental center of mass. The moment arm for the entire arm segment with respect to the shoulder is therefore the perpendicular distance between the weight vector (acting at the arm's center of gravity) and the shoulder (Figure 7-15). When the elbow is in flexion, the effects

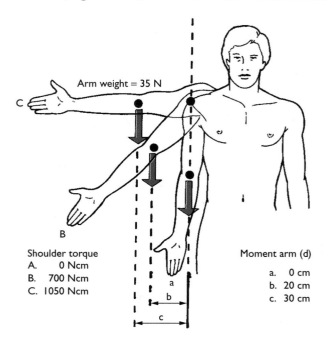

Arm weight = 35 N

Shoulder torque
A.   0 Ncm
B.   700 Ncm
C.   1050 Ncm

Moment arm (d)
a.   0 cm
b.   20 cm
c.   30 cm

FIGURE 7-15

The torque created at the shoulder by the weight of the arm is the product of arm weight and the perpendicular distance between the arm's center of gravity and the shoulder (the arm's moment arm). Adapted from Chaffin DB and Andersson GBJ, *Occupational Biomechanics* (2nd ed), New York, 1991, John Wiley & Sons.

The torque created at
the shoulder by each arm
segment is the product of
the segment's weight and
the segment's moment arm.
Adapted from Chaffin DB and
Andersson GBJ, *Occupational
Biomechanics* (2nd ed), New York,
1991, John Wiley & Sons.

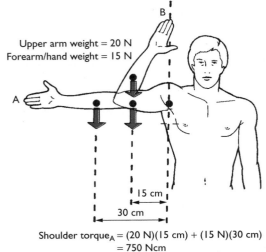

Upper arm weight = 20 N
Forearm/hand weight = 15 N

15 cm

30 cm

Shoulder torque$_A$ = (20 N)(15 cm) + (15 N)(30 cm)
= 750 Ncm
Shoulder torque$_B$ = (20 N)(15 cm) + (15 N)(15 cm)
= 525 Ncm

of the upper arm and the forearm/hand segments must be analyzed separately (Figure 7-16). Sample Problem 7.1 illustrates the effect of arm position on shoulder loading.

Although the weight of the arm is only approximately 5% of body weight, the length of the horizontally extended arm creates large segment moment arms and therefore large torques that must be countered by the shoulder muscles. When these muscles contract to support the extended arm, the glenohumeral joint sustains compressive forces estimated to reach 50% of body weight (61). Although this load is reduced by about half when the elbow is maximally flexed due to the shortened moment arms of the forearm and hand, this can place a rotational torque on the humerus that requires the activation of additional shoulder muscles (Figure 7-17).

Because of the effect of arm position on shoulder loading, ergonomists recommend that workers seated at a desk or a table attempt to position the arms with 20° or less of abduction and 25° or less of flexion (7). Workers who are required to hold the arms in a sustained position overhead are particularly susceptible to degenerative tendinitis in the biceps and supraspinatus (25).

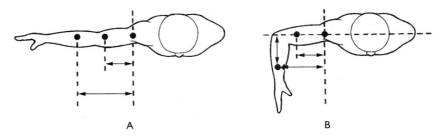

A

B

FIGURE 7-17   **A.** The weight of the arm segments creates a frontal plane torque at the shoulder, with moment arms as shown.
B. The weight of the upper arm segment creates a frontal plane torque at the shoulder. The weight of the forearm/hand creates both frontal plane and sagittal plane torques at the shoulder, with moment arms as shown. Adapted from Chaffin DB and Andersson GBJ, *Occupational Biomechanics* (2nd ed), New York, 1991, John Wiley & Sons.

SAMPLE PROBLEM 7.1

Using the simplifying assumptions of Poppen and Walker (38), a free body diagram of the arm and shoulder can be constructed as shown below. If the weight of the arm is 33 N, the moment arm for the total arm segment is 30 cm, and the moment arm for the deltoid muscle (F_m) is 3 cm, how much force must be supplied by the deltoid to maintain the arm in this position? What is the magnitude of the horizontal component of the joint reaction force (R_h)?

Known

$$wt = 33 \text{ N}$$
$$d_{wt} = 30 \text{ cm}$$
$$d_m = 3 \text{ cm}$$

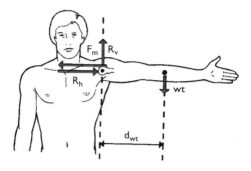

Solution

The torque at the shoulder created by the muscle force must equal the torque at the shoulder created by arm weight, yielding a net shoulder torque of zero.

$$\sum T_s = 0$$
$$\sum T_s = (F_m)(d_m) - (wt)(d_{wt})$$
$$0 = (F_m)(3 \text{ cm}) - (33 \text{ N})(30 \text{ cm})$$
$$0 = (F_m)(3 \text{ cm}) - (33 \text{ N})(30 \text{ cm})$$
$$F_m = \frac{(33 \text{ N})(30 \text{ cm})}{3 \text{ cm}}$$
$$F_m = 330 \text{ N}$$

Since the horizontal component of joint reaction force (R_h) and F_m are the only two horizontal forces present, and since the arm is stationary, these forces must be equal and opposite. The magnitude of R_h is therefore the same as the magnitude of F_m.

$$R_h = 330 \text{ N}$$

Note: Both components of the joint reaction force are directed through the joint center, and so have a moment arm of zero with respect to the center of rotation.

Muscles that attach to the humerus at small angles with respect to the glenoid fossa contribute primarily to shear as opposed to compression at the joint. These muscles serve the important role of stabilizing the humerus in the glenoid fossa against the contractions of powerful muscles that might otherwise dislocate the joint. For example, when the arm is elevated, the deltoid and rotator cuff muscles act in tandem, with the deltoid producing an upward shear force that is counteracted by the downward shear produced by the rotator cuff (58) (Figure 7-18). Maximum shear force has been found to be present at the glenohumeral joint when the arm is elevated approximately 60° (79).

FIGURE 7-18

Abduction of the humerus requires the cooperative action of the deltoid and the rotator cuff muscles. Because the vertical components of muscle force largely cancel each other, the oppositely directed horizontal components produce rotation of the humerus.

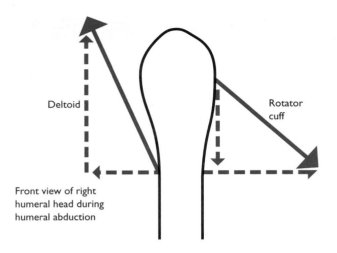

Deltoid

Rotator cuff

Front view of right humeral head during humeral abduction

• *When a glenohumeral joint dislocation occurs, the supporting soft tissues are often stretched beyond their elastic limits, thus predisposing the joint to subsequent dislocations.*

COMMON INJURIES OF THE SHOULDER

The shoulder is susceptible to both traumatic and overuse types of injuries, including 8–13% of all sport-related injuries (25, 50).

Dislocations

The glenohumeral joint is the most commonly dislocated joint in the body (6). The loose structure of the glenohumeral joint enables extreme mobility but provides little stability, and dislocations may occur in anterior, posterior, and inferior directions. The strong coracohumeral ligament usually prevents displacement in the superior direction. Glenohumeral dislocations typically occur when the humerus is abducted and externally rotated, with anterior-inferior dislocations more common than those in other directions. Factors that predispose the joint to dislocations include inadequate size of the glenoid fossa, anterior tilt of the glenoid fossa, inadequate retroversion of the humeral head, and deficits in the rotator cuff muscles (67).

Glenohumeral dislocation may result from sustaining a large external force during an accident, such as in cycling, or during participation in a contact sport such as wrestling or football. Unfortunately, once the joint has been dislocated, the stretching of the surrounding collagenous tissues beyond their elastic limits commonly predisposes it to subsequent dislocations. Glenohumeral capsular laxity may also be present due to genetic factors. Individuals with this condition should strengthen their shoulder muscles before athletic participation (24).

Dislocations or separations of the acromioclavicular joint are also common among wrestlers and football players. When a rigidly outstretched arm sustains the force of a full-body fall, either acromioclavicular separation or fracture of the clavicle is likely to result.

Rotator Cuff Damage

A common injury among workers and athletes who engage in forceful overhead movements typically involving abduction or flexion along with medial rotation is *rotator cuff impingement syndrome,* also known as subacromial impingement syndrome, or shoulder impingement syndrome. This is the most common disorder of the shoulder, with progressive loss of function and disability (52). The cause is progressive pressure on the rotator cuff tendons by the surrounding bone and soft tissue structures. Symptoms include hypermobility of the anterior shoulder capsule, hypomobility of the posterior capsule, excessive external rotation coupled with limited internal rotation of the humerus, and general ligamentous laxity of the glenohumeral joint (80). This can result in inflammation of the underlying tendons and bursae or, in severe cases, rupture of one of the rotator cuff tendons. The muscle most commonly affected is the supraspinatus, possibly because its blood supply is the most susceptible to pressure (68). This condition is accompanied by pain and tenderness in the superior and anterior shoulder regions, and sometimes by associated shoulder weakness. The symptoms are exacerbated by rotary movements of the humerus, especially those involving elevation and internal rotation.

Activities that may promote the development of shoulder impingement syndrome include throwing (particularly an implement like the javelin), serving in tennis, and swimming (especially the freestyle, backstroke, and butterfly) (3, 59, 72). Among competitive swimmers, the syndrome is known as *swimmer's shoulder*. Reports indicate shoulder pain complaints in up to 50% of competitive swimmers (40). Older golfers also frequently develop impaired rotator cuff function secondary to degenerative changes such as osteophyte formation that impinges upon the subacromial space (4).

Anatomical factors believed to predispose a person to impingement syndrome include a flat acromion with only a small inclination, bony spurs at the acromioclavicular joint secondary to osteoarthritis, and a superiorly positioned humeral head (20, 63, 77). A number of theories have been proposed regarding the biomechanical causes of rotator cuff problems. The impingement theory suggests that a genetic factor results in the formation of too narrow a space between the acromion process of the scapula and the head of the humerus. In this situation, the rotator cuff and associated bursae are pinched between the acromion, the acromioclavicular ligament, and the humeral head each time the arm is elevated, with the resulting friction causing irritation and wear. An alternative theory proposes that the major factor is inflammation of the supraspinatus tendon caused by repeated overstretching of the muscle–tendon unit. When the rotator cuff tendons become stretched and weakened, they cannot perform their normal function of holding the humeral head in the glenoid fossa. Consequently, the deltoid muscles pull the humeral head up too high during abduction, resulting in impingement and subsequent wear and tear on the rotator cuff.

The problem is relatively common among swimmers. Research has shown that during the recovery phase of swimming, when the arm is elevated above the shoulder while being medially rotated, the serratus anterior rotates the scapula so that the supraspinatus, infraspinatus, and middle deltoid may freely abduct the humerus (56). Each arm is in this position during an average of approximately 12% of the stroke cycle among collegiate male swimmers (81). It has been hypothesized that if the serratus, which develops nearly maximum tension to accomplish this task, becomes fatigued, the scapula may not be rotated sufficiently to abduct the humerus freely, and impingement may develop (56). Swimming technique appears to be related to the likelihood of developing shoulder impingement in swimmers, with a large amount of medial rotation of the arm during the pull phase, late initiation of lateral rotation of the arm during the overhead phase, and breathing exclusively on one side all implicated as contributing factors (65, 82).

Overarm sport activities such as pitching in baseball often result in overuse injuries of the shoulder.

Rotational Injuries

Tears of the labrum, the rotator cuff muscles, and the biceps brachii tendon are among the injuries that may result from repeated, forceful rotation at the shoulder. Throwing, serving in tennis, and spiking in volleyball are examples of forceful rotational movements. If the attaching muscles do not sufficiently stabilize the humerus, it can articulate with the glenoid labrum rather than with the glenoid fossa, contributing to wear on the labrum. Most tears are located in the anterior-superior region of the labrum. Tears of the rotator cuff, primarily of the supraspinatus, have been attributed to the extreme tension requirements placed on the muscle group during the deceleration phase of a vigorous rotational activity. There is a region of low vascularity near the insertion of the supraspinatus tendon, which is the most common site of rotator cuff inflammation and tears (49). The fact that vascularization diminishes with aging may explain why rotator cuff problems occur more frequently in individuals over 40 years of age (49). Tears of the biceps brachii tendon at the site of its attachment to the glenoid fossa may result from the forceful development of tension in the biceps when it negatively accelerates the rate of elbow extension during throwing (51).

Other pathologies of the shoulder attributed to throwing movements are calcifications of the soft tissues of the joint and degenerative changes in the articular surfaces (62). Bursitis, the inflammation of one or more bursae, is another overuse syndrome, generally caused by friction within the bursa (64).

Subscapular Neuropathy

Subscapular neuropathy, or subscapular nerve palsy, most commonly occurs in athletes involved in overhead activities and weight lifting (16). It has been reported in volleyball, baseball, football, and racquetball players, as well as backpackers, gymnasts, and dancers. The condition arises from compression of the subscapular nerve, which occurs most commonly at the suprascapular notch. In volleyball players, it has been attributed to the repeated stretching of the nerve during extreme shoulder abduction and lateral rotation, such as occurs during the serving motion (23).

STRUCTURE OF THE ELBOW

Although the elbow is generally considered a simple hinge joint, it is actually categorized as a trochoginglymus joint that encompasses three articulations: the humeroulnar, humeroradial, and proximal radioulnar joints (2). All are enclosed in the same joint capsule, which is reinforced by the anterior and posterior radial collateral and ulnar collateral ligaments. Bony structure provides about half of the elbow's stability, with the remaining stability provided by the joint capsule and the ulnar and radial ligament complexes (53).

Humeroulnar Joint

humeroulnar joint
hinge joint in which the humeral trochlea articulates with the trochlear fossa of the ulna

●*The humeroulnar hinge joint is considered to be the elbow joint.*

The hinge joint at the elbow is the humeroulnar joint, where the ovular trochlea of the humerus articulates with the reciprocally shaped trochlear fossa of the ulna (Figure 7-19). Flexion and extension are the primary movements, although in some individuals, a small amount of hyperextension is allowed. The joint is most stable in the close-packed position of extension.

Humeroradial Joint

humeroradial joint
gliding joint in which the capitellum of the humerus articulates with the proximal end of the radius

The humeroradial joint is immediately lateral to the humeroulnar joint and is formed between the spherical capitellum of the humerus and the proximal end of the radius (Figure 7-19). Although the humeroradial articulation is classified as a gliding joint, the immediately adjacent humeroulnar joint restricts motion to the sagittal plane. In the close-packed position, the elbow is flexed at 90° and the forearm is supinated about 5°.

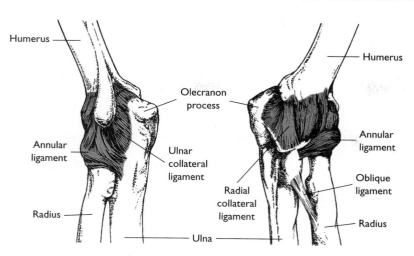

FIGURE 7-19

The major ligaments of the
elbow.

Proximal Radioulnar Joint

The annular ligament binds the head of the radius to the radial notch of
the ulna, forming the proximal radioulnar joint. This is a pivot joint, with
forearm pronation and supination occurring as the radius rolls medially
and laterally over the ulna (Figure 7-20). The close-packed position is at
5° of forearm supination.

Carrying Angle

The angle between the longitudinal axes of the humerus and the ulna
when the arm is in anatomical position is referred to as the *carrying angle*. The size of the carrying angle ranges from 10° to 15° in adults and
tends to be larger in females than in males. The carrying angle changes
with skeletal growth and is always greater on the side of the dominant

radioulnar joints
*the proximal and distal radioulnar
joints are pivot joints; the middle
radioulnar joint is a syndesmosis*

• *When pronation and supination
of the forearm occur, the radius
pivots around the ulna.*

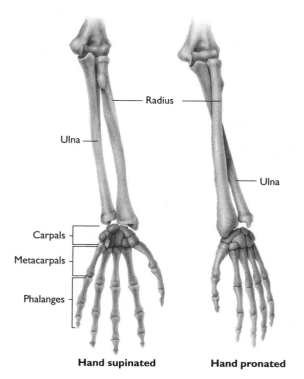

Hand supinated   **Hand pronated**

FIGURE 7-20

Bones of the elbow, forearm,
wrist, and hand. From Shier,
Butler, and Lewis, *Hole's Human
Anatomy and Physiology,* © 1996.
Reprinted by permission of The
McGraw-Hill Companies, Inc.

hand (57). No particular functional significance has been associated with the carrying angle.

MOVEMENTS AT THE ELBOW

Muscles Crossing the Elbow

Numerous muscles cross the elbow, including those that also cross the shoulder or extend into the hands and fingers. The muscles classified as primary movers of the elbow are summarized in Table 7-2.

Flexion and Extension

The muscles crossing the anterior side of the elbow are the elbow flexors (Figure 7-21). The strongest of the elbow flexors is the brachialis. Since

TABLE 7-2

Muscles of the Elbow

MUSCLE	PROXIMAL ATTACHMENT	DISTAL ATTACHMENT	PRIMARY ACTION(S)	INNERVATION
Biceps brachii		Radial tuberosity	Flexion, assists with supination	Musculocutaneous $(C_5–C_7)$
(Long head)	Upper rim of glenoid fossa			
(Short head)	Coracoid process of scapula			
Brachioradialis	Upper two-thirds of lateral supracondylar ridge of humerus	Styloid process of radius	Flexion, pronation from supinated position to neutral, supination from pronated position to neutral	Radial (C_5, C_6)
Brachialis	Anterior lower half of humerus	Anterior coronoid process of ulna	Flexion	Musculocutaneous (C_5, C_6)
Pronator teres		Lateral midpoint of radius	Pronation, assists with flexion	Median (C_6, C_7)
(Humeral head)	Medial epicondyle of humerus			
(Ulnar head)	Coronoid process of ulna			
Pronator quadratus	Lower fourth of anterior ulna	Lower fourth of anterior radius	Pronation	Anterior interosseous (C_8, T_1)
Triceps brachii		Olecranon process of ulna	Extension	Radial $(C_6–C_8)$
(Long head)	Just inferior to glenoid fossa			
(Lateral head)	Upper half of posterior humerus			
(Medial head)	Lower two-thirds of posterior humerus			
Anconeus	Posterior, lateral epicondyle of humerus	Lateral olecranon and posterior ulna	Assists with extension	Radial (C_7, C_8)
Supinator	Lateral epicondyle of humerus and adjacent ulna	Lateral upper third of radius	Supination	Posterior interosseou (C_5, C_6)

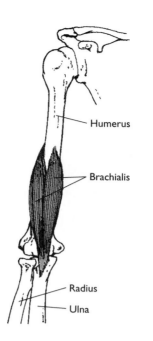

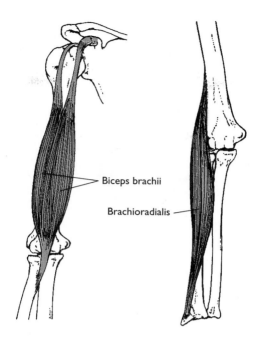

FIGURE 7-21

The major flexor muscles of the elbow.

the distal attachment of the brachialis is the coronoid process of the ulna, the muscle is equally effective when the forearm is in supination or pronation.

Another elbow flexor is the biceps brachii, with both long and short heads attached to the radial tuberosity by a single common tendon. The muscle contributes effectively to flexion when the forearm is supinated, because it is slightly stretched. When the forearm is pronated, the muscle is less taut and consequently less effective.

The brachioradialis is a third contributor to flexion at the elbow. This muscle is most effective when the forearm is in a neutral position (midway between full pronation and full supination) because of its distal attachment to the base of the styloid process on the lateral radius. In this position, the muscle is in slight stretch, and the radial attachment is centered in front of the elbow joint.

The major extensor of the elbow is the triceps, which crosses the posterior aspect of the joint (Figure 7-22). Although the three heads have separate proximal attachments, they attach to the olecranon process of the ulna through a common distal tendon. Even though the distal attachment is relatively close to the axis of rotation at the elbow, the size and strength of the muscle make it effective as an elbow extensor. The relatively small anconeus muscle, which courses from the posterior surface of the lateral epicondyle of the humerus to the lateral olecranon and posterior proximal ulna, also assists with extension. Research has shown that the lateral and medial heads of the triceps contribute 70–90% of the extension moment, with approximately 15% contributed by the anconeus (83). The long head of the triceps, crossing both the elbow and the shoulder, was found to contribute significantly less than the other muscles (83).

Pronation and Supination

Pronation and supination of the forearm involve rotation of the radius around the ulna. There are three radioulnar articulations: the proximal, middle, and distal radioulnar joints. Both the proximal and the distal

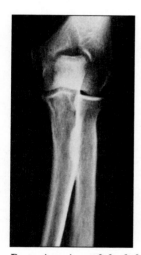

Posterior view of the left elbow. From Shier, Butler, and Lewis, Hole's Human Anatomy and Physiology, © 1996. Reprinted by permission of The McGraw-Hill Companies, Inc.

FIGURE 7-22

The major extensor muscles of the elbow.

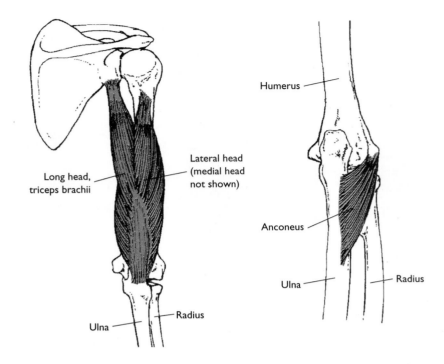

joints are pivot joints, and the middle radioulnar joint is a syndesmosis at which an elastic, interconnecting membrane permits supination and pronation but prevents longitudinal displacement of the bones.

The major pronator is the pronator quadratus, which attaches to the distal ulna and radius (Figure 7-23). When pronation is resisted or rapid, the pronator teres crossing the proximal radioulnar joint assists.

As the name suggests, the supinator is the muscle primarily responsible for supination (Figure 7-24). It is attached to the lateral epicondyle of the humerus and to the lateral proximal third of the radius. When the elbow is in flexion, tension in the supinator lessens, and the biceps assists with supination. When the elbow is flexed to 90° or less, the biceps is positioned to serve as a supinator.

FIGURE 7-23

The major pronator muscle is the pronator quadratus.

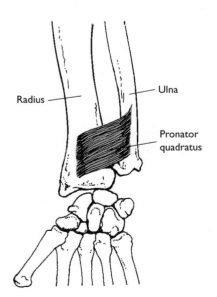

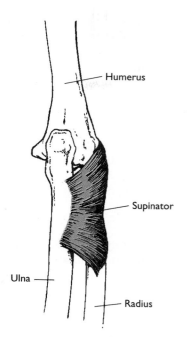

FIGURE 7-24

The major supinator muscle is the supinator.

LOADS ON THE ELBOW

Although the elbow is not considered to be a weight-bearing joint, it regularly sustains large loads during daily activities. For example, research shows that the compressive load at the elbow reaches an estimated 300 N (67 lb) during activities such as dressing and eating, 1700 N (382 lb) when the body is supported by the arms when rising from a chair, and 1900 N (427 lb) when the individual is pulling a table across the floor (34). Even greater loads are present during the execution of selected sport skills. During baseball pitching, the elbow undergoes a valgus torque of as much as 64 N-m, with muscle force as large as 1000 N required to prevent dislocation (47). The amount of valgus torque generated is most closely related to the pitcher's body weight (66). During the execution of gymnastic skills such as the handspring and the vault, the elbow functions as a weight-bearing joint. Research indicates that maximal isometric flexion when the elbow is fully extended can produce joint compression forces of as much as two times body weight (29).

Since the attachment of the triceps tendon to the ulna is closer to the elbow joint center than the attachments of the brachialis on the ulna and the biceps on the radius, the extensor moment arm is shorter than the flexor moment arm. This means that the elbow extensors must generate more force than the elbow flexors to produce the same amount of joint torque. This translates to larger compression forces at the elbow during extension than during flexion when movements of comparable speed and force requirements are executed. Sample Problem 7.2 illustrates the relationship between moment arm and torque at the elbow.

Because of the shape of the olecranon process, the triceps moment arm also varies with the position of the elbow. As shown in Figure 7-25, the triceps moment arm is larger when the arm is fully extended than when it is flexed past 90°.

SAMPLE PROBLEM 7.2

How much force must be produced by the brachioradialis and biceps (F_m) to maintain the 15 N forearm and hand in the position shown below, given moment arms of 5 cm for the muscles and 15 cm for the forearm/hand weight? What is the magnitude of the joint reaction force?

Known

$$wt = 15 \text{ N}$$
$$d_{wt} = 15 \text{ cm}$$
$$d_m = 5 \text{ cm}$$

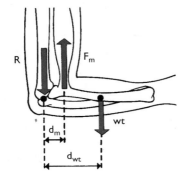

Solution

The torque at the elbow created by the muscle force must equal the torque at the elbow created by forearm/hand weight, yielding a net elbow torque of zero.

$$\sum T_e = 0$$
$$\sum T_e = (F_m)(d_m) - (wt)(d_{wt})$$
$$0 = (F_m)(5 \text{ cm}) - (15 \text{ N})(15 \text{ cm})$$

$$F_m = \frac{(15 \text{ N})(15 \text{ cm})}{5 \text{ cm}}$$

$$\boxed{F_m = 45 \text{ N}}$$

Since the arm is stationary, the sum of all of the acting vertical forces must be equal to zero. In writing the force equation, it is convenient to regard upward as the positive direction.

$$\sum F_v = 0$$
$$\sum F_v = F_m - wt - R$$
$$\sum F_v = 45 \text{ N} - 15 \text{ N} - R$$

$$\boxed{R = 30 \text{ N}}$$

FIGURE 7-25

Because of the shape of the olecranon process of the ulna, the moment arm for the triceps tendon is shorter when the elbow is in flexion.

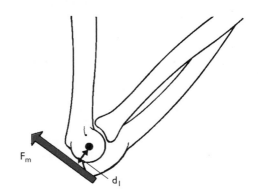

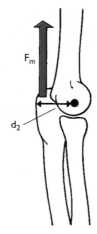

COMMON INJURIES OF THE ELBOW

Although the elbow is a stable joint reinforced by large, strong ligaments, the large loads placed on the joint during daily activities and sport participation render it susceptible to dislocations and overuse injuries.

Sprains and Dislocations

Forced hyperextension of the elbow can cause posterior displacement of the coronoid process of the ulna with respect to the trochlea of the humerus. Such displacement stretches the ulnar collateral ligament, which may rupture (sprain) anteriorly.

Continued hyperextension of the elbow can cause the distal humerus to slide over the coronoid process of the ulna, resulting in dislocation. Although dislocations of the elbow do not occur as frequently as dislocations of the glenohumeral joint, elbow dislocations occurring in individuals under age 30 are most likely to arise during participation in sports (37). The mechanism involved typically is a fall on an outstretched hand or a forceful, twisting blow. The subsequent stability of a once-dislocated elbow is impaired, particularly if the dislocation was accompanied by humeral fracture or rupture of the ulnar collateral ligament (37). Because of the large number of nerves and blood vessels passing through the elbow, elbow dislocations are of particular concern.

Elbow dislocations in young children age 1–3 are sometimes referred to as "nursemaid's elbow" or "pulled elbow." Adults should avoid lifting or swinging young children by the hands, wrists, or forearms, as this type of injury can result.

Overuse Injuries

With the exception of the knee, the elbow is the joint most commonly affected by overuse injuries (35). Stress injuries to the collagenous tissues at the elbow are progressive. The first symptoms are inflammation and swelling, followed by scarring of the soft tissues. If the condition progresses further, calcium deposits accumulate and ossification of the ligaments ensues.

Lateral epicondylitis involves inflammation or microdamage to the tissues on the lateral side of the distal humerus, including the tendinous attachment of the extensor carpi radialis brevis and possibly that of the extensor digitorum. Although a host of factors may contribute to the development of the condition, overuse of the wrist extensors is cited as a major culprit (26).

Because of the relatively high incidence of lateral epicondylitis among tennis players, the injury is commonly referred to as *tennis elbow*. A reported 30–40% of tennis players develop lateral epicondylitis, with onset typically in players age 35–50 (35). The amount of force to which the lateral aspect of the elbow is subjected during tennis play increases with poor technique and improper equipment. For example, hitting off-center shots and using an overstrung racquet increase the amount of force transmitted to the elbow (26). Activities such as swimming, fencing, and hammering can contribute to lateral epicondylitis as well.

Medial epicondylitis, which has been called *Little Leaguer's elbow*, is the same type of injury to the tissues on the medial aspect of the distal humerus. During pitching, the valgus strain imparted to the medial aspect of the elbow during the initial stage, when the trunk and shoulder are

epicondylitis
inflammation and sometimes microrupturing of the collagenous tissues on either the lateral or the medial side of the distal humerus; believed to be an overuse injury

brought forward ahead of the forearm and hand, contributes to development of the condition. Valgus torque increases with late trunk rotation, reduced external rotation of the throwing shoulder, and increased elbow flexion (1). Medial epicondyle avulsion fractures have also been attributed to forceful terminal wrist flexion during the follow-through phase of the pitch (35). More commonly, however, throwing injuries to the elbow are chronic rather than acute. Injury or stretching of the ulnar collateral ligament can result in valgus instability, which, with repeated valgus overload during repetitive throwing, can provoke the development of bony changes that further exacerbate the problem (9, 19). Although uncommon among athletes in general, valgus instability is seen with a higher incidence in individuals who throw repetitively (32). Proper pitching mechanics in young pitchers can help prevent shoulder and elbow injuries by lowering internal rotation torque on the humerus and reducing the valgus load on the elbow (14).

Medial and lateral epicondylitis occur with about equal frequency in golfers, particularly amateurs (4, 70). Among right-handed golfers, lateral epicondylitis occurs more often on the left side, and medial epicondylitis is found more often on the right side (4). Lateral epicondylitis may be related to gripping the club with excessive pronation of the right hand, while medial epicondylitis appears to be associated with repeatedly striking the ground with the club (4).

STRUCTURE OF THE WRIST

radiocarpal joints
condyloid articulations between the radius and the three carpal bones

●*The radiocarpal joints make up the wrist.*

The wrist is composed of radiocarpal and intercarpal articulations (Figure 7-26). Most wrist motion occurs at the radiocarpal joint, a condyloid joint where the radius articulates with the scaphoid, the lunate, and the triquetrum. The joint allows sagittal plane motions (flexion, extension, and hyperextension) and frontal plane motions (radial deviation and ulnar deviation), as well as circumduction. Its close-packed position is

FIGURE 7-26

The bones of the wrist.

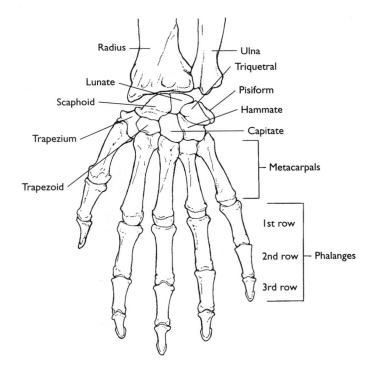

in extension with radial deviation. A cartilaginous disc separates the distal head of the ulna from the lunate and triquetral bones and the radius. Although this articular disc is common to both the radiocarpal joint and the distal radioulnar joint, the two articulations have separate joint capsules. The radiocarpal joint capsule is reinforced by the volar radiocarpal, dorsal radiocarpal, radial collateral, and ulnar collateral ligaments. The intercarpal joints are gliding joints that contribute little to wrist motion.

The fascia around the wrist is thickened into strong fibrous bands called retinacula, which form protective passageways through which tendons, nerves, and blood vessels pass. The flexor retinaculum protects the extrinsic flexor tendons and the median nerves where they cross the palmar side of the wrist. On the dorsal side of the wrist, the extensor retinaculum provides a passageway for the extrinsic extensor tendons.

retinacula
fibrous bands of fascia

MOVEMENTS OF THE WRIST

The wrist is capable of sagittal and frontal plane movements, as well as rotary motion (Figure 7-27). Flexion is motion of the palmar surface of the hand toward the anterior forearm. Extension is the return of the hand to anatomical position, and in hyperextension, the dorsal surface of the hand approaches the posterior forearm. Movement of the hand toward the thumb side of the arm is radial deviation, with movement in the opposite direction designated as ulnar deviation. Movement of the hand through all four directions produces circumduction. Because of the complex structure of the wrist, rotational movements at the wrist are also complex, with different axes of rotation and different mechanisms through which wrist motions occur (55).

Flexion

The muscles responsible for flexion at the wrist are the flexor carpi radialis and the powerful flexor carpi ulnaris (Figure 7-28). The palmaris longus, which is often absent in one or both forearms, contributes to flexion when present. All three muscles have proximal attachments on the medial epicondyle of the humerus. The flexor digitorum superficialis and flexor digitorum profundus can assist with flexion at the wrist when the fingers are completely extended, but when the fingers are in flexion, these muscles cannot develop sufficient tension due to active insufficiency.

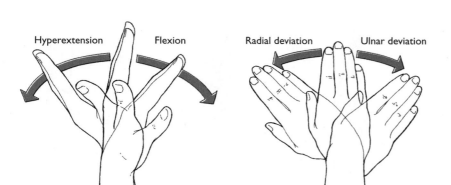

Hyperextension   Flexion   Radial deviation   Ulnar deviation

FIGURE 7-27

Movements occurring at the wrist.

FIGURE 7-28

The major flexor muscles of the wrist.

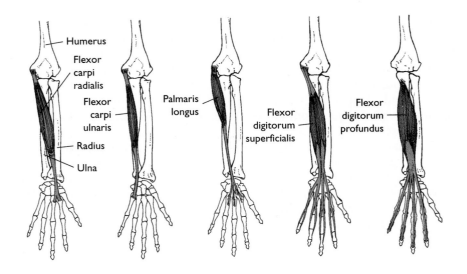

Extension and Hyperextension

Extension and hyperextension at the wrist result from contraction of the extensor carpi radialis longus, extensor carpi radialis brevis, and extensor carpi ulnaris (Figure 7-29). These muscles originate on the lateral epicondyle of the humerus. The other posterior wrist muscles may also assist with extension, particularly when the fingers are in flexion. Included in this group are the extensor pollicis longus, extensor indicis, extensor digiti minimi, and extensor digitorum (Figure 7-30).

Radial and Ulnar Deviation

Cooperative action of both flexor and extensor muscles produces lateral deviation of the hand at the wrist. The flexor carpi radialis and extensor carpi radialis longus and brevis contract to produce radial deviation, and the flexor carpi ulnaris and extensor carpi ulnaris cause ulnar deviation.

FIGURE 7-29

The major extensor muscles of the wrist.

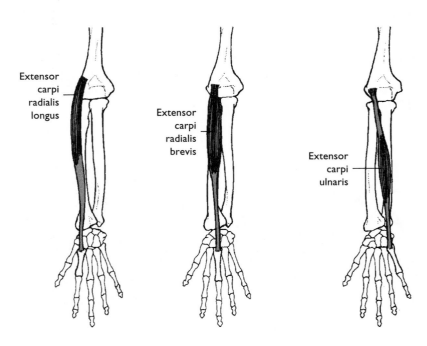

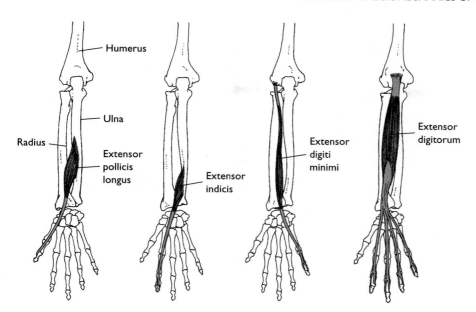

FIGURE 7-30

Muscles that assist with extension of the wrist.

STRUCTURE OF THE JOINTS OF THE HAND

A large number of joints are required to provide the extensive motion capabilities of the hand. Included are the carpometacarpal (CM), inter-metacarpal, metacarpophalangeal (MP), and interphalangeal (IP) joints (Figure 7-31). The fingers are referred to as digits one through five, with the first digit being the thumb.

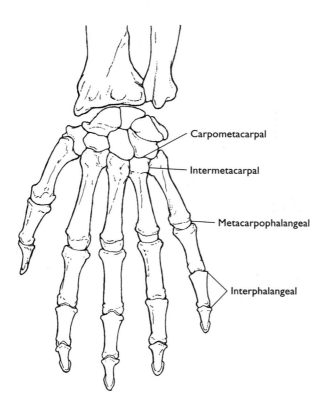

FIGURE 7-31

The bones of the hand.

Carpometacarpal and Intermetacarpal Joints

The carpometacarpal (CM) joint of the thumb, the articulation between the trapezium and the first metacarpal, is a classic saddle joint. The other CM joints are generally regarded as gliding joints. All carpometacarpal joints are surrounded by joint capsules, which are reinforced by the dorsal, volar, and interosseous CM ligaments. The irregular intermetacarpal joints share these joint capsules.

Metacarpophalangeal Joints

The metacarpophalangeal (MP) joints are the condyloid joints between the rounded distal heads of the metacarpals and the concave proximal ends of the phalanges. These joints form the knuckles of the hand. Each joint is enclosed in a capsule that is reinforced by strong collateral ligaments. A dorsal ligament also merges with the MP joint of the thumb. Close-packed positions of the MP joints in the fingers and thumb are full flexion and opposition, respectively.

FIGURE 7-32

Movements of the thumb.

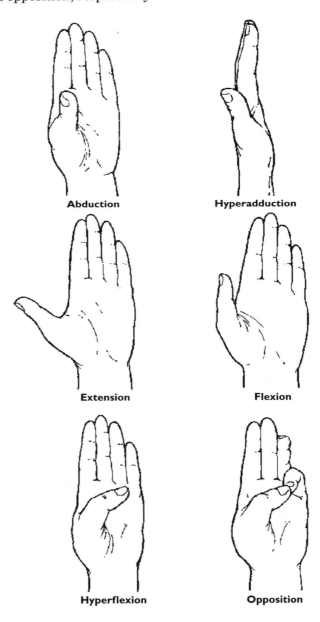

Interphalangeal Joints

The proximal and distal interphalangeal (IP) joints of the fingers and the single interphalangeal joint of the thumb are all hinge joints. An articular capsule joined by volar and collateral ligaments surrounds each IP joint. These joints are most stable in the close-packed position of full extension.

MOVEMENTS OF THE HAND

The carpometacarpal (CM) joint of the thumb allows a large range of movement similar to that of a ball-and-socket joint (Figure 7-32). Motion at CM joints two through four is slight due to constraining ligaments, with somewhat more motion permitted at the fifth CM joint.

The metacarpophalangeal (MP) joints of the fingers allow flexion, extension, abduction, adduction, and circumduction for digits two through five, with abduction defined as movement away from the middle finger and adduction being movement toward the middle finger (Figure 7-33). Because the articulating bone surfaces at the metacarpophalangeal joint of the thumb are relatively flat, the joint functions more as a hinge joint, allowing only flexion and extension.

The interphalangeal (IP) joints permit flexion and extension, and in some individuals, slight hyperextension. These are classic hinge joints. Due to passive tension in the extrinsic muscles, when the hand is relaxed and the wrist moves from full flexion to full extension, the distal interphalangeal joints go from approximately 12° to 31° of flexion, and the proximal IP joints go from about 19° to 70° of flexion (73).

A relatively large number of muscles are responsible for the many precise movements performed by the hand and fingers (Table 7-3). There

•The large range of movement of the thumb compared to that of the fingers is derived from the structure of the thumb's carpometacarpal joint.

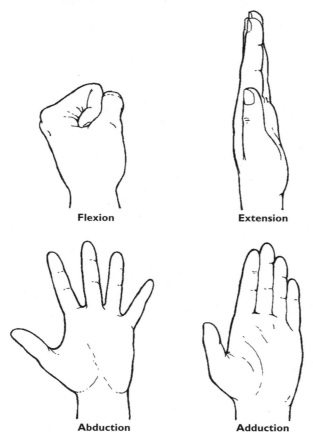

Flexion　　　**Extension**

Abduction　　**Adduction**

FIGURE 7-33

Movements of the fingers.

TABLE 7-3

Major Muscles of the Hand and Fingers

MUSCLE	PROXIMAL ATTACHMENT	DISTAL ATTACHMENT	PRIMARY ACTION(S)	INNERVATION
EXTRINSIC MUSCLES				
Extensor pollicis longus	Middle dorsal ulna	Dorsal distal phalanx of thumb	Extension at MP and IP joints of thumb, adduction at MP joint of thumb	Radial (C_7, C_8)
Extensor pollicis brevis	Middle dorsal radius	Dorsal proximal phalanx of thumb	Extension at MP and CM joints of thumb	Radial (C_7, C_8)
Flexor pollicis longus	Middle palmar radius	Palmar distal phalanx of thumb	Flexion at IP and MP joints of thumb	Median (C_8, T_1)
Abductor pollicis longus	Middle dorsal ulna and radius	Radial base of 1st metacarpal	Abduction at CM joint of thumb	Radial (C_7, C_8)
Extensor indicis	Distal dorsal ulna	Ulnar side of extensor digitorum tendon	Extension at MP joint of 2nd digit	Radial (C_7, C_8)
Extensor digitorum	Lateral epicondyle of humerus	Base of 2nd and 3rd phalanges, digits 2–5	Extension at MP, proximal and distal IP joints, digits 2–5	Radial (C_7, C_8)
Extensor digiti minimi	Proximal tendon of extensor digitorum	Tendon of extensor digitorum distal to 5th MP joint	Extension at 5th MP joint	Radial (C_7, C_8)
Flexor digitorum profundus	Proximal three-fourths of ulna	Base of distal phalanx, digits 2–5	Flexion at distal and proximal IP joints and MP joints, digits 2–5	Ulnar and median (C_8, T_1)
Flexor digitorum superficialis	Medial epicondyle of humerus	Base of middle phalanx, digits 2–5	Flexion at proximal IP and MP joints, digits 2–5	Median (C_7, C_8, T_1)
INTRINSIC MUSCLES				
Flexor pollicis brevis	Ulnar side, 1st metacarpal	Ulnar, palmar base of proximal phalanx of thumb	Flexion and adduction at MP joint of thumb	Median (C_8, T_1)
Abductor pollicis brevis	Trapezium and scaphoid bones	Radial base of 1st phalanx of thumb	Abduction at 1st CM joint	Median (C_8, T_1)
Opponens pollicis	Schaphoid bone	Radial side of 1st metacarpal	Flexion and abduction at CM joint of thumb	Median (C_8, T_1)
Adductor pollicis	Capitate, distal 2nd and 3rd metacarpals	Ulnar proximal phalanx of thumb	Adduction and flexion at CM joint of thumb	Ulnar (C_8, T_1)
Abductor digiti minimi	Pisiform bone	Ulnar base of proximal phalanx, 5th digit	Abduction and flexion at 5th MP joint	Ulnar (C_8, T_1)
Flexor digiti minimi brevis	Hamate bone	Ulnar base of proximal phalanx, 5th digit	Flexion at 5th MP joint	Ulnar (C_8, T_1)
Opponens digiti minimi	Hamate bone	Ulnar metacarpal of 5th metacarpal	Opposition at 5th CM joint	Ulnar (C_8, T_1)
Dorsal interossei (four muscles)	Sides of metacarpals, all digits	Base of proximal phalanx, all digits	Abduction at 2nd and 4th MP joints, radial and ulnar deviation at 3rd MP joint, flexion at MP joints 2–4	Ulnar (C_8, T_1)
Palmar interossei (three muscles)	2nd, 4th, and 5th metacarpals	Base of proximal phalanx, digits 2, 4, and 5	Adduction and flexion at MP joints, digits 2, 4, and 5	Ulnar (C_8, T_1)
Lumbricales (four muscles)	Tendons of flexor digitorum profundus, digits 2–5	Tendons of extensor digitorum, digits 2–5	Flexion at MP joints of digits 2–5	Median and ulnar (C_8, T_1)

are 9 extrinsic muscles that cross the wrist and 10 intrinsic muscles with both of their attachments distal to the wrist.

The extrinsic flexor muscles of the hand are more than twice as strong as the strongest of the extrinsic extensor muscles (78). This should come as little surprise, given that the flexor muscles of the hand are used extensively in everyday activities involving gripping, grasping, or pinching movements, while the extensor muscles rarely exert much force. Multidirectional force measurement for the index finger shows the highest force production in flexion with forces generated in extension, abduction, and adduction being about 38%, 98%, and 79%, respectively, of the flexion force (45). The strongest of the extrinsic flexor muscles are the flexor digitorum profundus and the flexor digitorum superficialis, collectively contributing over 80% of all flexion force (46).

extrinsic muscles
muscles with proximal attachments located proximal to the wrist and distal attachments located distal to the wrist

intrinsic muscles
muscles with both attachments distal to the wrist

COMMON INJURIES OF THE WRIST AND HAND

The hand is used almost continuously in daily activities and in many sports. Wrist sprains or strains are fairly common and are occasionally accompanied by dislocation of a carpal bone or the distal radius. These types of injuries often result from the natural tendency to sustain the force of a fall on the hyperextended wrist. It has been shown that falls from heights greater than 0.6 m can readily result in wrist fracture, since the peak force sustained exceeds the average fracture force for the distal radius (10). Fracture of the distal radius is the most common type of fracture in the population under 75 years of age and is second only to vertebral fractures among the elderly (10). Fractures of the scaphoid and lunate bones are relatively common for the same reason.

Certain hand/wrist injuries are characteristic of participation in a given sport. Examples are metacarpal (boxer's) fractures and mallet or drop finger deformity resulting from injury at the distal interphalangeal joints among football receivers and baseball catchers. Forced abduction of the thumb leading to ulnar collateral ligament injury often results from wrestling, football, hockey, and skiing (36). The most common injuries encountered in skateboarding and snowboarding are fractures of or close to the wrist (17). In sport rock climbing, 62% of all injuries are to the elbow, forearm, wrist, and hand, with many injuries specific to the handholds employed (30).

The wrist is the most frequently injured joint among golfers, with right-handed golfers tending to injure the left wrist (4). Both overuse injuries such as De Quervains disease (tendinitis of the extensor pollicis brevis and the abductor pollicis longus) and impact-related injuries are common (54). According to one study, golfers with overuse injuries of the wrist use a larger-than-average range of motion of the wrists during the swing (5).

Carpal tunnel syndrome is a fairly common disorder. The carpal tunnel is a passageway between the carpal bones and the flexor retinaculum on the palmar side of the wrist. Although the cause of this disorder in a given individual is often unknown, any swelling caused by acute or chronic trauma in the region can compress the median nerve, which passes through the carpal tunnel, thus bringing on the syndrome. Tendon and nerve movement during prolonged repetitive hand movement and incursion of the flexor muscles into the carpal tunnel during wrist extension have been hypothesized as causes for carpal tunnel syndrome (39, 75). Symptoms include pain and numbness along the median nerve, clumsiness of finger function, and eventually weakness and atrophy of the muscles supplied by the median nerve. Workers at tasks requiring large handgrip forces, repetitive movements, or use of vibrating tools are particularly susceptible to carpal tunnel syndrome (69, 76). Likewise, office workers who repeatedly rest the arms on the palmar surface of the wrists are vulnerable. Research indicates that modifications in keyboard design can dramatically affect tendon motion at the wrist, with promising implications for reducing the incidence of overuse injuries (74). The goal of workstation modifications in preventing carpal tunnel syndrome is enabling work with the wrist in neutral position (22, 44). Carpal tunnel syndrome has also been reported among athletes in badminton, baseball, cycling, gymnastics, field hockey, racquetball, rowing, skiing, squash, tennis, and rock climbing (16).

SUMMARY

The shoulder is the most complex joint in the human body, with four different articulations contributing to movement. The glenohumeral joint is a loosely structured ball-and-socket joint in which range of movement is substantial and stability is minimal. The sternoclavicular joint enables some movement of the bones of the shoulder girdle, clavicle, and scapula. Movements of the shoulder girdle contribute to optimal positioning of the glenohumeral joint for different humeral movements. Small movements are also provided by the acromioclavicular and coracoclavicular joints.

The humeroulnar articulation controls flexion and extension at the elbow. Pronation and supination of the forearm occur at the proximal and distal radioulnar joints.

The structure of the condyloid joint between the radius and the three carpal bones controls motion at the wrist. Flexion, extension, radial flexion, and ulnar flexion are permitted. The joints of the hand at which most movements occur are the carpometacarpal joint of the thumb, the metacarpophalangeal joints, and the hinges at the interphalangeal articulations.

INTRODUCTORY PROBLEMS

1. Construct a chart listing all muscles crossing the glenohumeral joint according to whether they are superior, inferior, anterior, or posterior to the joint center. Note that some muscles may fall into more than one category. Identify the action or actions performed by muscles in each of the four categories.
2. Construct a chart listing all muscles crossing the elbow joint according to whether they are medial, lateral, anterior, or posterior to the joint center with the arm in anatomical position. Note that some muscles may fall into more than one category. Identify the action or actions performed by muscles in each of the four categories.
3. Construct a chart listing all muscles crossing the wrist joint according to whether they are medial, lateral, anterior, or posterior to the joint center with the arm in anatomical position. Note that some muscles may fall into more than one category. Identify the action or actions performed by muscles in each of the four categories.
4. List the muscles that develop tension to stabilize the scapula during each of the following activities:
   a. Carrying a suitcase
   b. Water-skiing
   c. Performing a push-up
   d. Performing a pull-up

5. List the muscles used as agonists, antagonists, stabilizers, and neutralizers during the performance of a push-up.
6. Explain how the use of an overhand as compared to an underhand grip affects an individual's ability to perform a pull-up.
7. Select a familiar activity and identify the muscles of the upper extremity that are used as agonists during the activity.
8. Using the diagram in Sample Problem 7.1 as a model, calculate the tension required in the deltoid with a moment arm of 3 cm from the shoulder, given the following weights and moment arms for the upper arm (u), forearm (f), and hand (h) segments: wt_u = 19 N, wt_f = 11 N, wt_h = 4 N, d_u = 12 cm, d_f = 40 cm, d_h = 64 cm. (Answer: 308 N)
9. Which of the three segments in Problem 8 creates the largest torque about the shoulder when the arm is horizontally extended? Explain your answer and discuss the implications for positioning the arm for shoulder-level tasks.
10. Solve Sample Problem 7.2 with the addition of a 10 kg bowling ball held in the hand at a distance of 35 cm from the elbow. (Remember that kg is a unit of mass, not weight!) (Answer: F_m = 732 N, R5619 N)

ADDITIONAL PROBLEMS

1. Identify the sequence of movements that occur at the scapulothoracic, shoulder, elbow, and wrist joints during the performance of an overhand throw.
2. Which muscles are most likely to serve as agonists to produce the movements identified in your answer to Problem 1?
3. Select a familiar racket sport and identify the sequence of movements that occur at the shoulder, elbow, and wrist joints during the execution of forehand and backhand strokes.
4. Which muscles are most likely to serve as agonists to produce the movements identified in your answer to Problem 3?
5. Select five resistance-based exercises for the upper extremity, and identify which muscles are the primary movers and which muscles assist during the performance of each exercise.
6. Discuss the importance of the rotator cuff muscles as stabilizers of the glenohumeral joint and movers of the humerus.
7. Discuss possible mechanisms of rotator cuff injury. Include in your discussion the implications of the force–length relationship for muscle (described in Chapter 6).
8. How much tension (F_m) must be supplied by the triceps to stabilize the arm against an external force (F_e) of 200 N, given d_m = 2 cm and d_e = 25 cm? What is the magnitude of the joint reaction force (R)? (Since the forearm is vertical, its weight does not produce torque at the elbow.) (Answer: F_m = 2500 N, R = 2700 N)

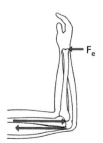

9. What is the length of the moment arm between the dumbbell and the shoulder when the extended 50 cm arm is positioned at a 60° angle? (Answer: 43.3 cm)

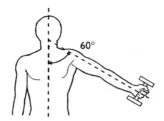

10. The medial deltoid attaches to the humerus at an angle of 15°. What are the sizes of the rotary and stabilizing components of muscle force when the total muscle force is 500 N? (Answer: rotary component = 129 N, stabilizing component = 483 N)

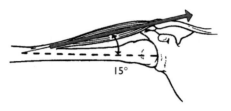

LABORATORY EXPERIENCES

1. Study anatomical models of the shoulder, elbow, and wrist. Be able to locate and identify the major bones, muscle attachments, and ligaments. (The bones are also viewable on the *Dynamic Human* CD, Skeletal System, 3-D Viewer: Thoracic.)

Bones articulating at the glenohumeral joint: _____

_____

Bones articulating at the sternoclavicular joint: _____

_____

Bones articulating at the acromioclavicular joint: _____

_____

Bones articulating at the coracoclavicular joint: _____

_____

Ligaments crossing the glenohumeral joint: _____

_____

Ligaments crossing the sternoclavicular joint: _____

_____

Ligaments crossing the acromioclavicular joint: _____

_____

Ligaments crossing the coracoclavicular joint: _____

_____

Muscles crossing the glenohumeral joint: _____

_____

_____

_____

_____

_____

2. With a partner, use a goniometer to measure wrist range of motion (ROM) in flexion and hyperextension, with the fingers both completely flexed and completely extended. Explain your results.

Flexion ROM with fingers in full flexion: _____

Hyperextension ROM with fingers in full flexion: _____

Flexion ROM with fingers in full extension: _____

Hyperextension ROM with fingers in full extension: _____

Explanation: _____

_____

_____

_____

_____

_____

_____

3. Perform pull-ups using an overhand grip and an underhand grip. Explain which is easier in terms of muscle function.

Which is easier? _____

Explanation: _____

_____

_____

_____

_____

_____

_____

4. Perform push-ups using wide, medium, and narrow hand placements. Explain which is easiest and which is hardest in terms of muscle function.

Which is easiest? _____

Which is hardest? _____

Explanation: _____

_____

_____

_____

_____

_____

5. With a partner, use a goniometer to measure both active and passive ROM at the shoulders of the dominant and nondominant arms in flexion, hyperextension, abduction, and horizontal abduction. Explain your results.

Dominant arm active ROM in flexion: _____

Dominant arm active ROM in hyperextension: _____

Dominant arm active ROM in abduction: _____

Dominant arm active ROM in horizontal abduction: _____

Dominant arm passive ROM in flexion: _____

Dominant arm passive ROM in hyperextension: _____

Dominant arm passive ROM in abduction: _____

Dominant arm passive ROM in horizontal abduction: _____

Nondominant arm active ROM in flexion: _____

Nondominant arm active ROM in hyperextension: _____

Nondominant arm active ROM in abduction: _____

Nondominant arm active ROM in horizontal abduction: _____

Nondominant arm passive ROM in flexion: _____

Nondominant arm passive ROM in hyperextension: _____

Nondominant arm passive ROM in abduction: _____

Nondominant arm passive ROM in horizontal abduction: _____

Explanation: _____

_____

_____

_____

_____

_____

_____

REFERENCES

1. Aguinaldo AL and Chambers H: Correlation of throwing mechanics with elbow valgus load in adult baseball pitchers, *Am J Sports Med,* 37:2043, 2009.

2. Alcid JG, Dugas JR, Wolf RS, and Andrews JR: Elbow injuries in the paediatric athlete, *Sports Med* 30:117, 2000.

3. Allegrucci M, Whitney SL, and Irrgang JJ: Clinical implications of secondary impingement of the shoulder in freestyle swimmers, *J Orthop Sports Phys Ther* 20:307, 1994.

4. Batt ME: Golfing injuries: an overview, *Sports Med* 16:64, 1993.

5. Calahan TD, et al: Biomechanics of the golf swing in players with pathological conditions of the forearm, wrist, and hand, *Am J Sports Med* 19:288, 1991.

6. Callanan M, Tzannes A, Hayes K, Paxinos A, Walton J, and Murrell GA: Shoulder instability: diagnosis and management, *Aust Fam Physician* 30:655, 2001.

7. Chaffin DB and Andersson GBJ: *Occupational biomechanics* (2nd ed), New York, 1991, John Wiley & Sons.

8. Chang YW, Highes RE, Su FC, Itoi E, and An KN: Prediction of muscle force involved in shoulder internal rotation, *J Shoulder Elbow Surg* 9:188, 2000.

9. Chen FS, Rokito AS, and Jobe FW: Medial elbow problems in the overhead-throwing athlete, *J Am Acad Orthop Surg* 9:99, 2001.

10. Chiu J and Robinovitvh SN: Prediction of upper extremity impact forces during falls on the outstretched hand, *J Biomech* 31:1169, 1998.

11. Colas F, Nevoux J, and Gagey O: The subscapular and subcoracoid bursae: Descriptive and functional anatomy, *J Shoulder Elbow Surg* 13:454, 2004.

12. Crosbie J, Kilbreath SL, and Dylke E: The kinematics of the scapulae and spine during a lifting task, *J Biomech* 43:1302, 2010.

13. David G, Magarey ME, Jones MA, Dvir Z, Türker KS, and Sharpe M: EMG and strength correlates of selected shoulder muscles during rotations of the glenohumeral joint, *Clin Biomech* 15:95, 2000.

14. Davis JT, Limpisvasti O, Fluhme D, Mohr KJ, Yocum LA, Elattrache NS, and Jobe FW: The effect of pitching biomechanics on the upper extremity in youth and adolescent baseball pitchers, *Am J Sports Med* 37:1484, 2009.

15. Dayanidhi S, Orlin M, Kozin S, Duff S, and Karduna A: Scapular kinematics during humeral elevation in adults and children, *Clin Biomech* 20:600, 2005.

16. Dimeff RJ: Entrapment neuropathies of the upper extremity, *Curr Sports Med Rep* 2: 255, 2003.

17. Dingerkus ML, Imhoff A, and Hipp E: Snowboard sports technique, injury pattern, prevention, *Fortschr Med* 115:26, 1997.

18. Doody SG, Freedman L, and Waterland JC: Shoulder movement during abduction in the scapular plane, *Arch Phys Med Rehabil* 51:595, 1970.

19. Edelson G, Kunos CA, Vigder F, and Obed E: Bony changes at the lateral epicondyle of possible significance in tennis elbow syndrome, *J Shoulder Elbow Surg* 10:158, 2001.

20. Edelson G and Teitz C: Internal impingement in the shoulder: J Shoulder Elbow Surg 9:308, 2000.

21. Endo K, Yukata K, and Yasui N: Influence of age on scapulo-thoracic orientation, *Clin Biomech* 19:1009, 2004.

22. Fagarasanu M, Kumar S, and Narayan Y: Measurement of angular wrist neutral zone and forearm muscle activity, *Clin Biomech* 19:671, 2004.

23. Ferretti A, Cerullo G, and Russo G: Subscapular neuropathy in volleyball players, *J Bone Joint Surg* 69-A:260, 1987.

24. Goldberg B and Boiardo R: Profiling children for sports participation, *Clin Sports Med* 3:153, 1984.

25. Hagberg M: Shoulder pain pathogenesis. In Hadler N, ed: *Clinical concepts in regional musculoskeletal illness,* Orlando, 1988, Grune and Stratton.

26. Henning EM, Rosenbaum D, and Milani TL: Transfer of tennis racket vibrations onto the human forearm, *Med Sci Sports Exerc* 24:1134, 1992.

27. Hogfors C, et al.: Biomechanical model of the human shoulder joint—II. The shoulder rhythm, *J Biomech* 24:699, 1991.

28. Holibka HH, Holibkova A, Laichman S, and Ruzickova K: Some peculiarities of the rotator cuff muscles, *Biomed Pap Med Fac Univ Palacky Olomouc Czech Repub* 147:233, 2003.

29. Hui FC, Chao EY, and An KN: Muscle and joint forces at the elbow during isometric lifting. *In Abstracts of the 24th annual orthopedic research society,* Dallas, 1978.

30. Holtzhausen LM and Noakes TD: Elbow, forearm, wrist, and hand injuries among sport rock climbers, *Clin J Sport Med* 6:196, 1996.

31. Hurschler C, Wülker N, and Mendila M: The effect of negative intraarticular pressure and rotator cuff force on glenohumeral translation during simulated active elevation, *Clin Biomech* 15:306, 2000.

32. Hyman J, Breazeala NM, and Altchek DW: Valgus instability of the elbow in athletes, *Clin Sports Med* 20:25, 2001.

33. Inman VT, Saunders, and Abbott LC: Observations on the function of the shoulder joint. 1994., *Clin Orthop Relat Res* 330:3, 1996.

34. Jazrawi LM, Rokito AS, Birdzell G, and Zuckerman JD: Biomechanics of the elbow. In Nordin M and Frankel VH, eds: *Basic biomechanics of the musculo-skeletal system* (3rd ed), Philadelphia, 2001, Lippincott Williams & Wilkins.

35. Jobe FW and Nuber G: Throwing injuries of the elbow, *Clin Sports Med* 5:621, 1986.

36. Johnson RE and Rust RJ: Sports related injury: an anatomic approach, part 2, *Minn Med* 68:829, 1985.

37. Josefsson PO and Nilsson BE: Incidence of elbow dislocation, *Acta Orthop Scand* 57:537, 1986.

38. Karduna AR, Williams GR, Illiams JL, and Iannotti JP: Kinematics of the glenohumeral joint: influences of muscle forces, ligamentous constraints, and articular geometry, *J Orthop Res* 14:986, 1996.

39. Keir PJ and Bach JM: Flexor muscle incursion into the carpal tunnel: a mechanism for increased carpal tunnel pressure? *Clin Biomech* 15:301, 2000.

40. Kent BE: Functional anatomy of the shoulder complex: a review, *J Am Phys Ther Assoc* 51:867, 1971.

41. Kon Y, Nishinaka N, Gamada K, Tsutsui H, and Banks SA: The influence of hand-held weight on the scapulohumeral rhythm, *J Shoulder Elbow Surg* 943:17, 2008.

42. Kuechle DK, Newman SR, Itoi E, Niebur GL, Morrey BF, and An KN: The relevance of the moment arm of shoulder muscles with respect to axial rotation of the glenohumeral joint in four positions, *Clin Biomech* 15:322, 2000.

43. Levy AS, Kelly BT, Lintner SA, Osbahr DC, and Speer KP: Function of the long head of the biceps at the shoulder: electromyographic analysis, *J Shoulder Elbow Surg* 10:250, 2001.

44. Li Z-M, Kuxhaus L, Fisk JA, and Christophel TH: Coupling between wrist flexion-extension and radial-ulnar deviation, *Clin Biomech* 20:177, 2005.

45. Li Z-M, Pfaeffle HJ, Sotereanos DG, Goitz RJ, and Woo S L-Y: Multi-directional strength and force envelope of the index finger, *Clin Biomech* 18:908, 2003.

46. Li Z-M, Zatsiorsky VM, and Latash ML: The effect of finger extensor mechanism on the flexor force during isometric tasks, *J Biomech* 34:1097, 2001.

47. Loftis J, Fleisig GS, Zheng N, and Andrews JR: Biomechanics of the elbow in sports, *Clin Sports Med* 23: 519, 2004.

48. Ludewig PM, Reynolds JF: The association of scapular kinematics and gleno-humeral joint pathologies, *J Orthop Sports Phys Ther* 39:90, 2009.

49. Lyons PM and Orwin JF: Rotator cuff tendinopathy and subacromial impingement syndrome, *Med Sci Sports Exerc* 30:S12, 1998.

50. McLeod WD and Andrews JR: Mechanisms of shoulder injuries, *Phys Ther* 66:1901, 1986.

51. Meister K: Internal impingement in the shoulder of the overhand athlete: pathophysiology, diagnosis, and treatment, *Am J Orthop* 29:433, 2000.

52. Michener LA, McClure PW, and Karduna AR: Anatomical and biomechanical mechanisms of subacromial impingement syndrome, *Clin Biomech* 18: 369, 2003.

53. Morrey BF: Applied anatomy and biomechanics of the elbow joint, *Instr Course Lect* 35:59, 1986.

54. Murray PM and Cooney WP: Golf-induced injuries of the wrist, *Clin Sports Med* 15:85, 1996.

55. Neu CP, Crisco JJ, and Wolfe SW: In vivo kinematic behavior of the radio-capitate joint during wrist flexion-extension and radio-ulnar deviation, *J Biomech* 34:1429, 2001.

56. Nuber GW et al: Fine wire electromyography analysis of muscles of the shoulder during swimming, *Am J Sports Med* 14:7, 1986.

57. Paraskevas G, Papadopoulos A, Papaziogas B, Spanidou S, Argiriadou H, and Gigis J: Study of the carrying angle of the human elbow joint in full extension: a morphometric analysis, *Surg Radiol Anat* 26:19, 2004.

58. Perry J and Glousman R: Biomechanics of throwing. In Nicholas JA and Hershman EB, eds: *The upper extremity in sports medicine,* St Louis, 1990, CV Mosby.

59. Plancher KD, Litchfield R, and Hawkins RJ: Rehabilitation of the shoulder in tennis players, *Clin Sports Med* 14:111, 1995.

60. Poncelet E, Demondion X, Lapegue F, Drizenko A, Cotton A, and Francke JP: Anatomic and biometric study of the acromioclavicular joint by ultrasound, *Surg Radiol Anat* 25: 439, 2003.

61. Poppen KN and Walker PS: Normal and abnormal motion of the shoulder, *J Bone Joint Surg* [Am] 58:195, 1976.

62. Prescher A: Anatomical basics, variations, and degenerative changes of the shoulder joint and shoulder girdle, *Eur J Rad* 35:88, 2000.

63. Rafii M et al: Athlete shoulder injuries: CT arthrographic findings, *Radiology* 162:559, 1987.

64. Renstrom P and Johnson RJ: Overuse injuries in sports, *Sports Med* 2:316, 1985.

65. Richardson AB, Jobe FW, and Collins HR: The shoulder in competitive swimming, *Am J Sports Med* 8:159, 1980.

66. Sabick MB, Torry MR, Lawton RL, and Hawkins RJ: Valgus torque in youth baseball pitchers: a biomechanical study, *J Shoulder Elbow Surg* 13:349, 2004.

67. Saha AK: Dynamic stability of the glenohumeral joint, *Acta Orthop Scand* 42:491, 1971.

68. Salter RB: *Textbook of disorders and injuries of the musculoskeletal system* (2nd ed.), Baltimore, 1983, Williams & Wilkins.

69. Shiri R, Miranda H, Heliövaara M, and Viikari-Juntura E: Physical work load factors and carpal tunnel syndrome: a population-based study, *Occup Environ Med* 66:368, 2009.

70. Stockard AR: Elbow injuries in golf, *J Am Osteopath Assoc* 101:509, 2001.

71. Stocker D, Pink M, and Jobe FW: Comparison of shoulder injury in collegiate- and master's-level swimmers, *Clin J Sport Med* 5:4, 1995.

72. Stockdijk M, EIlers PH, Nagels J, and Rozing PM: External rotation in the glenohumeral joint during elevation of the arm, *Clin Biomech* 18:296, 2003.

73. Su F-C, Chou YL, Yang CS, Lin GT, and An KN: Movement of finger joints induced by synergistic wrist motion, *Clin Biomech* 5:491, 2005.

74. Treaster DE and Marras WS: An assessment of alternate keyboards using finger motion, wrist motion and tendon travel, *Clin Biomech* 15:499, 2000.

75. Ugbolue UC, Hsu W-H, Goitz RJ, and Li Z-M: Tendon and nerve displacement at the wrist during finger movements, *Clin Biomech* 20:50, 2005.

76. van Rijn RM, Huisstede BM, Koes BW, Burdorf A: Associations between work-related factors and specific disorders at the elbow: a systematic literature review, *Rheumatology* (Oxford), 48:528, 2009.

77. Vaz S, Soyer J, Preis P, and Clarac JP: Subacromial impingement: influence of coracoacromial arch geometry on shoulder function, *Joint Bone Spine* 67:305, 2000.

78. Von Lanz T and Wachsmuth W: Functional anatomy. In Boyes JH, ed: *Bunnell's surgery of the hand* (5th ed), Philadelphia, 1970, JB Lippincott.

79. Walker PS and Poppen NK: Biomechanics of the shoulder joint during abduction in the plane of the scapula, *Bull Hosp Joint Dis* 38:107, 1977.

80. Wilk KE and Arrigo C: Current concepts in the rehabilitation of the athletic shoulder, *J Orthop Sports Phys Ther* 18:365, 1993.

81. Yani T and Hay JG: Shoulder impingement in front-crawl swimming, II: Analysis of stroking technique, *Med Sci Sports Exerc* 32:30, 2000.

82. Yani T, Hay JG, and Miller GF: Shoulder impingement in front-crawl swimming, I: A method to identify impingement, *Med Sci Sports Exerc* 32:21, 2000.

83. Zang LQ and Nuber GW: Moment distribution among human elbow extensor muscles during isometric and submaximal extension, *J Biomech* 33:145, 2000.

ANNOTATED READINGS

Andrews JR, Wilk KE, and Reinold MM: *The athlete's shoulder,* New York, 2008, Churchill Livingstone.
  Serves as a complete reference guide to athletic injuries of the shoulder.

Fleisig GS, Weber A, Hassell N, and Andrews JR: Prevention of elbow injuries in youth baseball pitchers, *Curr Sports Med Rep,* 8:250, 2009.
  Reviews the research on pitching mechanics as related to elbow injuries in youth baseball pitcers.

Neumann DA: *Kinesiology of the musculoskeletal system: Foundations for rehabilitation,* St. Louis, 2010, Mosby.
  Includes a major section on the upper extremity with separate chapters on the shoulder, elbow and forearm, wrist, and hand, with full color illustrations and detailed descriptions of anatomy, muscle attachments, and innervations.

Robinson P (ed.): *Essential radiology for sports medicine,* New York, 2010, Springer.
  Chapters on the shoulder, elbow, and wrist present detailed anatomical descriptions, common injury mechanisms, and radiographs illustrating these.

RELATED WEBSITES

Arthroscopy.com
http://www.arthroscopy.com/sports.htm
  Includes information and color graphics on the anatomy and function of the upper extremity.

E-hand.com: The Electronic Handbook of Hand Surgery
http://www.eatonhand.com/
  Presents a comprehensive list of links related to anatomy, injury, treatment, and research on the hand, including a clip art image gallery and hand surgery slide lecture archives.

Northern Rockies Orthopaedics Specialists
http://www.orthopaedic.com
  Provides links to pages describing descriptions of injuries, diagnostic tests, and surgical procedures for the shoulder, elbow, wrist, and hand.

Rothman Institute
http://www.rothmaninstitute.com/
  Includes information on common sport injuries to the shoulder and elbow.

Southern California Orthopaedic Institute
http://www.scoi.com
  Includes links to anatomical descriptions and labeled photographs of the shoulder, elbow, wrist, and hand.

University of Washington Orthopaedic Physicians
http://www.orthop.washington.edu
  Provides radiographs and information on common injuries and pathological conditions for the hand and wrist.

The Virtual Hospital: Joint Fluoroscopy
http://www.janela1.com/
  Provides detailed descriptions of anatomical features and includes multiple-view electric joint fluoroscopies of the elbow, lateral elbow, shoulder, and wrist.

The "Virtual" Medical Center: Anatomy & Histology Center
http://www.martindalecenter.com/MedicalAnatomy.html
  Contains numerous images, movies, and course links for human anatomy.

Wheeless' Textbook of Orthopaedics
http://www.wheelessonline.com/
  Provides comprehensive, detailed information, graphics, and related literature for all the joints.

KEY TERMS

acromioclavicular joint	irregular joint between the acromion process of the scapula and the distal clavicle
bursae	sacs secreting synovial fluid internally that lessen friction between soft tissues around joints
coracoclavicular joint	syndesmosis with the coracoid process of the scapula bound to the inferior clavicle by the coracoclavicular ligament
epicondylitis	inflammation and sometimes microrupturing of the collagenous tissues on either the lateral or the medial side of the distal humerus; believed to be an overuse injury
extrinsic muscles	muscles with proximal attachments located proximal to the wrist and distal attachments located distal to the wrist
glenohumeral joint	ball-and-socket joint in which the head of the humerus articulates with the glenoid fossa of the scapula
glenoid labrum	rim of soft tissue located on the periphery of the glenoid fossa that adds stability to the glenohumeral joint
humeroradial joint	gliding joint in which the capitellum of the humerus articulates with the proximal end of the radius
humeroulnar joint	hinge joint in which the humeral trochlea articulates with the trochlear fossa of the ulna
intrinsic muscles	muscles with both attachments distal to the wrist
radiocarpal joints	condyloid articulations between the radius and the three carpal bones
radioulnar joints	the proximal and distal radioulnar joints are pivot joints; the middle radioulnar joint is a syndesmosis
retinacula	fibrous bands of fascia
rotator cuff	band of tendons of the subscapularis, supraspinatus, infraspinatus, and teres minor, which attach to the humeral head
scapulohumeral rhythm	a regular pattern of scapular rotation that accompanies and facilitates humeral abduction
sternoclavicular joint	modified ball-and-socket joint between the proximal clavicle and the manubrium of the sternum

The Biomechanics of the Human Lower Extremity

After completing this chapter, you will be able to:

Explain how anatomical structure affects movement capabilities of lower-extremity articulations.

Identify factors influencing the relative mobility and stability of lower-extremity articulations.

Explain the ways in which the lower extremity is adapted to its weight-bearing function.

Identify muscles that are active during specific lower-extremity movements.

Describe the biomechanical contributions to common injuries of the lower extremity.

ONLINE LEARNING CENTER RESOURCES

www.mhhe.com/hall6e

Log on to our Online Learning Center (OLC) for access to these additional resources:

- Online Lab Manual
- Flashcards with definitions of chapter key terms
- Chapter objectives
- Chapter lecture PowerPoint presentation
- Self-scoring chapter quiz
- Additional chapter resources
- Web links for study and exploration of chapter-related topics

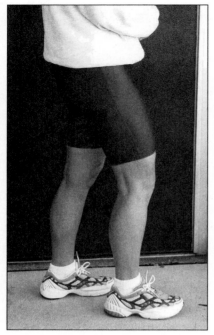

The lower extremity is well structured for its functions of weight bearing and locomotion.

•The hip is inherently more stable than the shoulder because of bone structure and the number and strength of the muscles and ligaments crossing the joint.

A lthough there are some similarities between the joints of the upper and the lower extremities, the upper extremity is more specialized for activities requiring large ranges of motion. In contrast, the lower extremity is well equipped for its functions of weight bearing and locomotion. Beyond these basic functions, activities such as kicking a field goal in football, performing a long jump or a high jump, and maintaining balance *en pointe* in ballet reveal some of the more specialized capabilities of the lower extremity. This chapter examines the joint and muscle functions that enable lower-extremity movements.

STRUCTURE OF THE HIP

The hip is a ball-and-socket joint (Figure 8-1). The ball is the head of the femur, which forms approximately two-thirds of a sphere. The socket is the concave acetabulum, which is angled obliquely in an anterior, lateral, and inferior direction. Joint cartilage covers both articulating surfaces. The cartilage on the acetabulum is thicker around its periphery, where it merges with a rim, or labrum, of fibrocartilage that contributes to the stability of the joint. Hydrostatic pressure is greater within the labrum than outside of it, contributing to lubrication of the joint (31). The acetabulum provides a much deeper socket than the glenoid fossa of the shoulder joint, and the bony structure of the hip is therefore much more stable or less likely to dislocate than that of the shoulder.

Several large, strong ligaments also contribute to the stability of the hip (Figure 8-2). The extremely strong iliofemoral or Y ligament and the pubofemoral ligament strengthen the joint capsule anteriorly, with posterior reinforcement from the ischiofemoral ligament. Tension in these major ligaments acts to twist the head of the femur into the acetabulum during hip extension, as when a person moves from a sitting to a standing position. Inside the joint capsule, the ligamentum teres supplies a direct attachment from the rim of the acetabulum to the head of the femur.

As with the shoulder joint, several bursae are present in the surrounding tissues to assist with lubrication. The most prominent are the iliopsoas

FIGURE 8-1

The bony structure of the hip.

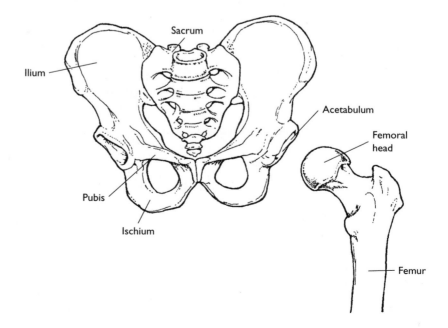

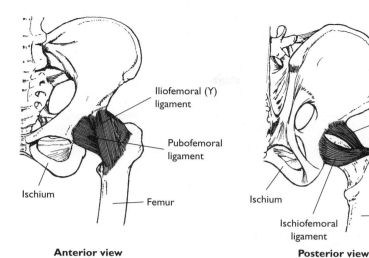

FIGURE 8-2

The ligaments of the hip.

bursa and the deep trochanteric bursa. The iliopsoas bursa is positioned between the iliopsoas and the articular capsule, serving to reduce the friction between these structures. The deep trochanteric bursa provides a cushion between the greater trochanter of the femur and the gluteus maximus at the site of its attachment to the iliotibial tract. The trochanteric bursae vary in number, position, and appearance among older individuals, suggesting that these bursae are formed to lessen friction between the greater trochanter and the gluteus maximus (24).

The femur is a major weight-bearing bone and is the longest, largest, and strongest bone in the body. Its weakest component is the femoral neck, which is smaller in diameter than the rest of the bone, and weak internally because it is primarily composed of trabecular bone. The femur angles medially downward from the hip during the support phase of walking and running, enabling single-leg support beneath the body's center of gravity.

MOVEMENTS AT THE HIP

Although movements of the femur are due primarily to rotation occurring at the hip joint, the pelvic girdle has a function similar to that of the shoulder girdle in positioning the hip joint for effective limb movement. Unlike the shoulder girdle, the pelvis is a single nonjointed structure, but it can rotate in all three planes of movement. The pelvis facilitates movement of the femur by rotating so that the acetabulum is positioned toward the direction of impending femoral movement. For example, posterior pelvic tilt, with the anterior superior iliac spine tilted backward with respect to the acetabulum, positions the head of the femur in front of the hipbone to enable ease of flexion. Likewise, anterior pelvic tilt promotes femoral extension, and lateral pelvic tilt toward the opposite side facilitates lateral movements of the femur. Movement of the pelvic girdle also coordinates with certain movements of the spine (see Chapter 9).

pelvic girdle
the two hip bones plus the sacrum, which can be rotated forward, backward, and laterally to optimize positioning of the hip joint

Muscles of the Hip

A number of large muscles cross the hip, further contributing to its stability. The locations and functions of the muscles of the hip are summarized in Table 8-1.

TABLE 8-1

Muscles of the Hip

MUSCLE	PROXIMAL ATTACHMENT	DISTAL ATTACHMENT	PRIMARY ACTION(S) ABOUT THE HIP	INNERVATION
Rectus femoris	Anterior inferior iliac spine (ASIS)	Patella	Flexion	Femoral (L_2–L_4)
Iliopsoas		Lesser trochanter of femur	Flexion	L_1 and femoral
(Iliacus)	Iliac fossa and adjacent sacrum			(L_2–L_4)
(Psoas)	12th thoracic and all lumbar vertebrae and lumbar discs			(L_1–L_3)
Sartorius	Anterior superior iliac spine	Upper medial tibia	Assists with flexion, abduction, lateral rotation	Femoral (L_2, L_3)
Pectineus	Pectineal crest of pubic ramus	Medial, proximal femur	Flexion, adduction, medial rotation	Femoral (L_2, L_3)
Tensor fascia latae	Anterior crest of ilium and ASIS	Iliotibial band	Assists with flexion, abduction, medial rotation	Superior gluteal (L_4–S_1)
Gluteus maximus	Posterior ilium, iliac crest, sacrum, coccyx	Gluteal tuberosity of femur and iliotibial band	Extension, lateral rotation	Inferior gluteal (L_5–S_2)
Gluteus medius	Between posterior and anterior gluteal lines on posterior ilium	Superior, lateral greater trochanter	Abduction, medial rotation	Superior gluteal (L_4–S_1)
Gluteus minimus	Between anterior and inferior gluteal lines on posterior ilium	Anterior surface of greater trochanter	Abduction, medial rotation	Superior gluteal (L_4–S_1)
Gracilis	Anterior, inferior pubic symphysis	Medial, proximal tibia	Adduction	Obturator (L_3, L_4)
Adductor magnus	Inferior ramus of pubis and ischium	Entire linea aspera	Adduction, lateral rotation	Obturator (L_3, L_4)
Adductor longus	Anterior pubis	Middle linea aspera	Adduction, assists with flexion	Obturator (L_3, L_4)
Adductor brevis	Inferior ramus of the pubis	Upper linea aspera	Adduction, lateral rotation	Obturator (L_3, L_4)
Semitendinosus	Medial ischial tuberosity	Proximal, medial tibia	Extension	Tibial (L_5–S_1)
Semimembranosus	Lateral ischial tuberosity	Proximal, medial tibia	Extension	Tibial (L_5–S_1)
Biceps femoris (long head)	Lateral ischial tuberosity	Posterior lateral condyle of tibia, head of fibula	Extension	Tibial (L_5–S_2)
The six outward rotators	Sacrum, ilium, ischium	Posterior greater trochanter	Outward rotation	(L_5–S_2)

Flexion

The six muscles primarily responsible for flexion at the hip are those crossing the joint anteriorly: the iliacus, psoas major, pectineus, rectus femoris, sartorius, and tensor fascia latae. Of these, the large iliacus and psoas major—often referred to jointly as the iliopsoas because of their common attachment to the femur—are the major hip flexors (Figure 8-3). Other major hip flexors are shown in Figure 8-4. Because the rectus femoris is

iliopsoas
the psoas major and iliacus muscles with a common insertion on the lesser trochanter of the femur

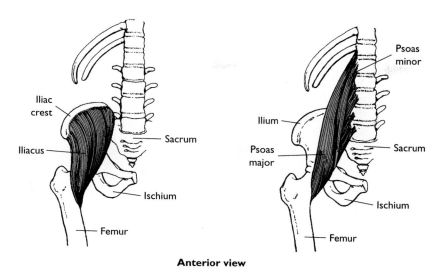

FIGURE 8-3

The iliopsoas complex is the major flexor of the hip.

Anterior view

a two-joint muscle active during both hip flexion and knee extension, it functions more effectively as a hip flexor when the knee is in flexion, as when a person kicks a ball. The thin, straplike sartorius, or *tailor's muscle*, is also a two-joint muscle. Crossing from the superior anterior iliac spine to the medial tibia just below the tuberosity, the sartorius is the longest muscle in the body.

Extension

The hip extensors are the gluteus maximus and the three hamstrings—the biceps femoris, semitendinosus, and semimembranosus (Figure 8-5). The gluteus maximus is a massive, powerful muscle that is usually active only when the hip is in flexion, as during stair climbing or cycling, or when extension at the hip is resisted (Figure 8-6). The hamstrings derive their name from their prominent tendons, which can readily be palpated on the posterior aspect of the knee. These two-joint muscles contribute to both extension at the hip and flexion at the knee, and are active during standing, walking, and running.

hamstrings
the biceps femoris, semimembranosus, and semitendinosus

● *Two-joint muscles function more effectively at one joint when the position of the other joint stretches the muscle slightly.*

FIGURE 8-4

Assistant flexor muscles of the hip.

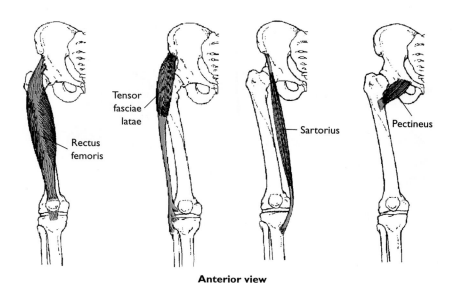

Rectus femoris

Tensor fasciae latae

Sartorius

Pectineus

Anterior view

FIGURE 8-5

The hamstrings are major hip extensors and knee flexors.

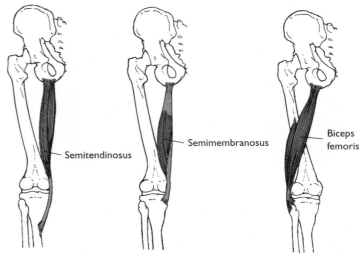

Semitendinosus

Semimembranosus

Biceps femoris

Posterior view

Abduction

The gluteus medius is the major abductor acting at the hip, with the gluteus minimus assisting. These muscles stabilize the pelvis during the support phase of walking and running and when an individual stands on one leg. For example, when body weight is supported by the right foot during walking, the right hip abductors contract isometrically and eccentrically to prevent the left side of the pelvis from being pulled downward by the weight of the swinging left leg. This allows the left leg to move freely through the swing phase. If the hip abductors are too weak to perform this function, then lateral pelvic tilt and dragging of the swing foot occurs with every step during gait. The hip abductors are also active during the performance of dance movements such as the *grande ronde jambe*.

• *Contraction of the hip abductors is required during the swing phase of gait to prevent the swinging foot from dragging.*

Adduction

The hip adductors are those muscles that cross the joint medially and include the adductor longus, adductor brevis, adductor magnus, and gracilis (Figure 8-7). The hip adductors are regularly active during the swing phase of the gait cycle to bring the foot beneath the body's center of gravity for placement during the support phase. The gracilis is a long, relatively weak strap muscle that also contributes to flexion of the leg at the

FIGURE 8-6

The three gluteal muscles.

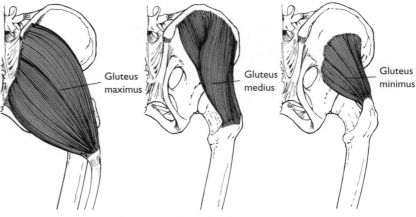

Gluteus maximus

Gluteus medius

Gluteus minimus

Posterior view

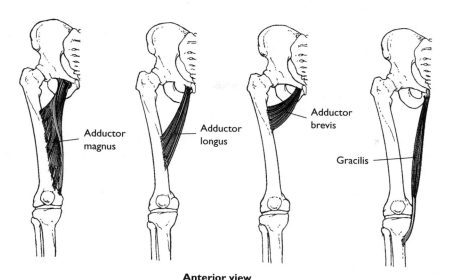

FIGURE 8-7

Adductor muscles of the hip.

Anterior view

knee. The other three adductor muscles also contribute to flexion and lateral rotation at the hip, particularly when the femur is medially rotated.

Medial and Lateral Rotation of the Femur

Although a number of muscles contribute to lateral rotation of the femur, six muscles function solely as lateral rotators. These are the piriformis, gemellus superior, gemellus inferior, obturator internus, obturator externus, and quadratus femoris (Figure 8-8). Although we tend to think of walking and running as involving strictly sagittal plane movement at the joints of the lower extremity, outward rotation of the femur also occurs with every step to accommodate the rotation of the pelvis.

The major medial rotator of the femur is the gluteus minimus, with assistance from the tensor fascia latae, semitendinosus, semimembranosus, and gluteus medius. Medial rotation of the femur is usually not a resisted motion requiring a substantial amount of muscular force. The medial rotators are weak in comparison to the lateral rotators, with the estimated strength of the medial rotators only approximately one-third that of the lateral rotators (57).

• During the gait cycle, lateral and medial rotation of the femur occur in coordination with pelvic rotation.

Horizontal Abduction and Adduction

Horizontal abduction and adduction of the femur occur when the hip is in 90° of flexion while the femur is either abducted or adducted. These actions require the simultaneous, coordinated actions of several muscles. Tension is required in the hip flexors for elevation of the femur. The hip abductors can then produce horizontal abduction and, from a horizontally abducted position, the hip adductors can produce horizontal adduction. The muscles located on the posterior aspect of the hip are more effective

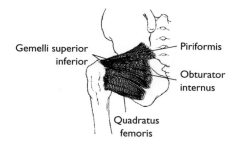

FIGURE 8-8

The lateral rotator muscles of the femur.

as horizontal abductors and adductors than the muscles on the anterior aspect, because the former are stretched when the femur is in 90° of flexion, whereas tension in the anterior muscles is usually reduced with the femur in this position.

LOADS ON THE HIP

The hip is a major weight-bearing joint that is never fully unloaded during daily activities (31). When body weight is evenly distributed across both legs during upright standing, the weight supported at each hip is one-half the weight of the body segments above the hip, or about one-third of total body weight. However, the total load on each hip in this situation is greater than the weight supported, because tension in the large, strong hip muscles further adds to compression at the joint (Figure 8-9).

Because of muscle tension, compression on the hip is approximately the same as body weight during the swing phase of walking (85). Data from instrumented femoral implants show hip joint loading to range around 238% of body weight (BW) during the support phase of walking, with values of approximately 251% BW and 260% BW during climbing and descending stairs, respectively (9). Calculated hip contact forces for running at 3.5 m/s, Alpine skiing on a flat slope, Alpine skiing on a steep slope, cross-country skiing with classical technique, and cross-country skiing with skating technique, are around 520% BW, 410% BW, 780% BW, 400% BW, and 460% BW, respectively (113). During side diving by the goalkeeper in soccer, forces acting on the hip range from 4.2 to 8.6 times body weight (100). Hip loading increases with the wearing of hard-soled as compared to soft-soled shoes (8). Carrying a load such as a suitcase of 25% body weight on one side produces a 167% increase in loading on the contralateral hip as compared to the hip on the loaded side (7).

As gait speed increases, the load on the hip increases during both swing and support phases. Hip loads during jogging can be reduced with a smooth gait pattern and soft heel strikes (8). In summary, body weight, impact forces translated upward through the skeleton from the foot, and muscle tension all contribute to this large compressive load on the hip, as demonstrated in Sample Problem 8.1.

Fortunately, the hip joint is well designed to bear the large loads it habitually sustains. Because the diameter of the humeral head is

• *During the support phase of walking, compression force at the hip reaches a magnitude of three to four times body weight.*

FIGURE 8-9

The major forces on the hip during static stance are the weight of the body segments above the hip (with one-half of the weight on each hip), tension in the hip abductor muscles (F_m), and the joint reaction force (R).

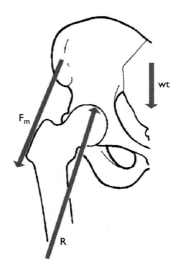

SAMPLE PROBLEM 8.1

How much compression acts on the hip during two-legged standing, given that the joint supports 250 N of body weight and the abductor muscles are producing 600 N of tension?

Known

$$wt = 250 \text{ N}$$
$$F_m = 600 \text{ N}$$

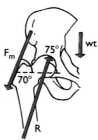

Graphic Solution

Since the body is motionless, all vertical force components must sum to zero and all horizontal force components must sum to zero. Graphically, this means that all acting forces can be transposed to form a closed force polygon (in this case, a triangle). The forces from the diagram of the hip above can be reconfigured to form a triangle.

If the triangle is drawn to scale (perhaps 1 cm = 100 N), the amount of joint compression can be approximated by measuring the length of the joint reaction force (R).

$$R \approx 840 \text{ N}$$

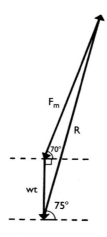

Mathematical Solution

The law of cosines can be used with the same triangle to calculate the length of R.

$$R^2 = F_m{}^2 + wt^2 - 2(F_m)(wt)\cos 160°$$
$$R^2 = 600 \text{ N}^2 + 250 \text{ N}^2 - 2(600 \text{ N})(250 \text{ N})\cos 160°$$
$$R = 839.3 \text{ N}$$

---

somewhat larger than the articulating surface of the acetabulum, contact between the two bones during initiation of weight bearing begins around the periphery. As loading increases, the contact area at the joint also increases, such that stress levels remain approximately constant (83).

Use of a crutch or cane on the side opposite an injured or painful hip is beneficial in that it serves to more evenly distribute the load between the legs throughout the gait cycle. During stance, a support opposite the painful hip reduces the amount of tension required of the powerful abductor muscles, thereby reducing the load on the painful hip. This reduction in load on the painful hip, however, increases the stress on the opposite hip.

COMMON INJURIES OF THE HIP

Fractures

Although the pelvis and the femur are large, strong bones, the hip is subjected to high, repetitive loads ranging from four to seven times body weight during locomotion (57). Fractures of the femoral neck frequently occur during the support phase of walking among elderly individuals with osteoporosis, a

●*Fracture of the femoral neck (broken hip) is a seriously debilitating injury that occurs frequently among elderly individuals with osteoporosis.*

condition of reduced bone mineralization and strength (see Chapter 4). These femoral neck fractures often result in loss of balance and a fall. A common misconception is that the fall always causes the fracture rather than the reverse, which may also be true. Hip fractures in the elderly are a serious health issue, with increased risk of death increasing five- to eight-fold during the first three months following the fracture (41). A high percentage of cortical bone in the proximal femur is protective against hip fractures (17). When the hipbones have good health and mineralization, they can sustain tremendous loads, as illustrated during many weight-lifting events. Research demonstrates that regular physical activity helps protect against the risk of hip fracture (52).

Contusions

The muscles on the anterior aspect of the thigh are in a prime location for sustaining blows during participation in contact sports. The resulting internal hemorrhaging and appearance of bruises vary from mild to severe. A relatively uncommon but potentially serious complication secondary to thigh contusions is acute compartment syndrome, in which internal bleeding causes a build-up of pressure in the muscle compartment causing compression of nerves, blood vessels, and muscle (55). If not treated, this can lead to tissue death due to lack of oxygenation as the blood vessels are compressed by the raised pressure within the compartment.

Strains

Since most daily activities do not require simultaneous hip flexion and knee extension, the hamstrings are rarely stretched unless exercises are performed for that specific purpose. The resulting loss of extensibility makes the hamstrings particularly susceptible to strain. Strains to these muscles most commonly occur during sprinting, particularly if the individual is fatigued and neuromuscular coordination is impaired. Researchers believe that hamstring strains typically occur during the late stance or late swing phases of gait as a result of an eccentric contraction (99, 122). These injuries are troubling for athletes, given their high incidence rate and slowness of healing, and with a recurrence rate of nearly one-third during the first year following return to athletic participation (45). Strains to the groin area are also relatively common among athletes in sports in which forceful thigh abduction movements may overstretch the adductor muscles. A study of professional ice hockey players showed that players were 17 times more likely to sustain adduction sprains if adductor strength was less than 80% of abduction strength (112). This strongly suggests that strengthening of the hip adductors can reduce the likelihood of adduction sprains.

STRUCTURE OF THE KNEE

The structure of the knee permits the bearing of tremendous loads, as well as the mobility required for locomotor activities. The knee is a large synovial joint, including three articulations within the joint capsule. The weight-bearing joints are the two condylar articulations of the tibiofemoral joint, with the third articulation being the patellofemoral joint. Although not a part of the knee, the proximal tibiofibular joint has soft-tissue connections that also slightly influence knee motion.

Tibiofemoral Joint

The medial and lateral condyles of the tibia and the femur articulate to form two side-by-side condyloid joints (Figure 8-10). These joints function

tibiofemoral joint
dual condyloid articulations between the medial and lateral condyles of the tibia and the femur, composing the main hinge joint of the knee

patellofemoral joints
articulation between the patella and the femur

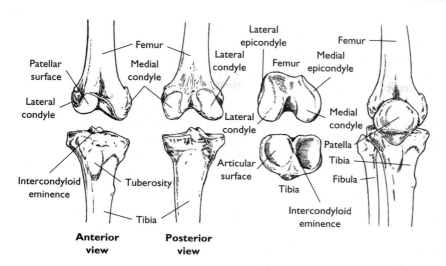

FIGURE 8-10

The bony structure of the tibiofemoral joint.

together primarily as a modified hinge joint because of the restricting ligaments, with some lateral and rotational motions allowed. The condyles of the tibia, known as the *tibial plateaus*, form slight depressions separated by a region known as the *intercondylar eminence*. Because the medial and lateral condyles of the femur differ somewhat in size, shape, and orientation, the tibia rotates laterally on the femur during the last few degrees of extension to produce "locking" of the knee. This phenomenon, known as the "screw-home" mechanism, brings the knee into the close-packed position of full extension. Because the curvatures of the tibial plateau are complex, asymmetric, and vary significantly from individual to individual, some knees are much more stable and resistant to injury than others (43).

●*The bony anatomy of the knee requires a small amount of lateral rotation of the tibia to accompany full extension.*

Menisci

The menisci, also known as semilunar cartilages after their half-moon shapes, are discs of fibrocartilage firmly attached to the superior plateaus of the tibia by the coronary ligaments and joint capsule (Figure 8-11). They are also joined to each other by the transverse ligament. The menisci are thickest at their peripheral borders, where fibers from the joint capsule solidly anchor them to the tibia (5). The medial semilunar disc is also directly attached to the medial collateral ligament. Medially, both menisci taper down to paper thinness, with the inner edges unattached to the bone.

menisci
cartilaginous discs located between the tibial and femoral condyles

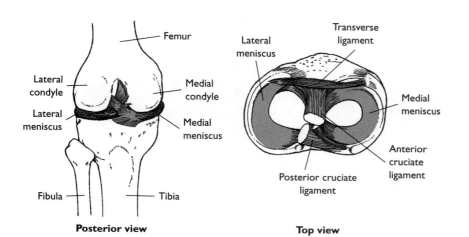

FIGURE 8-11

The menisci of the knee.

The menisci receive a rich supply of both blood vessels and nerves. The blood supply reaches the outer 10–33% of each meniscus, enabling inflammation, repair, and remodeling (39). The outer 66% of each meniscus is innervated, providing proprioceptive information regarding knee position, as well as the velocity and acceleration of knee movements (39).

The menisci deepen the articulating depressions of the tibial plateaus and assist with load transmission and shock absorption at the knee. The internal structure of the medial two-thirds of each meniscus is particularly well suited to resisting compression (5). The stress on the tibiofemoral joint can be an estimated three times higher during load bearing if the menisci have been removed (101). Injured knees, in which part or all of the menisci have been removed, may still function adequately but undergo increased wear on the articulating surfaces, significantly increasing the likelihood of the development of degenerative conditions at the joint. Knee osteoarthritis is frequently accompanied by meniscal tears. Whereas a meniscal tear can lead to the development of osteoarthritis over time, having osteoarthritis can also cause a spontaneous meniscal tear (26).

Ligaments

Many ligaments cross the knee, significantly enhancing its stability (Figure 8-12). The location of each ligament determines the direction in which it is capable of resisting the dislocation of the knee.

The medial and lateral collateral ligaments prevent lateral motion at the knee, as do the collateral ligaments at the elbow. They are also respectively referred to as the *tibial* and *fibular collateral ligaments,* after their distal attachments. Fibers of the medial collateral ligament complex merge with the joint capsule and the medial meniscus to connect the medial epicondyle of the femur to the medial tibia (92). The attachment is just below the pes anserinus, the common attachment of the semitendinosus, semimembranosus, and gracilis to the tibia, thereby positioning the ligament to resist medially directed shear (valgus) and rotational forces acting on the knee. The lateral collateral ligament courses from a few millimeters posterior to the ridge of the lateral epicondyle of the femur to the head of the fibula, contributing to lateral stability of the knee (73).

The anterior and posterior cruciate ligaments limit the forward and backward sliding of the femur on the tibial plateaus during knee flexion

●*The menisci distribute the load at the knee over a larger surface area and also help absorb shock.*

collateral ligaments
major ligaments that cross the medial and lateral aspects of the knee

cruciate ligaments
major ligaments that cross each other in connecting the anterior and posterior aspects of the knee

FIGURE 8-12

The ligaments of the knee.

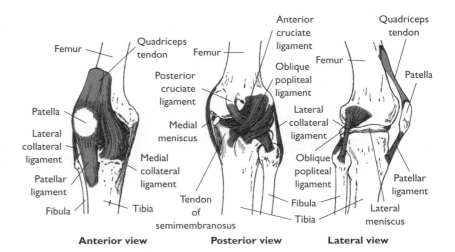

Anterior view   **Posterior view**   **Lateral view**

and extension, and also limit knee hyperextension. The name *cruciate* is derived from the fact that these ligaments cross each other; *anterior* and *posterior* refer to their respective tibial attachments. The anterior cruciate ligament stretches from the anterior aspect of the intercondyloid fossa of the tibia just medial and posterior to the anterior tibial spine in a superior, posterior direction to the posterior medial surface of the lateral condyle of the femur. The shorter and stronger posterior cruciate ligament runs from the posterior aspect of the tibial intercondyloid fossa in a superior, anterior direction to the lateral anterior medial condyle of the femur. These ligaments restrict the anterior and posterior sliding of the femur on the tibial plateaus during knee flexion and extension, and limit knee hyperextension.

Several other ligaments contribute to the integrity of the knee. The oblique and arcuate popliteal ligaments cross the knee posteriorly, and the transverse ligament connects the two semilunar discs internally. Another restricting tissue is the iliotibial band or tract, a broad, thickened band of the fascia lata with attachments to the lateral condyle of the femur and the lateral tubercle of the tibia.

iliotibial band
thick, strong band of tissue connecting the tensor fascia lata to the lateral condyle of the femur and the lateral tuberosity of the tibia

Patellofemoral Joint

The patellofemoral joint consists of the articulation of the triangularly shaped patella, encased in the patellar tendon, with the trochlear groove between the femoral condyles. The posterior surface of the patella is covered with articular cartilage, which reduces the friction between the patella and the femur.

The patella serves several biomechanical functions. Most notably, it increases the angle of pull of the quadriceps tendon on the tibia, thereby improving the mechanical advantage of the quadriceps muscles for producing knee extension by as much as 50% (38). It also centralizes the divergent tension from the quadriceps muscles that is transmitted to the patellar tendon. The patella also increases the area of contact between the patellar tendon and the femur, thereby decreasing patellofemoral joint contact stress. Finally, it also provides some protection for the anterior aspect of the knee and helps protect the quadriceps tendon from friction against the adjacent bones.

● *The patella improves the mechanical advantage of the knee extensors by as much as 50%.*

Joint Capsule and Bursae

The thin articular capsule at the knee is large and lax, encompassing both the tibiofemoral and the patellofemoral joints. A number of bursae are located in and around the capsule to reduce friction during knee movements. The suprapatellar bursa, positioned between the femur and the quadriceps femoris tendon, is the largest bursa in the body. Other important bursae are the subpopliteal bursa, located between the lateral condyle of the femur and the popliteal muscle, and the semimembranosus bursa, situated between the medial head of the gastrocnemius and semi-membranosus tendons.

Three other key bursae associated with the knee, but not contained in the joint capsule, are the prepatellar, superficial infrapatellar, and deep infrapatellar bursae. The prepatellar bursa is located between the skin and the anterior surface of the patella, allowing free movement of the skin over the patella during flexion and extension. The superficial infrapatellar bursa provides cushioning between the skin and the patellar tendon, and the deep infrapatellar bursa reduces friction between the tibial tuberosity and the patellar tendon.

MOVEMENTS AT THE KNEE

Muscles Crossing the Knee

Like the elbow, the knee is crossed by a number of two-joint muscles. The primary actions of the muscles crossing the knee are summarized in Table 8-2.

Flexion and Extension

Flexion and extension are the primary movements permitted at the tibiofemoral joint. For flexion to be initiated from a position of full extension, however, the knee must first be "unlocked." In full extension, the articulating surface of the medial condyle of the femur is longer than that of the lateral condyle, rendering motion almost impossible. The service of locksmith is provided by the popliteus, which acts to medially rotate the tibia with respect to the femur, enabling flexion to occur (Figure 8-13). As flexion proceeds, the femur must slide forward on the tibia to prevent rolling off the tibial plateaus. Likewise, the femur must slide backwards on the tibia during extension. Medial rotation of the tibia and anterior translation of the fibia on the tibial plateau have been shown to be coupled to

popliteus
muscle known as the unlocker of the knee because its action is lateral rotation of the femur with respect to the tibia

TABLE 8-2

Muscles of the Knee

MUSCLE	PROXIMAL ATTACHMENT	DISTAL ATTACHMENT	PRIMARY ACTION(S) ABOUT THE KNEE	INNERVATION
Rectus femoris	Anterior inferior iliac spine (ASIS)	Patella	Extension	Femoral (L_2–L_4)
Vastus lateralis	Greater trochanter and lateral linea aspera	Patella	Extension	Femoral (L_2–L_4)
Vastus intermedius	Anterior femur	Patella	Extension	Femoral (L_2–L_4)
Vastus medialis	Medial linea aspera	Patella	Extension	Femoral (L_2–L_4)
Semitendinosus	Medial ischial tuberosity	Proximal medial tibia at pes	Flexion, medial rotation	Sciatic (L_5–S_2)
Semimembranosus	Lateral ischial tuberosity	Proximal medial tibia	Flexion, medial rotation	Sciatic (L_5–S_2)
Biceps femoris		Posterior lateral condyle of tibia, head of fibula	Flexion, lateral rotation	Sciatic (L_5–S_2)
(Long head)	Ischial tuberosity			
(Short head)	Lateral linea aspera			
Sartorius	Anterior superior iliac spine	Proximal medial tibia at pes	Assists with flexion and lateral rotation of thigh	Femoral (L_2, L_3)
Gracilis	Anterior, inferior symphysis pubis	Proximal medial tibia at pes	Adduction of thigh, flexion of lower leg	Obturator (L_2, L_3)
Popliteus	Lateral condyle of the femur	Posterior medial tibia	Medial rotation, flexion	Tibial (L_4, L_5)
Gastrocnemius	Posterior medial and lateral femoral condyles	Tuberosity of calcaneus via Achilles tendon	Flexion	Tibial (S_1, S_2)
Plantaris	Distal posterior femur	Tuberosity of calcaneus	Flexion	Tibial (S_1, S_2)

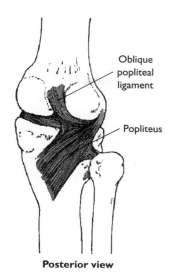

Posterior view

FIGURE 8-13

The popliteus muscle is the locksmith of the knee.

flexion at the knee, even when flexion is passive (22, 36, 56, 120). The exact nature of this coupling can vary between the knees of a given individual and is also influenced by loading at the knee (117). Both the ligaments of the knee and the shapes of the articular surfaces influence the patterns of these coupled motions at the knee (117).

The three hamstring muscles are the primary flexors acting at the knee. Muscles that assist with knee flexion are the gracilis, sartorius, popliteus, and gastrocnemius.

The quadriceps muscles, consisting of the rectus femoris, vastus lateralis, vastus medialis, and vastus intermedius, are the extensors of the knee (Figure 8-14). The rectus femoris is the only one of these muscles that also crosses the hip joint. All four muscles attach distally to the patellar tendon, which inserts on the tibia.

quadriceps
the rectus femoris, vastus lateralis, vastus medialis, and vastus intermedius

Rotation and Passive Abduction and Adduction

Rotation of the tibia relative to the femur is possible when the knee is in flexion and not bearing weight, with rotational capability greatest at approximately 90° of flexion. Tension development in the semimembranosus, semitendinosus, and popliteus produces medial rotation of the tibia,

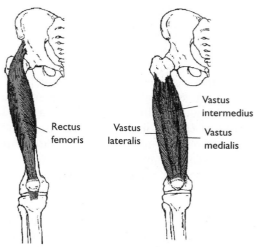

Anterior view

FIGURE 8-14

The quadriceps muscles extend the knee.

with the gracilis and sartorius assisting. The biceps femoris is solely responsible for lateral rotation of the tibia.

A few degrees of passive abduction and adduction are permitted at the knee. Abduction and adduction moments at the knee can also be actively generated by co-contraction of the muscles crossing the medial and lateral aspects of the knee to resist externally applied adduction and abduction moments (123). The primary contribution to these resistive moments comes from the co-contraction of the hamstrings and the quadriceps, with secondary contributions from the gracilis and the tensor fascia lata (67). The ability to resist abduction/adduction moments is relatively small, however, with only 11–14% of the experimentally applied moments resisted (67).

Patellofemoral Joint Motion

During flexion and extension at the tibiofemoral joint, the patella glides inferiorly and superiorly against the distal end of the femur with an excursion of approximately 7 cm. The path of the center of the patella is circular and uniplanar (51). Tracking of the patella against the femur is dependent on the direction of the net force produced by the attached quadriceps. The vastus lateralis tends to pull the patella laterally, while the vastus medialis oblique opposes the lateral pull of the vastus lateralis, keeping the patella centered in the patellofemoral groove. The medial and lateral quadriceps force components also tilt the patella in the sagittal and transverse planes (65). The iliotibial band also influences knee mechanics, and excessive tightness can cause maltracking of the patella (74).

LOADS ON THE KNEE

Because the knee is positioned between the body's two longest bony levers (the femur and the tibia), the potential for torque development at the joint is large. The knee is also a major weight-bearing joint.

Forces at the Tibiofemoral Joint

The tibiofemoral joint is loaded in both compression and shear during daily activities. Weight bearing and tension development in the muscles crossing the knee contribute to these forces, with compression dominating when the knee is fully extended.

● *The medial tibial plateau is well adapted for its weight-bearing function during stance, with greater surface area and thicker articular cartilage than the lateral plateau.*

Compressive force at the tibiofemoral joint has been reported to be slightly greater than three times body weight during the stance phase of gait, increasing up to approximately four times body weight during stair climbing (58). The medial tibial plateau bears most of this load during stance when the knee is extended, with the lateral tibial plateau bearing more of the much smaller loads imposed during the swing phase (121). Since the medial tibial plateau has a surface area roughly 60% larger than that of the lateral tibial plateau, the stress acting on the joint is less than if peak loads were distributed medially (58). The fact that the articular cartilage on the medial plateau is three times thicker than that on the lateral plateau also helps protect the joint from wear.

The menisci act to distribute loads at the tibiofemoral joint over a broader area, thus reducing the magnitude of joint stress. The menisci also directly assist with force absorption at the knee, bearing as much as an estimated 45% of the total load (104). Since the menisci help protect the articulating bone surfaces from wear, knees that have undergone meniscectomies are more likely to develop degenerative conditions.

Measurements of articular cartilage deformation on the tibial plateau during weight bearing show that stress at the joint is maximal from 180 to 120 degrees of flexion, with minimal stress at approximately 30 degrees of flexion (11). Comparison of front and back squat exercises shows no difference in overall muscle recruitment, but significantly less compressive force acting on the knee during the front squat (40). Among runners larger loads on the knee are related to reduced hamstring flexibility, greater body weight, greater weekly mileage, and greater muscular strength (75). Other general risk factors for development of knee osteoarthritis include a high body mass index and meniscal damage (93).

Forces at the Patellofemoral Joint

Compressive force at the patellofemoral joint has been found to be one-half of body weight during normal walking gait, increasing up to over three times body weight during stair climbing (89). As shown in Figure 8-15, patellofemoral compression increases with knee flexion during weight bearing. There are two reasons for this. First, the increase in knee flexion increases the compressive component of force acting at the joint. Second, as flexion increases, a larger amount of quadriceps tension is required to prevent the knee from buckling against gravity.

The squat exercise, known for being particularly stressful to the knee complex, produces a patellofemoral joint reaction force on the order of 7.6 times body weight (89). Patellofemoral joint reaction forces increase with the depth of the squat, although the variations in the magnitudes of these forces are not significantly different among shallow, medium, and deep squats (96). Training within the 0–50° knee flexion range is recommended for those who wish to minimize knee forces, however (28). The squat has been shown to be an effective exercise for use during rehabilitation following cruciate ligament or patellofemoral surgery (27). Sample Problem 8.2 illustrates the relationship between quadriceps force and patellofemoral joint compression.

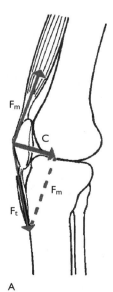

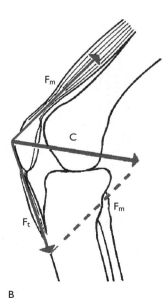

A

B

FIGURE 8-15

Compression at the patellofemoral joint is the vector sum of tension in the quadriceps and the patellar tendon. **A.** In extension, the compressive force is small because tension in the muscle group and tendon act nearly perpendicular to the joint. **B.** As flexion increases, compression increases because of changed orientation of the force vectors and increased tension requirement in the quadriceps to maintain body position.

SAMPLE PROBLEM 8.2

How much compression acts on the patellofemoral joint when the quadriceps exerts 300 N of tension and the angle between the quadriceps and the patellar tendon is (a) 160° and (b) 90°?

Known

$$F_m = 300 \text{ N}$$

Angle between F_m and F_t:

1. 160°
2. 90°

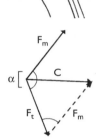

Graphic Solution

Vectors for F_m and F_t are drawn to scale (perhaps 1 cm: 100 N), with the angle between them first at 160° and then at 90°. The tip-to-tail method of vector composition is then used (see Chapter 3) to translate one of the vectors so that its tail is positioned on the tip of the other vector. The compression force is the resultant of F_m and F_t and is constructed with its tail on the tail of the original vector and its tip on the tip of the transposed vector.

The amount of joint compression can be approximated by measuring the length of vector C.

1. C ≈ 100 N
2. C ≈ 420 N

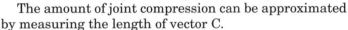

Mathematical Solution

The angle between F_t and transposed vector F_m is 180° minus the size of the angle between the two original vectors, or (a) 20° and (b) 90°. The law of cosines can be used to calculate the length of C.

1. $C^2 = F_m{}^2 + F_t{}^2 - 2(F_m)(F_t) \cos 20$
   $C^2 = 300 \text{ N}^2 + 300 \text{ N}^2 - 2(300 \text{ N})(300 \text{ N}) \cos 20$

   C = 104 N

2. $C^2 = F_m{}^2 + F_t{}^2 - 2(F_m)(F_t) \cos 90$
   $C^2 = 300 \text{ N}^2 + 300 \text{ N}^2 - 2(300 \text{ N})(300 \text{ N}) \cos 90$

   C = 424 N

Note: This problem illustrates the extent to which patellofemoral compression can increase due solely to changes in knee flexion.

Normally, there is also increased quadriceps force with increased knee flexion.

COMMON INJURIES OF THE KNEE AND LOWER LEG

The location of the knee between the long bones of the lower extremity, combined with its weight-bearing and locomotion functions, makes it susceptible to injury, particularly during participation in contact sports. A common injury mechanism involves the stretching or tearing of soft tissues on one side of the joint when a blow is sustained from the opposite side during weight bearing.

Anterior Cruciate Ligament Injuries

Injuries to the anterior cruciate ligament (ACL) are common in sports such as basketball and team handball, which involve pivoting and cutting, as well as in Alpine skiing, where a common mechanism involves catching a ski tip in the snow, with the skier simultaneously twisting and falling. Approximately 70% of ACL injuries are noncontact, with most of these being sustained when the femur is rotated on the planted leg with the knee close to full extension during cutting, landing, or stopping (13, 60). These kinds of activities involving sudden changes in direction combined with acceleration or deceleration of the body produce large rotational moments and varus/valgus forces at the knee, particularly when such movements are inadequately planned. The ACL is loaded when the net shear force at the knee is directed anteriorly (102). So, for ACL rupture to occur, there must be excess anterior translation or rotation of the femur on the tibia.

There is a striking gender disparity in the incidence of ACL injuries, with women sustaining significantly more than men. A number of hypotheses related to anatomical or neuromuscular factors have been advanced, but the reason for this disparity remains unknown. Research has shown that during running, cutting, and landing, women, as compared to men, tend to have less knee flexion, greater knee valgus angles, more hip abduction, greater quadriceps activation, less hamstring activation, and generally less variability in lower-extremity coordination patterns (32, 33, 49, 69, 72, 86). Some have advocated strengthening or stiffening of the hamstrings to protect the ACL, since female athletes tend to have greater quadriceps-to-hamstrings strength ratios than do male athletes (12). Research on the notion that tension in the hamstrings can reduce forces on the ACL has produced conflicting results, however (64, 105). Others have advocated strengthening the quadriceps to protect against ACL injuries, since the quadriceps provide the primary muscular restraint to anterior tibial translation during activities such as running and jumping (14). Yet other research has shown that women, as compared to men, have increased coactivation of the quadriceps during knee flexion, which may increase anterior tibial loads under dynamic conditions and increase the risk of ACL injury (116). Still others have encouraged training programs for female athletes focusing on landing and pivoting with increased knee flexion and not allowing medial or lateral sagging of the knees during cutting maneuvers to protect the ACL from excessive strain (25, 79). It has been demonstrated that sagittal plane knee forces cannot rupture the ACL during sidestep cutting, so during this maneuver, valgus loading is a more likely injury mechanism (71).

Following unrepaired ACL rupture, most individuals have difficulty with movements involving lateral or rotational loads at the knee, and 70–80% experience knee instability or "giving way" (103). There is also a notable lessening of flexion–extension range of motion at the knee during walking, which has been attributed to "quadriceps avoidance" (4). This does not appear to be related to a deficit in quadriceps strength, but instead may be attributed to the fact that quadriceps tension produces an anteriorly directed force on the tibia when the knee is near full extension. Individuals seem to adapt to the absence of the ACL by minimizing activation of the quadriceps when the knee is near full extension (4). There is also evidence of general impairment of neuromuscular control of the quadriceps, with the muscles remaining active during tasks not requiring quadriceps activation (119). Although some ACL-deficient individuals are able to stabilize their knees even during cutting and pivoting, for most, the absence of the ACL results in local instability of the knee, a change in the location of the center of rotation at the knee, a change in the area of tibiofemoral contact during gait, and altered joint kinetics, with subsequent onset of osteoarthritis (3, 22, 98, 107).

Surgical repair of ACL rupture involves reconstruction of the ligament using either the middle third of the patellar tendon, the semitendinosus, or the semitendinosus and gracilis (34). Problems that follow trauma to the knee, whether the trauma is injury- or surgery-induced, include notable weakness and loss of mass in the knee extensor muscles, dramatic reduction of joint range of motion, and impaired joint proprioception (109). The reasons for these changes are not understood; they may be neural or mechanical in origin or a product of deconditioning (46). One factor hypothesized to play a major role in precipitating these changes is muscle inhibition, or the inability to activate all motor units of a muscle during maximal voluntary contraction (108). It has been shown that muscle inhibition can persist for an extended time and may be responsible for long-term strength deficits that alter joint kinetics and lead to osteoarthritis (50, 106, 108).

muscle inhibition
the inability to activate all motor units of a muscle during maximal voluntary contraction

Posterior Cruciate Ligament Injuries

Posterior cruciate ligament (PCL) injuries most commonly result from sport participation or motor vehicle accidents (20, 54). When the PCL is ruptured in isolation, with no damage to the other ligaments or to the menisci, the mechanism is usually hyperflexion of the knee with the foot plantar flexed (54). Impact with the dashboard during motor vehicle accidents, on the other hand, with direct force on the proximal anterior tibia, results in combined ligamentous damage in 95% of cases (54). Isolated PCL injuries are usually treated nonoperatively.

Medial Collateral Ligament Injuries

•*In contact sports, blows to the knee are most commonly sustained on the lateral side, with injury occurring to the stretched tissues on the medial side.*

Blows to the lateral side of the knee are much more common than blows to the medial side, because the opposite leg commonly protects the medial side of the joint. When the foot is planted and a lateral blow of sufficient force is sustained, the result is sprain or rupture of the medial collateral ligament (MCL). Modeling studies suggest that the muscles crossing the knee are able to resist approximately 17% of external medial and lateral loads on the knee, with the remaining 83% sustained by the ligaments and other soft tissues (66). However, direct measurements of strain to the knee ligaments indicate that external tibial torque is more dangerous than medially directed tibial force for the MCL (48). In contact sports such as football, the MCL is more frequently injured, while both medial and lateral collateral ligament sprains occur among wrestlers.

Prophylactic Knee Bracing

To prevent knee ligament injuries, especially during contact sports, some athletes wear prophylactic knee braces. The wearing of such braces by healthy individuals has been a contentious issue since the American Academy of Orthopaedics issued a position statement against their use in 1987. Research on this topic has shown that knee braces can protect the ACL against anterior and torsional loads on the tibia by significantly reducing the strain present in the ligament under these conditions (10). Knee braces can also contribute 20–30% added resistance against lateral blows to the knee, with custom-fitted braces providing the best protection (1). A possible concern, however, is that knee braces act to change the pattern of lower-extremity muscle activity during gait, with less work performed at the knee and more at the hip (23). Other problems that appear to affect some athletes more than others and may be brace-specific include reduced sprinting speed and earlier onset of fatigue (1).

Meniscus Injuries

Because the medial collateral ligament attaches to the medial meniscus, stretching or tearing of the ligament can also result in damage to the meniscus. A torn meniscus is the most common knee injury, with damage to the medial meniscus occurring approximately 10 times as frequently as damage to the lateral meniscus. This is the case partly because the medial meniscus is more securely attached to the tibia, and therefore less mobile than the lateral meniscus. In knees that have undergone ACL rupture, the normal stress distribution is disrupted such that the force on the medial meniscus is doubled (84). In the absence of ACL reconstruction, this results in an increased incidence of medial meniscal tears, although loading of the meniscus returns to normal if the ACL is reconstructed (2, 47). A torn meniscus is problematic in that the unattached cartilage often slips from its normal position, interfering with normal joint mechanics. Symptoms include pain, which is sometimes accompanied by intermittent bouts of locking or buckling of the joint.

Iliotibial Band Friction Syndrome

The tensor fascia lata develops tension to assist with stabilization of the pelvis when the knee is in flexion during weight bearing. This can produce friction of the posterior edge of the iliotibial band (ITB) against the lateral condyle of the femur around the time of footstrike, primarily during foot contact with the ground (82). The result is inflammation of the distal portion of the ITB, as well as the knee joint capsule under the ITB, with accompanying symptoms of pain and tenderness over the lateral aspect of the knee (59). This condition is an overuse syndrome that affects approximately 2–12% of runners and can also affect cyclists (91). Both training errors and anatomical malalignments within the lower extremity increase the risk of ITB syndrome. Training factors in running include excessive running in the same direction on a track, greater-than-normal weekly mileage, and downhill running (35). Runners who had previously sustained iliotibial band syndrome displayed characteristic gait anomalies, including greater peak knee internal rotation and greater peak hip adduction compared to normal runners (30). Both of these kinematic differences place added stress on the iliotibial band. Improper seat height, as well as greater-than-normal weekly mileage, predisposes cyclists to the syndrome (29). Predisposing malalignments include excessive femoral anteversion, increased Q-angle (see below), lateral tibial torsion, tibial genu varum or valgum, subtalar varus, and excessive pronation (61).

Breaststroker's Knee

A condition of pain and tenderness localized on the medial aspect of the knee is often associated with performance of the whip kick, the kick

FIGURE 8-16

The likelihood of acquiring breaststroker's knee is much lower if the angle of hip abduction is between 37° and 42° at the beginning of the propulsive phase of the kick.

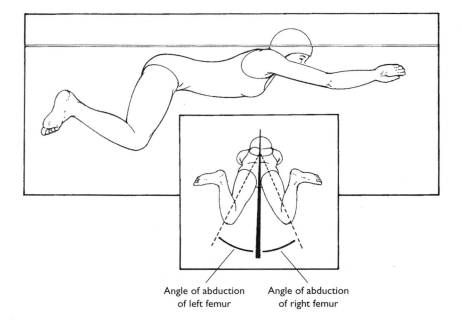

Angle of abduction of left femur   Angle of abduction of right femur

used with the breaststroke. The forceful whipping together of the lower legs that provides the propulsive thrust of the kick often forces the lower leg into slight abduction at the knee, with subsequent irritation to the medial collateral ligament and the medial border of the patellar tract. A survey of 391 competitive swimmers revealed incidences of knee pain among 73% of the breaststroke specialists and 48% of nonbreaststrokers (115). In a study of breaststroke kinematics, it was found that angles of hip abduction of less than 37° or greater than 42° at the initiation of the kick resulted in a dramatically increased incidence of knee pain (115) (Figure 8-16).

Patellofemoral Pain Syndrome

Painful patellofemoral joint motion involves anterior knee pain during and after physical activity, particularly activities requiring repeated flexion at the knee, such as running, ascending and descending stairs, and squatting. It is more common among females than among males. This syndrome has been attributed to a number of possible causes, including anatomical malalignment(s), imbalance between the VMO and the VL in strength or activation timing, decreased quadriceps and hamstring strength, increased hip external rotator strength, and overactivity (15, 68, 78, 86, 87, 93).

●*Painful lateral deviation in patellar tracking could be caused by weakness of the vastus medialis oblique.*

While the causes of this disorder are unknown, most research attention has focused on the relationship between the VMO and the VL. Weakness of the VMO relative to the VL has been shown to be associated with a lateral shift of the patella, particularly early in the range of knee flexion (87, 95). Research has also documented that individuals with patellofemoral pain display reduced patellofemoral joint loads with strengthening of the VMO (80). Such individuals also demonstrate onset of VL activation prior to VMO activation during stair stepping (21). An experiment in which the VMO and VL were electrically stimulated showed a respective increase and decrease in patellofemoral joint loading when VMO activation was delayed (80).

An anatomical factor hypothesized to contribute to lower-extremity malalignment that could trigger patellofemoral pain is an excessively large Q-angle, the angle formed between the anterior superior iliac spine, the center of the patella, and the tibial tuberosity (44). The Q-angle provides an approximation of the angle of pull of the quadriceps on the patella, and it has been hypothesized that a large Q-angle could lead to lateral patellar dislocation or increased lateral patellofemoral contact pressures (77). To date, however, research has not documented a relationship between Q-angle and incidence of patellofemoral pain. The one anatomical factor that has been found to be related to patellar maltracking is a shallow intercondylar groove (88).

Q-angle
angle formed between the anterior superior iliac spine, the center of the patella, and the tibial tuberosity

Shin Splints

Generalized pain along the anterolateral or posteromedial aspect of the lower leg is commonly known as *shin splints*. This is a loosely defined overuse injury, often associated with running or dancing, that may involve microdamage to muscle attachments on the tibia and/or inflammation of the periosteum. Common causes of the condition include running or dancing on a hard surface and running uphill. A change in workout conditions or rest usually alleviates shin splints.

•Shin splints *is a generic term often ascribed to any pain emanating from the anterior aspect of the lower leg.*

STRUCTURE OF THE ANKLE

The ankle region includes the distal tibiofibular, tibiotalar, and fibulotalar joints (Figure 8-17). The distal tibiofibular joint is a syndesmosis where dense fibrous tissue binds the bones together. The joint is supported by the anterior and posterior tibiofibular ligaments, as well as by the crural interosseous tibiofibular ligament. Most motion at the ankle occurs at the tibiotalar hinge joint, where the convex surface of the superior talus articulates with the concave surface of the distal tibia. All three articulations are enclosed in a joint capsule that is thickened on the medial side and extremely thin on the posterior side. Three ligaments—the anterior and posterior talofibular, and the calcaneofibular—reinforce the joint

FIGURE 8-17

The bony structure of the ankle.

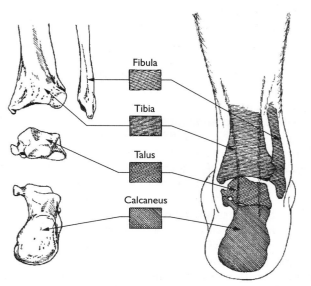

Fibula

Tibia

Talus

Calcaneus

Posterior view

FIGURE 8-18

The ligaments of the ankle.

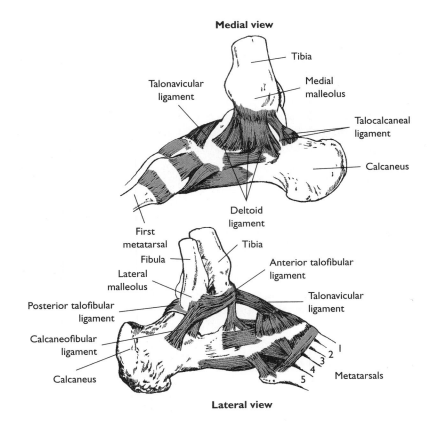

Medial view

Tibia

Talonavicular
ligament

Medial
malleolus

Talocalcaneal
ligament

Calcaneus

Deltoid
ligament

First
metatarsal

Tibia

Fibula

Anterior talofibular
ligament

Lateral
malleolus

Posterior talofibular
ligament

Talonavicular
ligament

Calcaneofibular
ligament

Metatarsals

Calcaneus

Lateral view

capsule laterally. The four bands of the deltoid ligament contribute to joint stability on the medial side. The ligamentous structure of the ankle is displayed in Figure 8-18.

MOVEMENTS AT THE ANKLE

● *The malleoli serve as pulleys to channel muscle tendons anterior or posterior to the axis of rotation at the ankle; those anterior to the malleoli are dorsiflexors and those posterior to the malleoli serve as plantar flexors.*

The axis of rotation at the ankle is essentially frontal, although it is slightly oblique and its orientation changes somewhat as rotation occurs at the joint. Motion at the ankle occurs primarily in the sagittal plane, with the ankle functioning as a hinge joint with a moving axis of rotation during the stance phase of gait (63). Ankle flexion and extension are termed *dorsiflexion* and *plantar flexion*, respectively (see Chapter 2). During passive motion, the articular surfaces and ligaments govern joint kinematics, with the articular surfaces sliding upon each other without appreciable tissue deformation (63).

The medial and lateral malleoli serve as pulleys to channel the tendons of muscles crossing the ankle either posterior or anterior to the axis of rotation, thereby enabling their contributions to either dorsiflexion or plantar flexion.

The tibialis anterior, extensor digitorum longus, and peroneus tertius are the prime dorsiflexors of the foot. The extensor hallucis longus assists in dorsiflexion (Figure 8-19).

The major plantar flexors are the two heads of the powerful two-joint gastrocnemius and the soleus, which lies beneath the gastrocnemius (Figure 8-20). Assistant plantar flexors include the tibialis posterior,

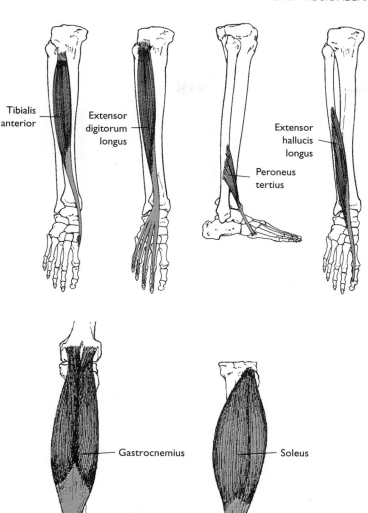

FIGURE 8-19

The dorsiflexors of the ankle.

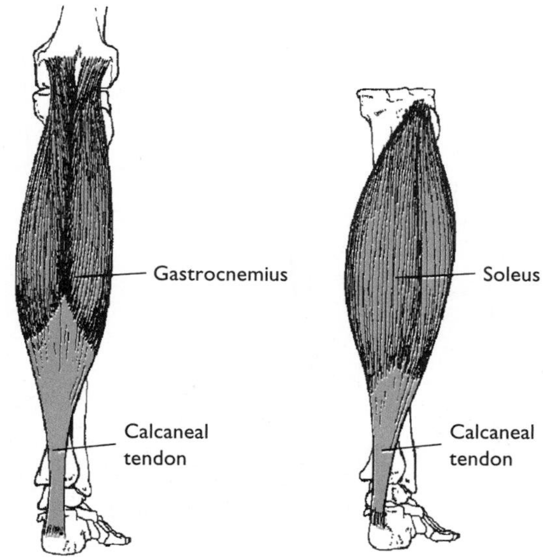

FIGURE 8-20

The major plantar flexors of the ankle.

peroneus longus, peroneus brevis, plantaris, flexor hallucis longus, and flexor digitorum longus (Figure 8-21).

STRUCTURE OF THE FOOT

Like the hand, the foot is a multibone structure. It contains a total of 26 bones with numerous articulations (Figure 8-22). Included are the subtalar and midtarsal joints and several tarsometatarsal, intermetatarsal, metatarsophalangeal, and interphalangeal joints. Together, the bones and joints of the foot provide a foundation of support for the upright body and help it adapt to uneven terrain and absorb shock.

Subtalar Joint

As the name suggests, the subtalar joint lies beneath the talus, where anterior and posterior facets of the talus articulate with the sustentaculum tali on the superior calcaneus. Four talocalcaneal ligaments join the

FIGURE 8-21

Muscles with tendons passing posterior to the malleoli assist with plantar flexion of the ankle.

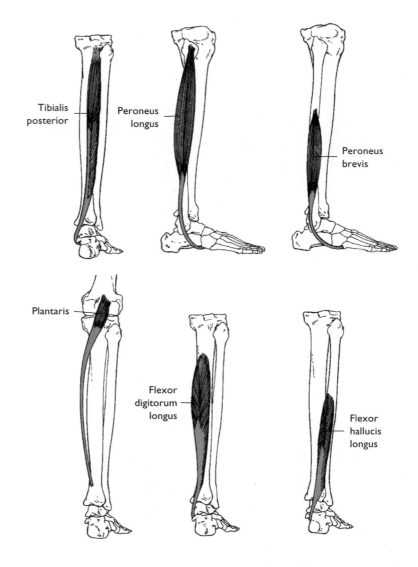

FIGURE 8-22

The foot is composed of numerous articulating bones.

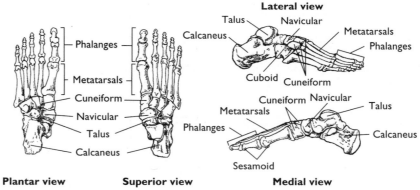

talus and the calcaneus. The joint is essentially uniaxial, with an alignment slightly oblique to the conventional descriptive planes of motion.

Tarsometatarsal and Intermetatarsal Joints

Both the tarsometatarsal and the intermetatarsal joints are nonaxial, with the bone shapes and the restricting ligaments permitting only gliding movements. These joints enable the foot to function as a semirigid unit or to adapt flexibly to uneven surfaces during weight bearing.

Metatarsophalangeal and Interphalangeal Joints

The metatarsophalangeal and interphalangeal joints are similar to their counterparts in the hand, with the former being condyloid joints and the latter being hinge joints. Numerous ligaments provide reinforcement for these joints. The toes function to smooth the weight shift to the opposite foot during walking and help maintain stability during weight bearing by pressing against the ground when necessary. The first digit is referred to as the *hallux,* or "great toe."

Plantar Arches

The tarsal and metatarsal bones of the foot form three arches. The medial and lateral longitudinal arches stretch from the calcaneus to the metatarsals and tarsals. The transverse arch is formed by the bases of the metatarsal bones.

Several ligaments and the plantar fascia support the plantar arches. The spring ligament is the primary supporter of the medial longitudinal arch, stretching from the sustentaculum tali on the calcaneus to the inferior navicular. The long plantar ligament provides the major support for the lateral longitudinal arch, with assistance from the short plantar ligament. Thick, fibrous, interconnected bands of connective tissue known as the plantar fascia extend over the plantar surface of the foot, assisting with support of the longitudinal arch (Figure 8-23). When muscle tension is present, the muscles of the foot, particularly the tibialis posterior, also contribute support to the arches and joints as they cross them.

As the arches deform during weight bearing, mechanical energy is stored in the stretched tendons, ligaments, and plantar fascia. Additional energy is stored in the gastrocnemius and soleus as they develop eccentric tension. During the push-off phase, the stored energy in all of these elastic structures is released, contributing to the force of push-off and actually reducing the metabolic energy cost of walking or running.

plantar fascia
thick bands of fascia that cover the plantar aspect of the foot

●*During weight bearing, mechanical energy is stored in the stretched ligaments, tendons, and plantar fascia of the arches of the foot, as well as in eccentrically contracting muscles. This stored energy is released to assist with push-off of the foot from the surface.*

FIGURE 8-23

The plantar fascia.

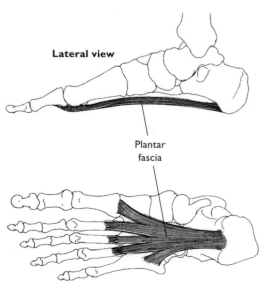

Lateral view

Plantar fascia

Plantar view

MOVEMENTS OF THE FOOT

Muscles of the Foot

The locations and primary actions of the major muscles of the ankle and foot are summarized in Table 8-3. As with the muscles of the hand, extrinsic muscles are those crossing the ankle, and intrinsic muscles have both attachments within the foot.

Toe Flexion and Extension

Flexion involves the curling under of the toes. Flexors of the toes include the flexor digitorum longus, flexor digitorum brevis, quadratus plantae, lumbricals, and interossei. The flexor hallucis longus and brevis produce flexion of the hallux. Conversely, the extensor hallucis longus, extensor digitorum longus, and extensor digitorum brevis are responsible for extension of the toes.

TABLE 8-3

Muscles of the Ankle and Foot

MUSCLE	PROXIMAL ATTACHMENT	DISTAL ATTACHMENT	PRIMARY ACTION(S)	INNERVATION
Tibialis anterior	Upper two-thirds of lateral tibia	Medial surface of first cuneiform and first metatarsal	Dorsiflexion, inversion	Deep peroneal $(L_4–S_1)$
Extensor digitorum longus	Lateral condyle of tibia, head of fibula, upper two-thirds of anterior fibula	Second and third phalanges of four lesser toes	Toe extension, dorsiflexion, eversion	Deep peroneal $(L_4–S_1)$
Peroneus tertius	Lower third anterior fibula	Dorsal surface of fifth metatarsal	Dorsiflexion, eversion	Deep peroneal $(L_4–S_1)$
Extensor hallucis longus	Middle two-thirds of medial anterior fibula	Dorsal surface of distal phalanx of great toe	Dorsiflexion, inversion, hallux extension	Deep peroneal $(L_4–S_1)$
Gastrocnemius	Posterior medial and lateral condyles of femur	Tuberosity of calcaneus via Achilles tendon	Plantar flexion	Tibial $(S_1–S_2)$
Plantaris	Distal, posterior femur	Tuberosity of calcaneus via Achilles tendon	Assists with plantar flexion	Tibial $(S_1–S_2)$
Soleus	Posterior proximal fibula and proximal two-thirds of posterior tibia	Tuberosity of calcaneus via Achilles tendon	Plantar flexion	Tibial $(S_1–S_2)$
Peroneus longus	Head and upper two-thirds of lateral fibula	Lateral surface of first cuneiform and first metatarsal	Plantar flexion, eversion	Superficial peroneal $(L_4–S_1)$
Peroneus brevis	Distal two-thirds lateral	Tuberosity of fifth fibula	Plantar flexion, eversion metatarsal	Superficial peroneal $(L_4–S_1)$
Flexor digitorum longus	Middle third of posterior tibia	Distal phalanx of four lesser toes	Plantar flexion, inversion, toe flexion	Tibial $(L_5–S_1)$
Flexor hallucis longus	Middle two-thirds of posterior fibula	Distal phalanx of great toe	Plantar flexion, inversion, toe flexion	Tibial $(L_4–S_2)$
Tibialis posterior	Posterior upper two-thirds tibia and fibula and interosseous membrane	Cuboid, navicular, and second to fifth metatarsals	Plantar flexion, inversion	Tibial $(L_5–S_1)$

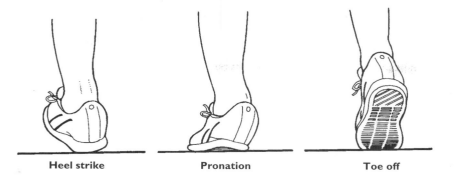

Heel strike   Pronation   Toe off

FIGURE 8-24

Rearfoot movement during running. Adapted from Nigg BM et al: Factors influencing kinetic and kinematic variables in running, in BM Nigg (ed): *Biomechanics of running shoes,* Champaign, 1986, Human Kinetics Publishers.

Inversion and Eversion

Rotational movements of the foot in medial and lateral directions are termed *inversion* and *eversion,* respectively (see Chapter 2). These movements occur largely at the subtalar joint, although gliding actions among the intertarsal and tarsometatarsal joints also contribute. Inversion results in the sole of the foot turning inward toward the midline of the body. The tibialis posterior and tibialis anterior are the main muscles involved. Turning the sole of the foot outward is termed *eversion.* The muscles primarily responsible for eversion are the peroneus longus and the peroneus brevis, both with long tendons coursing around the lateral malleolus. The peroneus tertius assists.

Pronation and Supination

During walking and running, the foot and ankle undergo a cyclical sequence of movements (Figure 8-24). As the heel contacts the ground, the rear portion of the foot typically inverts to some extent. When the foot rolls forward and the forefoot contacts the ground, plantar flexion occurs (114). The combination of inversion, plantar flexion, and adduction of the foot is known as supination (see Chapter 2). While the foot supports the weight of the body during midstance, there is a tendency for eversion and abduction to occur as the foot moves into dorsiflexion. These movements are known collectively as pronation. Pronation serves to reduce the magnitude of the ground reaction force sustained during gait by increasing the time interval over which the force is sustained (18).

supination
combined conditions of plantar flexion, inversion, and adduction

pronation
combined conditions of dorsiflexion, eversion, and abduction

LOADS ON THE FOOT

Impact forces sustained during gait increase with body weight and with gait speed in accordance with Newton's third law of motion (see Chapter 3). The vertical ground reaction force applied to the foot during running is bimodal, with an initial impact peak followed almost immediately by a propulsive peak, as the foot pushes off against the ground (Figure 8-25). As running speed increases from 3.0 m/s (8:56 minutes per mile) to 5.0 m/s (5:22 minutes per mile), impact forces range from 1.6 to 2.3 times body weight and propulsive forces range from 2.5 to 2.8 times body weight (78).

The structures of the foot are anatomically linked such that the load is evenly distributed over the foot during weight bearing. Approximately 50% of body weight is distributed through the subtalar joint to the calcaneus, with the remaining 50% transmitted across the metatarsal heads (97). The head of the first metatarsal sustains twice the load borne by each of the other metatarsal heads (97). A factor that influences this loading pattern,

Vertical ground reaction force during running with an initial impact peak followed almost immediately by a propulsive peak.

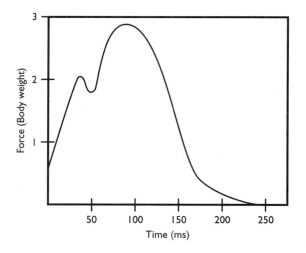

however, is the architecture of the foot. A pes planus (relatively flat arch) condition tends to reduce the load on the forefoot, with pes cavus (relatively high arch) significantly increasing the load on the forefoot (117, 118).

COMMON INJURIES OF THE ANKLE AND FOOT

Because of the crucial roles played by the ankle and foot during locomotion, injuries to this region can greatly limit mobility. Injuries of the lower extremity, especially those of the foot and ankle, may result in weeks or even months of lost training time for athletes, particularly runners. Among dancers, the foot and the ankle are the most common sites of both chronic and acute injuries.

Ankle Injuries

●*Ankle sprains usually occur on the lateral side because of weaker ligamentous support than is present on the medial side.*

Ankle sprains are the most common of all sport- and dance-related injuries (94). Because the joint capsule and ligaments are stronger on the medial side of the ankle, inversion sprains involving stretching or rupturing of the lateral ligaments are much more common than eversion sprains of the medial ligaments (16). In fact, the bands of the deltoid ligament are so strong that excessive eversion is more likely to result in a fracture of the distal fibula than in rupturing of the deltoid ligament. The ligaments most commonly injured are the anterior and posterior talofibular ligaments and the calcaneofibular ligament. Because of the protection by the opposite limb on the medial side, fractures in the ankle region also occur more often on the lateral than on the medial side. Repeated ankle sprains can result in functional ankle instability, which is characterized by significantly altered patterns of ankle and knee movement. Researchers hypothesize that this is related to altered motor control of the foot/ankle complex, which unfortunately predisposes the ankle to further injury (18).

Bracing and taping are two prophylactic measures often used to protect ankles from sprain during sport participation. Ankle braces are designed to preload the ankle and maintain it in a neutral position, thereby counteracting inversionally directed rotation (111). There are contradictory reports in the scientific literature regarding the efficacy of both bracing and taping of the ankle. One reported disadvantage of both bracing and taping of the ankle is a reduction in postural control, causing less stability and more touching down on the foot of the supported ankle (6).

Overuse Injuries

Achilles tendinitis involves inflammation and sometimes microrupturing of tissues in the Achilles tendon, typically accompanied by swelling. Two possible mechanisms for tendinitis have been proposed (90). The first is that repeated tension development results in fatigue and decreased flexibility in the muscle, increasing tensile load on the tendon even during relaxation of the muscle. The second theory is that repeated loading actually leads to failure or rupturing of the collagen threads in the tendon. Achilles tendinitis is usually associated with running and jumping activities and is extremely common among theatrical dancers. It has also been reported in skiers. Complete rupturing of the Achilles tendon occurs almost exclusively in male skiers, although incidence of the injury has decreased with the advent of high, rigid ski boots and effective release bindings (81).

Repetitive stretching of the plantar fascia can result in plantar fasciitis, a condition characterized by microtears and inflammation of the plantar fascia near its attachment to the calcaneus. The symptoms are pain in the heel and/or arch of the foot. The condition is the fourth most common cause of pain among runners, and also occurs with some frequency among basketball players, tennis players, gymnasts, and dancers (62). Anatomical factors believed to contribute to the development of plantar fasciitis include pes planus (flat foot), a rigid cavus (high-arch) foot, and a tight Achilles tendon, all of which reduce the foot's shock-absorbing capability (110).

Another factor linked to overuse injuries of the lower extremity is excessive pronation (19). Although walking normally involves approximately 6–8° of subtalar pronation, individuals with pes planus undergo 10–12° of pronation (90). Because pronation also causes a compensatory inward rotation of the tibia, excessive pronation can result in increased stress within the plantar fascia and Achilles tendon. Excessive pronation has been associated with running injuries, including shin splints, chondromalacia, plantar fasciitis, and Achilles tendinitis. Excessive pronation has been documented among 60% of one group of injured runners (53).

Stress fractures (see Chapter 4) occur relatively frequently in the bones of the lower extremity. Among a group of 320 athletes with bone-scan-positive stress fractures, the bone most frequently injured was the tibia (49.1%), followed by the tarsals (25.3%), metatarsals (8.8%), femur (7.2%), fibula (6.6%), and pelvis (1.6%) (70). The sites most frequently injured in the older athletes in the group were the femur and the tarsals, with the fibula and tibia most often injured among the younger athletes. Among runners, factors associated with stress fractures include training errors, forefoot striking (toe–heel gait), running on hard surfaces such as concrete, improper footwear, and alignment anomalies of the trunk and/or lower extremity (37). Female runners with a history of tibial stress fracture displayed greater peak hip adduction and peak rear foot eversion angles as compared to other runners, suggesting that this kinematic pattern may be predisposing for the injury (76). Stress fractures among dancers occur most frequently to the second and third metatarsals, and appear to be related to dancing on overly hard surfaces (42). Dancing *en pointe* is particularly stressful to the second metatarsal, because tension in the peroneus longus and tibialis posterior needed for maintaining the *en pointe* position places the stressed second metatarsal in traction (42). Stress fractures among women runners, dancers, and gymnasts may be related to decreased bone mineral density secondary to oligomenorrhea (see Chapter 4).

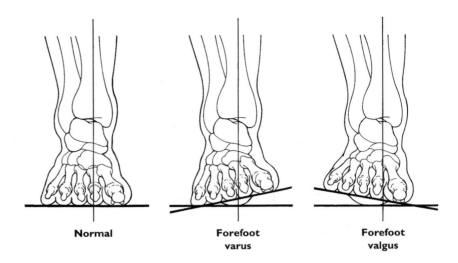

FIGURE 8-26

Varus and valgus conditions of the forefoot.

Normal   Forefoot varus   Forefoot valgus

Alignment Anomalies of the Foot

varus
condition of inward deviation in alignment from the proximal to the distal end of a body segment

valgus
condition of outward deviation in alignment from the proximal to the distal end of a body segment

Varus and valgus conditions (inward and outward lateral deviation, respectively, of a body segment) can occur in all of the major links of the lower extremity. These may be congenital or may arise from an imbalance in muscular strength.

In the foot, varus and valgus conditions can affect the forefoot, the rearfoot, and the toes. Forefoot varus and forefoot valgus refer to inversion and eversion misalignments of the metatarsals, and rearfoot varus and valgus involve inversion and eversion misalignments at the subtalar joint (Figure 8-26). Hallux valgus is a lateral deviation of the big toe often caused by wearing women's shoes with pointed toes (Figure 8-27).

Varus and valgus conditions in the tibia and the femur can alter the kinematics and kinetics of joint motion, because they cause added tensile stress on the stretched side of the affected joint. For example, a combination of femoral varus and tibial valgus (a knock-knee condition) places added tension on the medial aspect of the knee (Figure 8-28). In contrast, the bow-legged condition of femoral valgus and tibial varus stresses the lateral aspect of the knee and is therefore a predisposing factor for iliotibial band friction syndrome. Unfortunately, lateral misalignments at one joint of the lower extremity are typically accompanied by compensatory

FIGURE 8-27

Wearing shoes with pointed toes can cause hallux valgus.

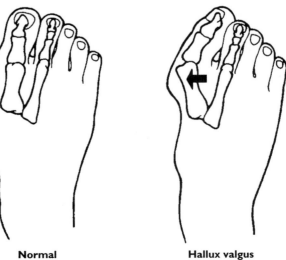

Normal   Hallux valgus

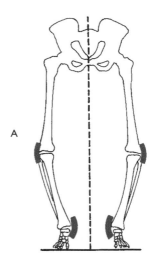

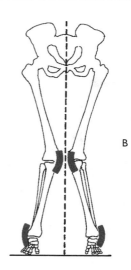

FIGURE 8-28

Areas of strain are highlighted. **A.** Femoral valgus and tibial varus. **B.** Femoral varus and tibial valgus.

misalignments at other lower-extremity joints because of the nature of joint loading during weight bearing.

Depending on the cause of the misalignment problem, correctional procedures may involve exercises to strengthen or stretch specific muscles and ligaments of the lower extremity, as well as the use of orthotics, specially designed inserts worn inside the shoe to provide added support for a portion of the foot.

• *Misalignment at a lower extremity joint typically results in compensatory misalignments at one or more other joints because of the lower extremity's weightbearing function.*

Injuries Related to High and Low Arch Structures

Arches that are higher or lower than the normal range have been found to influence lower-extremity kinematics and kinetics, with implications for injury. Specifically, as compared to those with normal arches, high-arched runners have been found to exhibit increased vertical loading rate, with related higher incidences of ankle sprains, plantar fascitis, iliotibial band friction syndrome, and fifth metatarsal stress fractures (115, 116). Low-arched runners, as compared to those with normal arches, have been found to exhibit increased range of motion and velocity in rearfoot eversion, as well as an increased ratio of eversion to tibial internal rotation (116). These kinematic alterations were found to result in increased incidences of general knee pain, patellar tendinitis, and patellar fascitis (115).

SUMMARY

The lower extremity is well adapted to its functions of weight bearing and locomotion. This is particularly evident at the hip, where the bony structure and several large, strong ligaments provide considerable joint stability. The hip is a typical ball-and-socket joint, with flexion, extension, abduction, adduction, horizontal abduction, horizontal adduction, medial and lateral rotation, and circumduction of the femur permitted.

The knee is a large, complex joint composed of two side-by-side condyloid articulations. Medial and lateral menisci improve the fit between the articulating bone surfaces and assist in absorbing forces transmitted across the joint. Because of differences in the sizes, shapes, and orientations of the medial and lateral articulations, medial rotation of the tibia accompanies full knee extension. A number of ligaments cross the knee and restrain its mobility. The primary movements allowed at the knee are flexion and extension, although some rotation of the tibia is also possible when the knee is in flexion and not bearing weight.

The ankle includes the articulations of the tibia and the fibula with the talus. This is a hinge joint that is reinforced both laterally and medially by ligaments. Movements at the ankle joint are dorsiflexion and plantar flexion.

Like the hand, the foot is composed of numerous small bones and their articulations. Movements of the foot include inversion and eversion, abduction and adduction, and flexion and extension of the toes.

INTRODUCTORY PROBLEMS

1. Construct a chart listing all muscles crossing the hip joint according to whether they are anterior, posterior, medial, or lateral to the joint center. Note that some muscles may fall into more than one category. Identify the action or actions performed by the muscles listed in the four categories.

2. Construct a chart listing all muscles crossing the knee joint according to whether they are anterior, posterior, medial, or lateral to the joint center. Note that some muscles may fall into more than one category. Identify the action or actions performed by the muscles listed in the four categories.

3. Construct a chart listing all muscles crossing the ankle joint according to whether they are anterior or posterior to the joint center. Note that some muscles may fall into more than one category. Identify the action or actions performed by the muscles listed in the four categories.

4. Compare the structure of the hip (including bones, ligaments, and muscles) to the structure of the shoulder. What are the relative advantages and disadvantages of the two joint structures?

5. Compare the structure of the knee (including bones, ligaments, and muscles) to the structure of the elbow. What are the relative advantages and disadvantages of the two joint structures?

6. Describe sequentially the movements of the lower extremity that occur during the activity of kicking a ball. Identify the agonist muscle groups for each of these movements.

7. Describe sequentially the movements of the lower extremity that occur during performance of a vertical jump. Identify the agonist muscle groups for each of these movements.

8. Describe sequentially the movements of the lower extremity that occur during the activity of rising from a seated position. Identify the agonist muscle groups for each of these movements.

9. Use the diagram in Sample Problem 8.1 as a model to determine the magnitude of the reaction force at the hip when tension in the hip abductors is 750 N and 300 N of body weight is supported. (Answer: 1037 N)

10. Use the diagram in Sample Problem 8.2 as a model to determine how much compression acts on the patellofemoral joint when the quadriceps exert 400 N of tension and the angle between the quadriceps and the patellar tendon is (a) 140° and (b) 100°. (Answer: (a) 273.6 N, (b) 514.2 N)

ADDITIONAL PROBLEMS

1. Explain the roles of two-joint muscles in the lower extremity, using the rectus femoris and the gastrocnemius as examples. How does the orientation of the limbs articulating at one joint influence the action of a two-joint muscle at the other joint?

2. Explain the sequencing of joint actions in the ankle and foot during the support phase of gait.
3. Describe the sequencing of contractions of the major muscle groups of the lower extremity during the gait cycle, indicating when contractions are concentric and eccentric.
4. Which muscles of the lower extremity are called on more for running uphill than for running on a level surface? For running downhill as compared to running on a level surface? Explain why.
5. The squat exercise with a barbell is sometimes performed with the heels elevated by a block of wood. Explain what effect this has on the function of the major muscle groups involved.
6. Explain why compression at the hip is higher than compression at the knee despite the fact that the knee supports more body weight during stance than does the hip.
7. Construct a free body diagram that demonstrates how the use of a cane can alleviate compression on the hip.
8. Explain why lifting a heavy load with one or both knees in extreme flexion is dangerous. What structure(s) are placed at risk?
9. Explain how excessive pronation predisposes the individual to stress-related injuries of the Achilles tendon and plantar fascia.
10. What compensations during gait are likely to be made by individuals with genu valgum and genu varus?

Laboratory Experiences

1. Study anatomical models of the hip, knee, and ankle, or use the *Dynamic Human* CD to locate and identify the major bones and muscle attachments. (On the *Dynamic Human* CD, click on Skeletal, Anatomy, Gross Anatomy, Lower Limbs, Femur, Patella, Tibia/Fibula, and then Foot.)

Bones Articulating at the Hip

Bone	Muscle Attachments
_____	_____
_____	_____

Bones Articulating at the Knee

Bone	Muscle Attachments
_____	_____
_____	_____
_____	_____
_____	_____

Bones Articulating at the Ankle

Bone	Muscle Attachments
_____	_____
_____	_____
_____	_____

2. Study anatomical models of the hip, knee, and ankle, or use the *Dynamic Human* CD to locate and identify the major muscles and their attachment sites. (On the *Dynamic Human* CD, click on muscular, anatomy, body regions, thigh, and then lower leg.)

Muscles of the Thigh

Muscle	Attachment Sites
_____	_____
_____	_____
_____	_____
_____	_____
_____	_____
_____	_____
_____	_____

_____   _____

_____   _____

_____   _____

_____   _____

_____   _____

Muscles of the Lower Leg

Muscle   **Attachment Sites**

_____   _____

_____   _____

_____   _____

_____   _____

_____   _____

_____   _____

_____   _____

_____   _____

_____   _____

3. Using the *Dynamic Human* CD, click on Skeletal, Clinical Concepts, and then Joint Disorders. Also review the relevant material in the chapter, and write a paragraph explaining what kinds of activities and injuries can lead to osteoarthritis of the knee.

_____

_____

_____

_____

_____

_____

_____

_____

4. From the side view, video a volunteer walking at a slow pace, a normal pace, and a fast pace. Review the video several times, and construct a chart that characterizes the differences in lower extremity kinematics among the three trials. Explain what differences in muscle activity are associated with the major kinematic differences.

Slow Pace   **Normal Pace**   **Fast Pace**

_____   _____   _____

_____   _____   _____

_____   _____   _____

_____   _____   _____

_____   _____   _____

_____   _____   _____

_____   _____   _____

5. From the side view, video a volunteer rising from a seated position. Review the video several times, and construct a list indicating the sequencing and timing of the major joint actions and the associated activity of major muscle groups.

Joint Actions   **Major Muscle Group(s)**

_____   _____

_____   _____

_____   _____

_____   _____

REFERENCES

1. Albright JP, Saterbak A, and Stokes J: Use of knee braces in sport. Current recommendations, *Sports Med* 20:281, 1995.
2. Allen CR, Wong EK, Livesay GA, Sakane M, Fu FH, and Woo SL: Importance of the medial meniscus in the anterior cruciate ligment-deficient knee, *J Orthop Res* 18:109, 2000.
3. Andriacchi TP and Dyrby CO: Interactions between kinematics and loading during walking for the normal and ACL deficient knee, *J Biomech* 38:293, 2005.
4. Andriacchi TP, Natarajan RN, and Hurwitz DE: Musculoskeletal dynamics, locomotion, and clinical applications. In Mow VC and Hayes WC, eds: *Basic orthopaedic biomechanics* (2nd ed.), Philadelphia, 1997, Lippincott-Raven.
5. Beaupre A et al: Knee menisci: correlation between microstructure and biomechanics, *Clin Orthop* 208:72, 1986.
6. Bennell KL and Goldie PA: The differential effects of external ankle support on postural control, *J Orthop Sports Phys Ther* 20:287, 1994.
7. Bergmann G, Graichen A, and Rohlmann A: Hip joint forces during load carrying, *Clin Orthop* 335:190, 1997.
8. Bergmann G, Kniggendorf H, Graichen F, and Rohlmann A: Influence of shoes and heel strike on the loading of the hip joint, *J Biomech* 28:817, 1995.
9. Bergmann G et al: Hip contact forces and gait patterns from routine activities, *J Biomech* 34:859, 2001.
10. Beynnon BD and Fleming BC: Anterior cruciate ligament strain in-vivo: A review of previous work, *J Biomech* 31:519, 1998.
11. Bingham JT, Papannagari R, Van de Velde SK, Gross C, Gill TJ, Felson DT, Rubash HE, and Li G: In vivo cartilage contact deformation in the healthy human tibiofemoral joint, *Rheumatology* (Oxford), 47:1622, 2008.
12. Blackburn JT, Riemann BL, and Guskiewicz KM: Sex comparison of extensibility, passive and active stiffness of the knee flexors, *Clin Biomech* 19:36, 2004.
13. Boden BP, Dean GS, Feagin JA Jr, and Garrett WE Jr: Mechanisms of anterior cruciate ligament injury, *Orthop* 23:573, 2000.
14. Bodor M: Quadriceps protects the anterior cruciate ligament, *J Orthop Res* 19:629, 2001.
15. Boling MC, Padua DA, Marshall SW, Guskiewicz K, Pyne S, and Beutler A: A prospective investigation of biomechanical risk factors for patellofemoral pain syndrome: the Joint Undertaking to Monitor and Prevent ACL Injury (JUMP-ACL) cohort, *Am J Sports Med*, 37:2108, 2009.
16. Browning D and Potter MB: Lateral ankle springs: The 5-minute sports medicine consult. Philadelphia: Lippincott Williams & Wilkins; 2001:14–15.
17. Bryan R, Nair PB, and Taylor M: Use of a statistical model of the whole femur in a large scale, multi-model study of femoral neck fracture risk, *J Biomech* 42:2171, 2009.
18. Caulfield B and Garrett M: Changes in ground reaction force during jump landing in subjects with functional instability of the ankle joint, *Clin Biomech* 19:617, 2004.
19. Clarke TE, Frederick EC, and Hamill C: The effects of shoe design parameters on rearfoot control in running, *Med Sci Sports Exerc* 15:376, 1983.
20. Cosgarea AJ and Jay PR: Posterior cruciate ligament injuries: evaluation and management, *J Am Acad Orthop Surg* 9:297, 2001.
21. Cowan SM: Delayed onset of electromyographic activity of the vastus medialis obliquus relative to vastus lateralis in subjects with pattellofemoral pain syndrome, *Arch Phys Med Rehabil* 82:183, 2001.
22. Dennis DA, Mahfouz MR, Komistek RD, and Hoff W: In vivo determination of normal and anterior cruciate ligament-deficient knee kinematics, *J Biomech* 38:241, 2005.
23. DeVita P, Torry M, Glover KL, and Speroni DL: A functional knee brace alters joint torque and power patterns during walking and running, *J Biomech* 29:583, 1996.
24. Dunn T, Heller CA, McCarthy SW, and Dos Remedios C: Anatomical study of the "trochanteric bursa," *Clin Anat* 16:233, 2003.

25. Ebstrup JF and Bojsen-Moller F: Anterior cruciate ligament injury in indoor ball games, *Scand J Med Sci Sports* 10:114, 2000.

26. Englund M, Guermazi A, and Lohmander LS: The meniscus in knee osteoarthritis, *Rheum Dis Clin North Am*, 35:579, 2009.

27. Escamilla RF: Knee biomechanics of the dynamic squat exercise, *Med Sci Sports Exerc* 33:127, 2001.

28. Escamilla RF et al: Effects of technique variations on knee biomechanics during the squat and leg press, *Med Sci Sports Exerc* 33:1552, 2001.

29. Farrell KC, Reisinger KD, and Tillman MD: Force and repetition in cycling: possible implications for iliotibial band friction syndrome, *Knee* 10:103, 2003.

30. Ferber R, Noehren B, Hamill J, and Davis IS: Competitive female runners with a history of iliotibial band syndrome demonstrate atypical hip and knee kinematics, *J Orthop Sports Phys Ther*, 40:52, 2010.

31. Ferguson SJ, Bryant JT, Ganz R, and Ito K: An in vitro investigation of the acetabular labral seal in hip joint mechanics, *J Biomech* 36:171, 2003.

32. Ford KR, Myer GD, Toms HE, and Hewett TE: Gender differences in the kinematics of unanticipated cutting in young athletes, *Med Sci Sports Exerc* 37:124, 2005.

33. Ford KR, Myer GD, and Hewett TE: Valgus knee motion during landing in high school female and male basketball players, *Med Sci Sports Exerc* 35:1745, 2003.

34. Frank CB and Jackson DW: Current concepts review: the science of reconstruction of the anterior cruciate ligament, *J Bone Joint Surg* 79-A:1556, 1997.

35. Fredrickson M and Wolf C: Iliotibial band syndrome in runners: innovations in treatment, *Sports Med* 35:451, 2005.

36. Freeman MAR and Pinskerova V: The movement of the normal tibio-femoral joint, *J Biomech* 38:197, 2005.

37. Frey C: Footwear and stress fractures, *Clin Sports Med* 16:249, 1997.

38. Grabiner MD, Koh TJ, and Draganich LF: Neuromechanics of the patellofemoral joint, *Med Sci Sports Exerc* 26:10, 1994.

39. Gray JC: Neural and vascular anatomy of the menisci of the human knee, *J Orthop Sports Phys Ther* 29:23, 1999.

40. Gullett JC, Tillman MD, Gutierrez GM, and Chow JW: A biomechanical comparison of back and front squats in healthy trained individuals, *J Strength Cond Res* 23:284, 2009.

41. Haentjens P, Magaziner J, Colón-Emeric CS, Vanderschueren D, Milisen K, Velkeniers B, and Boonen S: Meta-analysis: excess mortality after hip fracture among older women and men, *Ann Intern Med* 152:380, 2010.

42. Harrington T et al: Overuse ballet injury of the base of the second metatarsal, *Am J Sports Med* 21:591, 1993.

43. Hashemi J, Chandrashekar N, Gill B, Beynnon BD, Slauterbeck JR, Schutt RC Jr, Mansouri H, and Dabezies E: The geometry of the tibial plateau and its influence on the biomechanics of the tibiofemoral joint, *J Bone Joint Surg Am* 90:2724, 2008.

44. Heiderscheit BC, Hamill J, and Van Emmerik REA: Q-angle influences on the variability of lower extremity coordination during running, *Med Sci Sports Exerc* 31:1313, 1999.

45. Heiderscheit BC, Sherry MA, Silder A, Chumanov ES, and Thelen DG.: Hamstring strain injuries: recommendations for diagnosis, rehabilitation, and injury prevention, *J Orthop Sports Phys Ther* 40:67, 2010.

46. Heimstra LA, Webber S, MacDonald PB, and Kriellaars DJ: Knee strength deficits after hamstring tendon and patellar tendon anterior cruciate ligament reconstruction, *Med Sci Sports Exerc* 32:1472, 2000.

47. Hollis JM, Pearsall AW 4th, and Niciforos PG: Change in meniscal strain with anterior cruciate ligament injury and after reconstruction, *Am J Sports Med* 28:700, 2000.

48. Hull ML, Berns, GS, Varma H, and Patterson HA: Strain in the medial collateral ligament of the human knee under single and combined loads, *J Biomech* 29:199, 1996.

49. Hurd WJ, Chmielewski TL, Axe MJ, Davis I, and Snyder-Mackler L: Differences in normal and perturbed walking kinematics between male and female athletes, *Clin Biomech* 19:465, 2004.

50. Hurley MV, Jones DW, and Newham DJ: Arthrogenic quadriceps inhibition and rehabilitation of patients with extensive traumatic knee injuries, *Clin Sci* 86:305, 1994.

51. Iranpour F, Merican AM, Baena FR, Cobb JP, and Amis AA: Patellofemoral joint kinematics: the circular path of the patella around the trochlear axis, *J Orthop Res* 28:589, 2010.

52. Jaglal SB, Kreiger N, and Darlington G: Past and recent physical activity and risk of hip fracture, *Am J Epidemiol* 138:107, 1993.

53. James SL, Bates BT, and Osternig LR: Injuries to runners, *Am J Sports Med* 6:40, 1978.

54. Janousek AT, Jones DG, Clatworthy M, Higgins LD, and Fu FH: Posterior cruciate ligament injuries of the knee joint, *Sports Med* 28:429, 1999.

55. Joglekar SB, Rehman S: Delayed onset thigh compartment syndrome secondary to contusion, *Orthopedics*, 32: ii, 2009.

56. Johal P, Williams A, Wragg P, Hunt D, and Gedroyc W: Tibio-femoral movement in the living knee: a study of weight bearing and non-weight bearing knee kinematics using "interventional" MRI, *J Biomech* 38:269, 2005.

57. Johnston RC: Mechanical considerations of the hip joint, *Arch Surg* 107:411, 1973.

58. Kettlekamp DB and Jacobs AW: Tibiofemoral contact area determination and implications, *J Bone Joint Surg* 54A:349, 1972.

59. Khaund R and Flynn SH: Iliotibial band syndrome: a common source of knee pain, *Am Fam Physician* 71:1545, 2005.

60. Kirkendall DT and Garrett WE Jr: The anterior cruciate ligament enigma: injury mechanisms and prevention, *Clin Orthop* 372:64, 2000.

61. Krivickas LS: Anatomical factors associated with overuse sports injuries, *Sports Med* 24:132, 1997.

62. Leach RE, Seavey MS, and Salter DK: Results of surgery in athletes with plantar fascitis, *Foot Ankle* 7:156, 1986.

63. Leardini A, O'Connor JJ, Catani F, and Giannini S: A geometric model of the human ankle joint, *J Biomech* 32:585, 1999.

64. Li G, Rudy TW, Sakane M, Kanamori A, Ma CB, and Woo SL-Y: The importance of quadriceps and hamstring muscle loading on knee kinematics and in-situ forces in the ACL, *J Biomech* 32:395, 1999.

65. Lin F, Wang Guangzhi, Koh JK, Hendrix RW, and Zhang L-Q: In vivo and noninvasive three-dimensional patellar tracking induced by individual heads of quadriceps, *Med Sci Sports Exerc* 36:93, 2004.

66. Lloyd DG and Buchanan TS: A model of load sharing between muscles and soft tissues at the human knee during static tasks, *J Biomech Eng* 118:367, 1996.

67. Lloyd DG and Buchanan TS: Strategies of muscular support of varus and valgus isometric loads at the human knee, *J Biomech* 34:1257, 2001.

68. Makhsous M, Lin F, Koh J, Nuber G, and Zhang L-Q: In vivo and noninvasive load sharing among the vasti in patellar malalignment, *Med Sci Sports Exerc* 36:1768, 2004.

69. Malinzak RA, Colby Sm, Kirkendall DT, Yu B, and Garrett WE: A comparison of knee joint motion patterns between men and women in selected athletic tasks, *Clin Biomech* 16:438, 2001.

70. Matheson GO et al: Stress fractures in athletes, *Am J Sports Med* 15:46, 1987.

71. McLean SG, Lipfert SW, and van den Bogert AJ: Effect of gender and defensive opponent on the biomechanics of sidestep cutting, *Med Sci Sports Exer* 36:6, 2004.

72. McGibbon CA, Krebs DE, and RW Mann: In vivo hip pressures during cane and load-carrying gait, *Arthritis Care Res* 10:300, 1997.

73. Meister BR, Michael Sp, Moyer RA, Kelly JD, and Schneck CD: Anatomy and kinematics of the lateral collateral ligament of the knee, *Am J Sports Med* 28:869, 2000.

74. Merican AM, Amis AA: Iliotibial band tension affects patellofemoral and tibiofemoral kinematics, *J Biomech* 42:1539, 2009.

75. Messier SP, Legault C, Schoenlank CR, Newman JJ, Martin DF, and DeVita P: Risk factors and mechanisms of knee injury in runners, *Med Sci Sports Exerc* 40:1873, 2008.

76. Milner CE, Hamill J, Davis IS: Distinct hip and rearfoot kinematics in female runners with a history of tibial stress fracture, *J Orthop Sports Phys Ther* 40:59, 2010.

77. Mizuno Y et al: Q-angle influences tibiofemoral and patellofemoral kinematics, *J Orthop Res* 19:834, 2001.

78. Munro CF, Miller DI, and Fuglevand AJ: Ground reaction forces in running: a reexamination, *J Biomech* 20:147, 1987.

79. Myer GD, Ford KR, Palumbo JP, and Hewett TE: Neuromuscular traiing improves performance and lower-extremity biomechanics in female athletes, *J Strength Cond Res* 19:51, 2005.

80. Neptune RR, Wright IC, and van den Bogert AJ: The influence of orthotic devices and vastus medialis strength and timing on patellofemoral loads during running, *Clin Biomech* 15:611, 2000.

81. Oden RR: Tendon injuries about the ankle resulting from skiing, *Clin Orthop* 216:63, 1987.

82. Orchard JW, Fricker PA, Abud AT, and Mason BR: Biomechanics of iliotibial band friction syndrome in runners, *Am J Sports Med* 24:375, 1996.

83. Panjabi MM and White AA: *Biomechanics in the musculoskeletal system*, New York, Churchill Livingstone, 2001.

84. Papageoriou CD, Gil JE, Kanamori A, Fenwick JA, Woo SL, and Fu FH: The biomechanical interdependence between the anterior cruciate ligament replacement graft and the medial meniscus, *Am J Sports Med* 29:226, 2001.

85. Paul JP and McGrouther DA: Forces transmitted at the hip and knee joint of normal and disabled persons during a range of activities, *Acta Orthop Belg*, Suppl. 41:78, 1975.

86. Pollard CD, Heiderscheit BC, van Emmerik REA, and Hamill J: Gender differences in lower extremity coupling variability during an unanticipated cutting maneuver, *J Appl Biomech* 21:143, 2005.

87. Powers CM: Patellar kinematics, I: The influence of vastus muscle activity in subjects with and without patellofemoral pain, *Phys Ther* 80:956, 2000.

88. Powers CM: Patellar kinematics, II: The influence of the depth of the trochlear groove in subjects with and without patellofemoral pain, *Phys Ther* 80:965, 2000.

89. Reilly DT and Martens M: Experimental analysis of the quadriceps muscle force and patello-femoral joint reaction force for various activities, *Acta Orthop Scand* 43:126, 1972.

90. Renstrom P and Johnson RJ: Overuse injuries in sports: a review, *Sports Med* 2:316, 1985.

91. Richards DP, Barber A, and Troop RL: Iliotibial band Z-lengthening, *Arthroscopy* 19:326, 2003.

92. Robinson JR, Bull AMJ, and Amis A: Structural properties of the medial collateral ligament complex of the human knee, *J Biomech* 38:1067, 2005.

93. Roemer FW, Zhang Y, Niu J, Lynch JA, Crema MD, Marra MD, Nevitt MC, Felson DT, Hughes LB, El-Khoury GY, Englund M, and Guermazi A; Multicenter Osteoarthritis Study Investigators: Tibiofemoral joint osteoarthritis: risk factors for MR-depicted fast cartilage loss over a 30-month period in the multicenter osteoarthritis study, *Radiology* 252:772, 2009.

94. Safran MR, Benedetti RS, Bartolozzi AR III, and Mandelbaum BR: Lateral ankle sprains: a comprehensive review, I: Etiology, pathoanatomy, histopathogenesis, and diagnosis, *Med Sci Sports Exerc* 31, S429, 1999.

95. Sakai N, Luo ZP, Rand JA, and An KN: The influence of weakness in the vastus medialis oblique muscle on the patellofemoral joint: an in vitro biomechanical study, *Clin Biomech* 15:335, 2000.

96. Salem GJ and Powers CM: Patellofemoral joint kinetics during squatting in collegiate women athletes, *Clin Biomech* 16:424, 2001.

97. Sammarco GJ and Hockenbury RT: Biomechanics of the foot. In Nordin M and Frankel VH, eds: *Basic biomechanics of the musculoskeletal system* (3rd ed.), Philadelphia, 2001, Lippincott Williams & Wilkins.

98. Scarvell JM, Smith PN, Refshauge KM, Galloway H, and Woods K: Comparison of kinematics in the healthy and ACL injured knee using MRI, *J Biomech* 38:255, 2005.

99. Schache AG, Wrigley TV, Baker R, and Pandy MG: Biomechanical response to hamstring muscle strain injury, *Gait Posture* 29:332, 2009.

100. Schmitt KU, Schlittler M, and Boesiger P: Biomechanical loading of the hip during side jumps by soccer goalkeepers, *J Sports Sci* 28:53, 2010.

101. Seedhom BB, Dowson D, and Wright V: The load-bearing function of the menisci: a preliminary study. In Ingwerson OS et al, eds: *The knee joint: recent advances in basic research and clinical aspects,* Amsterdam, 1974, Excerpta Medica.

102. Shelburne KB, Pandy MG, Anderson FC, and Torry MR: Pattern of anterior cruciate ligament fore in normal walking, *J Biomech* 37:797, 2004.

103. Shelton WR, Barrett GR, and Dukes A: Early season anterior cruciate ligament tears: a treatment dilemma, *Am J Sports Med* 25:656, 1997.

104. Shrive NG, O'Connor JJ, and Goodfellow JW: Load-bearing in the knee joint, *Clin Orthop* 131:279, 1978.

105. Simonsen EB et al: Can the hamstring muscles protect the anterior cruciate ligament during a side-cutting maneuver? *Scand J Med Sci Sports* 10:78, 2000.

106. Sneyers CJ, Lysens R, Feys H, and Andries R: Influence of malalignment of feet on the plantar pressure pattern in running, *Foot Ankle Int* 16:624, 1995.

107. Stergiou N, Moraiti C, Giakas G, Ristanis S, and Anastasios DG: The effect of walking speed on the stability of the anterior cruciate ligament deficient knee, *Clin Biomech* 19:957, 2004.

108. Suter E and Herzog W: Does muscle inhibition after knee injury increase the risk of osteoarthritis? *Exerc and Sport Sci Rev* 28:15, 2000.

109. Suter E, Herzog W, De Souza K, and Bray R: Inhibition of the quadriceps muscles in patients with anterior knee pain, *J Appl Biomech* 14:360, 1998.

110. Tanner SM and Harvey JS: How we manage plantar fasciitis, *Physician Sportsmed* 16:39, 1988.

111. Thonnard JL, Bragard D, Willems PA, and Plaghki L: Stability of the braced ankle: a biomechanical investigation, *Am J Sports Med* 24:356, 1996.

112. Tyler TF, Nicholas SJ, Campbell RJ, and McHugh MP: The association of hip strength and flexibility with the incidence of adductor muscle strains in professional ice hockey players, *Am J Sports Med* 29:124, 2001.

113. van den Bogert AJ, Read L, and Nigg BM: An analysis of hip joint loading during walking, running, and skiing, *Med Sci Sports Exerc* 31:131, 1999.

114. Van Ingen Schenau GJ et al: A new skate allowing powerful plantar flexion improves performance, *Med Sci Sports Exerc* 28:531, 1996.

115. Vizsolyi P et al: Breaststroker's knee, *Am J Sports Med* 15:63, 1987.

116. White KK, Lee SS, Cutuk A, Hargens AR, and Pedowitz RA: EMG power spectra of intercollegiate athletes and anterior cruciate ligament injury risk in females, *Med Sci Sports Exerc* 35:371, 2003.

117. Williams DS III, McClay IA, and Hamill J: Arch structure and injury patterns in runners, *Clin Biomech* 16:341, 2001.

118. Williams DS III, McClay IA, Hamill J, and Buchanan TS: Lower extremity kinematic and kinetic differences in runners with high and low arches, *J Appl Biomech* 17:153, 2001.

119. Williams GN, Barrance PJ, Snyder-Mackler L, and Buchanan TS: Altered quadriceps control in people with anterior cruciate ligament deficiency, *Med Sci Sports Exerc* 36:1089, 2004.

120. Wilson DR, Feikes JD, Zavatsky AB, and O'Connor JJ: The components of passive knee movement are coupled to flexion angle, *J Biomech* 33:465, 2000.

121. Winby CR, Lloyd DG, Besier TF, and Kirk TB: Muscle and external load contribution to knee joint contact loads during normal gait, *J Biomech* 42(14):2294, 2009.

122. Yu B, Queen RM, Abbey AN, Liu Y, Moorman CT, and Garrett WE: Hamstring muscle kinematics and activation during overground sprinting, *J Biomech* 41:3121, 2008.

123. Zhang L-Q and Wang G: Dynamic and static control of the human knee joint in abduction-adduction, *J Biomech* 34:1107, 2001.

ANNOTATED READINGS

Galloway J and Hannaford D: *Running injuries: Treatment and prevention*, New York, 2009, Meyer & Meyer Fachverlag und Buchhandel GmbH.
  Includes information on prevention and treatment of numerous running-related injuries.

McDougall: *Born to run: A hidden tribe, superathletes, and the greatest race the world has never seen*, New York, 2009, Knopf.
  Describes practices of the Tarahumara Indians of Mexico's Copper Canyons, who run hundreds of miles without rest while enjoying uncanny health and serenity.

Perry J and Burnfield J: *Gait analysis: Normal and pathological function* (2nd ed.), Philadelphia, 2010, Slack.
  Comprises a definitive, comprehensive text on analyzing gait.

Puleo J and Milroy P: *Running anatomy*, Champaign, IL, 2010, Human Kinetics.
  Presents scientifically based advice on training for running.

RELATED WEB SITES

Arthroscopy.com
http://www.arthroscopy.com/sports.htm
  Includes information and color graphics on the foot and ankle and on knee surgery.

Biomechanics Laboratory University of Essen
http://www.uni-essen.de/~qpd800/research.html
  Includes links to animations of a barefoot pressure isobarograph, barefoot walking, in-shoe pressures in two running shoes, and others.

M & M Orthopaedics
http://mmortho.com/
  Includes information on ACL reconstruction, plantar fascitis, clubfoot, and hip and knee replacement, with QuickTime movies of the inside of a knee.

Medical Multimedia Group
http://www.medicalmultimediagroup.com/
  Includes descriptive information and video clips related to lower extremity injuries.

Northern Rockies Orthopaedics Specialists
http://www.orthopaedic.com
  Provides links to pages describing descriptions of injuries, diagnostic tests, and surgical procedures for the shoulder, elbow, wrist, and hand.

Rothman Institute
http://www.rothmaninstitute.com/
  Includes information on common sport injuries to the hip, knee, foot, and ankle.

Southern California Orthopaedic Institute
http://www.scoi.com
  Includes links to anatomical descriptions and labeled photographs of the hip, knee, ankle, and toe.

University of Washington Orthopaedic Physicians
http://www.orthop.washington.edu
  Provides radiographs and information on common injuries and pathological conditions for the hip, knee, ankle, and foot.

Wheeless' Textbook of Orthopaedics
http://www.wheelessonline.com/
  Provides comprehensive, detailed information, graphics, and related literature for all the joints.

KEY TERMS

collateral ligaments	major ligaments that cross the medial and lateral aspects of the knee
cruciate ligaments	major ligaments that cross each other in connecting the anterior and posterior aspects of the knee
hamstrings	the biceps femoris, semimembranosus, and semitendinosus
iliopsoas	the psoas major and iliacus muscles with a common insertion on the lesser trochanter of the femur
iliotibial band	thick, strong band of tissue connecting the tensor fascia lata to the lateral condyle of the femur and the lateral tuberosity of the tibia
menisci	cartilaginous discs located between the tibial and femoral condyles
muscle inhibition	inability to activate all motor units of a muscle during maximal voluntary contraction
patellofemoral joint	articulation between the patella and the femur
pelvic girdle	the two hip bones plus the sacrum, which can be rotated forward, backward, and laterally to optimize positioning of the hip joint
plantar fascia	thick bands of fascia that cover the plantar aspect of the foot
popliteus	muscle known as the unlocker of the knee because its action is lateral rotation of the femur with respect to the tibia
pronation	combined conditions of dorsiflexion, eversion, and abduction
Q-angle	angle formed between the anterior superior iliac spine, the center of the patella, and the tibial tuberosity
quadriceps	the rectus femoris, vastus lateralis, vastus medialis, and vastus intermedius
supination	combined conditions of plantar flexion, inversion, and adduction
tibiofemoral joint	dual condyloid articulations between the medial and lateral condyles of the tibia and the femur, composing the main hinge joint of the knee
valgus	condition of outward deviation in alignment from the proximal to the distal end of a body segment
varus	condition of inward deviation in alignment from the proximal to the distal end of a body segment

The Biomechanics of the Human Spine

After completing this chapter, you will be able to:

Explain how anatomical structure affects movement capabilities of the spine.

Identify factors influencing the relative mobility and stability of different regions of the spine.

Explain the ways in which the spine is adapted to carry out its biomechanical functions.

Explain the relationship between muscle location and the nature and effectiveness of muscle action in the trunk.

Describe the biomechanical contributions to common injuries of the spine.

The spine is a complex and functionally significant segment of the human body. Providing the mechanical linkage between the upper and lower extremities, the spine enables motion in all three planes, yet still functions as a bony protector of the delicate spinal cord. To many researchers and clinicians, the lumbar region of the spine is of particular interest because low back pain is a major medical and socioeconomic problem in modern times.

STRUCTURE OF THE SPINE

Vertebral Column

The spine consists of a curved stack of 33 vertebrae divided structurally into five regions (Figure 9-1). Proceeding from superior to inferior, there are 7 cervical vertebrae, 12 thoracic vertebrae, 5 lumbar vertebrae, 5 fused sacral vertebrae, and 4 small, fused coccygeal vertebrae. There may be one extra vertebra or one less, particularly in the lumbar region.

Because of structural differences and the ribs, varying amounts of movement are permitted between adjacent vertebrae in the cervical, thoracic, and lumbar portions of the spine. Within these regions, two adjacent

FIGURE 9-1

A. Left lateral and **B.** posterior views of the major regions of the spine. From Shier, Butler, and Lewis, *Hole's Human Anatomy and Physiology,* © 1996. Reprinted by permission of The McGraw-Hill Companies, Inc.

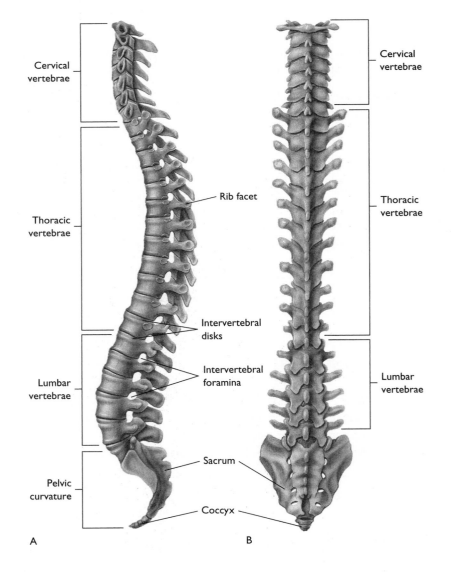

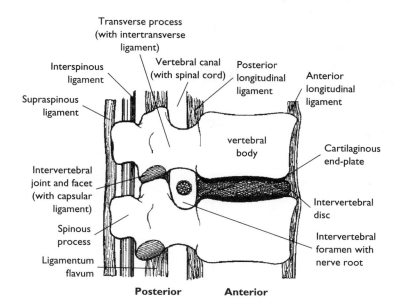

Transverse process
(with intertransverse
ligament)

Vertebral canal
(with spinal cord)

Posterior
longitudinal
ligament

Anterior
longitudinal
ligament

Interspinous
ligament

Supraspinous
ligament

vertebral
body

Cartilaginous
end-plate

Intervertebral
joint and facet
(with capsular
ligament)

Intervertebral
disc

Spinous
process

Intervertebral
foramen with
nerve root

Ligamentum
flavum

Posterior   **Anterior**

FIGURE 9-2

The motion segment,
composed of two adjacent
vertebrae and the associated
soft tissues, is the functional
unit of the spine.

vertebrae and the soft tissues between them are known as a motion segment. The motion segment is considered the functional unit of the spine (Figure 9-2).

Each motion segment contains three joints. The vertebral bodies separated by the intervertebral discs form a symphysis type of amphiarthrosis. The right and left facet joints between the superior and inferior articular processes are diarthroses of the gliding type that are lined with articular cartilage.

Vertebrae

A typical vertebra consists of a body, a hollow ring known as the *neural arch,* and several bony processes (Figure 9-3). The vertebral bodies serve as the primary weight-bearing components of the spine. The neural arches and posterior sides of the bodies and intervertebral discs form a protective passageway for the spinal cord and associated blood vessels known as the vertebral canal. From the exterior surface of each neural arch, several bony processes protrude. The spinous and transverse processes serve as outriggers to improve the mechanical advantage of the attached muscles.

The first two cervical vertebrae are specialized in shape and function. The first cervical vertebra, known as the *atlas,* provides a reciprocally shaped receptacle for the condyles of the occiput of the skull. The atlanto-occipital joint is extremely stable, with flexion/extension of about 14–15° permitted, but with virtually no motion occurring in any other plane (17). A large range of axial rotation is provided at the next joint between the atlas and the second cervical vertebrae, the axis. Motion at the atlanto-axial joint averages around 75° of rotation, 14° of extension, and 24° of lateral flexion (17).

There is a progressive increase in vertebral size from the cervical region down through the lumbar region (Figure 9-3). The lumbar vertebrae, in particular, are larger and thicker than the vertebrae in the superior regions of the spine. This serves a functional purpose, since when the body is in an upright position each vertebra must support the weight of not only the arms and head but all the trunk positioned above it. The increased surface area of the lumbar vertebrae reduces the amount of stress

motion segment
two adjacent vertebrae and the associated soft tissues; the functional unit of the spine

●*The spine may be viewed as a triangular stack of articulations, with symphysis joints between vertebral bodies on the anterior side and two gliding diarthrodial facet joints on the posterior side.*

●*Although all vertebrae have the same basic shape, there is a progressive superior-inferior increase in the size of the vertebral bodies and a progression in the size and orientation of the articular processes.*

●*The orientation of the facet joints determines the movement capabilities of the motion segment.*

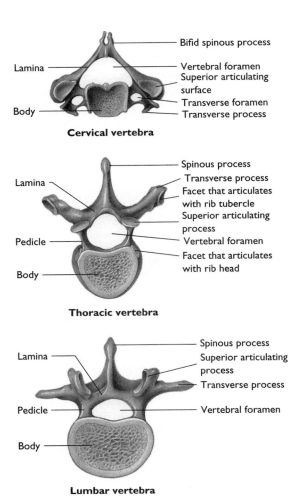

to which these vertebrae would otherwise be subjected. The weight-bearing surface area of the intervertebral disc also increases with the weight supported in all mammals (129).

The size and angulation of the vertebral processes vary throughout the spinal column (Figure 9-4). This changes the orientation of the facet joints, which limit range of motion in the different spinal regions. In addition to channeling the movement of the motion segment, the facet joints assist in load bearing. The facet joints and discs provide about 80% of the spine's ability to resist rotational torsion and shear, with half of this contribution from the facet joints (43, 61). The facet joints also sustain up to approximately 30% of the compressive loads on the spine, particularly when the spine is in hyperextension (Figure 9-5) (72). Contact forces are largest at the L5-S1 facet joints (77). Recent studies suggest that 15–40% of chronic low back pain emanates from the facet joints (11).

Intervertebral Discs

The articulations between adjacent vertebral bodies are symphysis joints with intervening fibrocartilaginous discs that act as cushions. Healthy intervertebral discs in an adult account for approximately one-fourth of the height of the spine. When the trunk is erect, the differences in the anterior and posterior thicknesses of the discs produce the lumbar, thoracic, and cervical curves of the spine.

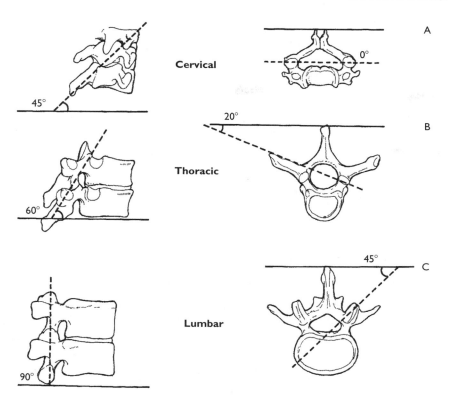

Cervical

Thoracic

Lumbar

FIGURE 9-4

Approximate orientations of the facet joints. **A.** Lower cervical spine, with facets oriented 45° to the transverse plane and parallel to the frontal plane. **B.** Thoracic spine, with facets oriented 60° to the transverse plane and 20° to the frontal plane. **C.** Lumbar spine, with facets oriented 90° to the transverse plane and 45° to the frontal plane.

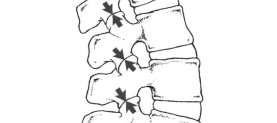

Hyperextension

FIGURE 9-5

Hyperextension of the lumbar spine creates compression at the facet joints.

FIGURE 9-6

In the intervertebral disc, the annulus fibrosus, made up of laminar layers of criss-crossed collagen fibers, surrounds the nucleus pulposus.

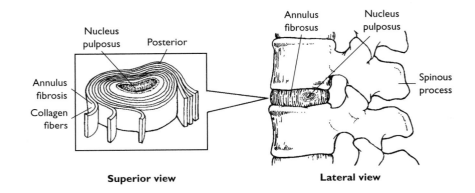

Superior view   Lateral view

annulus fibrosus

thick, fibrocartilaginous ring that forms the exterior of the intervertebral disc

nucleus pulposus

colloidal gel with a high fluid content, located inside the annulus fibrosus of the intervertebral disc

The intervertebral disc incorporates two functional structures: A thick outer ring composed of fibrous cartilage called the annulus fibrosus, or annulus, surrounds a central gelatinous material known as the nucleus pulposus, or nucleus (Figure 9-6). The annulus consists of about 90 concentric bands of collagenous tissue that are bonded together. The collagen fibers of the annulus crisscross vertically at about 30° angles to each other, making the structure more sensitive to rotational strain than to compression, tension, and shear (43). These fibers, which are crucial for the mechanical functioning of the disc, display changes in organization and orientation with loading of the disc, as well as with disc degeneration (64, 90). The nucleus of a young, healthy disc is approximately 90% water, with the remainder being collagen and proteoglycans, specialized materials that chemically attract water (98). The extremely high fluid content of the nucleus makes it resistant to compression.

Mechanically, the annulus acts as a coiled spring whose tension holds the vertebral bodies together against the resistance of the nucleus pulposus, and the nucleus pulposus acts like a ball bearing composed of an incompressible gel (Figure 9-7) (74). During flexion and extension, the vertebral bodies roll over the nucleus while the facet joints guide the movements. As shown in Figure 9-8, spinal flexion, extension, and lateral flexion produce compressive stress on one side of the discs and tensile stress on the other, whereas spinal rotation creates shear stress in the

FIGURE 9-7

Mechanically, the annulus fibrosus behaves as a coiled spring, holding the vertebral bodies together, whereas the nucleus pulposus acts like a ball bearing that the vertebrae roll over during flexion/extension and lateral bending.

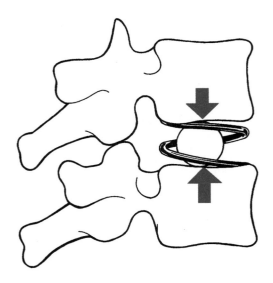

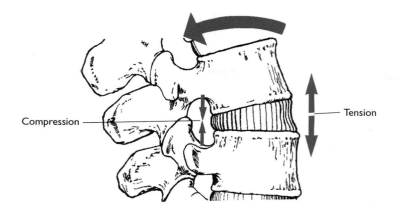

FIGURE 9-8

When the spine bends, a tensile load is created on one side of the discs, and a compressive load is created on the other.

discs (Figure 9-9) (72). Stress on the discs is significantly higher with flexion as compared to rotation, with bending stress approximately 450 times greater than the twisting stress from the same angle of bending or twisting of the annulus fibers (26). During daily activities, compression is the most common form of loading on the spine.

When a disc is loaded in compression, it tends to simultaneously lose water and absorb sodium and potassium until its internal electrolyte concentration is sufficient to prevent further water loss (70). When this chemical equilibrium is achieved, internal disc pressure is equal to the external pressure (8). Continued loading over a period of several hours results in a further slight decrease in disc hydration (1). For this reason, the spine undergoes a height decrease of up to nearly 2 cm over the course of a day, with approximately 54% of this loss occurring during the first 30 minutes after an individual gets up in the morning (101).

Once pressure on the discs is relieved, the discs quickly reabsorb water, and disc volumes and heights are increased (70). Astronauts experience a temporary increase in spine height of approximately 5 cm while free from the influence of gravity (88). On earth, disc height and volume are typically greatest when a person first arises in the morning. Because increased disc volume also translates to increased spinal stiffness, there appears to be a heightened risk of disc injury early in the morning (38). Measurements of spinal shrinkage following activities performed for one hour immediately after rising in the morning yielded average values of −7.4 mm for standing, −5.0 mm for sitting, −7.9 mm for walking, −3.7 mm for cycling, and +0.4 mm for lying down (121). Body positions that allow rehydration and height increase in the discs are spinal hyperextension in the prone position and trunk flexion in the supine position (93).

The intervertebral discs have a blood supply up to about the age of 8 years, but after that the discs must rely on a mechanically based means

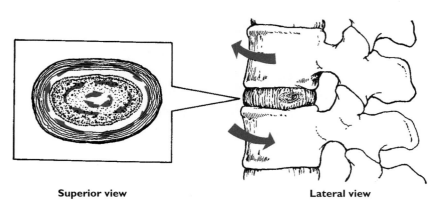

Superior view   **Lateral view**

FIGURE 9-9

Spinal rotation creates shear stress within the discs, with the greatest shear around the periphery of the annulus.

• *It is important not to maintain any one body position for too long, since the intervertebral discs rely on body movement to pump nutrients in and waste products out.*

for maintaining a healthy nutritional status. Intermittent changes in posture and body position alter internal disc pressure, causing a *pumping action* in the disc. The influx and outflux of water transports nutrients in and flushes out metabolic waste products, basically fulfilling the same function that the circulatory system provides for vascularized structures within the body. Maintaining even an extremely comfortable fixed body position over a long period curtails this pumping action and can negatively affect disc health. Research has shown that there is a zone of optimal loading frequency and magnitude that promotes disc health (125).

Injury and aging irreversibly reduce the water-absorbing capacity of the discs, with a concomitant decrease in shock-absorbing capability. Magnetic resonance imaging (MRI) studies show degenerative changes to be the most common at L5-S1, the disc subjected to the most mechanical stress by virtue of its position (104). However, the fluid content of all discs begins to diminish around the second decade of life (8). A typical geriatric disc has a fluid content that is reduced by approximately 35% (128). As this normal degenerative change occurs, abnormal movements occur between adjacent vertebral bodies, and more of the compressive, tensile, and shear loads on the spine must be assumed by other structures—particularly the facets and joint capsules. Results include reduced height of the spinal column, often accompanied by degenerative changes in the spinal structures that are forced to assume the loads of the discs. Postural alterations may also occur. The normal lordotic curve of the lumbar region may be reduced as an individual attempts to relieve compression on the facet joints by maintaining a posture of spinal flexion (128). Factors such as habitual smoking and exposure to vibration can negatively affect disc nutrition, while regular exercise can improve it (98).

• *Although most of the load sustained by the spine is borne by the symphysis joints, the facet joints may play a role, particularly when the spine is in hyperextension and when disc degeneration has occurred.*

Ligaments

A number of ligaments support the spine, contributing to the stability of the motion segments (Figure 9-10). The powerful anterior longitudinal ligament and the weaker posterior longitudinal ligament connect the vertebral bodies in the cervical, thoracic, and lumbar regions. The supraspinous

• *The enlarged cervical portion of the supraspinous ligament is the ligamentum nuchae, or ligament of the neck.*

FIGURE 9-10

The major ligaments of the spine. (The intertransverse ligament is not visible in this medial section through the spine.)

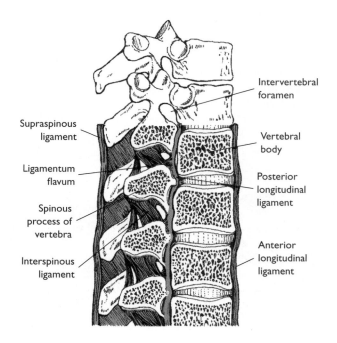

Supraspinous ligament

Ligamentum flavum

Spinous process of vertebra

Interspinous ligament

Intervertebral foramen

Vertebral body

Posterior longitudinal ligament

Anterior longitudinal ligament

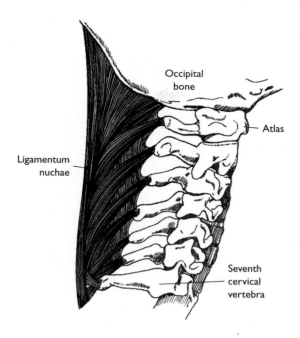

Occipital bone

Atlas

Ligamentum nuchae

Seventh cervical vertebra

FIGURE 9-11

The supraspinous ligament is well developed in the cervical region, where it is referred to as the *ligamentum nuchae.*

ligamentum flavum
yellow ligament that connects the laminae of adjacent vertebrae; distinguished by its elasticity

prestress
stress on the spine created by tension in the resting ligaments

primary spinal curves
curves that are present at birth

secondary spinal curves
curves that do not develop until the weight of the body begins to be supported in sitting and standing positions

ligament attaches to the spinous processes throughout the length of the spine. This ligament is prominently enlarged in the cervical region, where it is referred to as the *ligamentum nuchae,* or *ligament of the neck* (Figure 9-11). Adjacent vertebrae have additional connections between spinous processes, transverse processes, and laminae, supplied respectively by the interspinous ligaments, the intertransverse ligaments, and the ligamenta flava.

Another major ligament, the ligamentum flavum, connects the laminae of adjacent vertebrae. Although most spinal ligaments are composed primarily of collagen fibers that stretch only minimally, the ligamentum flavum contains a high proportion of elastic fibers, which lengthen when stretched during spinal flexion and shorten during spinal extension. The ligamentum flavum is in tension even when the spine is in anatomical position, enhancing spinal stability. This tension creates a slight, constant compression in the intervertebral discs, referred to as prestress.

Spinal Curves

As viewed in the sagittal plane, the spine contains four normal curves. The thoracic and sacral curves, which are concave anteriorly, are present at birth and are referred to as primary curves. The lumbar and cervical curves, which are concave posteriorly, develop from supporting the body in an upright position after young children begin to sit up and stand. Since these curves are not present at birth, they are known as the secondary spinal curves. Although the cervical and thoracic curves change little during the growth years, the curvature of the lumbar spine increases approximately 10% between the ages of 7 and 17 (124). Spinal curvature (posture) is influenced by heredity, pathological conditions, an individual's mental state, and the forces to which the spine is habitually subjected. Mechanically, the curves enable the spine to absorb more shock without injury than if the spine were straight.

As discussed in Chapter 4, bones are constantly modeled or shaped in response to the magnitudes and directions of the forces acting on them. Similarly, the four spinal curves can become distorted when the spine is habitually subjected to asymmetrical forces.

Note the relatively flat spine of this 3-year-old. The lumbar curve does not reach full development until approximately age 17.

lordosis
extreme lumbar curvature

Exaggeration of the lumbar curve, or lordosis, is often associated with weakened abdominal muscles and anterior pelvic tilt (Figure 9-12). Causes of lordosis include congenital spinal deformity, weakness of the abdominal muscles, poor postural habits, and overtraining in sports requiring repeated lumbar hyperextension, such as gymnastics, figure skating, javelin throwing, and swimming the butterfly stroke. Because lordosis places added compressive stress on the posterior elements of the spine, some have hypothesized that excessive lordosis is a risk factor for low back pain development. Limited range of motion in hip extension is associated with exaggerated lumbar lordosis (45). Obesity causes reduced range of motion of the entire spine and pelvis, and obese individuals resultingly display increased anterior pelvic tilt and an associated increased lumbar lordosis (123). Similarly, increased anterior pelvic tilt and increased lordosis are greater during running than during walking (45).

kyphosis
extreme thoracic curvature

Another abnormality in spinal curvature is kyphosis (exaggerated thoracic curvature) (Figure 9-12). The incidence of kyphosis has been estimated to be as high as 8% in the general population, with equal distribution across genders (2). Kyphosis can result from a congenital abnormality, a pathology such as osteoporosis, or Scheuermann's disease, in which one or more wedge-shaped vertebrae develop because of abnormal epiphyseal plate behavior. Scheuermann's disease typically develops in individuals between the ages of 10 and 16 years, which is the period of most rapid growth of the thoracic spine (35). Both genetic and biomechanical factors are believed to play a role (7). The condition has been called *swimmer's back* because it is frequently seen in adolescents who have trained heavily with the butterfly stroke (119). Scheuermann's disease is not limited to swimmers, however, with research showing a strong association between incidence of this pathology and cumulative training time in any sport (130). Treatment for mild cases may consist of exercises to strengthen the posterior thoracic muscles, although bracing or surgical corrections are used in more severe cases. Kyphosis also often develops in elderly women with osteoporosis, as discussed in Chapter 4. Both the thoracic vertebrae and the intervertebral discs in the region develop a characteristic wedge shape (110).

scoliosis
lateral spinal curvature

Lateral deviation or deviations in spinal curvature are referred to as scoliosis (Figure 9-12). The lateral deformity is coupled with rotational deformity of the involved vertebrae, with the condition ranging from mild

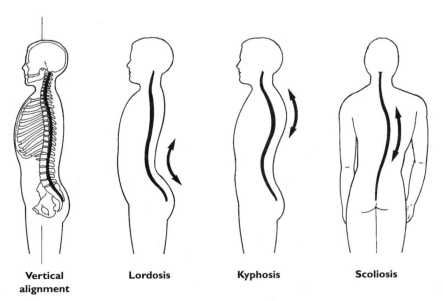

FIGURE 9-12

Abnormal spinal curvatures.

Vertical   Lordosis   Kyphosis   Scoliosis
alignment

to severe. Scoliosis may appear as either a C- or an S-curve involving the thoracic spine, the lumbar spine, or both.

A distinction is made between structural and nonstructural scoliosis. Structural scoliosis involves inflexible curvature that persists even with lateral bending of the spine. Nonstructural scoliotic curves are flexible and are corrected with lateral bending.

Scoliosis results from a variety of causes. Congenital abnormalities and selected cancers can contribute to the development of structural scoliosis. Nonstructural scoliosis may occur secondary to a leg length discrepancy or local inflammation. Small lateral deviations in spinal curvature are relatively common and may result from a habit such as carrying books or a heavy purse on one side of the body every day. Approximately 70–90% of all scoliosis, however, is termed *idiopathic,* which means that the cause is unknown (133). Idiopathic scoliosis is most commonly diagnosed in children between the ages of 10 and 13 years, but can be seen at any age. It is present in 2–4% of children between 10 and 16 years of age and is more common in females (100). Low bone mineral density is typically associated with idiopathic scoliosis and may play a causative role in its development (28).

Symptoms associated with scoliosis vary with the severity of the condition. Mild cases may be nonsymptomatic and may self-correct with time. A growing body of evidence supports the effectiveness of appropriate stretching and strengthening exercises for resolving the symptoms and appearance of mild to moderate scoliosis (133). Severe scoliosis, however, which is characterized by extreme lateral deviation and localized rotation of the spine, can be painful and deforming, and is treated with bracing and/or surgery. As is the case with kyphosis, both the vertebrae and the intervertebral discs in the affected region(s) assume a wedge shape (114).

MOVEMENTS OF THE SPINE

As a unit, the spine allows motion in all three planes of movement, as well as circumduction. Because the motion allowed between any two adjacent vertebrae is small, however, spinal movements always involve a number of motion segments. The range of motion (ROM) allowed at each motion segment is governed by anatomical constraints that vary through the cervical, thoracic, and lumbar regions of the spine.

●*The movement capabilities of the spine as a unit are those of a ball-and-socket joint, with movement in all three planes, as well as circumduction, allowed.*

Female gymnasts undergo extreme lumbar hyperextension during many commonly performed skills. Photo courtesy of Royalty-Free/ CORBIS.

Flexion, Extension, and Hyperextension

The ROM for flexion/extension of the motion segments is considerable in the cervical and lumbar regions, with representative values as high as 17° at the C5-C6 vertebral joint and 20° at L5-S1. In the thoracic spine, however, due to the orientation of the facets, the ROM increases from only approximately 4° at T1-T2 to approximately 10° at T11-T12 (127).

It is important not to confuse spinal flexion with hip flexion or anterior pelvic tilt, although all three motions occur during an activity such as touching the toes (Figure 9-13). Hip flexion consists of anteriorly directed sagittal plane rotation of the femur with respect to the pelvic girdle (or vice versa), and anterior pelvic tilt is anteriorly directed movement of the anterior superior iliac spine with respect to the pubic symphysis, as discussed in Chapter 8. Just as anterior pelvic tilt facilitates hip flexion, it also promotes spinal flexion.

Extension of the spine backward past anatomical position is termed *hyperextension*. The ROM for spinal hyperextension is considerable in the cervical and lumbar regions. Lumbar hyperextension is required in the execution of many sport skills, including several swimming strokes, the high jump and pole vault, and numerous gymnastic skills. For example, during the execution of a back handspring, the curvature normally present in the lower lumbar region may increase twentyfold (53).

Lateral Flexion and Rotation

Frontal plane movement of the spine away from anatomical position is termed *lateral flexion*. The largest ROM for lateral flexion occurs in the cervical region, with approximately 9–10° of motion allowed at C4-C5. Somewhat less lateral flexion is allowed in the thoracic region, where the ROM between adjacent vertebrae is about 6°, except in the lower segments, where lateral flexion capability may be as high as approximately 8–9°. Lateral flexion in the lumbar spine is also on the order of 6°, except at L5-S1, where it is reduced to only about 3° (127).

Spinal rotation in the transverse plane is again freest in the cervical region of the spine, with up to 12° of motion allowed at C1-C2. It is next freest in the thoracic region, where approximately 9° of rotation is permitted among the upper motion segments. From T7-T8 downward, the range of rotational capability progressively decreases, with only about 2° of motion allowed in the lumbar spine due to the interlocking of the articular

FIGURE 9-13

When the trunk is flexed, the first 50–60° of motion occurs in the lumbar spine, with additional motion resulting from anterior pelvic tilt.

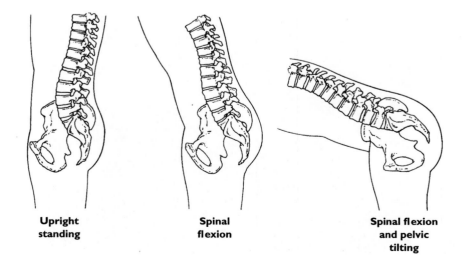

Upright standing   **Spinal flexion**   **Spinal flexion and pelvic tilting**

processes there. At the lumbosacral joint, however, rotation on the order of 5° is allowed (127). Since the structure of the spine causes lateral flexion and rotation to be coupled, rotation is accompanied by slight lateral flexion to the same side, although this motion is not observable with the naked eye.

Studies of activities of daily living have quantified ROMs in the cervical and lumbar regions of the spine. Among these activities, backing up a car was found to require the most motion in the cervical region, with approximately 32% of sagittal, 26% of lateral, and 92% of rotational motion capability involved (13). Similarly, the task requiring the greatest lumbar motion was picking up an object from the floor (14). ROM in the cervical spine has been found to decrease linearly with increasing age, with a loss in passive range of motion of about 0.5 degrees per year (103).

MUSCLES OF THE SPINE

The muscles of the neck and trunk are named in pairs, with one on the left and the other on the right side of the body. These muscles can cause lateral flexion and/or rotation of the trunk when they act unilaterally, and trunk flexion or extension when acting bilaterally. The primary functions of the major muscles of the spine are summarized in Table 9-1.

Anterior Aspect

The major anterior muscle groups of the cervical region are the prevertebral muscles, including the rectus capitis anterior, rectus capitis lateralis, longus capitis, and longus colli, and the eight pairs of hyoid muscles (Figures 9-14 and 9-15). Bilateral tension development by these muscles results in flexion of the head, although the main function of the hyoid muscles appears to be to move the hyoid bone during the act of swallowing. Unilateral tension development in the prevertebrals contributes to lateral flexion of the head toward the contracting muscles or to rotation of the head away from the contracting muscles, depending on which other muscles are functioning as neutralizers.

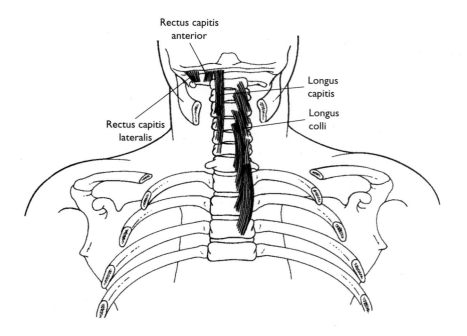

FIGURE 9-14

Anterior muscles of the cervical region.

TABLE 9-1

Muscles of the Spine

MUSCLE	PROXIMAL ATTACHMENT	DISTAL ATTACHMENT	PRIMARY ACTION(S) ABOUT THE HIP	INNERVATION
Prevertebral muscles (rectus capitis anterior, rectus capitis lateralis, longus capitis, longus coli)	Anterior aspect of occipital bone and cervical vertebrae	Anterior surfaces of cervical and first three thoracic vertebrae	Flexion, lateral flexion, rotation to opposite side	Cervical nerves (C_1–C_6)
Rectus abdominis	Costal cartilage of ribs 5–7	Pubic crest	Flexion, lateral flexion	Intercostal nerves (T_6–T_{12})
External oblique	External surface of lower eight ribs	Linea alba and anterior iliac crest	Flexion, lateral flexion, rotation to opposite side	Intercostal nerves (T_7–T_{12})
Internal oblique	Linea alba and lower four ribs	Inguinal ligament, iliac crest, lumbodorsal fascia	Flexion, lateral flexion, rotation to same side	Intercostal nerves (T_7–T_{12}, L_1)
Splenius (capitis and cervicis)	Mastoid process of temporal bone, transverse processes of first three cervical vertebrae	Lower half of ligamentum nuchae, spinous processes of seventh cervical and upper six thoracic vertebrae	Extension, lateral flexion, rotation to same side	Middle and lower cervical nerves (C_4–C_8)
Suboccipitals (obliquus capitus superior and inferior, rectus capitis posterior major and minor)	Occipital bone, transverse process of first cervical vertebra	Posterior surfaces of first two cervical vertebrae	Extension, lateral flexion, rotation to same side	Suboccipital nerve (C_1)
Erector spinae (spinalis, longissimus, iliocostalis)	Lower part of ligamentum nuchae, posterior cervical, thoracic, and lumbar spine, lower nine ribs, iliac crest, posterior sacrum	Mastoid process of temporal bone, posterior cervical, thoracic, and lumbar spine, twelve ribs	Extension, lateral flexion, rotation to opposite side	Spinal nerves (T_1–T_{12})
Semispinalis (capitis, cervicis, thoracis)	Occipital bone, spinous processes of thoracic vertebrae 2–4	Transverse presses of thoracic and seventh cervical vertebrae	Extension, lateral flexion, rotation to opposite side	Cervical and thoracic spinal nerves (C_1–T_{12})
Deep spinal muscles (multifidi, rotatores, interspinales, intertransversarii, levatores costarum)	Posterior processes of all vertebrae, posterior sacrum	Spinous and transverse processes and laminae of vertebrae below those of proximal attachment	Extension, lateral flexion, rotation to opposite side	Spinal and intercostal nerves (T_1–T_{12})
Sternocleidomastoid	Mastoid process of temporal bone	Superior sternum, inner third of clavicle	Flexion of neck, extension of head, lateral flexion, rotation to opposite side	Accessory nerve and C_2 spinal nerve
Levator scapulae	Transverse processes of first four cervical vertebrae	Vertebral border of scapula	Lateral flexion	Spinal nerves (C_3–C_4), dorsal scapular nerve (C_3–C_5)
Scaleni (scalenus anterior, medius, posterior)	Transverse processes of cervical vertebrae	Upper two ribs	Flexion, lateral flexion	Cervical nerves (C_3–C_7)
Quadratus lumborum	Last rib, transverse processes of first four lumbar vertebrae	Iliolumbar ligament, adjacent iliac crest	Lateral flexion	Spinal nerves (T_{12}–L_4)
Psoas major	Sides of twelfth thoracic and all lumbar vertebrae	Lesser trochanter of the femur	Flexion	Femoral nerve (L_1–L_3)

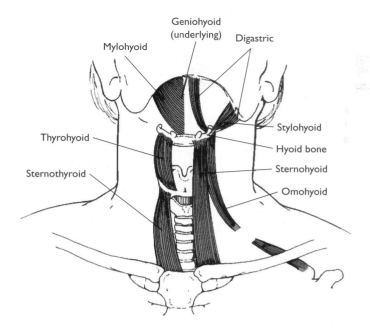

FIGURE 9-15
The hyoid muscles.

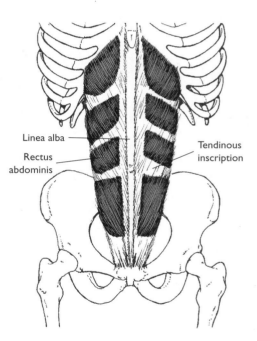

FIGURE 9-16
The rectus abdominis.

The main abdominal muscles are the rectus abdominis, the external obliques, and the internal obliques (Figures 9-16, 9-17, and 9-18). Functioning bilaterally, these muscles are the major spinal flexors and also reduce anterior pelvic tilt. Unilateral tension development by the muscles produces lateral flexion of the spine toward the tensed muscles. Tension development in the internal obliques causes rotation of the spine toward the same side. Tension development by the external obliques results in rotation toward the opposite side. If the spine is fixed, the internal obliques produce pelvic rotation toward the opposite side, with the external obliques producing rotation of the pelvis toward the same side. These muscles also form the major part of the abdominal wall, which protects the internal organs of the abdomen.

The rectus abdominis is a prominent abdominal muscle.

FIGURE 9-17

The external obliques.

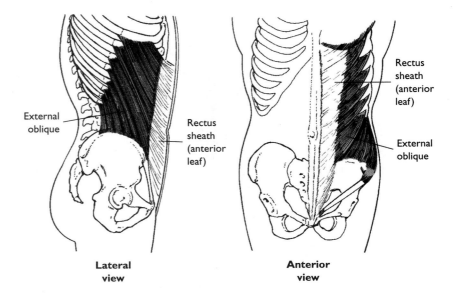

Lateral view | Anterior view

FIGURE 9-18

The internal obliques.

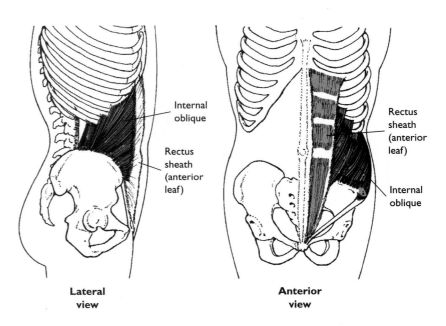

Lateral view | Anterior view

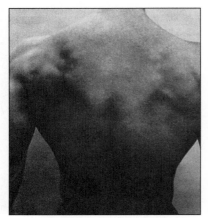

The superficial muscles of the posterior trunk.

Posterior Aspect

The splenius capitis and splenius cervicis are the primary cervical extensors (Figure 9-19) (89). Bilateral tension development in the four suboccipitals— the rectus capitis posterior major and minor and the obliquus capitis superior and inferior—assist (Figure 9-20). When these posterior cervical muscles develop tension on one side only, they laterally flex or rotate the head toward the side of the contracting muscles.

The posterior thoracic and lumbar region muscle groups are the massive erector spinae (sacrospinalis), the semispinalis, and the deep spinal muscles. As shown in Figure 9-21, the erector spinae group includes the spinalis, longissimus, and iliocostalis muscles. The semispinalis, with its capitis, cervicis, and thoracis branches, is shown in Figure 9-22. The deep spinal muscles, including the multifidi, rotatores, interspinales, intertransversarii,

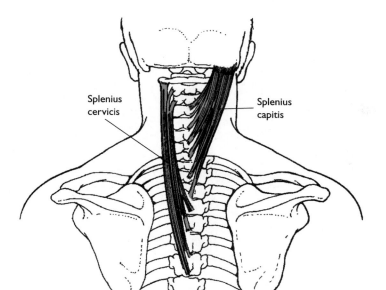

FIGURE 9-19

The major cervical extensors.

Splenius
cervicis

Splenius
capitis

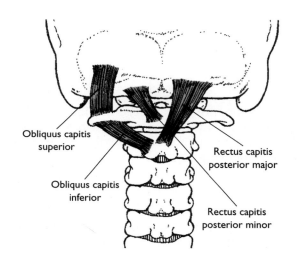

FIGURE 9-20

The suboccipital muscles.

Obliquus capitis
superior

Rectus capitis
posterior major

Obliquus capitis
inferior

Rectus capitis
posterior minor

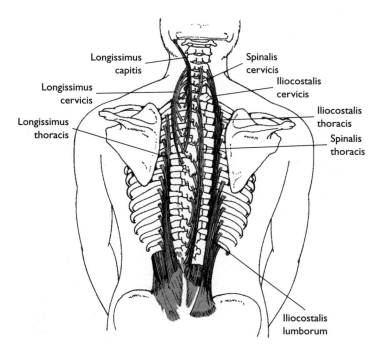

FIGURE 9-21

The erector spinae group.

Longissimus
capitis

Spinalis
cervicis

Longissimus
cervicis

Iliocostalis
cervicis

Longissimus
thoracis

Iliocostalis
thoracis

Spinalis
thoracis

Iliocostalis
lumborum

FIGURE 9-22

The semispinalis group.

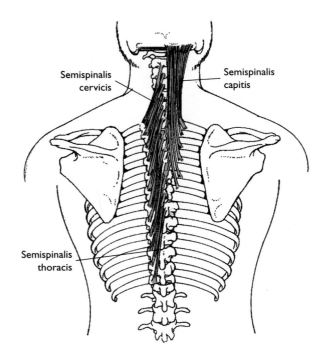

●*The prominent erector spinae muscle group—the major extensor and hyperextensor of the trunk—is the muscle group of the trunk most often strained.*

and levatores costarum, are represented in Figure 9-23. The muscles of the erector spinae group are the major extensors and hyperextensors of the trunk. All posterior trunk muscles contribute to extension and hyperextension when contracting bilaterally and to lateral flexion when contracting unilaterally.

Lateral Aspect

Muscles on the lateral aspect of the neck include the prominent sternocleidomastoid, the levator scapulae, and the scalenus anterior, posterior,

FIGURE 9-23

The deep spinal muscles.

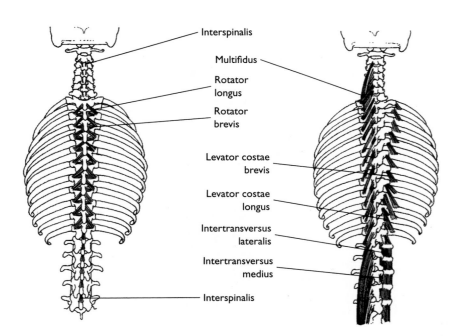

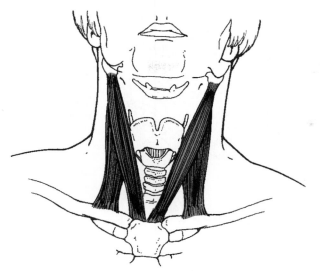

Sternocleidomastoid

FIGURE 9-24

The sternocleidomastoid.

and medius (Figures 9-24, 9-25, and 9-26). Bilateral tension development in the sternocleidomastoid may result in either flexion of the neck or extension of the head, with unilateral contraction producing lateral flexion to the same side or rotation to the opposite side. The levator scapulae can also contribute to lateral flexion of the neck when contracting unilaterally with the scapula stabilized. The three scalenes assist with flexion and lateral flexion of the neck, depending on whether tension development is bilateral or unilateral.

In the lumbar region, the quadratus lumborum and psoas major are large, laterally oriented muscles (Figures 9-27 and 9-28). These muscles function bilaterally to flex and unilaterally to laterally flex the lumbar spine.

● *Many muscles of the neck and trunk cause lateral flexion when contracting unilaterally but either flexion or extension when contracting bilaterally.*

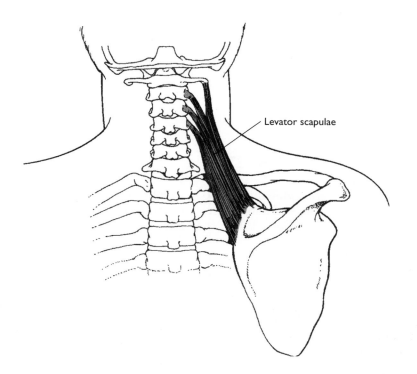

Levator scapulae

FIGURE 9-25

The levator scapulae.

FIGURE 9-26

The scaleni muscles.

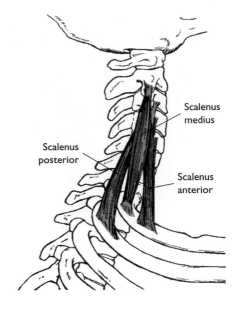

FIGURE 9-27

The quadratus lumborum.

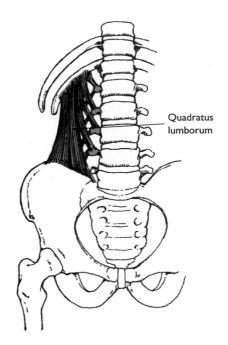

LOADS ON THE SPINE

Forces acting on the spine include body weight, tension in the spinal ligaments, tension in the surrounding muscles, intraabdominal pressure, and any applied external loads. When the body is in an upright position, the major form of loading on the spine is axial. In this position, body weight, the weight of any load held in the hands, and tension in the surrounding ligaments and muscles all contribute to spinal compression.

During erect standing, the total-body center of gravity is anterior to the spinal column, placing the spine under a constant forward-bending moment (Figure 9-29). To maintain body position, this torque must be counteracted by tension in the back extensor muscles. As the trunk or the arms are flexed, the increasing moment arms of these body segments

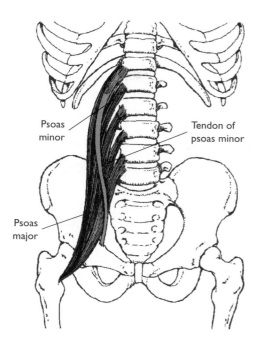

FIGURE 9-28

The psoas muscle.

Psoas minor

Tendon of psoas minor

Psoas major

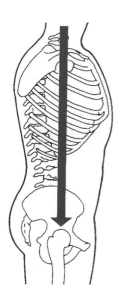

FIGURE 9-29

Because the line of gravity for the head, trunk, and upper extremity passes anterior to the vertebral column, a forward torque acts on the spine.

contribute to increasing flexor torque and increasing compensatory tension in the back extensor muscles (Figure 9-30). Because the spinal muscles have extremely small moment arms with respect to the vertebral joints, they must generate large forces to counteract the torques produced about the spine by the weights of body segments and external loads (see Sample Problem 9.1). Consequently, the major force acting on the spine is usually that derived from muscle activity (105). In comparison to the load present during upright standing, compression on the lumbar spine increases with sitting, increases more with spinal flexion, and increases still further with a slouched sitting position (Figure 9-31). During sitting, the pelvis rotates backward and the normal lumbar lordosis tends to flatten, resulting in increased loading on the intervertebral discs (126). A slumped sitting posture increases disc loading even more (126). Ergonomically designed chairs that provide lumbar support and enable slight forward tilting of

● *Because the spinal muscles have small moment arms, they must generate large forces to counteract the flexion torques produced by the weight of body segments and external loads.*

FIGURE 9-30

The back muscles, with a moment arm of approximately 6 cm, must counter the torque produced by the weights of body segments plus any external load. This illustrates why it is advisable to lift and carry heavy objects close to the trunk.

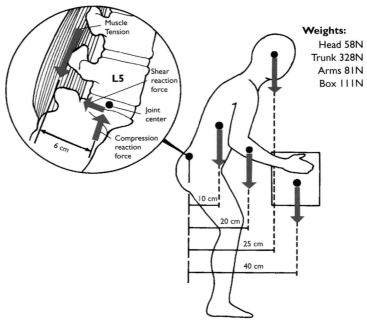

Weights:
Head 58N
Trunk 328N
Arms 81N
Box 111N

Torque at L5,S1 vertebral joint created by body segments and load:
$$T = (328 \text{ N})(10 \text{ cm}) + (81 \text{ N})(20 \text{ cm}) + (58 \text{ N})(25 \text{ cm}) + (111 \text{ N})(40 \text{ cm})$$
$$= 10{,}790 \text{ Ncm}$$

SAMPLE PROBLEM 9.1

How much tension must be developed by the erector spinae with a moment arm of 6 cm from the L5-S1 joint center to maintain the body in a lifting position with segment moment arms as specified? (Segment weights are approximated for a 600 N (135 lb) person.)

Known

SEGMENT	WT	MOMENT ARM
head	50 N	22 m
trunk	280 N	12 cm
arms	65 N	25 cm
box	100 N	42 cm
F_m		6 cm

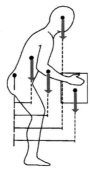

Solution

When the body is in a static position, the sum of the torques acting at any point is zero. At L5-S1:

$$\Sigma T_{L5,S1} = 0$$
$$0 = (F_m)(6 \text{ cm}) - [(50 \text{ N})(22 \text{ cm}) + (280 \text{ N})(12 \text{ cm})$$
$$+ (65 \text{ N})(25 \text{ cm}) + (100 \text{ N})(42 \text{ cm})]$$
$$0 = (F_m)(6 \text{ cm}) - 10{,}285 \text{ Ncm}$$

$$\boxed{F_m = 1714 \text{ N}}$$

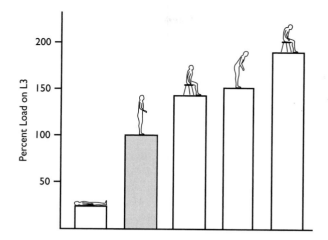

FIGURE 9-31

The load on the third lumbar disc during upright standing (100%) is markedly reduced in a supine position, but increases for each of the other positions shown.
Adapted from Nachemson A: Towards a better understanding of back pain: a review of the mechanics of the lumbar disc, Rheumatol Rehabil, 14:129, 1975.

the seat such that more weight is supported by the thighs have been shown to reduce the load on the spine (76).

Pressure within the intervertebral discs changes significantly with body position and loading, but is relatively consistent through the different regions of the spine (97). During static loading, the discs deform over time, transferring more of the load to the facet joints (30). After 30 minutes of dynamic repetitive spinal flexion, such as might occur with a lifting task, the general stiffness of the spine is decreased, and deformation of the discs in combination with elongation of the spinal ligaments results in altered loading patterns that may predispose the individual to low back pain (95).

During erect standing, body weight also loads the spine in shear. This is particularly true in the lumbar spine, where shear creates a tendency for vertebrae to displace anteriorly with respect to adjacent inferior vertebrae (Figure 9-32). Because very few of the fibers of the major spinal extensors lie parallel to the spine, as tension in these muscles increases, both compression and shear on the vertebral joints and facet joints increase (73). Fortunately, however, the shear component produced by muscle tension in the lumbar region is directed posteriorly, so that it partially counteracts the anteriorly directed shear produced by body weight (99). Shear is

●*Body weight produces shear as well as compression on the lumbar spine.*

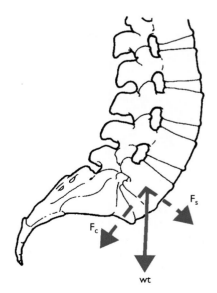

FIGURE 9-32

Body weight during upright standing produces shear (F_s) as well as compression (F_c) components on the lumbar spine. (Note that the vector sum of F_s and F_c is wt.)

a dominant force on the spine during flexion as well as during activities requiring backward lean of the trunk, such as rappelling and trapezing during sailing (54). Although the relative significance of compression and shear on the spine is poorly understood, excessive shear stress is believed to contribute to disc herniation.

Tension in the trunk extensors increases with spinal flexion until the spine approaches full flexion, when it abruptly disappears. This has been shown to occur at 57% of maximum hip flexion and at 84% of maximum vertebral flexion (51). At this point, the posterior spinal ligaments completely support the flexion torque. The quiescence of the spinal extensors at full flexion is known as the flexion relaxation phenomenon. Unfortunately, when the spine is in full flexion, tension in the interspinous ligament contributes significantly to anterior shear force and increases facet joint loading (79). During repeated trunk flexion and extension movements over time, the flexion relaxation period is lengthened, which reduces lumbar stability and may predispose the individual to low back pain (91).

When the spine undergoes lateral flexion and axial twisting, a more complex pattern of trunk muscle activation is required than for flexion and extension. Researchers estimate that, whereas 50 Nm of extension torque places 800 N of compression on the L4-L5 vertebral joint, 50 Nm of lateral flexion and axial twisting torques respectively generate 1400 N and 2500 N of compression on the joint (81). Information derived from biomechanical modeling of the spine suggests that tension in antagonist trunk muscles produces a significant part of these increased loads (79). Asymmetrical frontal plane loading of the trunk also increases both compressive and shear loads on the spine because of the added lateral bending moment (39, 82).

Another factor affecting spinal loading is body movement speed. It has been shown that executing a lift in a rapid, jerking fashion dramatically increases compression and shear forces on the spine, as well as tension in the paraspinal muscles (52). This is one of the reasons that resistance-training exercises should always be performed in a slow, controlled fashion.

flexion relaxation phenomenon
condition in which, when the spine is in full flexion, the spinal extensor muscles relax and the flexion torque is supported by the spinal ligaments

•*Lateral flexion and rotation create much larger spinal loads than do flexion and extension.*

Lifting while twisting should be avoided because it places about three times more stress on the back than lifting in the sagittal plane.

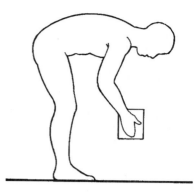

FIGURE 9-33

It is advisable to maintain normal lumbar curvature rather than allowing the lumbar spine to flex when lifting, as discussed in the text.

Maximizing the smoothness of the motion pattern of the external load acts to minimize the peaks in the compressive force on the lumbosacral joint (60). However, when lifting moderate loads that are awkwardly positioned, skilled workers may be able to reduce spinal loading by initially jerking the load in close to the body and then transferring momentum from the rotating trunk to the load (see Chapter 14) (81).

The old adage to lift with the legs and not with the back refers to the advisability of minimizing trunk flexion and thereby minimizing the torque generated on the spine by body weight. However, either the physical constraints of the lifting task or the added physiological cost of leg-lifting as compared to back-lifting (49) often make this advice impractical. Research suggests that a more important focus of attention for people performing lifts may be maintaining the normal lumbar curve, rather than either increasing lumbar lordosis or allowing the lumbar spine to flex (Figure 9-33) (79). Maintaining a normal or slightly flattened lumbar curve enables the active lumbar extensor muscles to partially offset the anterior shear produced by body weight (as discussed), and uniformly loads the lumbar discs rather than placing a tensile load on the posterior annulus of these discs (106). A lordotic lumbar posture, alternatively, increases loading of the posterior annulus and facet joints, while full lumbar flexion changes the line of action of the lumbar extensor muscles, such that they cannot

Many activities of daily living are stressful to the low back. The constraints of the automobile trunk make it difficult to lift with the spine erect.

Intraabdominal pressure, which often increases during lifting, contributes to the stiffness of the lumbar spine to help prevent buckling.

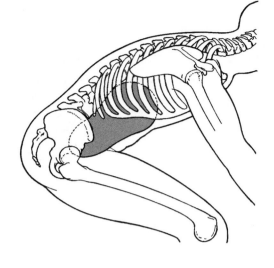

intraabdominal pressure

pressure inside the abdominal cavity; believed to help stiffen the lumbar spine against buckling

effectively counteract anterior shear (36, 80). Anterior shear load on the lumbar spine has been associated with increased risk of back injury (80).

A factor once believed to alleviate compression on the lumbar spine is intraabdominal pressure. Researchers hypothesized that intraabdominal pressure works like a balloon inside the abdominal cavity to support the adjacent lumbar spine by creating a tensile force that partially offsets the compressive load (10) (Figure 9-34). This was corroborated by the observation that intraabdominal pressure increases just prior to the lifting of a heavy load (52). More recently, however, scientists have discovered that pressure in the lumbar discs actually increases when intraabdominal pressure increases (84). It now appears that increased intraabdominal pressure may help to stiffen the trunk to prevent the spine from buckling under compressive loads (31). When an unstable load is assumed, there is also increased coactivation of antagonistic trunk muscles to contribute to spinal stiffening (122). Research has shown that increases in intraabdominal pressure generate proportional increases in trunk extensor moment (58). This is of value in performing a task such as lifting, because the spinal extensor muscles must generate sufficient extensor moment to overcome the flexion moment generated by the forward lean of the trunk, as well as by the load in front of the body being lifted.

Carrying a loaded book bag or backpack loads the spine, with heavier loads resulting in postural adjustments including forward trunk and head lean and reduced lumbar lordosis. Research shows that placing the load low within the backpack and limiting the load to no more than 15% body weight minimizes these postural adaptations (19, 33).

COMMON INJURIES OF THE BACK AND NECK

Low Back Pain

Low back pain is an extremely prevalent problem, with up to 85% of people experiencing low back pain at some time during their lives and more than half of the population having had back pain (5). That is, at any given time, approximately 35% of people have not yet had back pain, but will have it in the future. Low back pain is second only to the common cold in causing absences from the workplace, and back injuries are the most frequent and the most expensive of all worker's compensation claims in the

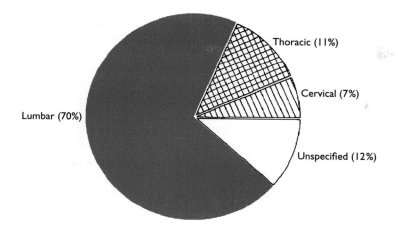

Lumbar (70%)

Thoracic (11%)

Cervical (7%)

Unspecified (12%)

FIGURE 9-35

The majority of back injuries that result in lost work time involve the lumbar region.

United States (29, 86). Moreover, the incidence of low back pain in the United States has been steadily increasing for more than the past decade (46). This is likely a direct result of the increasing incidence of overweight and obesity, which is significantly associated with low back pain in both men and women and at all ages (56, 57). Most back injuries also involve the lumbar or low back region (Figure 9-35).

Although psychological and social components are a factor in some low back pain cases, mechanical stress typically plays a significant causal role in the development of low back pain (47). Perhaps because of their predominance in occupations involving the handling of heavy materials, men experience low back pain about four times more frequently than do women (71). However, some female-dominated groups, such as nurses' aides, register higher rates of low back injury than do male workers in general (111).

The incidence of low back pain in children is nearly 30% (40). This incidence increases with age and by age 16 approaches that found in adults, with back pain more common in boys than in girls (22). In contrast to the situation for adults, however, low back pain in children is associated with increased physical activity and stronger back flexor muscles (87). The main causes of low back pain in children are believed to be musculotendinous strains and ligamentous sprains (87).

Not surprisingly, athletes have a much higher incidence of low back pain than do nonathletes, with over 9% of college athletes in different sports receiving treatment for low back pain (55, 79, 85). In some sports, such as gymnastics, as many as 85% of the participants experience low back pain (116, 117).

High incidences of low back pain have been found in workers who sit for prolonged periods and in those unable to sit at all during the work day (75). High-risk occupations for the development of low back pain, in order of frequency, include laborers, truck drivers, garbage collectors, warehouse workers, mechanics, nursing aides, materials handlers, lumber workers, practical nurses, and construction laborers (110). Cigarette smokers have an increased incidence of low back pain compared to non-smokers, possibly due to the contributions of habitual smoking to disc degeneration (44). Accidental low back pain is often associated with working in an unnatural posture, with sudden and unexpected motions, and with working single-handed (3, 131). Work involving dynamic motion in multiple planes is also associated with significantly increased risk for developing low back pain (34).

As discussed in Chapter 3, loading patterns that injure biological tissues may involve one or a few repetitions of a large load or numerous repetitions of a small load. Repeated loading, such as that which occurs in industrial work, in exercise performance, and in jobs such as truck driving that involve vibration, can all produce low back pain.

A factor recently implicated in our evolving understanding of low back injuries is the relative stability of the spine. In the absence of contraction of the surrounding musculature, buckling of the lumbar spine occurs under compressive loads as small as 4 N (41). Although bone structure, intervertebral discs, and ligaments all contribute to spinal stability, it is the surrounding muscles that have been shown to be the primary contributors to spinal stability (41). Recent research suggests that coactivation of spinal agonists and antagonists serves to increase the stiffness of the motion segments and enhance spinal stability (48, 50). Only relatively modest levels of coactivation of the paraspinal and abdominal wall muscles are needed to provide adequate spinal stability for daily living tasks, however (32). Accordingly, McGill advocates training the trunk muscles for endurance rather than for strength as a prophylactic for low back injury (78). Fatigue of the spinal extensor muscles with concomitantly reduced force output has been shown to increase the bending moment on the lumbar spine and also to reduce the ability of these muscles to protect the spine (36, 37).

Although injuries and some known pathologies may cause low back pain, 60% of low back pain is idiopathic, or of unknown origin. The risk factors for chronic disability due to low back pain include involvement of the spinal nerve roots, substantial functional disability, widespread pain, and previous injury with an extended absence from work (120). However, most low back pain is self-limiting, with 75% of cases resolving within three weeks and a recovery rate of over 90% within two months, whether treatment occurs or not (25).

Despite this fact, clinicians often recommend abdominal exercises as both a prophylactic and a treatment for low back pain. The general rationale for such prescriptions is that weak abdominal muscles may not contribute sufficiently to spinal stability. However, sit-up-type exercises, even when performed with the knees in flexion, generate compressive loads on the lumbar spine of well over 3000 N (78). According to one clinical report, the use of sit-up exercises appears to have actually contributed to low back pain development among a group of 29 exercisers (83). Partial curl-up exercises have also been advocated as providing strong abdominal muscle challenge, with minimal spinal compression (9).

Soft-Tissue Injuries

Contusions, muscle strains, and ligament sprains collectively compose the most common injury of the back, accounting for up to 97% of all back pain in the general population (4). These types of injuries typically result from either sustaining a blow or overloading the muscles, particularly those of the lumbar region. Painful spasms and knotlike contractions of the back muscles may also develop as a sympathetic response to spinal injuries, and may actually be only symptoms of the underlying problem. Researchers believe that a biochemical mechanism is responsible for these sympathetic muscle spasms (21), which act as a protective mechanism to immobilize the injured area (102).

Acute Fractures

Differences in the causative forces determine the type of vertebral fracture incurred. Transverse or spinous process fractures may result from extremely forceful contraction of the attached muscles or from the sustenance of a hard blow to the back of the spine, which may occur during participation in contact sports such as football, rugby, soccer, basketball, hockey, and lacrosse (102). The most common cause of cervical fractures is indirect trauma involving force applied to the head or trunk rather than to the cervical region itself (23). The most common mechanism for catastrophic cervical injuries is an axial force to the top of the head with the neck in slight flexion (118). Fractures of the cervical vertebrae frequently result from impacts to the head when people dive into shallow water or engage in gymnastics or trampolining activities without appropriate supervision (108, 109). In wrestling, there are three common cervical injury scenarios: (a) The wrestler's arms are in a hold, and he is unable to prevent himself from landing on his head; (b) the wrestler attempts to roll, but is landed on by his opponent's full weight, causing a hyperflexion injury of the neck; and (c) the wrestler lands on the top of his head and sustains an axial compression force (15, 16).

Large compressive loads (such as those encountered in the sport of weight lifting or in the handling of heavy materials) can cause fractures of the vertebral end plates. High levels of impact force may result in anterior compression fractures of the vertebral bodies. This type of injury is generally associated with vehicular accidents, although it can also result from contact with the boards in ice hockey, head-on blocks or tackles in football, collisions between baseball catchers and base runners, or impacts during tobogganing, snowmobiling, and hot air ballooning. When a snowmobile drops 4 ft, the forces generated far exceed the level known to cause a compression fracture (67).

Because one function of the spine is to protect the spinal cord, acute spinal fractures are extremely serious, with possible outcomes including paralysis and death. Unfortunately, increased participation in leisure activities is associated with an increased prevalence of spinal injuries (107). Whenever a spinal fracture is a possibility, only trained personnel should move the victim. Spinal stress injuries can also result from repeated sustenance of relatively modest forces, such as heading the ball in soccer. Degenerative changes in the cervical spine of soccer players have been shown to occur 10–20 years earlier than in the general population, with changes more pronounced among those with more years of playing time (132).

Fractures of the ribs are generally caused by blows received during accidents or participation in contact sports. Rib fractures are extremely painful because pressure is exerted on the ribs with each inhalation. Damage to the underlying soft tissues is a potentially serious complication with this type of injury.

Stress Fractures

The most common type of vertebral fracture is a stress fracture of the pars interarticularis, the region between the superior and inferior articular facets, which is the weakest portion of the neural arch (Figure 9-36). A fracture of the pars is termed spondylolysis, with the severity ranging from hairline fracture to complete separation of bone. Although some pars defects may be congenital, they are also known to be caused by mechanical stress. One mechanism of injury appears to involve repeated axial loading of the lumbar spine when it is hyperextended. The prevalence of lumbar spondylolysis in the general population is 11.5%, with approximately three times as many males as females affected (65).

A bilateral separation in the pars interarticularis, called spondylolisthesis, results in the anterior displacement of a vertebra with respect to the vertebra below it (Figure 9-36). The most common site of this injury is the lumbosacral joint, with 90% of the slips occurring at this level. Spondylolisthesis is often initially diagnosed in children between the ages of

spondylolysis
presence of a fracture in the pars interarticularis of the vertebral neural arch

spondylolisthesis
complete bilateral fracture of the pars interarticularis, resulting in the anterior slippage of the vertebra

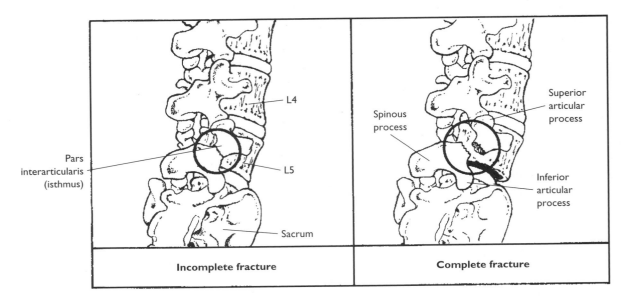

Incomplete fracture

Complete fracture

FIGURE 9-36 Stress fractures of the pars interarticularis may occur unilaterally or bilaterally and may or may not result in complete separation.

10 and 15 years, and is more commonly seen in boys. Risk factors for spondylolysis and spondylolisthesis in children include genetics, the growth spurt, repetitive stresses, and participation in sports for more than 15 hours per week (40). The prevalence of spondylolysthesis increases from the fifth through the eighth decades of life and affects approximately three times as many females as males (65).

Unlike most stress fractures, spondylolysis and spondylolisthesis do not typically heal with time, but tend to persist, particularly when there is no interruption in sport participation. Individuals whose sports or positions require repeated hyperextension of the lumbar spine are prime candidates for stress-related spondylolysis. Those particularly susceptible to this pathology include female gymnasts, interior football linemen, and weight lifters, with increased incidences also found among volleyball players, pole-vaulters, wrestlers, figure skaters, dancers, and rowers (92). Degenerative spondylolisthesis is also a common condition of the aging spine, although the causative factors for this condition are unknown (12). A high incidence of spondylolysis has been found in human remains from circa A.D. 1200 to 1500, a period characterized by the use of stone pillars for the construction of houses (6).

Disc Herniations

The source of approximately 1–5% of back pain is a herniated disc, which consists of the protrusion of part of the nucleus pulposus from the annulus. Disc herniations may be either traumatic or stress related, but only occur in discs that show signs of previous degeneration (62). The most common sites of protrusions are between the fifth and sixth and the sixth and seventh cervical vertebrae, and between the fourth and fifth lumbar vertebrae and the fifth lumbar vertebra and the first sacral vertebra (68, 69). Most occur on the posterior or posterior-lateral aspect of the disc (8).

Although the disc itself is not innervated and therefore is incapable of generating a sensation of pain, sensory nerves do supply the anterior and posterior longitudinal ligaments, the vertebral bodies, and the articular cartilage of the facet joints (24). If the herniation presses on one or more

• *Stress-related fractures of the pars interarticularis, the weakest section of the neural arch, are unusually common among participants in sports involving repeated hyperextension of the lumbar spine.*

• *The term* slipped disc *is often used to refer to a herniated or prolapsed disc. It is a misnomer, because the discs as intact units do not slip around.*

of these structures, on the spinal cord, or on a spinal nerve, pain or numbness may result.

Disc herniations do not cause a significant decrease in intervertebral space height, and so cannot be effectively detected through X-rays. The average decrease in the height of a disc that has undergone a herniation is less than that reported in normal daily disc height variation (59). Most lumbar disc herniations are treated conservatively, meaning without surgery. Many individuals have disc herniations without being aware of it, since there are no symptoms. For those who do present with symptoms of pain or minor neural deficits, treatment includes medications, physical therapy, and sometimes lumbar injection (63). More serious cases that fail to respond to conservative treatment options are treated with surgical interventions, often lumbar discectomy.

Whiplash Injuries

Whiplash injury to the cervical region is relatively common, with a reported incidence of 4 per 1000 individuals (42). Such injuries typically occur during automobile collisions, with the neck undergoing sudden acceleration and deceleration. Shear force and extension moment at the junction between the cervical and thoracic spines are the underlying mechanism causing neck motion and potential injury (113). Research suggests that the cervical vertebrae go through a sequence of abnormal motion, although there are different variations to the pattern depending on the direction and speed of the impact, as well as possibly the gender of the individual (20, 66). Generally, however, the cervical spine assumes an S-shape, with the upper segments in flexion and the lower segments in extension (27). The cervical muscles contract rapidly in such situations, with potential for forced eccentric tension development (20). Both upper and lower cervical injuries are possible (94). Symptoms of whiplash injury include neck pain; muscle pain; pain or numbness radiating from the neck to the shoulders, arms, or hands; and, in 50–60% of cases, accompanying headache (18, 96, 115). Women appear to be at greater risk for whiplash injury than men due to less stiffness of the cervical structures (112).

SUMMARY

The spine is composed of 33 vertebrae that are divided structurally into five regions: cervical, thoracic, lumbar, sacral, and coccygeal. Although most vertebrae adhere to a characteristic shape, there is a progression in vertebral size and in the orientation of the articular facets throughout the spinal column.

Within the cervical, thoracic, and lumbar regions, each pair of adjacent vertebrae with the intervening soft tissues is called a motion segment. Three joints interconnect the vertebrae of each motion segment. The intervertebral discs function as shock absorbers at the weight-bearing vertebral joints. The right and left pairs of superior and inferior facet joints significantly influence the movement capabilities of the motion segments at different levels of the spine.

The muscles of the neck and trunk are named in pairs, with one on the left and the other on the right side of the body. These muscles can cause lateral flexion and/or rotation of the trunk when they act unilaterally, and trunk flexion or extension when acting bilaterally.

Forces acting on the spine include body weight, tension in the spinal ligaments, tension in the surrounding muscles, intraabdominal pressure, and any applied external loads. Because the spinal muscles have extremely small moment arms with respect to the vertebral joints, they must generate large forces to counteract the torques produced about the spine by the weights of body segments and external loads.

Because the spine serves as the protector of the spinal cord, spinal injuries are serious. Low back pain is a major modern-day health problem and a leading cause of lost working days.

INTRODUCTORY PROBLEMS

1. What regions of the spine contribute the most to flexion? Hyperextension? Lateral flexion? Rotation?
2. Construct a chart listing the muscles of the cervical region of the trunk according to whether they are anterior, posterior, medial, or lateral to the joint center. Note that some muscles may fall into more than one category. Identify the action or actions performed by the muscles in each category.
3. Construct a chart listing the muscles of the thoracic region of the trunk according to whether they are anterior, posterior, medial, or lateral to the joint center. Note that some muscles may fall into more than one category. Identify the action or actions performed by the muscles in each category.
4. Construct a chart listing the muscles of the lumbar region of the trunk according to whether they are anterior, posterior, medial, or lateral to the joint center. Note that some muscles may fall into more than one category. Identify the action or actions performed by the muscles in each category.
5. How would trunk movement capability be affected if the lumbar region were immovable?
6. What are the postural consequences of having extremely weak abdominal muscles?
7. Weight training is used in conjunction with conditioning for numerous sports. What would you advise regarding spinal posture during weight training?
8. What exercises strengthen the muscles on the anterior, lateral, and posterior aspects of the trunk?
9. Why should twisting be avoided when performing a lift?
10. Solve Sample Problem 9.1 using the following data:

SEGMENT	WT	MOMENT ARM
Head	50 N	22 cm
Trunk	280 N	12 cm
Arms	65 N	25 cm
Box	100 N	42 cm

ADDITIONAL PROBLEMS

1. Explain how pelvic movements facilitate spinal movements.
2. What exercises should be prescribed for individuals with scoliosis? Lordosis? Kyphosis?

3. Compare and contrast the major muscles that serve as agonists during performances of straight-leg and bent-knee sit-ups. Should sit-ups be prescribed as an exercise for a low back pain patient? Explain why or why not.

4. Why do individuals who work at a desk all day often develop low back pain?

5. Explain why maintaining a normal lumbar curve is advantageous during lifting.

6. Formulate a theory explaining why osteoporosis is often associated with increased thoracic kyphosis.

7. Formulate a theory explaining why a loss in spinal flexibility of approximately 50% is a result of the aging process.

8. What are the consequences of the loss in intervertebral disc hydration that accompanies the aging process?

9. What spinal exercises are appropriate for senior citizens? Provide a rationale for your choices.

10. Make up a problem similar to (but different from) Sample Problem 9.1. Show a free body diagram and a solution for your problem.

LABORATORY EXPERIENCES

1. Study an anatomical model of the spine, or use the *Dynamic Human* CD to locate and identify the major bones and muscle attachments. (On the *Dynamic Human* CD, click on Skeletal, Anatomy, Gross Anatomy, Vertebral Column, Thoracic Cage, and then Pelvic Girdle.)

Bones of the Vertebral Column

_____   _____

_____   _____

_____   _____

_____   _____

Bones and Muscle Attachments of the Thoracic Cage

Bone   **Muscle Attachments**

_____   _____

_____   _____

_____   _____

_____   _____

_____   _____

Bones and Muscle Attachments of the Pelvic Girdle

Bone   **Muscle Attachments**

_____   _____

_____   _____

_____   _____

_____   _____

_____   _____

_____   _____

_____   _____

_____   _____

_____   _____

_____   _____

_____   _____

_____   _____

2. Study an anatomical model of the trunk, or use the *Dynamic Human* CD to locate and identify the major muscles of the trunk and their attachment sites. (On the *Dynamic Human* CD, click on Muscular, Anatomy, Body Regions, and then Abdomen and Back.)

Muscles of the Anterior Trunk

Muscle	Attachment Sites
_____	_____
_____	_____
_____	_____
_____	_____
_____	_____
_____	_____
_____	_____
_____	_____
_____	_____
_____	_____

Muscles of the Posterior Trunk

Muscle	Attachment Sites
_____	_____
_____	_____
_____	_____
_____	_____
_____	_____

3. Using a skeleton or an anatomical model of the spine, carefully study the differences in vertebral size and shape among the cervical, thoracic, and lumbar regions. Construct a chart that characterizes the differences between regions, and write a paragraph explaining how the form of the vertebrae in each region is related to their functions.

Cervical	Thoracic	Lumbar
_____	_____	_____
_____	_____	_____
_____	_____	_____

———————————— ———————————— ————————————

———————————— ———————————— ————————————

———————————— ———————————— ————————————

Explanation: _____

_____

_____

_____

_____

4. If available, use the *Dynamic Human* CD, and click on Skeletal, Clinical Concepts, and then Herniated Disc. After also reviewing the material in the chapter, explain what kinds of activities are most likely to contribute to a disc herniation.

Explanation: _____

_____

_____

_____

_____

5. From a side view, video a volunteer lifting objects of light, medium, and heavy weights. What differences do you observe in lifting kinematics? Write a short explanation of your findings.

Explanation: _____

_____

_____

_____

_____

REFERENCES

1. Adams MA and Hutton WC: The effect of posture on the fluid content of lumbar intervertebral discs, *Spine* 8:665, 1983.
2. Ali RM, Green DW, and Patel TC: Scheuermann's kyphosis, *Curr Opin Pediatr* 11:70, 1999.
3. Allread WG, Marras WS, and Parnianpour M: Trunk kinematics of one-handed lifting, and the effects of asymmetry and load weight, *Ergonomics* 39:322, 1996.
4. An HS, Jenis LG, and Vaccaro AR: Adult spine trauma. In Beaty JH and Rosemont IL (eds.): Orthopaedic knowledge update six. *American Academy of Orthopaedic Surgeons,* 1999:653.
5. Andersson GBJ: Epidemiology of low back pain, *Acta Orthop Scand* 69 (Suppl 281):28, 1998.
6. Arriaza BT: Spondylolysis in prehistoric human remains from Guam and its possible etiology, *Am J Phys Anthropol* 104:393, 1997.
7. Ashton-Miller JA: Thoracic hyperkyphosis in the young athlete: a review of the biomechanical issues, *Curr Sports Med Rep* 3:47, 2004.
8. Ashton-Miller JA and Schultz AB: Biomechanics of the human spine and trunk, *Exerc Sport Sci Rev* 16:169, 1988.
9. Axler CT and McGill SM: Low back loads over a variety of abdominal exercises: searching for the safest abdominal challenge, *Med Sci Sports Exerc* 29:804, 1997.
10. Bartelink DL: The role of abdominal pressure in relieving the pressure on the lumbar intervertebral discs, *J Bone Joint Surg* 39B:718, 1957.
11. Beresford ZM, Kendall RW, and Willick SE: Lumbar facet syndromes, *Curr Sports Med Rep* 9:50, 2010.
12. Berlemann U, Jeszensky DJ, Buhler DW, and Harms J: The role of lordosis, vertebral end-plate inclination, disc height, and facet orientation in degenerative spondylolisthesis, *J Spinal Disord* 12:68, 1999.
13. Bible JE, Biswas D, Miller CP, Whang PG, and Grauer JN: Normal functional range of motion of the cervical spine during 15 activities of daily living, *J Spinal Disord Tech* 23:15, 2010.
14. Bible JE, Biswas D, Miller CP, Whang PG, and Grauer JN: Normal functional range of motion of the lumbar spine during 15 activities of daily living, *J Spinal Disord Tech* 23:106, 2010.
15. Boden BP, Lin W, Young, M, and Mueller FO: Catastrophic injuries in wrestlers, *Am J Sports Med* 11:17, 2001.
16. Boden BP and Prior C: Catastrophic spine injuries in sports, *Curr Sports Med Rep* 4:45, 2005.
17. Bogduk N and Mercer S: Biomechanics of the cervical spine, I: Normal kinematics, *Clin Biomech* 15:633, 2000.
18. Bono G et al: Whiplash injuries: clinical picture and diagnostic workup, *Clin Exp Rheumatol* 18:S23, 2000.
19. Brackley HM, Stevenson JM, and Selinger JC: Effect of backpack load placement on posture and spinal curvature in prepubescent children, *Work* 32:351, 2009.
20. Brault JR, Siegmund GP, and Wheeler JB: Cervical muscle response during whiplash: evidence of a lengthening muscle contraction, *Clin Biomech* 15:426, 2000.
21. Brennan GP et al: Physical characteristics of patients with herniated intervertebral lumbar discs, *Spine* 12:699, 1987.
22. Burton AK: Low back pain in children and adolescents: to treat or not? *Bull Hosp Jt Dis* 55:127, 1996.
23. Byun HS, Cantos EL, and Patel PP: Severe cervical injury due to break dancing: a case report, *Orthopedics* 9:550, 1986.
24. Cailliet R: *Low back pain syndrome* (3rd ed.), Philadelphia, 1981, FA Davis.
25. Carpenter DM and Nelson BW: Low back strengthening for the prevention and treatment of low back pain, *Med Sci Sports Exerc* 31:18, 1999.
26. Chaudhry H, Ji Z, Shenoy N, and Findley T: Viscoelastic stresses on anisotropic annulus fibrosus of lumbar disk under compression, rotation and flexion in manual treatment, *J Bodyw Mov Ther* 13:182, 2009.

27. Chen HB, Yang KH, and Wang ZG: Biomechanics of whiplash injury, *Chin J Traumatol* 12:305, 2009.

28. Cheng JC, Tang SP, Guo X, Chan CW, and Qin L: Osteopenia in adolescent idiopathic scoliosis: a histomorphometric study, *Spine* 26:E19, 2001.

29. Cherkin DC et al: An international comparison of back surgery rates, *Spine* 19:1201, 1994.

30. Cheung J T-M, Zhang M, and Chow D H-K: Biomechanical responses of the intervertebral joints to static and vibrational loading: a finite element study, *Clin Biomech* 18:790, 2003.

31. Cholewicki J, Juluru K, and McGill SM: Intra-abdominal pressure mechanism for stabilizing the lumbar spine, *J Biomech* 32:13, 1999.

32. Cholewicki J and McGill SM: Mechanical stability of the in vivo lumbar spine: implications for injury and chronic low back pain, *Clin Biomech* 11:1, 1996.

33. Connolly BH, Cook B, Hunter S, Laughter M, Mills A, Nordtvedt N, and Bush A: Effects of backpack carriage on gait parameters in children, *Pediatr Phys Ther* 20:347, 2008.

34. Davis KG and Marras WS: The effects of motion on trunk biomechanics, *Clin Biomech* 15:703, 2000.

35. DiMeglio A and Bonnel F: *La rachis en croissance*. Paris, 1990, Springer-Verlag.

36. Dolan P and Adams MA: Recent advances in lumbar spinal mechanics and their significance for modelling, *Clin Biomech* 16:S8, 2001.

37. Dolan P and Adams MA: Repetitive lifting tasks fatigue the back muscles and increase the bending moment acting on the lumbar spine, *J Biomech* 31:713, 1998.

38. Dolan P, Benjamin E, and Adams M: Diurnal changes in bending and compressive stresses acting on the lumbar spine, *J Bone JT Surg* (Suppl) 75B:22, 1993.

39. Drury CG et al: Symmetric and asymmetric manual materials handling, II: Biomechanics, *Ergonomics*, 32:565, 1989.

40. Duggleby T and Kumar S: Epidemiology of juvenile low back pain: a review, *Disabil Rehabil* 19:505, 1997.

41. Ebenbichler GR, Oddsson LIE, Kollmitzer J, and Erim Z: Sensory-motor control of the lower back: implications for rehabilitation, *Med Sci Sports Exerc* 33:1889, 2001.

42. Eck JC, Hodges SD, and Humphreys SC: Whiplash: a review of a commonly misunderstood injury, *Am J Med* 110:651, 2001.

43. Farfan H: *Mechanical disorders of the low back,* Philadelphia, 1973, Lea & Febiger.

44. Fogelholm RR and Alho AV: Smoking and intervertebral disc degeneration, *Med Hypotheses* 56:537, 2001.

45. Franz JR, Paylo KW, Dicharry J, Riley PO, and Kerrigan DC: Changes in the coordination of hip and pelvis kinematics with mode of locomotion, *Gait Posture*, 29:494, 2009.

46. Freburger JK, Holmes GM, Agans RP, Jackman AM, Darter JD, Wallace AS, Castel LD, Kalsbeek WD, and Carey TS: The rising prevalence of chronic low back pain, *Arch Intern Med* 169:251, 2009.

47. Frymoyer JW and Pope M: The role of trauma in low back pain: a review, *J Trauma* 18:628, 1978.

48. Gardner-Morse MG and Stokes IAF: Trunk stiffness increases with steady-state effort, *J Biomech* 34:457, 2001.

49. Garg A and Herrin G: Stoop or squat: a biomechanical and metabolic evaluation, *Am Inst Ind Eng Trans* 11:293, 1979.

50. Granata KP and Orishimo KF: Response of trunk muscle coactivation to changes in spinal stability, *J Biomech* 34:1117, 2001.

51. Gupta A: Analyses of myoelectrical silence of erectors spinae, *J Biomech* 34:491, 2001.

52. Hall, SJ: Effect of attempted lifting speed on forces and torque exerted on the lumbar spine, *Med Sci Sports Exerc* 17:440, 1985.

53. Hall SJ: Mechanical contribution to lumbar stress injuries in female gymnasts, *Med Sci Sports Exerc* 18:599, 1986.

54. Hall SJ, Kent JA, and Dickinson, VR: Comparative assessment of novel sailing trapeze harness designs. *Int J Sport Biomech* 5:289, 1989.

55. Hangai M, Kaneoka K, Okubo Y, Miyakawa S, Hinotsu S, Mukai N, Sakane M, and Ochiai N: Relationship between low back pain and competitive sports activities during youth, *Am J Sports Med* 38:791, 2010.

56. Heliovaara M: Body height, obesity, and risk of herniated lumbar intervertebral disc, *Spine* 12:469, 1987.

57. Heuch I, Hagen K, Heuch I, Nygaard Ø, and Zwart JA: The impact of body mass index on the prevalence of low back pain: the HUNT study, *Spine* (Philadelphia, PA, 1976), 35:764, 2010.

58. Hodges PW, Cresswell AG, Daggfeldt K, and Thorstensson A: In vivo measurement of the effect of intra-abdominal pressure on the human spine, *J Biomech* 34:347, 2001.

59. Holodny AI, Kisza PS, Contractor S, and Liu WC: Does a herniated nucleus pulposus contribute significantly to a decrease in height of the intervertebral disc? Quantitative volumetric MRI, *Neuroradiology* 42:451, 2000.

60. Hsiang SM and McGorry: Three different lifting strategies for controlling the motion patterns of the external load, *Ergonomics* 40:928, 1997.

61. Hutton WC, Stott JRR, and Cyron BM: Is spondylolysis a fatigue fracture? *Spine* 2:202, 1977.

62. Iencean SM: Lumbar intervertebral disc herniation following experimental intradiscal pressure increase, *Acta Neurochir* 142:669, 2000.

63. Jegede KA, Ndu A, and Grauer JN: Contemporary management of symptomatic lumbar disc herniations, *Orthop Clin North Am* 41:217, 2010.

64. Johnson S, Halliwell M, Jones M, and McNally D: Visualisation of collagen fibre bundles in the intact intervertebral disc, *J Biomech* 34:S12, 2001.

65. Kalichman L, Kim DH, Li L, Guermazi A, Berkin V, and Hunter DJ: Spondylolysis and spondylolisthesis: prevalence and association with low back pain in the adult community-based population, *Spine* (Philadelphia, PA, 1976), 34:199, 2009.

66. Kaneoka K, Ono K, Inami S, and Hayashi K: Motion analysis of cervical vertebrae during whiplash loading, *Spine* 15:763, 1999.

67. Keene JS: Thoracolumbar fractures in winter sports, *Clin Orthop* 216:39, 1987.

68. Kelsey JL et al: Acute prolapsed lumbar intervertebral disc: an epidemiological study with special reference to driving automobiles and cigarette smoking, *Spine* 9:608, 1984.

69. Kelsey JL et al: An epidemiological study of acute prolapsed cervical intervertebral disc, *J Bone Joint Surg* 66A:907, 1984.

70. Kraemer J, Kolditz D, and Gowin R: Water and electrolyte content of human intervertebral discs under variable load, *Spine* 10:69, 1985.

71. Kuwashima A, Aizawa Y, Nakamura K, Taniguchi S, and Watanabe M: National survey on accidental low back pain in workplace, *Ind Health* 35:187, 1997.

72. Lindh M: Biomechanics of the lumbar spine. In Nordin M and Frankel VH: *Basic biomechanics of the musculoskeletal system,* Philadelphia, 1989, Lea & Febiger.

73. Macintosh JE and Bogduk N: The morphology of the lumbar erector spinae, *Spine* 12:658, 1987.

74. Macnab I and McCulloch J: *Backache* (2nd ed.), Baltimore, 1990, Williams & Wilkins.

75. Magora A: Investigation of the relationship between low back pain and occupation, *Indust Med* 41:5, 1972.

76. Makhsous M, Lin F, Bankard J, Hendrix RW, Hepler M, and Press J: Biomechanical effects of sitting with adjustable ischial and lumbar support on occupational low back pain: evaluation of sitting load and back muscle activity, *BMC Musculoskelet Disord* 5:10, 2009.

77. McGill SM: Estimation of force and extensor moment contributions of the disc and ligaments at L4/L5, *Spine* 13:1395, 1988.

78. McGill SM: Low back stability: From formal description to issues for performance and rehabilitation, *Exerc and Sport Sciences Rev* 29:26, 2001.

79. McGill SM: A myoelectrically based dynamic three-dimensional model to predict loads on lumbar spine tissues during lateral bending, *J Biomech* 25:395, 1992.

80. McGill SM, Hughson RL, and Parks K: Changes in lumbar lordosis modify the role of the extensor muscles, *Clin Biomech* 15:777, 2000.

81. McGill SM and Norman RW: Low back biomechanics in industry: the prevention of injury through safer lifting. In Grabiner MD: *Current issues in biomechanics,* Champaign, IL, 1993, Human Kinetics.

82. Mital A and Kromodihardjo S: Kinetic analysis of manual lifting activities, II: Biomechanical analysis of task variables, *Int J Ind Ergonomics* 1:91, 1986.

83. Mutoh Y et al: The relation between sit-up exercises and the occurrence of low back pain. In Matsui H and Kobayashi K, eds: *Biomechanics VIII-A,* Champaign, IL, 1983, Human Kinetics.

84. Nachemson A, Andersson GBJ, and Schultz AB: Valsalva manoeuvre biomechanics: effects on lumbar trunk loads of elevated intra-abdominal pressure, *Spine* 11:476, 1986.

85. Nadler SF, Wu KD, Galski T, and Feinberg JH: Low back pain in college athletes: a prospective study correlating lower extremity overuse or acquired ligamentous laxity with low back pain, *Spine* 23:828, 1998.

86. Neal C: The assessment of knowledge and application of proper body mechanics in the workspace, *Orthop Nurs* 16:66, 1997.

87. Newcomer K and Sinaki M: Low back pain and its relationship to back strength and physical activity in children, *Acta Paediatr* 85:1433, 1996.

88. Nixon J: Intervertebral disc mechanics: a review, *J World Soc Med* 79:100, 1986.

89. Nolan JP and Sherk HH: Biomechanical evaluation of the extensor musculature of the cervical spine, *Spine* 13:9, 1988.

90. O'Connell GD, Guerin HL, Elliott DM: Theoretical and uniaxial experimental evaluation of human annulus fibrosus degeneration, *J Biomech Eng* 131:111007, 2009.

91. Olson MW, Li L, and Solomonow M: Flexion-relaxation response to cyclic lumbar flexion, *Clin Biomech* 19:769, 2004.

92. Omey ML, Micheli LJ, and Gerbino PG 2nd: Idiopathic scoliosis and spondylolysis in the female athlete: tips for treatment, *Clin Orthop* 372:74, 2000.

93. Owens SC, Brismée JM, Pennell PN, Dedrick GS, Sizer PS, and James CR: Changes in spinal height following sustained lumbar flexion and extension postures: a clinical measure of intervertebral disc hydration using stadiometry, *J Manipulative Physiol Ther* 32:358, 2009.

94. Panjabi MM, Pearson AM, Ito S, Ivancic PC, and Wang J-L: Cervical spine curvature during simulated whiplash, *Clin Biomech* 19:1, 2004.

95. Parkinson RJ, Beach TAC, and Callaghan JP: The time-varying response of the in vivo lumbar spine to dynamic repetitive flexion, *Clin Biomech* 19:330, 2004.

96. Pearce JM: Headaches in the whiplash syndrome, *Spinal Cord* 39:228, 2001.

97. Polga DJ, Beaubien BP, Kallemeier PM, Schellhas KP, Lew WD, Buttermann GR, and Wood KB: Measurement of in vivo intradiscal pressure in healthy thoracic intervertebral discs, *Spine* 29:1320, 2004.

98. Pope MH, Frymoyer JW, and Lehman TR: Structure and function of the lumbar spine. In Pope MH, et al, eds: *Occupational low back pain: Assessment, treatment and prevention,* St. Louis, 1991, Mosby Year Book.

99. Potvin JR, McGill SM, and Norman RW: Trunk muscle and lumbar ligament contributions to dynamic lifts with varying degrees of trunk flexion, *Spine* 16:1099, 1991.

100. Reamy BV and Slakey JB: Adolescent idiopathic scoliosis: review and current concepts, *Am Fam Physician* 64:111, 2001.

101. Reilly T, Tynell A, and Troup JDG: Circadian variation in human stature, *Chronobiology Int* 1:121, 1984.

102. Rovere GD: Low back pain in athletes, *Physician Sportsmed* 15:105, 1987.

103. Salo PK, Häkkinen AH, Kautiainen H, and Ylinen JJ: Quantifying the effect of age on passive range of motion of the cervical spine in healthy working-age women, *J Orthop Sports Phys Ther* 39:478, 2009.

104. Savage RA, Whitehouse GH, and Roberts N: The relationship between magnetic resonance imaging appearance of the lumbar spine and low back pain, age and occupation in males, *Eur Spine* 6:106, 1997.

105. Schultz AB: Biomechanical analyses of loads on the lumbar spine. In Weinstein JN and Wiesel SW, eds: *The lumbar spine,* Philadelphia, 1990, WB Saunders.
106. Shirazi-Adl A and Parnianpour M: Effect of changes in lordosis on mechanics of the lumbar spine–lumbar curve fitting, *J Spinal Disord* 12:436, 1999.
107. Silver JR: Spinal injuries as a result of sporting accidents, *Paraplegia* 25:16, 1987.
108. Silver JR, Silver DD, and Godfrey JJ: Injuries of the spine sustained during gymnastic activities, *Br Med J* 293:861, 1986.
109. Silver JR, Silver DD, and Godfrey JJ: Trampolining injuries of the spine, *Injury* 17:117, 1986.
110. Snook S, Fine L, and Silverstein B: Musculoskeletal disorders. In Levy B and Wegman D, eds: *Occupational health: recognizing and preventing work-related disease,* Boston, 1988, Little, Brown.
111. Spengler D et al: Back injury in industry: a retrospective study, *Spine* 11:241, 1986.
112. Stemper BD, Yoganandan N, and Pintar FA: Gender dependent cervical spine segmental kinematics during whiplash, *J Biomech* 36:1281, 2003.
113. Stemper BD, Yoganandan N, and Pintar FA: Kinetics of the head-neck complex in low-speed rear impact, *Biomed Sci Instrum* 39:245, 2003.
114. Stokes IA and Aronsson DD: Disc and vertebral wedging in patients with progressive scoliosis, *J Spinal Disord* 14:317, 2001.
115. Suissa S, Harder S, and Veilleux M: The relation between initial symptoms and signs and the prognosis of whiplash, *Eur Spine J* 10:44, 2001.
116. Sward L, Hellstrom M, Jacobsson B, and Peterson L: Back pain and radiologic changes in the thoraco lumbar spine of athletes, *Spine* 15:24, 1990.
117. Tall RL and DeVault W: Spinal injury in sport: epidemiologic considerations, *Clin Sports Med* 12:441, 1993.
118. Torg JS, Guille JT, and Jaffe S: Current concepts review: injuries to the cervical spine in American football players, *J Bone Joint Surg Am* 84:112, 2002.
119. Tsirikos AI: Scheuermann's kyphosis: an update, *J Surg Orthop Adv* 18:122, 2009.
120. Turner JA, Franklin G, Fulton-Kehoe D, Sheppard L, Stover B, Wu R, Gluck JV, and Wickizer TM: ISSLS prize winner: early predictors of chronic work disability: a prospective, population-based study of workers with back injuries, *Spine* (Philadelphia, PA, 1976), 33:2809, 2008.
121. van Deursen LL, van Deursen DL, Snijders CJ, and Wilke HJ: Relationship between everyday activities and spinal shrinkage, *Clin Biomech* 20:547, 2005.
122. van Dieën JH, Kingma I, and van der Bug JCE: Evidence for a role of antagonistic cocontraction in controlling trunk stiffness during lifting, *J Biomech* 36: 1829, 2003.
123. Vismara L, Menegoni F, Zaina F, Galli M, Negrini S, and Capodaglio P: Effect of obesity and low back pain on spinal mobility: a cross sectional study in women, *J Neuroeng Rehabil* 18;7, 2010.
124. Voutsinas SA and MacEwan GD: Sagittal profiles of the spine, *Clin Orthop* 210:235, 1986.
125. Walsh AJL and Lotz JC: Biological response of the intervertebral disc to dynamic loading, *J Biomech* 37:329, 2004.
126. Watanabe S, Eguchi A, Kobara K, and Ishida H: Influence of trunk muscle co-contraction on spinal curvature during sitting reclining against the backrest of a chair, *Electromyogr Clin Neurophysiol* 48:359, 2008.
127. White AA and Panjabi MM: *Clinical biomechanics of the spine,* Philadelphia, 1978, JB Lippincott.
128. Wiesel SW, Bernini P, and Rothman RH: *The aging lumbar spine,* Philadelphia, 1982, WB Saunders.
129. Wilder DG, Krag MH, and Pope MH: *Atlas of mammalian lumbar vertebrae—pictorial and dimensional information*, Springfield, 1991, Charles C. Thomas.
130. Wojtys EM, Ashton-Miller JA, Huston L, and Moga P: The association between athletic training time and the sagittal curvature of the immature spine, *Am J Sports Med* 28:490, 2000.
131. Yamamoto S: A new trend in the study of low back pain in workplaces, *Ind Health* 35:173, 1997.

132. Yildiran KA, Senkoylu A, and Korkusuz F: Soccer causes degenerative changes in the cervical spine, *Eur Spine J* 13:76, 2004.
133. Zarzycka M, Rozek K, and Zarzycki M: Alternative methods of conservative treatment of idiopathic scoliosis, *Ortop Traumatol Rehabil* 11:396, 2009.

ANNOTATED READINGS

DiNubile N and Scali B: *Framework for the lower back* (2nd ed.), New York, 2010, Rodale Books.
Describes practical approaches for preventing and managing back pain. Written by physicians for the lay public.
Key J: *Back pain—a movement problem: A clinical approach incorporating relevant research and practice,* New York, 2010, Churchill Livingstone.
Presents a physiotherapy perspective for diagnosis and treatment of back disorders.
Kusumi K and Dunwoodie SL (eds.): *The genetics and development of scoliosis,* New York, 2010, Springer.
Discusses current research-based knowledge on development and treatment of scoliosis.
McGill SM: *Low back disorders* (2nd ed.), Champaign, IL, 2007, Human Kinetics.
Presents comprehensive information on prevention and rehabilitation of low back pain based on biomechanics research.

RELATED WEBSITES

M & M Orthopaedics
http://mmortho.com/
Includes information on spinal anatomy, laminotomy, microdiscectomy, and scoliosis.
North American Spine Society
http://www.spine.org/
Provides links to information on topics such as obesity and low back pain, artificial intervertebral discs, and herniated cervical and lumbar discs.
Rothman Institute
http://www.rothmaninstitute.com/index.cfm?rothman=omg&fuseaction=content.homePage
Includes information on spinal anatomy and problems such as scoliosis, spinal stenosis, and degenerative disc disease.
Southern California Orthopaedic Institute
http://www.scoi.com
Includes links to anatomical descriptions and labeled photographs of the spine, as well as laminotomy and microdiscectomy and scoliosis.
The Mount Sinai Medical Center
http://www.mountsinai.org/
Provides links to information about cervical, thoracic, and lumbar spinal anatomy, as well as pathological diseases and conditions of the spine.
University of Washington Orthopaedic Physicians
http://www.orthop.washington.edu
Provides radiographs and information on common injuries and pathological conditions for the spine.
Wheeless' Textbook of Orthopaedics
http://www.wheelessonline.com
Provides comprehensive, detailed information, graphics, and related literature for all the joints.

KEY TERMS

annulus fibrosus	thick, fibrocartilaginous ring that forms the exterior of the intervertebral disc
flexion relaxation phenomenon	condition in which, when the spine is in full flexion, the spinal extensor muscles relax and the flexion torque is supported by the spinal ligaments
intraabdominal pressure	pressure inside the abdominal cavity; believed to help stiffen the lumbar spine against buckling

kyphosis	extreme curvature in the thoracic region of the spine
ligamentum flavum	yellow ligament that connects the laminae of adjacent vertebrae; distinguished by its elasticity
lordosis	extreme curvature in the lumbar region of the spine
motion segment	two adjacent vertebrae and the associated soft tissues; the functional unit of the spine
nucleus pulposus	colloidal gel with a high fluid content, located inside the annulus fibrosus of the intervertebral disc
prestress	stress on the spine created by tension in the resting ligaments
primary spinal curves	curves that are present at birth
scoliosis	lateral spinal curvature
secondary spinal curves	the cervical and lumbar curves, which do not develop until the weight of the body begins to be supported in sitting and standing positions
spondylolisthesis	complete bilateral fracture of the pars interarticularis, resulting in anterior slippage of the vertebra
spondylolysis	presence of a fracture in the pars interarticularis of the vertebral neural arch

Linear Kinematics of Human Movement

After completing this chapter, you will be able to:

Discuss the interrelationships among kinematic variables.

Correctly associate linear kinematic quantities with their units of measure.

Identify and describe the effects of factors governing projectile trajectory.

Explain why the horizontal and vertical components of projectile motion are analyzed separately.

Distinguish between average and instantaneous quantities and identify the circumstances under which each is a quantity of interest.

Select and use appropriate equations to solve problems related to linear kinematics.

ONLINE LEARNING CENTER RESOURCES

www.mhhe.com/hall6e

Log on to our Online Learning Center (OLC) for access to these additional resources:

- Online Lab Manual
- Flashcards with definitions of chapter key terms
- Chapter objectives
- Chapter lecture PowerPoint presentation
- Self-scoring chapter quiz
- Additional chapter resources
- Web links for study and exploration of chapter-related topics

W hy is a sprinter's acceleration close to zero in the middle of a race? How does the size of a dancer's foot affect the performance time that a choreographer must allocate for jumps? At what angle should a discus or a javelin be thrown to achieve maximum distance? Why does a ball thrown horizontally hit the ground at the same time as a ball dropped from the same height? These questions all relate to the kinematic characteristics of a pure form of movement: linear motion. This chapter introduces the study of human movement mechanics with a discussion of linear kinematic quantities and projectile motion.

LINEAR KINEMATIC QUANTITIES

kinematics

the form, pattern, or sequencing of movement with respect to time

Kinematics is the geometry, pattern, or form of motion with respect to time. Kinematics, which describes the appearance of motion, is distinguished from kinetics, the forces associated with motion. Linear kinematics involves the shape, form, pattern, and sequencing of linear movement through time, without particular reference to the forces that cause or result from the motion.

Careful kinematic analyses of performance are invaluable for clinicians, physical activity teachers, and coaches. When people learn a new motor skill, a progressive modification of movement kinematics reflects the learning process. This is particularly true for young children, whose movement kinematics changes with the normal changes in anthropometry and neuromuscular coordination that accompany growth. Likewise, when a patient rehabilitates an injured joint, the therapist or clinician looks for the gradual return of normal joint kinematics.

Kinematics spans both qualitative and quantitative forms of analysis. For example, qualitatively describing the kinematics of a soccer kick entails identifying the major joint actions, including hip flexion, knee extension, and possibly plantar flexion at the ankle. A more detailed qualitative kinematic analysis might also describe the precise sequencing and timing of body segment movements, which translates to the degree of skill evident on the part of the kicker. Although most assessments of human movement are carried out qualitatively through visual observation, quantitative analysis is also sometimes appropriate. Physical therapists,

Movement kinematics is also referred to as form *or* technique. Photo courtesy of Getty Images.

for example, often measure the range of motion of an injured joint to help determine the extent to which range of motion exercises may be needed. When a coach measures an athlete's performance in the shot put or long jump, this too is a quantitative assessment.

Sport biomechanists often quantitatively study the kinematic factors that characterize an elite performance or the biomechanical factors that may limit the performance of a particular athlete. Researchers have discovered that elite sprinters, for example, develop both greater horizontal and vertical velocity coming off the starting blocks as compared to well-trained nonelite sprinters (24). During both the approach for the volleyball spike jump and the takeoff in the ski jump it is subtleties in approach kinematics that influence the height of the spike jump and the length of the ski jump (10, 27).

Most biomechanical studies of human kinematics, however, are performed on nonelite subjects. Kinematic research has shown that toddlers exhibit different strategies in walking than adults, exhibiting more kinematic variability and exploratory behavior as compared to adults (5). In collaboration with adapted physical education specialists, biomechanists have documented the characteristic kinematic patterns associated with relatively common disabling conditions such as cerebral palsy, Down syndrome, and stroke. Biomechanists studying wheelchair propulsion have shown that in order to propel a wheelchair up a slope, paraplegics increase forward trunk lean and employ increased stroke frequency coupled with decreased stroking speed (2).

Distance and Displacement

Units of distance and displacement are units of length. In the metric system, the most commonly used unit of distance and displacement is the **meter (m)**. A kilometer (km) is 1000 m, a centimeter (cm) is $\frac{1}{100}$ m, and a millimeter (mm) is $\frac{1}{1000}$ m. In the English system, common units of length are the inch, the foot (0.30 m), the yard (0.91 m), and the mile (1.61 km).

Distance and displacement are assessed differently. Distance is measured along the path of motion. When a runner completes 1½ laps around a 400 m track, the distance that the runner has covered is equal to 600 (400 + 200) m. **Linear displacement** is measured in a straight line from position 1 to position 2, or from initial position to final position. At the end of 1½ laps around the track, the runner's displacement is the length of the straight imaginary line that transverses the field, connecting the runner's initial position to the runner's final position halfway around the track (see Introductory Problem 1). At the completion of 2 laps around the track, the distance run is 800 m. Because initial and final positions are the same, however, the runner's displacement is zero. When a skater moves around a rink, the distance the skater travels may be measured along the tracks left by the skates. The skater's displacement is measured along a straight line from initial to final positions on the ice (Figure 10-1).

Another difference is that distance is a scalar quantity while displacement is a vector quantity. Consequently, the displacement includes more than just the length of the line between two positions. Of equal importance is the *direction* in which the displacement occurs. The direction of a displacement relates the final position to the initial position. For example, the displacement of a yacht that has sailed 900 m on a tack due south would be identified as 900 m to the south.

The direction of a displacement may be indicated in several different, equally acceptable ways. Compass directions such as south and northwest and the terms *left/right, up/down,* and *positive/negative* are all appropriate

meter
the most common international unit of length, on which the metric system is based

linear displacement
change in location, or the directed distance from initial to final location

●*The metric system is the predominant standard of measurement in every major country in the world except the United States.*

The distance a skater travels may be measured from the track on the ice. The skater's displacement is measured in a straight line from initial position to final position.

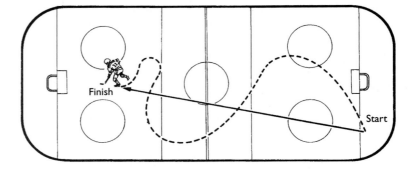

labels. The positive direction is typically defined as upward and/or to the right, with negative regarded as downward and/or to the left. This enables indication of direction using plus and minus signs. The most important thing is to be consistent in using the system or convention adopted for indicating direction in a given context. It would be confusing to describe a displacement as 500 m north followed by 300 m to the right.

Either distance or displacement may be the more important quantity of interest depending on the situation. Many 5 km and 10 km racecourses are set up so that the finish line is only a block or two from the starting line. Participants in these races are usually interested in the number of kilometers of distance covered or the number of kilometers left to cover as they progress along the racecourse. Knowledge of displacement is not particularly valuable during this type of event. In other situations, however, displacement is more important. For example, triathlon competitions may involve a swim across a lake. Because swimming in a perfectly straight line across a lake is virtually impossible, the actual distance a swimmer covers is always somewhat greater than the width of the lake (see Sample Problem 10.1). However, the course is set up so that the identified length of the swim course is the length of the displacement between the entry and exit points on the lake.

Displacement magnitude and distance covered can be identical. When a cross-country skier travels down a straight path through the woods, both distance covered and displacement are equal. However, any time the path of motion is not rectilinear, the distance traveled and the size of the displacement will differ.

• *Distance covered and displacement may be equal for a given movement. Or distance may be greater than displacement, but the reverse is never true.*

Speed and Velocity

Two quantities that parallel distance and linear displacement are speed and linear velocity. These terms are often used synonymously in general conversation, but in mechanics, they have precise and different meanings. *Speed,* a scalar quantity, is defined as the distance covered divided by the time taken to cover it:

linear velocity
the rate of change in location

$$\text{speed} = \frac{\text{length (or distance)}}{\text{change in time}}$$

Velocity (v) is the change in position, or displacement, that occurs during a given period of time:

$$\text{v} = \frac{\text{change in position (or displacement)}}{\text{change in time}}$$

Because the Greek capital letter delta (Δ) is commonly used in mathematical expressions to mean "change in," a shorthand version of the relationship

SAMPLE PROBLEM 10.1

A swimmer crosses a lake that is 0.9 km wide in 30 minutes. What was his average velocity? Can his average speed be calculated?

Known

After reading the problem carefully, the next step is to sketch the problem situation, showing all quantities that are known or may be deduced from the problem statement:

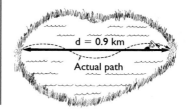

$$t = 30 \text{ min } (0.5 \text{ hr})$$

d = 0.9 km

Actual path

Solution

In this situation, we know that the swimmer's displacement is 0.9 km. However, we know nothing about the exact path that he may have followed. The next step is to identify the appropriate formula to use to find the unknown quantity, which is velocity:

$$v = \frac{d}{t}$$

The known quantities can now be filled in to solve for velocity:

$$v = \frac{0.9 \text{ km}}{0.5 \text{ hr}}$$

$$= 1.8 \text{ km/hr}$$

Speed is calculated as distance divided by time. Although we know the time taken to cross the lake, we do not know, nor can we surmise from the information given, the exact distance covered by the swimmer. Therefore, his speed cannot be calculated.

---

expressed follows, with t representing the amount of time elapsed during the velocity assessment:

$$v = \frac{\Delta \text{ position}}{\Delta \text{ time}} = \frac{d}{\Delta t}$$

Another way to express change in position is position$_2$ − position$_1$, in which position$_1$ represents the body's position at one point in time and position$_2$ represents the body's position at a later point:

$$\text{velocity} = \frac{\text{position}_2 - \text{position}_1}{\text{time}_2 - \text{time}_1}$$

Because velocity is based on displacement, it is also a vector quantity. Consequently, description of velocity must include an indication of both the direction and the magnitude of the motion. If the direction of the motion is positive, velocity is positive; if the direction is negative, velocity is negative. A change in a body's velocity may represent a change in its speed, movement direction, or both.

● *Displacement and velocity are vector equivalents of the scalar quantities distance and speed.*

FIGURE 10-2

The velocity of a swimmer in a river is the vector sum of the swimmer's velocity and the velocity of the current.

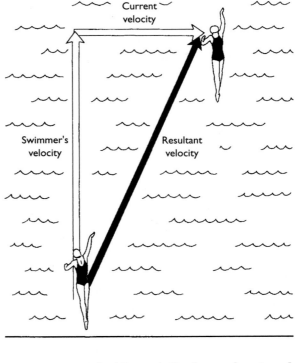

Units of speed and velocity are always units of length divided by units of time.

Running speed is the product of stride length and stride frequency. Photo courtesy of Karl Weatherly/Getty Images.

Whenever two or more velocities act, the laws of vector algebra govern the ultimate speed and direction of the resultant motion. For example, the path actually taken by a swimmer crossing a river is determined by the vector sum of the swimmer's speed in the intended direction and the velocity of the river's current (Figure 10-2). Sample Problem 10.2 provides an illustration of this situation.

Units of speed and velocity are units of length divided by units of time. In the metric system, common units for speed and velocity are meters per second (m/s) and kilometers per hour (km/hr). However, any unit of length divided by any unit of time yields an acceptable unit of speed or velocity. For example, a speed of 5 m/s can also be expressed as 5000 mm/s or 18,000 m/hr. It is usually most practical to select units that will result in expression of the quantity in the smallest, most manageable form.

For human gait, speed is the product of stride length and stride frequency. Adults in a hurry tend to walk with both longer stride lengths and faster stride frequency than they use under more leisurely circumstances.

During running, a kinematic variable such as stride length is not simply a function of the runner's body height, but is also influenced by muscle fiber composition, footwear, level of fatigue, injury history, and the inclination (grade) and stiffness of the running surface (7). Runners traveling at a slow pace tend to increase velocity primarily by increasing stride length. At faster running speeds, recreational runners rely more on increasing stride frequency to increase velocity (Figure 10-3). In cross-country skiing, as speed increases, stride rate increases and stride length tends to decrease (19). Overstriding, or using an overly long stride length, should be avoided in both running and skiing, since it is a risk factor for hamstring strains.

Those who run regularly for exercise usually prefer a given stride frequency over a range of slow-to-moderate running speeds. One reason for this may be related to running economy—the oxygen consumption required for performing a given task. Most runners tend to choose a

SAMPLE PROBLEM 10.2

A swimmer orients herself perpendicular to the parallel banks of a river. If the swimmer's velocity is 2 m/s and the velocity of the current is 0.5 m/s, what will be her resultant velocity? How far will she actually have to swim to get to the other side if the banks of the river are 50 m apart?

Solution
A diagram is drawn showing vector representations of the velocities of the swimmer and the current:

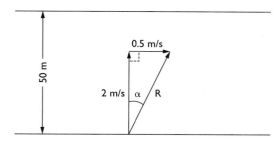

The resultant velocity can be found graphically by measuring the length and the orientation of the vector resultant of the two given velocities:

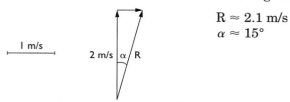

$$R \approx 2.1 \text{ m/s}$$
$$\alpha \approx 15°$$

The resultant velocity can also be found using trigonometric relationships. The magnitude of the resultant velocity may be calculated using the Pythagorean theorem:

$$R^2 = (2 \text{ m/s})^2 + (0.5 \text{ m/s})^2$$
$$R^2 = \sqrt{(2\text{m/s})^2} + \sqrt{(0.5 \text{ m/s})^2}$$
$$= 2.06 \text{ m/s}$$

The direction of the resultant velocity may be calculated using the cosine relationship:

$$R \cos \alpha = 2 \text{ m/s}$$
$$(2.06 \text{ m/s}) \cos \alpha = 2 \text{ m/s}$$

$$\alpha = \arccos\left(\frac{2 \text{ m/s}}{2.06 \text{ m/s}}\right)$$

$$= 14°$$

If the swimmer travels in a straight line in the direction of her resultant velocity, the cosine relationship may be used to calculate her resultant displacement:

$$D \cos \alpha = 50 \text{ m}$$
$$D \cos 14 = 50 \text{ m}$$
$$\boxed{D = 51.5 \text{ m}}$$

FIGURE 10-3

Changes in stride length and stride rate with running velocity. From Luhtanen P and Komi PV: Mechanical factors influencing running speed. In Asmussen E and Jorgensen K, eds: *Biomechanics VI-B,* Baltimore, 1978, University Park Press.

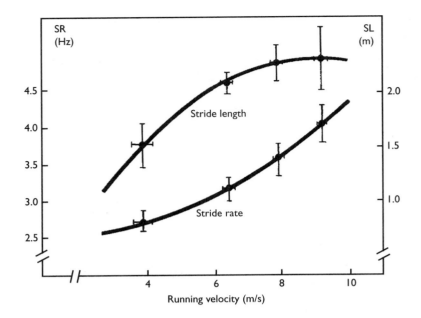

combination of stride length and stride frequency that minimizes the physiological cost of running. As discussed in Chapter 1, many species of animals do the same thing. Running on downhill and uphill surfaces tends to respectively increase and decrease running speed, with these differences primarily a function of increased and decreased stride length (20). The presence of fatigue, as would be expected near the end of a marathon event, tends to result in increased stride frequency and decreased stride length (11).

Since maximizing speed is the objective of all racing events, sport biomechanists have focused on the kinematic features that appear to accompany fast performances in running, skiing, skating, cycling, swimming, and rowing events. Research shows that elite 1500-m runners are distinguished from other skilled performers in the event by different hip kinematics, suggesting more efficient use of the hips during running (14). In cross-country skiing, use of the double-push technique is superior to other techniques for uphill skating speed in that it involves longer cycle lengths, a lower cycle rate, and longer recovery times (25). Elite long-distance skaters adopt the strategy of starting races fast and skating progressively slower, although completing all laps faster than their subelite competitors (18). Further research shows that competitive breaststrokers glide for longer periods than less-skilled swimmers (23, 26).

When racing performances are analyzed, comparisons are usually based on pace rather than speed or velocity. *Pace* is the inverse of speed. Rather than units of distance divided by units of time, pace is presented as units of time divided by units of distance. Pace is the time taken to cover a given distance and is commonly quantified as minutes per km or minutes per mile.

ACCELERATION

linear acceleration
the rate of change in linear velocity

We are well aware that the consequence of pressing down or letting up on the accelerator (gas) pedal of an automobile is usually a change in the automobile's speed (and velocity). Linear acceleration (a) is defined as the rate of change in velocity, or the change in velocity occurring over a given time interval (t):

$$a = \frac{\text{change in velocity}}{\text{change of time}} = \frac{\Delta v}{\Delta t}$$

Another way to express change in velocity is $v_2 - v_1$, where v_1 represents velocity at one point in time and v_2 represents velocity at a later point:

$$a = \frac{v_2 - v_1}{\Delta t}$$

Units of acceleration are units of velocity divided by units of time. If a car increases its velocity by 1 km/hr each second, its acceleration is 1 km/hr/s. If a skier increases velocity by 1 m/s each second, the acceleration is 1 m/s/s. In mathematical terms, it is simpler to express the skier's acceleration as 1 m/s squared (1 m/s^2). A common unit of acceleration in the metric system is m/s^2.

Acceleration is the rate of change in velocity, or the degree with which velocity is changing with respect to time. For example, a body accelerating in a positive direction at a constant rate of 2 m/s^2 is increasing its velocity by 2 m/s each second. If the body's initial velocity was zero, a second later its velocity would be 2 m/s, a second after that its velocity would be 4 m/s, and a second after that its velocity would be 6 m/s.

In general usage, the term *accelerating* means speeding up, or increasing in velocity. If v_2 is greater than v_1, acceleration is a positive number, and the body in motion may have speeded up during the period in question. However, because it is sometimes appropriate to label the direction of motion as positive or negative, a positive value of acceleration may not mean that the body is speeding up.

If the direction of motion is described in terms other than positive or negative, a positive value of acceleration does indicate that the body being analyzed has speeded up. For example, if a sprinter's velocity is 3 m/s on leaving the blocks and is 5 m/s a second later, calculation of the acceleration that has occurred will yield a positive number. Because $v_1 = 3$ m/s, $v_2 = 5$ m/s, and t = 1 s:

$$a = \frac{v_2 - v_1}{\Delta t}$$

$$= \frac{5 \text{ m/s} - 3 \text{ m/s}}{1 \text{ s}}$$

$$= 2 \text{ m/s}^2$$

Sliding into a base involves negative acceleration of the base runner. Photo courtesy of Royalty-Free/CORBIS.

Whenever the direction of motion is described in terms other than positive or negative, and v_2 is greater than v_1, the value of acceleration will be a positive number, and the object in question is speeding up.

Acceleration can also assume a negative value. As long as the direction of motion is described in terms other than positive or negative, negative acceleration indicates that the body in motion is slowing down, or that its velocity is decreasing. For example, when a base runner slides to a stop over home plate, acceleration is negative. If a base runner's velocity is 4 m/s when going into a 0.5 s slide that stops the motion, $v_1 = 4$ m/s, $v_2 = 0$, and $t = 0.5$ s. Acceleration may be calculated as the following:

$$a = \frac{v_2 - v_1}{t}$$
$$= \frac{0 - 4 \text{ m/s}}{0.5 \text{ s}}$$
$$= -8 \text{ m/s}^2$$

Whenever v_1 is greater than v_2 in this type of situation, acceleration will be negative. Sample Problem 10.3 provides another example of a situation involving negative acceleration.

Understanding acceleration is more complicated when one direction is designated as positive and the opposite direction is designated as negative. In this situation, a positive value of acceleration can indicate either that the object is speeding up in a positive direction or that it is slowing down in a negative direction (Figure 10-4).

Consider the case of a ball being dropped from a hand. As the ball falls faster and faster because of the influence of gravity, it is gaining speed—for example, 0.3 m/s to 0.5 m/s to 0.8 m/s. Because the downward direction is considered as the negative direction, the ball's velocity is actually −0.3 m/s to −0.5 m/s to −0.8 m/s. If $v_1 = -0.3$ m/s, $v_2 = -0.5$ m/s, and $t = 0.02$ s, acceleration is calculated as follows:

$$a = \frac{v_2 - v_1}{t}$$
$$= \frac{-0.5 \text{ m/s} - (-0.3 \text{ m/s})}{0.02 \text{ s}}$$
$$= -10 \text{ m/s}^2$$

In this situation, the ball is speeding up, yet its acceleration is negative because it is speeding up in a negative direction. If acceleration is negative, velocity may be either increasing in a negative direction or decreasing in a positive direction. Alternatively, if acceleration is positive, velocity may be either increasing in a positive direction or decreasing in a negative direction.

The third alternative is for acceleration to be equal to zero. Acceleration is zero whenever velocity is constant, that is, when v_1 and v_2 are the same. In the middle of a 100 m sprint, a sprinter's acceleration should be close to zero, because at that point the runner should be running at a constant, near-maximum velocity.

Acceleration and deceleration (the lay term for negative acceleration) have implications for injury of the human body, since changing velocity results from the application of force (see Chapter 12). The anterior cruciate ligament, which restricts the forward sliding of the femur on the tibial plateaus during knee flexion, is often injured when an athlete who is running decelerates rapidly or changes directions quickly.

It is important to remember that since acceleration is a vector quantity, changing directions, even while maintaining a constant speed, represents

•When acceleration is zero, velocity is constant.

SAMPLE PROBLEM 10.3

A soccer ball is rolling down a field. At t = 0, the ball has an instantaneous velocity of 4 m/s. If the acceleration of the ball is constant at −0.3 m/s², how long will it take the ball to come to a complete stop?

Known

After reading the problem carefully, the next step is to sketch the problem situation, showing all quantities that are known or given in the problem statement.

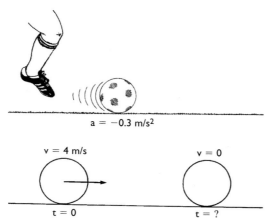

Solution

The next step is to identify the appropriate formula to use to find the unknown quantity:

$$a = \frac{v_2 - v_1}{t}$$

The known quantities can now be filled in to solve for the unknown variable (time):

$$-0.3 \text{ m/s}^2 = \frac{0 - 4 \text{ m/s}}{t}$$

Rearranging the equation, we have the following:

$$t = \frac{0 - 4 \text{ m/s}}{-0.3 \text{ m/s}^2}$$

Simplifying the expression on the right side of the equation, we have the solution:

$$t = 13.3 \text{ s}$$

---

a change in acceleration. The concept of angular acceleration, with direction constantly changing, is discussed in Chapter 11. The forces associated with change in acceleration based on change in direction must be compensated for by skiers and velodrome cyclists, in particular. That topic is discussed in Chapter 14.

Average and Instantaneous Quantities

It is often of interest to determine the velocity of acceleration of an object or body segment at a particular time. For example, the instantaneous

instantaneous
occurring during a small interval of time

FIGURE 10-4

Right is regarded as the positive direction, and *left* as the negative direction. Acceleration may be positive, negative, or equal to zero, based on the direction of the motion and the direction of the change in speed.

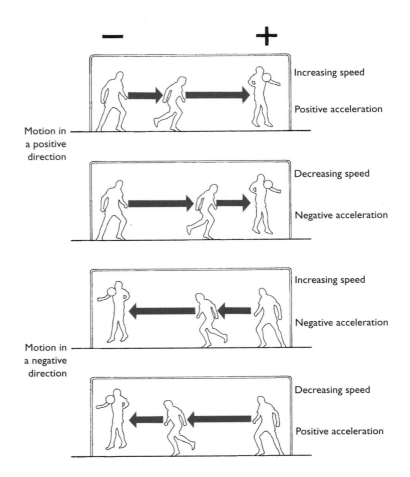

average
occurring over a designated time interval

projectile
body in free fall that is subject only to the forces of gravity and air resistance

The instantaneous velocity of the shot at the moment of release primarily determines the ultimate horizontal displacement of the shot. Photo courtesy of Digital Vision/Getty Images.

velocity of a shot or a discus at the moment the athlete releases it greatly affects the distance that the implement will travel. It is sometimes sufficient to quantify the average speed or velocity of the entire performance.

When speed and velocity are calculated, the procedures depend on whether the average or the instantaneous value is the quantity of interest. Average velocity is calculated as the final displacement divided by the total time. Average acceleration is calculated as the difference in the final and initial velocities divided by the entire time interval. Calculation of instantaneous values can be approximated by dividing differences in velocities over an extremely small time interval. With calculus, velocity can be calculated as the derivative of displacement, and acceleration as the derivative of velocity.

Selection of the time interval over which speed or velocity is quantified is important when analyzing the performance of athletes in racing events. Many athletes can maintain world record paces for the first one-half or three-fourths of the event, but slow during the last leg because of fatigue. Alternatively, some athletes may intentionally perform at a controlled pace during earlier segments of a race and then achieve maximum speed at the end. The longer the event is, the more information is potentially lost or concealed when only the final time or average speed is reported.

KINEMATICS OF PROJECTILE MOTION

Bodies projected into the air are projectiles. A basketball, a discus, a high jumper, and a sky diver are all projectiles as long as they are moving through the air unassisted. Depending on the projectile, different

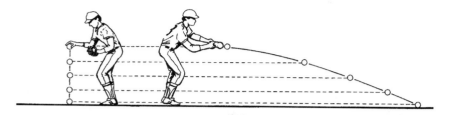

FIGURE 10-5

The vertical and horizontal components of projectile motion are independent. A ball hit horizontally has the same vertical component as a ball dropped with no horizontal velocity.

kinematic quantities are of interest. The resultant horizontal displacement of the projectile determines the winner of the contest in field events such as the shot put, discus throw, and javelin throw. High jumpers and pole-vaulters maximize ultimate vertical displacement to win events. Sky divers manipulate both horizontal and vertical components of velocity to land as close as possible to targets on the ground.

However, not all objects that fly through the air are projectiles. A projectile is a body in free fall that is subject only to the forces of gravity and air resistance. Therefore, objects such as airplanes and rockets do not qualify as projectiles, because they are also influenced by the forces generated by their engines.

Horizontal and Vertical Components

Just as it is more convenient to analyze general motion in terms of its linear and angular components, it is usually more meaningful to analyze the horizontal and vertical components of projectile motion separately. This is true for two reasons. First, the vertical component is influenced by gravity, whereas no force (neglecting air resistance) affects the horizontal component. Second, the horizontal component of motion relates to the distance the projectile travels, and the vertical component relates to the maximum height achieved by the projectile. Once a body has been projected into the air, its overall (resultant) velocity is constantly changing because of the forces acting on it. When examined separately, however, the horizontal and vertical components of projectile velocity change predictably.

Horizontal and vertical components of projectile motion are independent of each other. In the example shown in Figure 10-5, a baseball is dropped from a height of 1 m at the same instant that a second ball is horizontally struck by a bat at a height of 1 m, resulting in a line drive. Both balls land on the level field simultaneously, because the vertical components of their motions are identical. However, because the line drive also has a horizontal component of motion, it undergoes some horizontal displacement as well.

Influence of Gravity

A major factor that influences the vertical but not the horizontal component of projectile motion is the force of gravity, which accelerates bodies in a vertical direction toward the surface of the earth (Figure 10-6). Unlike aerodynamic factors that may vary with the velocity of the wind, gravitational force is a constant, unchanging force that produces a constant downward vertical acceleration. Using the convention that upward is positive and downward is negative, the acceleration of gravity is treated as a negative quantity (-9.81 m/s^2). This acceleration remains constant regardless of the size, shape, or weight of the projectile. The vertical component of the initial projection velocity determines the maximum vertical displacement achieved by a body projected from a given relative projection height.

The human body becomes a projectile during the airborne phase of a jump. Photo courtesy of Digital Vision/Getty Images.

● *The force of gravity produces a constant acceleration on bodies near the surface of the earth equal to approximately -9.81 m/s^2.*

Projectile trajectories without (A) and with (B) gravitational influence.

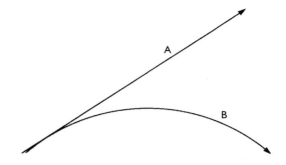

Figure 10-7 illustrates the influence of gravity on projectile flight in the case of a ball tossed into the air by a juggler. The ball leaves the juggler's hand with a certain vertical velocity. As the ball travels higher and higher, the magnitude of its velocity decreases because it is undergoing a negative acceleration (the acceleration of gravity in a downward direction). At the peak or apex of the flight, which is that instant between going up and coming down, vertical velocity is zero. As the ball falls downward, its speed progressively increases, again because of gravitational acceleration. Since the direction of motion is downward, the ball's velocity is becoming progressively more negative. If the ball is caught at the same height from which it was tossed, the ball's speed is exactly the same as its initial speed, although its direction is now reversed. Graphs of the vertical displacement, velocity, and acceleration of a tossed ball are shown in Figure 10-8.

apex
the highest point in the trajectory of a projectile

Influence of Air Resistance

If an object were projected in a vacuum (with no air resistance), the horizontal component of its velocity would remain exactly the same throughout the flight. However, in most real-life situations, air resistance affects the horizontal component of projectile velocity. A ball thrown with a given initial velocity in an outdoor area will travel much farther if it is thrown with a tailwind rather than into a headwind. Because the effects of air resistance are variable, however, it is customary to disregard air resistance in discussing and solving problems related to projectile motion since this allows treating the horizontal component of projectile motion as an unchanging (constant) quantity.

•*Neglecting air resistance, the horizontal speed of a projectile remains constant throughout the trajectory.*

The pattern of change in the vertical velocity of a projectile is symmetrical about the apex of the trajectory.

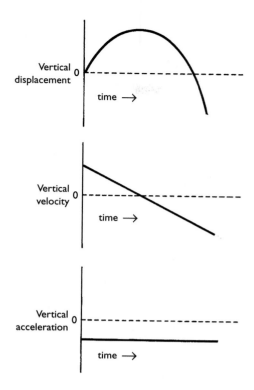

FIGURE 10-8

Vertical displacement, velocity, and acceleration graphs for a ball tossed into the air that falls to the ground. Note that velocity is in a positive (upward) direction but decreasing as the ball is going up. At the apex of the trajectory, in between going up and coming down, the ball's velocity is instantaneously zero. Then as the ball falls downward, velocity is in a negative (downward) direction but speed is increasing. Acceleration remains constant at -9.81 m/s^2 since gravity is the only force acting on the ball.

When a projectile drops vertically through the air in a typical real-life situation, its velocity at any point is also related to air resistance. A sky diver's velocity, for example, is much smaller after the opening of the parachute than before its opening.

FACTORS INFLUENCING PROJECTILE TRAJECTORY

Three factors influence the trajectory (flight path) of a projectile: (a) the angle of projection, (b) the projection speed, and (c) the relative height of projection (Figure 10-9) (Table 10-1). Understanding how these factors

trajectory
the flight path of a projectile

● *The three mechanical factors that determine a projectile's motion are projection angle, projection speed, and relative height of projection.*

FIGURE 10-9

Factors affecting the trajectory of a projectile include projection angle, projection speed, and relative height of projection.

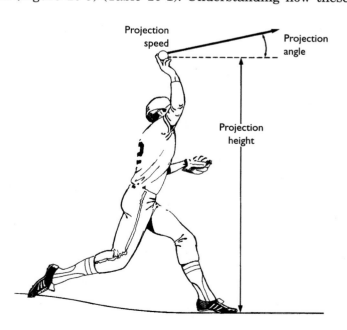

TABLE 10-1

Factors Influencing Projectile Motion (Neglecting Air Resistance)

VARIABLE	FACTORS OF INFLUENCE
Flight time	Initial vertical velocity Relative projection height
Horizontal displacement	Horizontal velocity Relative projection height
Vertical displacement	Initial vertical velocity Relative projection height
Trajectory	Initial speed Projection angle Relative projection height

interact is useful within the context of sport both for determining how to best project balls and other implements and for predicting how to best catch or strike projected balls.

Projection Angle

angle of projection
the direction at which a body is projected with respect to the horizontal

The angle of projection and the effects of air resistance govern the shape of a projectile's trajectory. Changes in projection speed influence the size of the trajectory, but trajectory shape is solely dependent on projection angle. In the absence of air resistance, the trajectory of a projectile assumes one of three general shapes, depending on the angle of projection. If the projection angle is perfectly vertical, the trajectory is also perfectly vertical, with the projectile following the same path straight up and then straight down again. If the projection angle is oblique (at some angle between 0° and 90°), the trajectory is *parabolic,* or shaped like a parabola. A parabola is symmetrical, so its right and left halves are mirror images of each other. A body projected perfectly horizontally (at an angle of 0°) will follow a trajectory resembling one-half of a parabola (Figure 10-10).

FIGURE 10-10

The effect of projection angle on projectile trajectory.

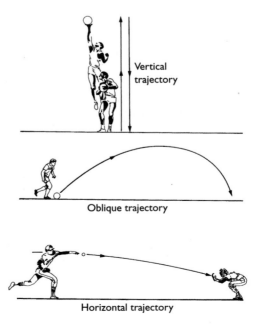

Vertical trajectory

Oblique trajectory

Horizontal trajectory

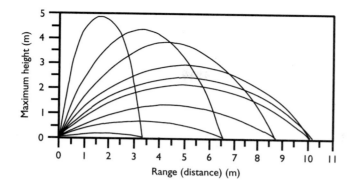

FIGURE 10-11

This scaled diagram shows the size and shape of trajectories for an object projected at 10 m/s.

Figure 10-11 displays scaled, theoretical trajectories for an object projected at different angles at a given speed. A ball thrown upward at a projection angle of 80° to the horizontal follows a relatively high and narrow trajectory, achieving more height than horizontal distance. A ball projected upward at a 10° angle to the horizontal follows a trajectory that is flat and long in shape.

Projection angle has direct implications for success in the sport of basketball, since a nearly vertical angle of entry into the basket allows a somewhat larger margin of error than a more horizontal angle of entry. Within 4.57 m of the basket, jump shot release angles are about 52–55°, providing a relatively vertical angle of entry, whereas shots taken from 6.40 m tend to be released at 48–50°, allowing for a minimum release speed, but a less vertical angle of entry (17). When shooting in close proximity to a defender, players tend to release the ball at a more vertical release angle and from a greater height than is the case when a player is open (21). Although the strategy behind this is typically to keep the shot from being blocked, it may also result in more accurate shooting.

In projection situations on a field, air resistance may, in reality, create irregularities in the shape of a projectile's trajectory. A typical modification in trajectory caused by air resistance is displayed in Figure 10-12. For purposes of simplification, the effects of aerodynamic forces will be disregarded in the discussion of projectile motion.

Projection Speed

When projection angle and other factors are constant, the projection speed determines the length or size of a projectile's trajectory. For example, when a body is projected vertically upward, the projectile's initial speed determines the height of the trajectory's apex. For a body that is projected at an oblique angle, the speed of projection determines both the height

Projection angle is particularly important in the sport of basketball. A common error among novice players is shooting the ball with too flat a trajectory. Photo courtesy of Blend Images/Getty Images.

projection speed
the magnitude of projection velocity

•*A projectile's range is the product of its horizontal speed and flight time.*

FIGURE 10-12

In real-life situations, air resistance causes a projectile to deviate from its theoretical parabolic trajectory.

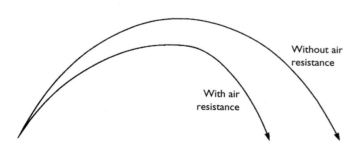

Without air resistance

With air resistance

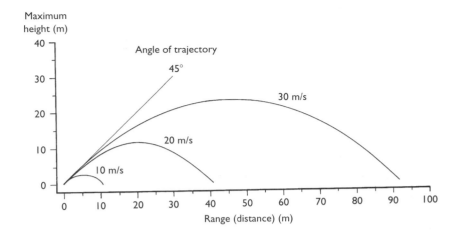

The effect of projection speed on projectile trajectory with projection angle held constant.

range
the horizontal displacement of a projectile at landing

and the horizontal length of the trajectory (Figure 10-13). The combined effects of projection speed and projection angle on the horizontal displacement, or range, of a projectile are shown in Table 10-2.

Performance in the execution of a vertical jump on a flat surface is entirely dependent on takeoff speed; that is, the greater the vertical velocity at takeoff, the higher the jump, and the higher the jump, the greater the amount of time the jumper is airborne (see margin). Elite beach volleyball players can jump higher and stay airborne longer when taking off from a solid surface than from sand because the instability of the sand produces a reduction in takeoff velocity (6).

The time required for the performance of a vertical jump can be an important issue for dance choreographers. The incorporation of vertical jumps into a performance must be planned carefully (12). If the tempo of the music necessitates that vertical jumps be executed within one-third of a second, the height of the jumps is restricted to approximately 12 cm. The choreographer must be aware that under these circumstances, most dancers do not have sufficient floor clearance to point their toes during jump execution.

Relative Projection Height

relative projection height
the difference between projection height and landing height

● *A projectile's flight time is increased by increasing the vertical component of projection velocity or by increasing the relative projection height.*

The third major factor influencing projectile trajectory is the **relative projection height** (Figure 10-14). This is the difference in the height from which the body is initially projected and the height at which it lands or stops. When a discus is released by a thrower from a height of 1½ m above the ground, the relative projection height is 1½ m, because the projection height is 1½ m greater than the height of the field on which the discus lands. If a driven golf ball becomes lodged in a tree, the relative projection height is negative, because the landing height is greater than the projection height. When projection velocity is constant, greater relative projection height translates to longer flight time and greater horizontal displacement of the projectile.

In the sport of diving, relative projection height is the height of the springboard or platform above the water. If a diver's center of gravity is elevated 1.5 m above the springboard at the apex of the trajectory, flight time is about 1.2 s from a 1 m board and 1.4 s from a 3 m board. This provides enough time for a skilled diver to complete 3 somersaults from a 1 m board and 3½-somersaults from a 3 m board (30). The implication is that a diver attempting to learn a 3½-somersault dive from the 3 m springboard should first be able to easily execute a 2½ somersault dive from the 1 m board.

VERTICAL JUMP HEIGHT (cm)	FLIGHT TIME (s)
5	0.2
11	0.3
20	0.4
31	0.5
44	0.6
60	0.7
78	0.8
99	0.9

PROJECTION SPEED (m/s)	PROJECTION ANGLE (°)	RANGE (m)
10	10	3.49
10	20	6.55
10	30	8.83
10	40	10.04
10	45	10.19
10	50	10.04
10	60	8.83
10	70	6.55
10	80	3.49
20	10	13.94
20	20	26.21
20	30	35.31
20	40	40.15
20	45	40.77
20	50	40.15
20	60	35.31
20	70	26.21
20	80	13.94
30	10	31.38
30	20	58.97
30	30	79.45
30	40	90.35
30	45	91.74
30	50	90.35
30	60	79.45
30	70	58.97
30	80	31.38

TABLE 10-2

The Effect of Projection Angle on Range (Relative Projection Height = 0)

Optimum Projection Conditions

In sporting events based on achieving maximum horizontal displacement or maximum vertical displacement of a projectile, the athlete's primary goal is to maximize the speed of projection. In the throwing events, another objective is to maximize release height, because greater relative projection height produces longer flight time, and consequently greater horizontal displacement of the projectile. However, it is generally not prudent for a thrower to sacrifice release speed for added release height.

The factor that varies the most, with both the event and the performer, is the optimum angle of projection. When relative projection height is zero, the angle of projection that produces maximum horizontal displacement

FIGURE 10-14

The relative projection height.

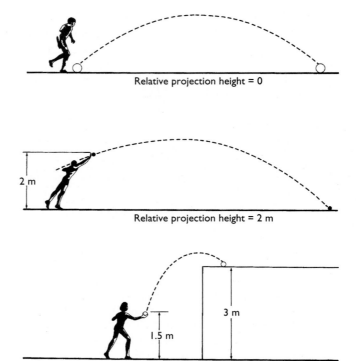

FIGURE 10-15

When projection speed is constant and aerodynamics are not considered, the optimum projection angle is based on the relative height of projection. When the relative projection height is zero, an angle of 45° is optimum. As the relative projection height increases, optimum projection angle decreases. As the relative projection height becomes increasingly negative, the optimum projection angle increases.

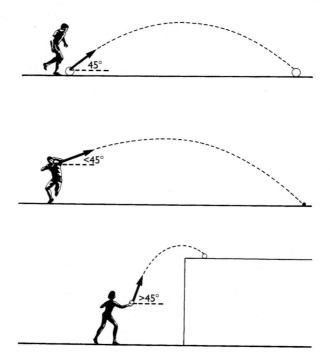

is 45°. As relative projection height increases, the optimum angle of projection decreases, and as relative projection height decreases, the optimum angle increases (Figure 10-15).

It is important to recognize that there are relationships among projection speed, height, and angle, such that when one is shifted closer to what would theoretically be optimal, another moves farther away from optimum. This is because humans are not machines, and human anatomy dictates certain constraints. Research has shown, for example, that the relationships among release speed, height, and angle for performance in the shot put are such that achievable release speed decreases with increasing release angle at 1.7 (m/s)/rad and decreases with increasing release height at 0.8 (m/s)/m (9). For both the shot put and the discus, however, biomechanists have found that the optimal angle of release is athlete-specific, ranging from 35 to 44 degrees among elite performers because of individual differences in the decrease of projection speed with increasing release angle (13, 15).

Likewise, when the human body is the projectile during a jump, high takeoff speed serves to constrain the projection angle that can be achieved (22). In the performance of the long jump, for example, because takeoff and landing heights are the same, the theoretically optimum angle of takeoff is 45° with respect to the horizontal. However, it has been estimated by Hay (8) that to obtain this theoretically optimum takeoff angle, long jumpers would decrease the horizontal velocity they could otherwise obtain by approximately 50%. Research has shown that success in the long jump, high jump, and pole vault is related to the athlete's ability to maximize horizontal velocity going into takeoff (1, 4, 28). The actual takeoff angles employed by elite long jumpers range from approximately 18° to 27° (8). Takeoff angles during all three phases of the triple jump are even smaller for elite performers than those used in the long jump (16). Performance in the triple jump is complicated by the fact that there is a direct trade-off between horizontal velocity and vertical velocity during the jumps (31). In the ski jump, where athletes have the advantage of a large relative height between takeoff and landing, takeoff angles are as small as 4.6–6.2° (29). In an event such as the high jump, in which the goal is to maximize vertical displacement, takeoff angles among skilled Fosbury Flop–style jumpers range from 40°–48° (3).

ANALYZING PROJECTILE MOTION

Because velocity is a vector quantity, the initial velocity of a projectile incorporates both the initial speed (magnitude) and the angle of projection (direction) into a single quantity. When the initial velocity of a projectile is resolved into horizontal and vertical components, the horizontal component has a certain speed or magnitude in a horizontal direction, and the vertical component has a speed or magnitude in a vertical direction (Figure 10-16). The magnitudes of the horizontal and vertical components are always quantified so that if they were added together through the process of vector composition, the resultant velocity vector would be equal in magnitude and direction to the original initial velocity vector. The horizontal and vertical components of initial velocity may be quantified both graphically and trigonometrically (see Sample Problem 10.4).

initial velocity
vector quantity incorporating both angle and speed of projection

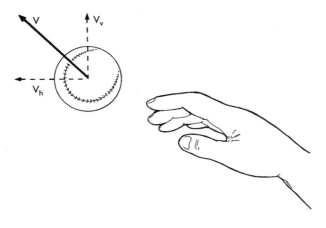

FIGURE 10-16

The vertical and horizontal components of projection velocity.

SAMPLE PROBLEM 10.4

A basketball is released with an initial speed of 8 m/s at an angle of 60°. Find the horizontal and vertical components of the ball's initial velocity, both graphically and trigonometrically.

Known

A diagram showing a vector representation of the initial velocity is drawn using a scale of 1 cm = 2 m/s:

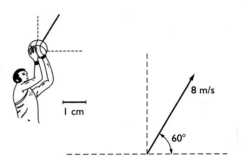

Solution

The horizontal component is drawn in along the horizontal line to a length that is equal to the length that the original velocity vector extends in the horizontal direction. The vertical component is then drawn in the same fashion in a direction perpendicular to the horizontal line:

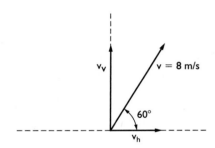

The lengths of the horizontal and vertical components are then measured:

length of horizontal component = 2 cm
length of vertical component = 3.5 cm

To calculate the magnitudes of the horizontal and vertical components, use the scale factor of 2 m/s/cm:

Magnitude of horizontal component:

$$v_h = 2 \text{ cm} \times 2 \text{ m/s/cm}$$

$$v_h = 4 \text{ m/s}$$

Magnitude of vertical component:

$$v_v = 3.5 \text{ cm} \times 2 \text{ m/s/cm}$$

$$v_v = 7 \text{ m/s}$$

To solve for v_h and v_v trigonometrically, construct a right triangle with the sides being the horizontal and vertical components of initial velocity and the initial velocity represented as the hypotenuse:

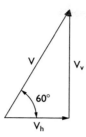

The sine and cosine relationships may be used to quantify the horizontal and vertical components:

$$v_h = (8 \text{ m/s}) (\cos 60)$$

$$v_h = 4 \text{ m/s}$$

$$v_v = (8 \text{ m/s})(\sin 60)$$

$$v_v = 6.9 \text{ m/s}$$

Note that the magnitude of the horizontal component is *always* equal to the magnitude of the initial velocity multiplied by the cosine of the projection angle. Similarly, the magnitude of the initial vertical component is *always* equal to the magnitude of the initial velocity multiplied by the sine of the projection angle.

For purposes of analyzing the motion of projectiles, it will be assumed that the horizontal component of projectile velocity is constant throughout the trajectory and that the vertical component of projectile velocity is constantly changing because of the influence of gravity (Figure 10-17). Since horizontal projectile velocity is constant, horizontal acceleration is equal to the constant of zero throughout the trajectory. The vertical acceleration of a projectile is equal to the constant -9.81 m/s^2.

- The vertical speed of a projectile is constantly changing because of gravitational acceleration.
- The horizontal acceleration of a projectile is always zero.

Equations of Constant Acceleration

When a body is moving with a constant acceleration (positive, negative, or equal to zero), certain interrelationships are present among the kinematic quantities associated with the motion of the body. These interrelationships may be expressed using three mathematical equations originally derived by Galileo, which are known as the laws of constant acceleration, or the laws of uniformly accelerated motion. Using the variable symbols d, v, a, and t (representing displacement, velocity, acceleration, and time, respectively) and

laws of constant acceleration
formulas relating displacement, velocity, acceleration, and time when acceleration is unchanging

FIGURE 10-17

The horizontal and vertical components of projectile velocity. Notice that the horizontal component is constant and the vertical component is constantly changing.

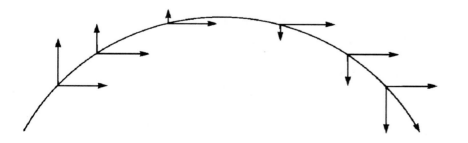

TABLE 10-3

Kinematic Variables

SYMBOL	MEANING	REPRESENTING IN EQUATIONS
d	Displacement	Change in position
v	Velocity	Rate of change in position
a	Acceleration	Rate of change in velocity
t	Time	Time interval
v_1	Initial or first velocity	Velocity at time 1
v_2	Later or final velocity	Velocity at time 2
v_v	Vertical velocity	Vertical component of total velocity
v_h	Horizontal velocity	Horizontal component of total velocity

with the subscripts $_1$ and $_2$ (representing first or initial and second or final points in time), the equations are the following:

$$v_2 = v_1 + at \qquad (1)$$

$$d = v_1 t + \tfrac{1}{2}at^2 \qquad (2)$$

$$v_2^2 = v_1^2 + 2ad \qquad (3)$$

Notice that each of the equations contains a unique combination of three of the four kinematic quantities: displacement, velocity, acceleration, and time. This provides considerable flexibility for solving problems in which two of the quantities are known and the objective is to solve for a third. The symbols used in these equations are listed in Table 10-3.

It is instructive to examine these relationships as applied to the horizontal component of projectile motion in which a = 0. In this case, each term containing acceleration may be removed from the equation. The equations then appear as the following:

$$v_2 = v_1 \qquad (1H)$$

$$d = v_1 t \qquad (2H)$$

$$v_2^2 = v_1^2 \qquad (3H)$$

Equations 1H and 3H reaffirm that the horizontal component of projectile velocity is a constant. Equation 2H indicates that horizontal displacement is equal to the product of horizontal velocity and time (see Sample Problem 10.5).

When the constant acceleration relationships are applied to the vertical component of projectile motion, acceleration is equal to -9.81 m/s^2, and the equations cannot be simplified by the deletion of the acceleration term. However, in analysis of the vertical component of projectile motion, the initial velocity (v_1) is equal to zero in certain cases. For example, when an object is dropped from a stationary position, the initial velocity of the object is zero. When this is the case, the equations of constant acceleration may be expressed as the following:

$$v_2 = at \qquad (1V)$$

$$d = \tfrac{1}{2}at^2 \qquad (2V)$$

$$v_2^2 = 2ad \qquad (3V)$$

When an object is dropped, equation 1V relates that the object's velocity at any instant is the product of gravitational acceleration and the amount of time the object has been in free fall. Equation 2V indicates

SAMPLE PROBLEM 10.5

The score was tied 20–20 in the final 1987 AFC playoff game between the Denver Broncos and the Cleveland Browns. During the first overtime period, Denver had the opportunity to kick a field goal, with the ball placed at a distance of 29 m from the goalposts. If the ball was kicked with the horizontal component of initial velocity being 18 m/s and a flight time of 2 s, was the kick long enough to make the field goal?

Known

$$v_h = 18 \text{ m/s}$$
$$t = 2 \text{ s}$$

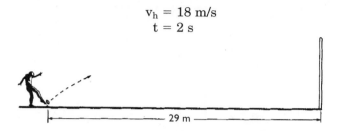

29 m

Solution

Equation 2H is selected to solve the problem, since two of the variables contained in the formula (v_h and t) are known quantities, and since the unknown variable (d) is the quantity we wish to find:

$$d_h = v_h t$$
$$d = (18 \text{ m/s}) (2 \text{ s})$$
$$d = 36 \text{ m}$$

The ball did travel a sufficient distance for the field goal to be good, and Denver won the game, advancing to Super Bowl XXI.

---

that the vertical distance through which the object has fallen can be calculated from gravitational acceleration and the amount of time the object has been falling. Equation 3V expresses the relationship between the object's velocity and vertical displacement at a certain time and gravitational acceleration.

It is useful in analyzing projectile motion to remember that at the apex of a projectile's trajectory, the vertical component of velocity is zero. If the goal is to determine the maximum height achieved by a projectile, v_2 in equation 3 may be set equal to zero:

$$0 = v_1^2 + 2ad \qquad (3A)$$

An example of this use of equation 3A is shown in Sample Problem 10.6. If the problem is to determine the total flight time, one approach is to calculate the time it takes to reach the apex, which is one-half of the total flight time if the projection and landing heights are equal. In this case, v_2 in equation 1 for the vertical component of the motion may be set equal to zero because vertical velocity is zero at the apex:

$$0 = v_1 + at \qquad (1A)$$

Sample Problem 10.7 illustrates this use of equation 1A.

When using the equations of constant acceleration, it is important to remember that they may be applied to the horizontal component of projectile motion or to the vertical component of projectile motion, but not to the resultant motion of the projectile. If the horizontal component of motion is being analyzed, a = 0, but if the vertical component is being analyzed, a = −9.81 m/s². The equations of constant acceleration and their special variations are summarized in Table 10-4.

SAMPLE PROBLEM 10.6

A volleyball is deflected vertically by a player in a game housed in a high school gymnasium where the ceiling clearance is 10 m. If the initial velocity of the ball is 15 m/s, will the ball contact the ceiling?

Known

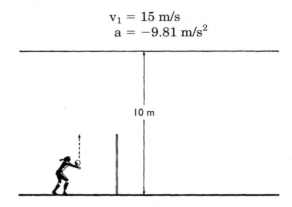

$v_1 = 15$ m/s
$a = -9.81$ m/s^2

10 m

Solution

The equation selected for use in solving this problem must contain the variable d for vertical displacement. Equation 2 contains d but also contains the variable t, which is an unknown quantity in this problem. Equation 3 contains the variable d, and, recalling that vertical velocity is zero at the apex of the trajectory, Equation 3A can be used to find d:

$$v_2^2 = v_1^2 + 2ad \qquad (3)$$
$$0 = v_1^2 + 2ad \qquad (3A)$$
$$0 = (15 \text{ m/s})^2 + (2)(-9.81 \text{ m/s}^2)d$$
$$(19.62 \text{ m/s}^2)d = 225 \text{ m}^2/\text{s}^2$$
$$\boxed{d = 11.47 \text{ m}}$$

Therefore, the ball has sufficient velocity to contact the 10 m ceiling.

SAMPLE PROBLEM 10.7

A ball is kicked at a 35° angle, with an initial speed of 12 m/s. How high and how far does the ball go?

Solution

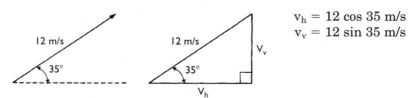

$v_h = 12 \cos 35$ m/s
$v_v = 12 \sin 35$ m/s

How high does the ball go?
Equation 1 cannot be used because it does not contain d. Equation 2 cannot be used unless t is known. Since vertical velocity is zero at the apex of the ball's trajectory, equation 3A is selected:

$$0 = v_1^2 + 2ad \tag{3A}$$
$$0 = (12 \sin 35 \text{ m/s})^2 + (2)(-9.81 \text{ m/s}^2)d$$
$$(19.62 \text{ m/s}^2)d = 47.37 \text{ m}^2/\text{s}^2$$

$$\boxed{d = 2.41 \text{ m}}$$

How far does the ball go?

Equation 2H for horizontal motion cannot be used because t for which the ball was in the air is not known. Equation 1A can be used to solve for the time it took the ball to reach its apex:

$$0 = v_1 + at \tag{1A}$$
$$0 = 12 \sin 35 \text{ m/s} + (-9.81 \text{ m/s}^2)t$$

$$t = \frac{6.88 \text{ m/s}}{9.81 \text{ m/s}^2}$$

$$t = 0.70 \text{ s}$$

Recalling that the time to reach the apex is one-half of the total flight time, total time is the following:

$$t = (0.70 \text{ s})(2)$$
$$t = 1.40 \text{ s}$$

Equation 2H can then be used to solve for the horizontal distance the ball traveled:

$$d_h = v_h t \tag{2H}$$
$$d_h = (12 \cos 35 \text{ m/s})(1.40 \text{ s})$$

$$\boxed{d_h = 13.76 \text{ m}}$$

TABLE 10-4

Formulas Relating to Projectile Motion

The Equations of Constant Acceleration

These equations may be used to relate linear kinematic quantities whenever acceleration (a) is a constant, unchanging value:

$$v_2 = v_1 + at \tag{1}$$
$$d = v_1 t + (\tfrac{1}{2})at^2 \tag{2}$$
$$v_2^2 = v_1^2 + 2ad \tag{3}$$

Special Case Applications of the Equations of Constant Acceleration

For the horizontal component of projectile motion, with $a = 0$:

$$d_h = v_h t \tag{2H}$$

For the vertical component of projectile motion, with $v_1 = 0$, as when the projectile is dropped from a static position:

$$v_2 = at \tag{1V}$$
$$d = (\tfrac{1}{2})at^2 \tag{2V}$$
$$v_2^2 = 2ad \tag{3V}$$

For the vertical component of projectile motion, with $v_2 = 0$, as when the projectile is at its apex:

$$0 = v_1 + at \tag{1A}$$
$$0 = v_1^2 + 2ad \tag{2A}$$

SUMMARY

Linear kinematics is the study of the form or sequencing of linear motion with respect to time. Linear kinematic quantities include the scalar quantities of distance and speed, and the vector quantities of displacement, velocity, and acceleration. Depending on the motion being analyzed, either a vector quantity or its scalar equivalent and either an instantaneous or an average quantity may be of interest.

A projectile is a body in free fall that is affected only by gravity and air resistance. Projectile motion is analyzed in terms of its horizontal and vertical components. The two components are independent of each other, and only the vertical component is influenced by gravitational force. Factors that determine the height and distance the projectile achieves are projection angle, projection speed, and relative projection height. The equations of constant acceleration can be used to quantitatively analyze projectile motion, with vertical acceleration being -9.81 m/s^2 and horizontal acceleration being zero.

INTRODUCTORY PROBLEMS

Start

160 m

Finish

Note: Some problems require vector algebra (see Chapter 3).

1. A runner completes 6½ laps around a 400 m track during a 12 min (720 s) run test. Calculate the following quantities:
   a. The distance the runner covered
   b. The runner's displacement at the end of 12 min
   c. The runner's average speed
   d. The runner's average velocity
   e. The runner's average pace
   (Answers: a. 2.6 km; b. 160 m; c. 3.6 m/s; d. 0.22 m/s; e. 4.6 min/km)
2. A ball rolls with an acceleration of -0.5 m/s^2. If it stops after 7 s, what was its initial speed? (Answer: 3.5 m/s)
3. A wheelchair marathoner has a speed of 5 m/s after rolling down a small hill in 1.5 s. If the wheelchair underwent a constant acceleration of 3 m/s^2 during the descent, what was the marathoner's speed at the top of the hill? (Answer: 0.5 m/s)
4. An orienteer runs 400 m directly east and then 500 m to the northeast (at a 45° angle from due east and from due north). Provide a graphic solution to show final displacement with respect to the starting position.
5. An orienteer runs north at 5 m/s for 120 s, and then west at 4 m/s for 180 s. Provide a graphic solution to show the orienteer's resultant displacement.
6. Why are the horizontal and vertical components of projectile motion analyzed separately?
7. A soccer ball is kicked with an initial horizontal speed of 5 m/s and an initial vertical speed of 3 m/s. Assuming that projection and landing heights are the same and neglecting air resistance, identify the following quantities:
   a. The ball's horizontal speed 0.5 s into its flight
   b. The ball's horizontal speed midway through its flight
   c. The ball's horizontal speed immediately before contact with the ground
   d. The ball's vertical speed at the apex of the flight
   e. The ball's vertical speed midway through its flight
   f. The ball's vertical speed immediately before contact with the ground

8. If a baseball, a basketball, and a 71.2 N shot were dropped simultaneously from the top of the Empire State Building (and air resistance was not a factor), which would hit the ground first? Why?
9. A tennis ball leaves a racket during the execution of a perfectly horizontal ground stroke with a speed of 22 m/s. If the ball is in the air for 0.7 s, what horizontal distance does it travel? (Answer: 15.4 m)
10. A trampolinist springs vertically upward with an initial speed of 9.2 m/s. How high above the trampoline will the trampolinist go? (Answer: 4.31 m)

ADDITIONAL PROBLEMS

1. Answer the following questions pertaining to the split times (in seconds) presented below for Ben Johnson and Carl Lewis during the 100 m sprint in the 1988 Olympic Games.

	Johnson	Lewis
10 m	1.86	1.88
20 m	2.87	2.96
30 m	3.80	3.88
40 m	4.66	4.77
50 m	5.55	5.61
60 m	6.38	6.45
70 m	7.21	7.29
80 m	8.11	8.12
90 m	8.98	8.99
100 m	9.83	9.86

   a. Plot velocity and acceleration curves for both sprinters. In what ways are the curves similar and different?
   b. What general conclusions can you draw about performance in elite sprinters?
2. Provide a trigonometric solution for Introductory Problem 4. (Answer: D = 832 m; ∠ = 25° north of due east)
3. Provide a trigonometric solution for Introductory Problem 5. (Answer: D = 937 m; ∠ = 50° west of due north)
4. A buoy marking the turn in the ocean swim leg of a triathlon becomes unanchored. If the current carries the buoy southward at 0.5 m/s, and the wind blows the buoy westward at 0.7 m/s, what is the resultant displacement of the buoy after 5 min? (Answer: 258m; ∠ = 54.5° west of due south)
5. A sailboat is being propelled westerly by the wind at a speed of 4 m/s. If the current is flowing at 2 m/s to the northeast, where will the boat be in 10 min with respect to its starting position? (Answer: D = 1.8 km; ∠ = 29° north of due west)
6. A Dallas Cowboy carrying the ball straight down the near sideline with a velocity of 8 m/s crosses the 50-yard line at the same time that the last Buffalo Bill who can possibly hope to catch him starts running from the 50-yard line at a point that is 13.7 m from the near sideline. What must the Bill's velocity be if he is to catch the Cowboy just short of the goal line? (Answer: 8.35 m/s)
7. A soccer ball is kicked from the playing field at a 45° angle. If the ball is in the air for 3 s, what is the maximum height achieved? (Answer: 11.0 m)
8. A ball is kicked a horizontal distance of 45.8 m. If it reaches a maximum height of 24.2 m with a flight time of 4.4 s, was the ball kicked

at a projection angle less than, greater than, or equal to 45°? Provide a rationale for your answer based on the appropriate calculations. (Answer: >45°)

9. A badminton shuttlecock is struck by a racket at a 35° angle, giving it an initial speed of 10 m/s. How high will it go? How far will it travel horizontally before being contacted by the opponent's racket at the same height from which it was projected? (Answer: $d_v = 1.68$ m; $d_h = 9.58$ m)

10. An archery arrow is shot with a speed of 45 m/s at an angle of 10°. How far horizontally can the arrow travel before hitting a target at the same height from which it was released? (Answer: 70.6 m)

LABORATORY EXPERIENCES

1. At the *Basic Biomechanics* Online Learning Center (www.mhhe.com/hall6e), go to Student Center, Chapter 10, Lab Manual, Lab 1, and then click on the Instantaneous Speed simulation. After viewing this simulation several times, answer the following questions.

   a. What variables are represented on the horizontal and vertical axes of the graph?

   Horizontal axis: _____ vertical axis: _____

   b. What variable is represented by the slope of the graph? _____

   c. Explain what the car is doing when the slope of the graph is zero (the graph is

   horizontal). _____

   _____

2. At the *Basic Biomechanics* Online Learning Center (www.mhhe.com/hall6e), go to Student Center, Chapter 10, Lab Manual, Lab 2, and then click on the Vector Addition in Two Dimensions simulation. After experimenting with this simulation several times, answer the following questions.

   a. What are the numbers shown at the top of the graph for vector A? _____

   _____

   _____

   b. What are the numbers shown at the top of the graph for vector B? _____

   _____

   _____

   c. What are the numbers shown at the top of the graph for vector C? _____

   _____

   _____

   d. What is the name of the vector operation that is shown? _____

   e. Which vector is the resultant? _____

3. Calculate the horizontal and vertical components of velocity for a ball projected at 15 m/s at the following angles:

	V_h	V_v
30°	_____	_____
40°	_____	_____
50°	_____	_____
60°	_____	_____

   At the *Basic Biomechanics* Online Learning Center, click on the Vectors simulation and verify that your answers are correct.

4. At the *Basic Biomechanics* Online Learning Center (www.mhhe.com/hall6e), click on Student Center, Chapter 10, Lab Manual, Lab 4, and then click on the Projectile Motion (falling arrow) simulation. After viewing this simulation several times, write an explanation.

_____

_____

_____

_____

5. At the *Basic Biomechanics* Online Learning Center (www.mhhe.com/hall6e), click on Student Center, Chapter 10, Lab Manual, Lab 5, and then click on the Projectile Motion (baseball) simulation. Click on the player to activate a throw. Your target is first base, which is 80 m from the thrower. After experimenting with this simulation several times, answer the following questions.

a. At what projection angle does a throw reach first base in the shortest time? _____

b. At what projection angle can a ball be thrown and reach first base with the least projection speed?

_____

c. What are the answers to questions a and b if the player is on the moon?

Angle for throw of shortest time: _____

Angle for throw with least speed: _____

Explain why these are different than on earth: _____

_____

_____

d. What are the answers to questions a and b if the player is on Mars?

Angle for throw of shortest time: _____

Angle for throw with least speed: _____

Explain why these are different than on earth: _____

_____

_____

REFERENCES

1. Chow JW and Hay JG: Computer simulation of the last support phase of the long jump, *Med Sci Sports Exerc* 37:115, 2005.
2. Chow JW, Millikan TA, Carlton LG, Chae WS, Lim YT, and Morse MI: Kinematic and electromyographic analysis of wheelchair propulsion on ramps of different slopes for young men with paraplegia, *Arch Phys Med Rehabil* 90:271, 2009.
3. Dapena J: Mechanics of translation in the Fosbury flop, *Med Sci Sports Exerc* 12:37, 1980.
4. Dapena J and Chung CS: Vertical and radial motions of the body during the take-off phase of high jumping, *Med Sci Sports Exerc* 20:290, 1988.
5. Dominici N, Ivanenko YP, Cappellini G, Zampagni ML, and Lacquaniti F: Kinematic strategies in newly walking toddlers stepping over different support surfaces, *J Neurophysiol* 103:1673, 2010.
6. Giatsis G, Kollias I, Panoutsakopoulos V, and Papaiakovou G: Biomechanical differences in elite beach-volleyball players in vertical squat jump on rigid and sand surface, *Sports Biomech* 3:145, 2004.
7. Hardin EC, van den Bogert AJ, and Hamill J: Kinematic adaptations during running: effects of footwear, surface, and duration, *Med Sci Sports Exerc* 36:838, 2004.
8. Hay JG: The biomechanics of the long jump, *Exerc Sport Sci Rev* 14:401, 1986.
9. Hubbard M, de Mestre NJ, and Scott J: Dependence of release variables in the shot put, *J Biomech* 34:449, 2001.
10. Janura M, Cabell L, Elfmark M, and Vaverka F: Kinematic characteristics of the ski jump inrun: a 10-year longitudinal study, *J Appl Biomech* 26:196, 2010.
11. Kyrolainen H et al: Effects of marathon running on running economy and kinematics, *Eur J Appl Physiol* 82:297, 2000.
12. Laws K: *The physics of dance,* New York, 1984, Schirmer Books.
13. Leigh S, Liu H, Hubbard M, and Yu B: Individualized optimal release angles in discus throwing, *J Biomech* 10;43:540, 2010.
14. Leskinen A, Häkkinen K, Virmavirta M, Isolehto J, and Kyröläinen H: Comparison of running kinematics between elite and national-standard 1500-m runners, *Sports Biomech* 8:1, 2009.
15. Linthorne NP: Optimum release angle in the shot put, *J Sports Sci* 19:359, 2001.
16. Miller JA and Hay JG: Kinematics of a world record and other world-class performances in the triple jump, *Int J Sport Biomech* 2:272, 1986.
17. Miller S and Bartlett R: The relationship between basketball shooting kinematics, distance and playing position, *J Sports Sci* 14:243, 1996.
18. Muehlbauer T, Panzer S, and Schindler C: Pacing pattern and speed skating performance in competitive long-distance events, *J Strength Cond Res* 24:114, 2010.
19. Nilsson J, Tveit P, and Eikrehagen O: Effects of speed on temporal patterns in classical style and freestyle cross-country skiing, *Sports Biomech* 3:85, 2004.
20. Paradisis GP and Cooke CB: Kinematic and postural characteristics of spring running on sloping surfaces, *J Sports Sci* 19:149, 2001.
21. Rojas FJ, Cepero M, Ona A, and Guitierrez M: Kinematic adjustments in basketball jump shot against an opponent, *Ergonomics* 43:1651, 2000.
22. Seyfarth A, Blickhan R, and van Leeuwen JL: Optimum take-off techniques and muscle design for long jump, *J Exp Biol* 203:741, 2000.
23. Seifert L, Leblanc H, Chollet D, and Delignières D: Inter-limb coordination in swimming: effect of speed and skill level, *Hum Mov Sci* 29:103, 2010.
24. Slawinski J, Bonnefoy A, Levêque JM, Ontanon G, Riquet A, Dumas R, and Chèze L: Kinematic and kinetic comparisons of elite and well-trained sprinters during sprint start, *J Strength Cond Res* 24:896, 2010.
25. Stöggl T, Kampel W, Müller E, and Lindinger S: Double-push skating versus V2 and V1 skating on uphill terrain in cross-country skiing, *Med Sci Sports Exerc* 42:187, 2010.

26. Thompson KG, Haljand R, MacLaren DP: An analysis of selected kinematic variables in national and elite male and female 100-m and 200-m breaststroke swimmers, *J Sports Sci* 18:421, 2000.

27. Wagner H, Tilp M, von Duvillard SP, and Mueller E: Kinematic analysis of volleyball spike jump, *Int J Sports Med* 30(10):760, 2009.

28. Wakai M and Linthorne NP: Optimum take-off angle in the standing long jump, *Hum Mov Sci* 24:81, 2005.

29. Watanabe K: Ski-jumping, alpine, cross-country, and nordic combination skiing. In Vaughan CL, ed: *Biomechanics of sport,* Boca Raton, FL, 1989, CRC Press.

30. Yeadon MR: Theoretical models and their application to aerial movement. In Van Gheluwe B and Atha J, eds: *Current research in sports biomechanics,* Basel, 1987, Karger.

31. Yu B: Horizontal-to-vertical velocity conversion in the triple jump, *J Sports Sci* 17:221, 1999.

ANNOTATED READINGS

Hay JG: Citius, altius, longius (faster, higher, longer): The biomechanics of jumping for distance, *J Biomech* 26:7, 1993.
Discusses current research related to performance in the long jump and triple jump.

McCoy RL: *Modern exterior ballistics: The launch and flight dynamics of symmetric projectiles,* New York, 1999, Schiffer.
Provides a historical perspective of early technological developments in the 19th century, including the first ballistic firing tables.

Saunders PU, Pyne DB, Telford RD, and Hawley JA: Factors affecting running economy in trained distance runners, *Sports Med* 34:465, 2004.
Reviews current knowledge about kinematic and other factors that may contribute to economy during running.

White C: *Projectile dynamics in sport: Principles and applications,* New York, 2010, Routledge.
Provides a comprehensive analysis of projection within the context of sports biomechanics, including data, worked equations, examples, and illustrations.

RELATED WEBSITES

Programming Example: Projectile Motion
http://www.cs.mtu.edu/~shene/COURSES/cs201/NOTES/chap02/projectile.html
Provides a documented, downloadable computer program for calculating horizontal and vertical displacements, resultant velocity, and direction of a projectile.

Projectile Motion
http://galileo.phys.virginia.edu/classes/109N/more_stuff/Applets/Projectile-Motion/jarapplet.html
Java applet that shows the trajectory of a projectile when the user enters the projection speed, angle, and height.

The Physics Classroom: Kinematics
http://www.physicsclassroom.com/Class/1DKin/1DKinTOC.html
High school–level tutorial on kinematics, including text and graphs.

The Physics Classroom: Projectile Motion
http://www.physicsclassroom.com/Class/vectors/U3L2a.html
High school–level tutorial on projectile motion, including animations.

The Physics of Projectile Motion
http://library.thinkquest.org/2779/
Provides textual information and historic drawings of the first accurate description of projectile motion by Galileo, plus links to projectile animations, including a projectile water balloon game with sound effects.

KEY TERMS

angle of projection	the direction at which a body is projected with respect to the horizontal
apex	the highest point in the trajectory of a projectile
average	occurring over a designated time interval
initial velocity	vector quantity incorporating both angle and speed of projection
instantaneous	occurring during a small interval of time
kinematics	the form, pattern, or sequencing of movement with respect to time
laws of constant acceleration	formulas relating displacement, velocity, acceleration, and time when acceleration is unchanging
linear acceleration	the rate of change in linear velocity
linear displacement	change in location, or the directed distance from initial to final location
linear velocity	the rate of change in location
meter	the most common international unit of length, on which the metric system is based
projectile	body in free fall that is subject only to the forces of gravity and air resistance
projection speed	the magnitude of projection velocity
range	the horizontal displacement of a projectile at landing
relative projection height	the difference between projection height and landing height
trajectory	the flight path of a projectile

Angular Kinematics of Human Movement

After completing this chapter, you will be able to:

Distinguish angular motion from rectilinear and curvilinear motion.

Discuss the relationships among angular kinematic variables.

Correctly associate angular kinematic quantities with their units of measure.

Explain the relationships between angular and linear displacement, angular and linear velocity, and angular and linear acceleration.

Solve quantitative problems involving angular kinematic quantities and the relationships between angular and linear kinematic quantities.

W hy is a driver longer than a 9-iron? Why do batters slide their hands up the handle of the bat to lay down a bunt but not to drive the ball? How does the angular motion of the discus or hammer during the windup relate to the linear motion of the implement after release?

These questions relate to angular motion, or rotational motion around an axis. The axis of rotation is a line, real or imaginary, oriented perpendicular to the plane in which the rotation occurs, like the axle for the wheels of a cart. In this chapter, we discuss angular motion, which, like linear motion, is a basic component of general motion.

OBSERVING THE ANGULAR KINEMATICS OF HUMAN MOVEMENT

Understanding angular motion is particularly important for the student of human movement, because most volitional human movement involves rotation of one or more body segments around the joints at which they articulate. Translation of the body as a whole during gait occurs by virtue of rotational motions taking place at the hip, knee, and ankle around imaginary mediolateral axes of rotation. During the performance of jumping jacks, both the arms and the legs rotate around imaginary anteroposterior axes passing through the shoulder and hip joints. The angular motion of sport implements such as golf clubs, baseball bats, and hockey sticks, as well as household and garden tools, is also often of interest.

As discussed in Chapter 2, clinicians, coaches, and teachers of physical activities routinely analyze human movement based on visual observation. What is actually observed in such situations is the angular kinematics of human movement. Based on observation of the timing and range of motion (ROM) of joint actions, the experienced analyst can make inferences about the coordination of muscle activity producing the joint actions and the forces resulting from those joint actions.

MEASURING ANGLES

As reviewed in Appendix A, an angle is composed of two sides that intersect at a vertex. Quantitative kinematic analysis can be achieved by projecting filmed images of the human body onto a piece of paper, with joint centers then marked with dots and the dots connected with lines representing the longitudinal axes of the body segments (Figure 11-1). A protractor can be used to make hand measurements of angles of interest from this representation, with the joint centers forming the vertices of the angles between adjacent body segments. (The procedure for measuring angles with a protractor is reviewed in Appendix A.) Videos and films of human movement can also be analyzed using this same basic procedure to evaluate the angles present at the joints of the human body and the angular orientations of the body segments. The angle assessments are usually done with computer software from stick figure representations of the human body constructed in computer memory.

Relative versus Absolute Angles

Assessing the angle at a joint involves measuring the angle of one body segment relative to the other body segment articulating at the joint. The relative angle at the knee is the angle formed between the longitudinal axis of the thigh and the longitudinal axis of the lower leg (Figure 11-2). When joint ROM is quantified, it is the relative joint angle that is measured.

relative angle
angle at a joint formed between the longitudinal axes of adjacent body segments

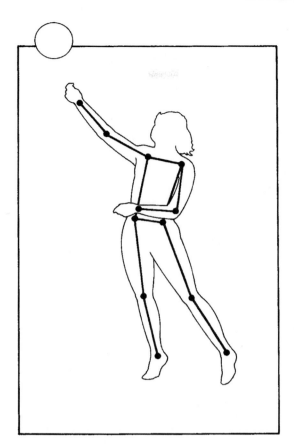

FIGURE 11-1

For the human body, joint centers form the vertices of body segment angles.

The convention used for measuring relative joint angles is that in anatomical reference position, all joint angles are at 0°. As discussed in Chapter 5, joint motion is then measured directionally. For example, when the extended arm is elevated 30° in front of the body in the sagittal plane, the arm is in 30° of flexion at the shoulder. When the leg is abducted at the hip, the ROM in abduction is likewise measured from 0° in anatomical reference position.

Other angles of interest are often the orientations of the body segments themselves. As discussed in Chapter 9, when the trunk is in flexion, the angle of inclination of the trunk directly affects the amount of force that must be generated by the trunk extensor muscles to support the trunk in the position assumed. The angle of inclination of a body segment, referred to as its absolute angle, is measured with respect to an absolute reference line, usually either horizontal or vertical. Figure 11-3 shows quantification of segment angles with respect to the right horizontal.

Tools for Measuring Body Angles

Goniometers are commonly used by clinicians for direct measurement of relative joint angles on a live human subject. A goniometer is essentially a protractor with two long arms attached. One arm is fixed so that it extends from the protractor at an angle of 0°. The other arm extends from the center of the protractor and is free to rotate. The center of the protractor is aligned over the joint center, and the two arms are aligned over the longitudinal axes of the two body segments that connect at the joint. The angle at the joint is then read at the intersection of the freely rotating

• Relative angles should consistently be measured on the same side of a given joint.

• The straight, fully extended position at a joint is regarded as 0°.

absolute angle
angular orientation of a body segment with respect to a fixed line of reference

• Absolute angles should consistently be measured in the same direction from a single reference—either horizontal or vertical.

FIGURE 11-2

Relative angles measured at joints are the angles between adjacent body segments.

arm and the protractor scale. The accuracy of the reading depends on the accuracy of the positioning of the goniometer. Knowledge of the underlying joint anatomy is essential for proper location of the joint center of rotation. Placing marks on the skin to identify the location of the center of rotation at the joint and the longitudinal axes of the body segments before aligning the goniometer is sometimes helpful, particularly if repeated measurements are being taken at the same joint.

Instant Center of Rotation

Quantification of joint angles is complicated by the fact that joint motion is often accompanied by displacement of one bone with respect to the

FIGURE 11-3

Angles of orientation of individual body segments are measured with respect to an absolute (fixed) line of reference.

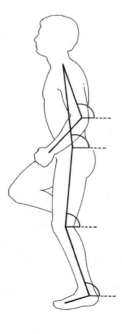

The relative angle at the knee (measured between adjacent body segments) and the absolute angle of the trunk (measured with respect to the right horizontal).

articulating bone at the joint. This phenomenon is caused by normal asymmetries in the shapes of the articulating bone surfaces. One example is the tibiofemoral joint, at which medial rotation and anterior displacement of the femur on the tibial plateau accompany flexion (Figure 11-4). As a result, the location of the exact center of rotation at the joint changes slightly when the joint angle changes. The center of rotation at a given joint angle, or at a given instant in time during a dynamic movement, is called the instant center. The exact location of the instant center for a given joint

instant center
precisely located center of rotation at a joint at a given instant in time

A goniometer is used to measure joint angles.

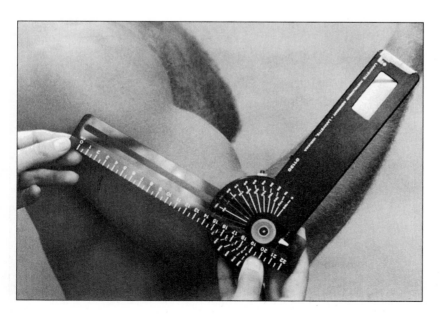

FIGURE 11-4

The path of the instant center at the knee during knee extension.

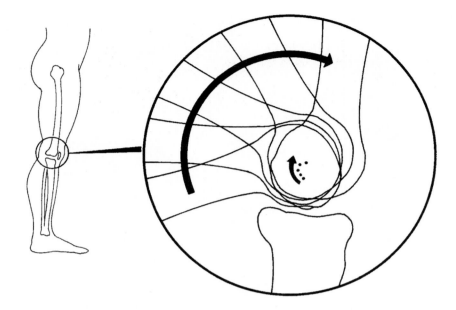

may be determined through measurements taken from roentgenograms (X-rays), which are usually taken at 10° intervals throughout the ROM at the joint (11). The instant center at the tibiofemoral joint of the knee shifts during angular movement at the knee due to the ellipsoid shapes of the femoral condyles (12).

ANGULAR KINEMATIC RELATIONSHIPS

The interrelationships among angular kinematic quantities are similar to those discussed in Chapter 10 for linear kinematic quantities. Although the units of measure associated with the angular kinematic quantities are different from those used with their linear counterparts, the relationships among angular units also parallel those present among linear units.

When the forearm returns to its original position at the completion of a curl exercise, the angular displacement at the elbow is zero.

Angular Distance and Displacement

Consider a pendulum swinging back and forth from a point of support. The pendulum is rotating around an axis passing through its point of support perpendicular to the plane of motion. If the pendulum swings through an arc of 60°, it has swung through an angular distance of 60°. If the pendulum then swings back through 60° to its original position, it has traveled an angular distance totaling 120° (60° + 60°). Angular distance is measured as the sum of all angular changes undergone by a rotating body.

The same procedure may be used for quantifying the angular distances through which the segments of the human body move. If the angle at the elbow joint changes from 90° to 160° during the flexion phase of a forearm curl exercise, the angular distance covered is 70°. If the extension phase of the curl returns the elbow to its original position of 90°, an additional 70° have been covered, resulting in a total angular distance of 140° for the complete curl. If 10 curls are performed, the angular distance transcribed at the elbow is 1400° (10 × 140°).

Just as with its linear counterpart, angular displacement is assessed as the difference in the initial and final positions of the moving body. If the angle at the knee of the support leg changes from 5° to 12° during the initial support phase of a running stride, the angular distance and the angular displacement at the knee are 7°. If extension occurs at the knee, returning the joint to its original 5° position, angular distance totals 14° (7° + 7°), but angular displacement is 0°, because the final position of the joint is the same as its original position. The relationship between angular distance and angular displacement is represented in Figure 11-5.

Like linear displacement, angular displacement is defined by both magnitude and direction. Since rotation observed from a side view occurs in either a clockwise or a counterclockwise direction, the direction of angular displacement may be indicated using these terms. The counterclockwise direction is conventionally designated as positive (+), and the clockwise direction as negative (−) (Figure 11-6). With the human body, it is also appropriate to indicate the direction of angular displacement with joint-related terminology such as flexion or abduction. However, there is no set relationship between the positive (counterclockwise) direction and either flexion or extension or any other movement at a joint. This is because when viewed from one side, flexion at a given joint such as the hip is positive, but when viewed from the opposite side it is negative. When biomechanists do motion capture studies with computer-linked cameras, the software quantifies joint motions in either positive

angular displacement
change in the angular position or orientation of a line segment

●*The counterclockwise direction is regarded as positive, and the clockwise direction is regarded as negative.*

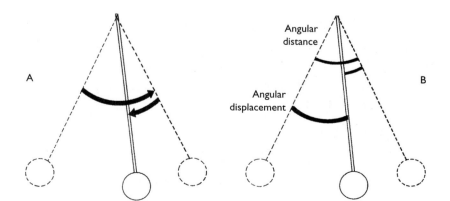

FIGURE 11-5

A. The path of motion of a swinging pendulum. **B.** The angular distance is the sum of all angular changes that have occurred; the angular displacement is the angle between the initial and final positions.

FIGURE 11-6

The direction of rotational motion is commonly identified as counterclockwise (or positive) versus clockwise (or negative).

Clockwise **−**

Counterclockwise **+**

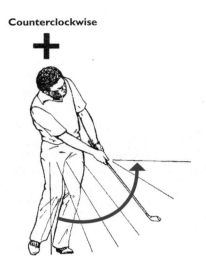

or negative directions. The researcher must then translate these values into flexion/extension or other joint motions depending on the camera view. More sophisticated software packages will do this translation with appropriate input from the researcher.

Three units of measure are commonly used to represent angular distance and angular displacement. The most familiar of these units is the degree. A complete circle of rotation transcribes an arc of 360°, an arc of 180° subtends a straight line, and 90° forms a right angle between perpendicular lines (Figure 11-7).

Another unit of angular measure sometimes used in biomechanical analyses is the radian. A line connecting the center of a circle to any point on the circumference of the circle is a radius. A radian is defined as the size of the angle subtended at the center of a circle by an arc equal in

radian

unit of angular measure used in angular-linear kinematic quantity conversions; equal to 57.3°

●*Pi (π) is a mathematical constant equal to approximately 3.14, which is the ratio of the circumference to the diameter of a circle.*

FIGURE 11-7

Angles measured in degrees.

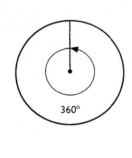

360°

180°

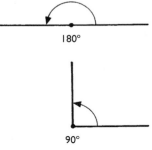

90°

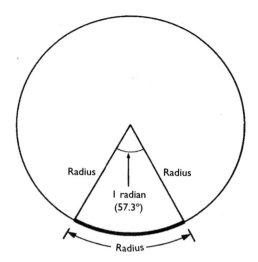

FIGURE 11-8

A radian is defined as the size of the angle subtended at the center of a circle by an arc equal in length to the radius of the circle.

length to the radius of the circle (Figure 11-8). One complete circle is an arc of 2π radians, or 360°. Because 360° divided by 2π is 57.3°, one radian is equivalent to 57.3°. Because a radian is much larger than a degree, it is a more convenient unit for the representation of extremely large angular distances or displacements. Radians are often quantified in multiples of pi (π).

The third unit sometimes used to quantify angular distance or displacement is the revolution. One revolution transcribes an arc equal to a circle. Dives and some gymnastic skills are often described by the number of revolutions the human body undergoes during their execution. The one-and-a-half forward somersault dive is a descriptive example. Figure 11-9 illustrates the way in which degrees, radians, and revolutions compare as units of angular measure.

Angular Speed and Velocity

Angular speed is a scalar quantity and is defined as the angular distance covered divided by the time interval over which the motion occurred:

$$\text{angular speed} = \frac{\text{angular distance}}{\text{change in time}}$$

$$\sigma = \frac{\phi}{\Delta t}$$

The lowercase Greek letter sigma (σ) represents angular speed, the lowercase Greek letter phi (ϕ) represents angular distance, and t represents time.

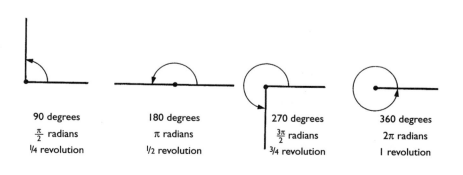

FIGURE 11-9

Comparison of degrees, radians, and revolutions.

angular velocity
rate of change in the angular position or orientation of a line segment

Angular velocity is calculated as the change in angular position, or the angular displacement, that occurs during a given period of time:

$$\text{angular velocity} = \frac{\text{change in angular position}}{\text{change in time}}$$

$$\omega = \frac{\Delta \text{ angular position}}{\Delta \text{ time}}$$

$$\text{angular velocity} = \frac{\text{angular displacement}}{\text{change in time}}$$

$$\omega = \frac{\theta}{\Delta t}$$

The lowercase Greek letter omega (ω) represents angular velocity, the capital Greek letter theta (θ) represents angular displacement, and t represents the time elapsed during the velocity assessment. Another way to express change in angular position is angular position$_2$ − angular position$_1$, in which angular position$_1$ represents the body's position at one point in time and angular position$_2$ represents the body's position at a later point:

$$\omega = \frac{\text{angular position}_2 - \text{angular position}_1}{\text{time}_2 - \text{time}_1}$$

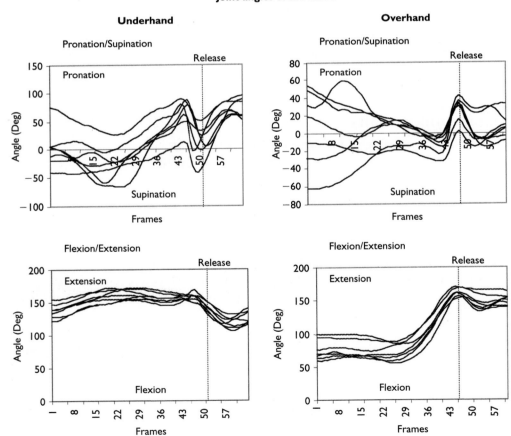

Joint angles of the elbow

Underhand

Overhand

FIGURE 11-10   Elbow angles exhibited by seven collegiate softball players during fast-pitch strikes.
Graphs courtesy of Kim Hudson and James Richards.

Because angular velocity is based on angular velocity, it must include an identification of the direction (clockwise or counterclockwise, negative or positive) in which the angular displacement on which it is based occurred.

Units of angular speed and angular velocity are units of angular distance or angular displacement divided by units of time. The unit of time most commonly used is the second. Units of angular speed and angular velocity are degrees per second (deg/s), radians per second (rad/s), revolutions per second (rev/s), and revolutions per minute (rpm).

Moving the body segments at a high rate of angular velocity is a characteristic of skilled performance in many sports. Angular velocities at the joints of the throwing arm in Major League Baseball pitchers have been reported to reach 2320 deg/s in elbow extension and 7240 deg/s in internal shoulder rotation (3). Interestingly, these values are also high in the throwing arms of youth pitchers, with 2230 deg/s in elbow extension and 6900 deg/s in internal rotation documented (3). What does change as pitchers advance to higher and higher levels of competition is that they tend to become more consistent with the kinematics of their pitching motions (4). However, this does not translate to better coordination or decreased risk of overuse injury (4). Comparison of different types of pitches thrown by collegiate baseball pitchers showed internal shoulder rotation values of 7550 deg/s for fastballs, 6680 deg/s for change-ups, 7120 deg/s for curveballs, and 7920 deg/s for sliders (2). Figures 11-10 to 11-13

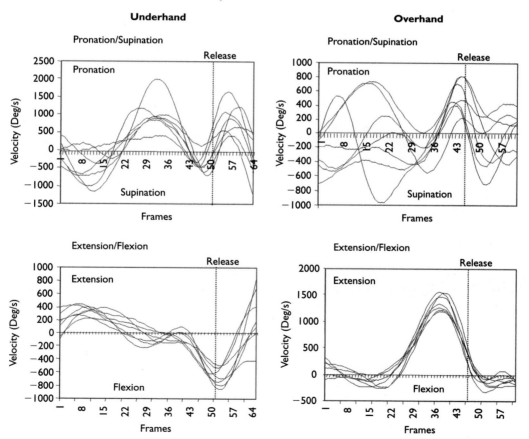

FIGURE 11-11   Elbow velocities exhibited by seven collegiate softball players during fast-pitch strikes. Graphs courtesy of Kim Hudson and James Richards.

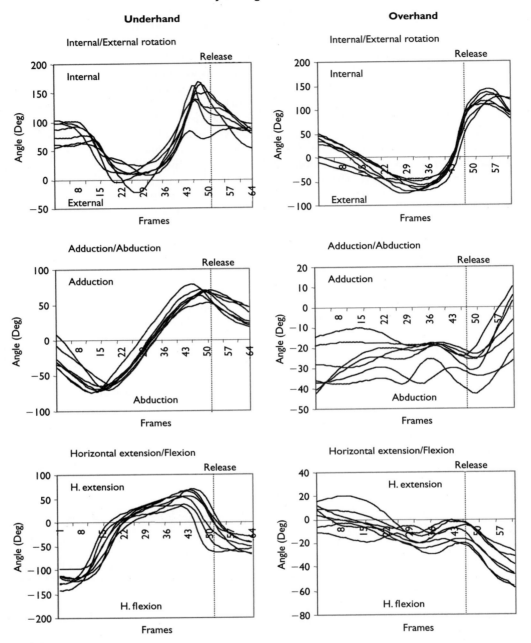

FIGURE 11-12 Shoulder angles exhibited by seven collegiate softball players during fast-pitch strikes. Graphs courtesy of Kim Hudson and James Richards.

display the patterns of joint angle and joint angular velocity at the elbow and shoulder during underhand and overhand throws executed by collegiate softball players. A study of world-class male and female tennis players has documented a sequential rotation of segmental rotations. Analysis of the cocked, preparatory position showed the elbow flexed to an average of 104° and the upper arm rotated to about 172° of external rotation at the shoulder. Proceeding from this position, there was a rapid sequence of segmental rotations, with averages of trunk tilt of 280 deg/s, upper torso

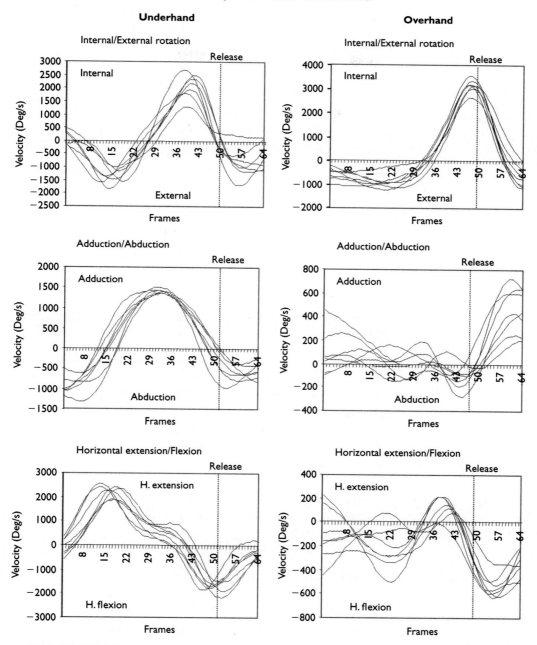

Joint velocities of the shoulder

Shoulder velocities exhibited by seven collegiate softball players during fast-pitch strikes. Graphs courtesy of Kim Hudson and James Richards.

rotation of 870 deg/s, pelvis rotation of 440 deg/s, elbow extension of 1510 deg/s, wrist flexion of 1950 deg/s, and shoulder internal rotation of 2420 deg/s for males and 1370 deg/s, for females (5). Angular velocity of the racquet during serves executed by professional male tennis players has been found to range from 1900 to 2200 deg/s (33.2 to 38.4 rad/s) just before ball impact (1).

As discussed in Chapter 10, when the human body becomes a projectile during the execution of a jump, the height of the jump determines the amount of time the body is in the air. When figure skaters perform a triple or quadruple axel, as compared to an axel or double axel, this means that either jump height or rotational velocity of the body must be greater. Measurements of these two variables indicate that it is the skater's angular velocity that increases, with skilled skaters rotating their bodies in excess of 5 rev/s while airborne during the triple axel (9, 10).

Angular Acceleration

angular acceleration

rate of change in angular velocity

Angular acceleration is the rate of change in angular velocity, or the change in angular velocity occurring over a given time. The conventional symbol for angular acceleration is the lowercase Greek letter alpha (α):

$$\text{angular acceleration} = \frac{\text{change in angular velocity}}{\text{change in time}}$$

$$\alpha = \frac{\Delta\omega}{\Delta t}$$

The calculation formula for angular acceleration is therefore the following:

$$\alpha = \frac{\omega_2 - \omega_1}{t_2 - t_1}$$

SAMPLE PROBLEM 11.1

A golf club is swung with an average angular acceleration of 1.5 rad/s^2. What is the angular velocity of the club when it strikes the ball at the end of a 0.8 s swing? (Provide an answer in both radian and degree-based units.)

Known

$$\alpha = 1.5 \text{ rad/s}^2$$
$$t = 0.8 \text{ s}$$

Solution

The formula to be used is the equation relating angular acceleration, angular velocity, and time:

$$\alpha = \frac{\omega_2 - \omega_1}{t}$$

Substituting in the known quantities yields the following:

$$1.5 \text{ rad/s}^2 = \frac{\omega_2 - \omega_1}{0.8 \text{ s}}$$

It may also be deduced that the angular velocity of the club at the beginning of the swing was zero:

$$1.5 \text{ rad/s}^2 = \frac{\omega_2 - 0}{0.8 \text{ s}}$$

$$(1.5 \text{ rad/s}^2)(0.8 \cdot \text{s}) = \omega_2 - 0$$

$$\boxed{\omega_2 = 1.2 \text{ rad/s}}$$

In degree-based units:

$$\omega_2 = (1.2 \text{ rad/s})(57.3 \text{ deg/rad})$$

$$\boxed{\omega_2 = 68.8 \text{ deg/s}}$$

	DISPLACEMENT	VELOCITY	ACCELERATION
Linear	meters	meters/second	meters/second^2
Angular	radians	radians/second	radians/second^2

TABLE 11-1

Common Units of Kinematic Measure

In this formula, ω_1 represents angular velocity at an initial point in time, ω_2 represents angular velocity at a second or final point in time, and t_1 and t_2 are the times at which velocity was assessed. Use of this formula is illustrated in Sample Problem 11.1.

Just as with linear acceleration, angular acceleration may be positive, negative, or zero. When angular acceleration is zero, angular velocity is constant. Just as with linear acceleration, positive angular acceleration may indicate either increasing angular velocity in the positive direction or decreasing angular velocity in the negative direction. Similarly, a negative value of angular acceleration may represent either decreasing angular velocity in the positive direction or increasing angular velocity in the negative direction.

Units of angular acceleration are units of angular velocity divided by units of time. Common examples are degrees per second squared (deg/s^2), radians per second squared (rad/s^2), and revolutions per second squared (rev/s^2). Units of angular and linear kinematic quantities are compared in Table 11-1.

•*Human movement rarely involves constant velocity or constant acceleration.*

Angular Motion Vectors

Because representing angular quantities using symbols such as curved arrows would be impractical, angular quantities are represented with conventional straight vectors, using what is called the right hand rule. According to this rule, when the fingers of the right hand are curled in the direction of an angular motion, the vector used to represent the motion is oriented perpendicular to the plane of rotation, in the direction the extended thumb points (Figure 11-14). The magnitude of the quantity may be indicated through proportionality to the vector's length.

right hand rule
procedure for identifying the direction of an angular motion vector

Average versus Instantaneous Angular Quantities

Angular speed, velocity, and acceleration may be calculated as instantaneous or average values, depending on the length of the time interval selected. The instantaneous angular velocity of a baseball bat at the

FIGURE 11-14

An angular motion vector is oriented perpendicular to the linear displacement (d) of a point on a rotating body.

instant of contact with a ball is typically of greater interest than the average angular velocity of the swing, because the former directly affects the resultant velocity of the ball.

RELATIONSHIPS BETWEEN LINEAR AND ANGULAR MOTION

Linear and Angular Displacement

The greater the radius is between a given point on a rotating body and the axis of rotation, the greater is the linear distance undergone by that point during an angular motion (Figure 11-15). This observation is expressed in the form of a simple equation:

$$s = r\phi$$

The curvilinear distance traveled by the point of interest s is the product of r, the point's radius of rotation, and ϕ, the angular distance through which the rotating body moves, which is quantified in radians.

For this relationship to be valid, two conditions must be met: (a) The linear distance and the radius of rotation must be quantified in the same units of length, and (b) angular distance must be expressed in radians. Although units of measure are normally balanced on opposite sides of an equal sign when a valid relationship is expressed, this is not the case here. When the radius of rotation (expressed in meters) is multiplied by angular displacement in radians, the result is linear displacement in meters. Radians disappear on the right side of the equation in this case because, as may be observed from the definition of the radian, the radian serves as a conversion factor between linear and angular measurements.

Linear and Angular Velocity

The same type of relationship exists between the angular velocity of a rotating body and the linear velocity of a point on that body at a given instant in time. The relationship is expressed as the following:

$$v = r\omega$$

radius of rotation

distance from the axis of rotation to a point of interest on a rotating body

FIGURE 11-15

The larger the radius of rotation (r), the greater the linear distance (s) traveled by a point on a rotating body.

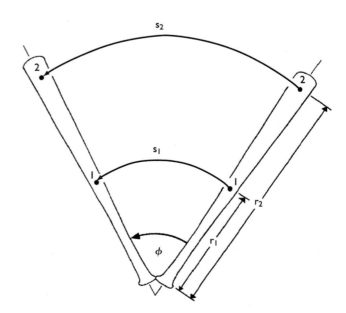

The linear (tangential) velocity of the point of interest is v, r is the radius of rotation for that point, and ω is the angular velocity of the rotating body. For the equation to be valid, angular velocity must be expressed in radian-based units (typically rad/s), and velocity must be expressed in the units of the radius of rotation divided by the appropriate units of time. Radians are again used as a linear-angular conversion factor, and are not balanced on opposite sides of the equals sign:

$$m/s = (m)\,(rad/s)$$

The use of radian-based units for conversions between linear and angular velocities is shown in Sample Problem 11.2.

During several sport activities, an immediate performance goal is to direct an object such as a ball, shuttlecock, or hockey puck accurately, while imparting a relatively large amount of velocity to it with a bat, club, racquet, or stick. In baseball batting, the initiation of the swing and the angular velocity of the swing must be timed precisely to make contact with the ball and direct it into fair territory. A 40 m/s pitch reaches the batter 0.41 s after leaving the pitcher's hand. It has been estimated that a difference of 0.001 s in the time of initiation of the swing can determine whether the ball is directed to center field or down the foul line, and that a swing initiated 0.003 s too early or too late will result in no contact with the ball (6). Similarly, there is a very small window of time during which gymnasts on the high bar can release from the bar to execute a skillful dismount. For high-bar finalists in the 2000 Olympic Games in Sydney, the release window was an average of 0.055 s (7).

Timing is important in the execution of a ground stroke in tennis. If the ball is contacted too soon or too late, it may be hit out of bounds. Photo © Royalty Free/CORBIS.

● *Skilled performances of high-velocity movements are characterized by precisely coordinated timing of body segment rotations.*

SAMPLE PROBLEM 11.2

Two baseballs are consecutively hit by a bat. The first ball is hit 20 cm from the bat's axis of rotation, and the second ball is hit 40 cm from the bat's axis of rotation. If the angular velocity of the bat was 30 rad/s at the instant that both balls were contacted, what was the linear velocity of the bat at the two contact points?

Known

$$r_1 = 20 \text{ cm}$$
$$r_2 = 40 \text{ cm}$$
$$\omega_1 = \omega_2 = 30 \text{ rad/s}$$

Solution
The formula to be used is the equation relating linear and angular velocities:

$$v = r\omega$$

For ball 1:

$$v_1 = (0.20 \text{ m})\,(30 \text{ rad/s})$$

$$v_1 = 6 \text{ m/s}$$

For ball 2:

$$v_2 = (0.40 \text{ m})\,(30 \text{ rad/s})$$

$$v_2 = 12 \text{ m/s}$$

The greater the angular velocity of a baseball bat, the farther a struck ball will travel, other conditions being equal.

With all other factors held constant, the greater the radius of rotation at which a swinging implement hits a ball, the greater the linear velocity imparted to the ball. In golf, longer clubs are selected for longer shots, and shorter clubs are selected for shorter shots. However, the magnitude of the angular velocity figures as heavily as the length of the radius of rotation in determining the linear velocity of a point on a swinging implement. Little Leaguers often select long bats, which increase the potential radius of rotation if a ball is contacted, but are also too heavy for the young players to swing as quickly as shorter, lighter bats. The relationship between the radius of rotation of the contact point between a striking implement and a ball and the subsequent velocity of the ball is shown in Figure 11-16.

It is important to recognize that the linear velocity of a ball struck by a bat, racket, or club is *not* identical to the linear velocity of the contact point on the swinging implement. Other factors, such as the directness of the hit and the elasticity of the impact, also influence ball velocity.

Linear and Angular Acceleration

The acceleration of a body in angular motion may be resolved into two perpendicular linear acceleration components. These components are directed along and perpendicular to the path of angular motion at any point in time (Figure 11-17).

The component directed along the path of angular motion takes its name from the term *tangent*. A tangent is a line that touches, but does not cross, a curve at a single point. The tangential component, known as tangential acceleration, represents the change in linear speed for a body traveling on a curved path. The formula for tangential acceleration is the following:

tangential acceleration
component of acceleration of a body in angular motion directed along a tangent to the path of motion; represents change in linear speed

$$a_t = \frac{v_2 - v_1}{t}$$

Tangential acceleration is a_t, v_1 is the tangential linear velocity of the moving body at an initial time, v_2 is the tangential linear velocity of the moving body at a second time, and t is the time interval over which the velocities are assessed.

Top view

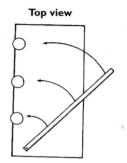

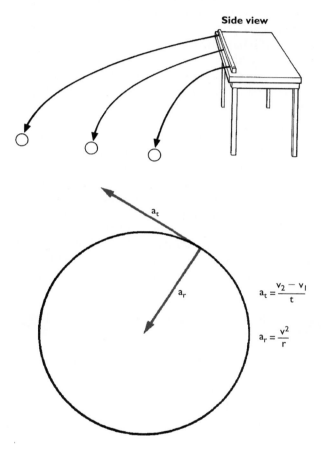

$$a_t = \frac{v_2 - v_1}{t}$$

$$a_r = \frac{v^2}{r}$$

FIGURE 11-16

A simple experiment in which a rotating stick strikes three balls demonstrates the significance of the radius of rotation.

FIGURE 11-17

Tangential and radial acceleration vectors shown relative to a circular path of motion.

When a ball is thrown, the ball follows a curved path as it is accelerated by the muscles of the shoulder, elbow, and wrist. The tangential component of ball acceleration represents the rate of change in the linear speed of the ball. Because the speed of projection greatly affects a projectile's range, tangential velocity should be maximum just before ball release if the objective is to throw the ball fast or far. Once ball release occurs, tangential acceleration is zero, because the thrower is no longer applying a force.

The relationship between tangential acceleration and angular acceleration is expressed as follows:

$$a_t = r\alpha$$

Linear acceleration is a_t, r is the radius of rotation, and α is angular acceleration. The units of linear acceleration and the radius of rotation must be compatible, and angular acceleration must be expressed in radian-based units for the relationship to be accurate.

•*At the instant that a thrown ball is released, its tangential and radial accelerations become equal to zero, because a thrower is no longer applying force.*

radial acceleration

component of acceleration of a body in angular motion directed toward the center of curvature; represents change in direction

Although the linear speed of an object traveling along a curved path may not change, its direction of motion is constantly changing. The second component of angular acceleration represents the rate of change in direction of a body in angular motion. This component is called radial acceleration, and it is always directed toward the center of curvature. Radial acceleration may be quantified by using the following formula:

$$a_r = \frac{v^2}{r}$$

Radial acceleration is a_r, v is the tangential linear velocity of the moving body, and r is the radius of rotation. An increase in linear velocity or a decrease in the radius of curvature increases radial acceleration. Thus, the smaller the radius of curvature (the tighter the curve) is, the more difficult it is for a cyclist to negotiate the curve at a high velocity (see Chapter 14).

During execution of a ball throw, the ball follows a curved path because the thrower's arm and hand restrain it. This restraining force causes radial acceleration toward the center of curvature throughout the motion. When the thrower releases the ball, radial acceleration no longer exists, and the implement follows the path of the tangent to the curve at that instant. The timing of release is therefore critical: If release occurs too soon or too late, the ball will be directed to the left or the right rather than straight ahead. Sample Problem 11.3 demonstrates the effects of the tangential and radial components of acceleration.

Both tangential and radial components of motion can contribute to the resultant linear velocity of a projectile at release. For example, during somersault dismounts from the high bar in gymnastics routines, although the primary contribution to linear velocity of the body's center of gravity is generally from tangential acceleration, the radial component can contribute up to 50% of the resultant velocity (8). The size of the contribution from the radial component, and whether the contribution is positive or negative, varies with the performer's technique.

SAMPLE PROBLEM 11.3

A windmill-style softball pitcher executes a pitch in 0.65 s. If her pitching arm is 0.7 m long, what are the magnitudes of the tangential and radial accelerations on the ball just before ball release, when tangential ball speed is 20 m/s? What is the magnitude of the total acceleration on the ball at this point?

Known

$$t = 0.65 \text{ s}$$
$$r = 0.7 \text{ m}$$
$$v_2 = 20 \text{ m/s}$$

Solution

To solve for tangential acceleration, use the following formula:

$$a_t = \frac{v_2 - v_1}{t}$$

Substitute in what is known and assume that $v_1 = 0$:

$$a_t = \frac{20 \text{ m/s} - 0}{0.65 \text{ s}}$$

$$a_t = 30.8 \text{ m/s}^2$$

To solve for radial acceleration, use the following formula:

$$a_r = \frac{v^2}{r}$$

Substitute in what is known:

$$a_r = \frac{(20 \text{ m/s})^2}{0.7\text{m}}$$

$$a_r = 571.4 \text{ m/s}^2$$

To solve for total acceleration, perform vector composition of tangential and radial acceleration. Since tangential and radial acceleration are oriented perpendicular to each other, the Pythagorean theorem can be used to calculate the magnitude of total acceleration.

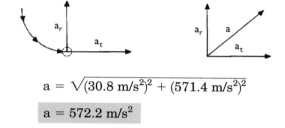

$$a = \sqrt{(30.8 \text{ m/s}^2)^2 + (571.4 \text{ m/s}^2)^2}$$

$$a = 572.2 \text{ m/s}^2$$

SUMMARY

An understanding of angular motion is an important part of the study of biomechanics, because most volitional motion of the human body involves the rotation of bones around imaginary axes of rotation passing through the joint centers at which the bones articulate. The angular kinematic quantities—angular displacement, angular velocity, and angular acceleration—possess the same interrelationships as their linear counterparts, with angular displacement representing change in angular position, angular velocity defined as the rate of change in angular position, and angular acceleration indicating the rate of change in angular velocity during a given time. Depending on the selection of the time interval, either average or instantaneous values of angular velocity and angular acceleration may be quantified.

Angular kinematic variables may be quantified for the relative angle formed by the longitudinal axes of two body segments articulating at a joint, or for the absolute angular orientation of a single body segment with respect to a fixed reference line. Different instruments are available for direct measurement of angles on a human subject.

INTRODUCTORY PROBLEMS

1. The relative angle at the knee changes from 0° to 85° during the knee flexion phase of a squat exercise. If 10 complete squats are performed, what is the total angular distance and the total angular displacement undergone at the knee? (Provide answers in both degrees and radians.) (Answer: ϕ = 1700°, 29.7 rad; θ = 0)

2. Identify the angular displacement, the angular velocity, and the angular acceleration of the second hand on a clock over the time interval in which it moves from the number 12 to the number 6. Provide answers

in both degree- and radian-based units. (Answer: $\theta = -180°$, $-\pi$ rad; $\omega = -6$ deg/s, $-\pi/30$ rad/s; $\alpha = 0$)

3. How many revolutions are completed by a top spinning with a constant angular velocity of 3π rad/s during a 20 s time interval? (Answer: 30 rev)

4. A kicker's extended leg is swung for 0.4 s in a counterclockwise direction while accelerating at 200 deg/s². What is the angular velocity of the leg at the instant of contact with the ball? (Answer: 80 deg/s, 1.4 rad/s)

5. The angular velocity of a runner's thigh changes from 3 rad/s to 2.7 rad/s during a 0.5 s time period. What has been the average angular acceleration of the thigh? (Answer: -0.6 rad/s², -34.4 deg/s²)

6. Identify three movements during which the instantaneous angular velocity at a particular time is the quantity of interest. Explain your choices.

7. Fill in the missing corresponding values of angular measure in the table below.

DEGREES	RADIANS	REVOLUTIONS
90	?	?
?	1	?
180	?	?
?	?	1

8. Measure and record the following angles for the drawing shown below:

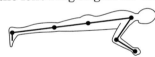

   a. The relative angle at the shoulder
   b. The relative angle at the elbow
   c. The absolute angle of the upper arm
   d. The absolute angle of the forearm
   Use the right horizontal as your reference for the absolute angles.

9. Calculate the following quantities for the diagram shown below:
   a. The angular velocity at the hip over each time interval
   b. The angular velocity at the knee over each time interval
   Would it provide meaningful information to calculate the average angular velocities at the hip and knee for the movement shown? Provide a rationale for your answer.

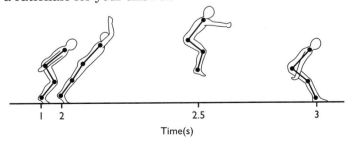

Time(s)

10. A tennis racquet swung with an angular velocity of 12 rad/s strikes a motionless ball at a distance of 0.5 m from the axis of rotation. What is the linear velocity of the racquet at the point of contact with the ball? (Answer: 6 m/s)

ADDITIONAL PROBLEMS

1. A 1.2 m golf club is swung in a planar motion by a right-handed golfer with an arm length of 0.76 m. If the initial velocity of the golf ball is 35 m/s, what was the angular velocity of the left shoulder at the point of ball contact? (Assume that the left arm and the club form a straight line, and that the initial velocity of the ball is the same as the linear velocity of the club head at impact.) (Answer: 17.86 rad/s)

2. David is fighting Goliath. If David's 0.75 m sling is accelerated for 1.5 s at 20 rad/s^2, what will be the initial velocity of the projected stone? (Answer: 22.5 m/s)

3. A baseball is struck by a bat 46 cm from the axis of rotation when the angular velocity of the bat is 70 rad/s. If the ball is hit at a height of 1.2 m at a 45° angle, will the ball clear a 1.2 m fence 110 m away? (Assume that the initial linear velocity of the ball is the same as the linear velocity of the bat at the point at which it is struck.) (Answer: No, the ball will fall through a height of 1.2 m at a distance of 105.7 m.)

4. A polo player's arm and stick form a 2.5 m rigid segment. If the arm and stick are swung with an angular speed of 1.0 rad/s as the player's horse gallops at 5 m/s, what is the resultant velocity of a motionless ball that is struck head-on? (Assume that ball velocity is the same as the linear velocity of the end of the stick.) (Answer: 7.5 m/s)

5. Explain how the velocity of the ball in Problem 4 would differ if the stick were swung at a 30° angle to the direction of motion of the horse.

6. List three movements for which a relative angle at a particular joint is important and three movements for which the absolute angle of a body segment is important. Explain your choices.

7. A majorette in the Rose Bowl Parade tosses a baton into the air with an initial angular velocity of 2.5 rev/s. If the baton undergoes a constant acceleration while airborne of -0.2 rev/s^2 and its angular velocity is 0.8 rev/s when the majorette catches it, how many revolutions does it make in the air? (Answer: 14 rev)

8. A cyclist enters a curve of 30 m radius at a speed of 12 m/s. As the brakes are applied, speed is decreased at a constant rate of 0.5 m/s^2. What are the magnitudes of the cyclist's radial and tangential accelerations when his speed is 10 m/s? (Answer: $a_r = 3.33$ m/s^2; $a_t = -0.5$ m/s^2)

9. A hammer is being accelerated at 15 rad/s^2. Given a radius of rotation of 1.7 m, what are the magnitudes of the radial and tangential components of acceleration when tangential hammer speed is 25 m/s? (Answer: $a_r = 367.6$ m/s^2; $a_t = 25.5$ m/s^2)

10. A speed skater increases her speed from 10 m/s to 12.5 m/s over a period of 3 s while coming out of a curve of 20 m radius. What are the magnitudes of her radial, tangential, and total accelerations as she leaves the curve? (Remember that a_r and a_t are the vector components of total acceleration.) (Answer: $a_r = 7.81$ m/s^2; $a_t = 0.83$ m/s^2; $a = 7.85$ m/s^2)

LABORATORY EXPERIENCES

1. Perform the experiment shown in Sample Problem 11.1 in this chapter. Record the linear distances traveled by the three balls, and write a brief explanation.

Distances: _____

Explanation: _____

_____

_____

2. With a partner, use a goniometer to measure range of motion for wrist flexion and hyperextension, for ankle plantar flexion and dorsiflexion, and for shoulder flexion and hyperextension. Provide an explanation for differences in these ranges of motion between your partner and yourself.

	Yourself	**Your Partner**
Wrist flexion	_____	_____
Wrist hyperextension	_____	_____
Plantar flexion	_____	_____
Dorsiflexion	_____	_____
Shoulder flexion	_____	_____
Shoulder hyperextension	_____	_____

Explanation:_____

_____

_____

3. Observe a young child executing a kick or a throw. Write a brief description of the angular kinematics of the major joint actions. What features distinguish the performance from that of a reasonably skilled adult?

Hip kinematics: _____

_____

_____

Knee kinematics:_____

_____

_____

Foot/ankle kinematics:_____

_____

_____

4. Working in a small group, observe from a side view two volunteers performing simultaneous maximal vertical jumps. Either video (and replay) the jumps or have the subjects repeat the jumps several times. Write a comparative description of the angular kinematics of the jumps, including both relative and absolute angles of importance. Does your description suggest a reason one jump is higher than the other?

Performer 1   **Performer 2**

_____   _____

_____   _____

_____   _____

_____   _____

_____   _____

_____   _____

_____   _____

_____   _____

_____   _____

_____   _____

_____   _____

_____   _____

5. Tape a piece of tracing paper over the monitor of a videocassette recorder. Using the single-frame advance button, draw at least three sequential stick figure representations of a person performing a movement of interest. (If the movement is slow, you may need to skip a consistent number of frames between tracings.) Use a protractor to measure the angle present at one major joint of interest on each figure. Given 1/30 s between adjacent video pictures, calculate the angular velocity at the joints between pictures 1 and 2 and between pictures 2 and 3. Record your answers in both degree- and radian-based units.

Joint selected: _____

Angle 1: _____ angle 2: _____ angle 3: _____

Number of frames skipped between tracings: _____

Calculation:

REFERENCES

1. Elliott BC: Tennis strokes and equipment. In Vaughan CL, ed: *Biomechanics of sport,* Boca Raton, FL, 1989, CRC Press.
2. Escamilla RF, Fleisig GS, Barrentine SW, Zheng N, and Andrews JR: Kinematic comparisons of throwing different types of baseball pitches, *J Appl Biomech* 14:1, 1998.
3. Fleisig GS, Barrentine SW, Zheng N, Escamilla RF, and Andrews JR: Kinematic and kinetic comparison of baseball pitching among various levels of development, *J Biomech* 32:1371, 1999.
4. Fleisig G, Chu Y, Weber A, and Andrews J: Variability in baseball pitching biomechanics among various levels of competition, *Sports Biomech* 8:10, 2009.
5. Fleisig G, Nicholls R, Elliott B, and Escamilla R: Kinematics used by world class tennis players to produce high-velocity serves, *Sports Biomech* 2:51, 2003.
6. Gutman D: The physics of foul play, *Discover* 70, April 1988.
7. Hiley MJ and Yeadon MR: Maximal dismounts from high bar, *J Biomech* 38:2221, 2005.
8. Kerwin DG, Yeadon MR, and Harwood MJ: High bar release in triple somersault dismounts, *J Appl Biomech* 9:279, 1993.
9. King DL: Performing triple and quadruple figure skating jumps: implications for training, *Can J Appl Physiol* 30:743, 2005.
10. King DL, Smith SL, Brown MR, McCrory JL, Munkasy BA, and Scheirman GI: Comparison of split double and triple twists in pair figure skating, *Sports Biomech* 7:222, 2008.
11. Nordin M and Frankel VH: Biomechanics of the knee. In Nordin M and Frankel VH, eds: *Biomechanics of the musculoskeletal system* (3rd ed.), Philadelphia, 2001, Lippincott Williams & Wilkins.
12. Smith PN, Refshauge KM, and Scarvell JM: Development of the concepts of knee kinematics, *Arch Phys Med Rehabil* 84:1895, 2003.

ANNOTATED READINGS

Fleisig GS, Jameson EG, Dillman CJ, and Andrews JR: Biomechanics of overhead sports. In Garrett WE Jr. and Kirkendall DT: *Exercise and sport science,* Philadelphia, 2000, Lippincott Williams & Wilkins.
  Chapter reviews kinematic and other aspects of throwing and related movements.
Gregor RJ: Biomechanics of cycling. In Garrett WE Jr. and Kirkendall DT: *Exercise and sport science,* Philadelphia, 2000, Lippincott Williams & Wilkins.
  Chapter comprises a comprehensive review of kinematic and other aspects of cycling.
Hong Y and Bartlett R: *Routledge handbook of biomechanics and human movement science,* New York, 2008, Routledge.
  Includes a section on biomechanics of sports, including kinematic analyses.
Jemni M (ed.): The science of gymnastics, New York, 2010, Routledge.
  Presents a comprehensive overview of scientific knowledge related to gymnastics, including kinematics of performance.

RELATED WEBSITES

Circular Motion and Rotational Kinematics
http://cnx.org/content/m14014/latest/
  Presents an online tutorial with formulas for relationships among angular kinematic quantities.
The Exploratorium: Science of Baseball
http://www.exploratorium.edu/baseball/index.html
  Explains scientific concepts related to baseball pitching and hitting.
The Exploratorium: Science of Cycling
http://www.exploratorium.edu/cycling/index.html
  Explains scientific concepts related to bicycle wheels and gear ratios.
The Exploratorium's Science of Hockey
http://www.exploratorium.edu/hockey/index.html
  Explains scientific concepts related to hockey, including how to translate rotational motion of the arms into linear motion of the puck.

KEY TERMS

absolute angle	angular orientation of a body segment with respect to a fixed line of reference
angular acceleration	rate of change in angular velocity
angular displacement	change in the angular position or orientation of a line segment
angular velocity	rate of change in the angular position or orientation of a line segment
instant center	precisely located center of rotation at a joint at a given instant in time
radial acceleration	component of acceleration of a body in angular motion directed toward the center of curvature; represents change in direction
radian	unit of angular measure used in angular-linear kinematic quantity conversions; equal to 57.3°
radius of rotation	distance from the axis of rotation to a point of interest on a rotating body
relative angle	angle at a joint formed between the longitudinal axes of adjacent body segments
right hand rule	procedure for identifying the direction of an angular motion vector
tangential acceleration	component of acceleration of a body in angular motion directed along a tangent to the path of motion; represents change in linear speed

Linear Kinetics
of Human Movement

12

After completing this chapter, you will be able to:

Identify Newton's laws of motion and gravitation and describe practical illustrations of the laws.

Explain what factors affect friction and discuss the role of friction in daily activities and sports.

Define impulse and momentum and explain the relationship between them.

Explain what factors govern the outcome of a collision between two bodies.

Discuss the relationships among mechanical work, power, and energy.

Solve quantitative problems related to kinetic concepts.

ONLINE LEARNING CENTER RESOURCES

www.mhhe.com/hall6e

Log on to our Online Learning Center (OLC) for access to these additional resources:

- Online Lab Manual
- Flashcards with definitions of chapter key terms
- Chapter objectives
- Chapter lecture PowerPoint presentation
- Self-scoring chapter quiz
- Additional chapter resources
- Web links for study and exploration of chapter-related topics

What can people do to improve traction when walking on icy streets? Why do some balls bounce higher on one surface than on another? How can football linemen push larger opponents backward? In this chapter, we introduce the topic of kinetics with a discussion of some important basic concepts and principles relating to linear kinetics.

NEWTON'S LAWS

Sir Isaac Newton (1642–1727) discovered many of the fundamental relationships that form the foundation for the field of modern mechanics. These principles highlight the interrelationships among the basic kinetic quantities introduced in Chapter 3.

Law of Inertia

Newton's first law of motion is known as the *law of inertia*. This law states the following:

> A body will maintain a state of rest or constant velocity unless acted on by an external force that changes the state.

In other words, a motionless object will remain motionless unless there is a net force (a force not counteracted by another force) acting on it. Similarly, a body traveling with a constant speed along a straight path will continue its motion unless acted on by a net force that alters either the speed or the direction of the motion.

It seems intuitively obvious that an object in a static (motionless) situation will remain motionless barring the action of some external force. We assume that a piece of furniture such as a chair will maintain a fixed position unless pushed or pulled by a person exerting a net force to cause its motion. When a body is traveling with a constant velocity, however, the enactment of the law of inertia is not so obvious, because, in most situations, external forces do act to reduce velocity. For example, the law of inertia implies that a skater gliding on ice will continue gliding with the same speed and in the same direction, barring the action of an external force. But in reality, friction and air resistance are two forces normally present that act to slow skaters and other moving bodies.

A skater has a tendency to continue gliding with constant speed and direction because of inertia. Photo courtesy of Karl Weatherly/Getty Images.

Law of Acceleration

Newton's second law of motion is an expression of the interrelationships among force, mass, and acceleration. This law, known as the *law of acceleration,* may be stated as follows for a body with constant mass:

> A force applied to a body causes an acceleration of that body of a magnitude proportional to the force, in the direction of the force, and inversely proportional to the body's mass.

When a ball is thrown, kicked, or struck with an implement, it tends to travel in the direction of the line of action of the applied force. Similarly, the greater the amount of force applied, the greater the speed the ball has. The algebraic expression of the law is a well-known formula that expresses the quantitative relationships among an applied force, a body's mass, and the resulting acceleration of the body:

$$F = ma$$

Thus, if a 1 kg ball is struck with a force of 10 N, the resulting acceleration of the ball is 10 m/s^2. If the ball has a mass of 2 kg, the application of the same 10 N force results in an acceleration of only 5 m/s^2.

Newton's second law also applies to a moving body. When a defensive football player running down the field is blocked by an opposing player, the velocity of the defensive player following contact is a function of the player's original direction and speed and the direction and magnitude of the force exerted by the offensive player.

Law of Reaction

The third of Newton's laws of motion states that every applied force is accompanied by a reaction force:

> For every action, there is an equal and opposite reaction.

In terms of forces, the law may be stated as follows:

> When one body exerts a force on a second, the second body exerts a reaction force that is equal in magnitude and opposite in direction on the first body.

When a person leans with a hand against a rigid wall, the wall pushes back on the hand with a force that is equal and opposite to that exerted by the hand on the wall. The harder the hand pushes against the wall, the greater is the amount of pressure felt across the surface of the hand where it contacts the wall. Another illustration of Newton's third law of motion is found in Sample Problem 12.1.

During gait, every contact of a foot with the floor or ground generates an upward reaction force. Researchers and clinicians measure and study these ground reaction forces (GRFs) in analyzing differences in gait patterns across the life span and among individuals with handicapping conditions. Researchers have studied the GRFs that are sustained with every footfall during running to investigate factors related to both performance and running-related injuries. The magnitude of the vertical component of the GRF during running on a level

In accordance with Newton's third law of motion, ground reaction forces are sustained with every footfall during running. Photo courtesy of Digital Vision/Getty Images.

SAMPLE PROBLEM 12.1

A 90 kg ice hockey player collides head-on with an 80 kg player. If the first player exerts a force of 450 N on the second player, how much force is exerted by the second player on the first?

Known

$$m_1 = 90 \text{ kg}$$
$$m_2 = 80 \text{ kg}$$
$$F_1 = 450 \text{ N}$$

Solution
This problem does not require computation. According to Newton's third law of motion, for every action, there is an equal and opposite reaction. If the force exerted by the first player on the second has a magnitude of 450 N and a positive direction, then the force exerted by the second player on the first has a magnitude of 450 N and a negative direction.

$$-450 \text{ N}$$

FIGURE 12-1

Typical ground reaction force patterns for rearfoot strikers and others. Runners may be classified as rearfoot, midfoot, or forefoot strikers according to the portion of the shoe that usually contacts the ground first.

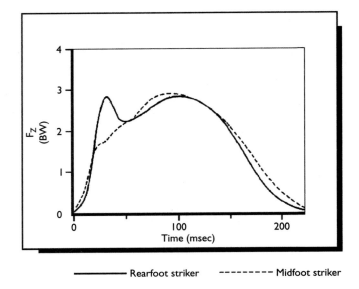

——————— Rearfoot striker   - - - - - - - Midfoot striker

surface is generally two to three times the runner's body weight, with the pattern of force sustained during ground contact varying with running style. Runners are classified as rearfoot, midfoot, or forefoot strikers, according to the portion of the shoe first making contact with the ground. Typical vertical GRF patterns for rearfoot strikers and others are shown in Figure 12-1.

Other factors influencing GRF patterns include running speed, running duration, knee flexion angle at contact, stride length, fatigue, footwear, surface stiffness, surface smoothness, light intensity, and grade (8). During treadmill running at 3 m/s at a grade of −9°, impacts perpendicular to the surface increase on the order of 54% compared to level running, significantly increasing the potential for stress-related injury (12). The presence of fatigue, on the other hand, slightly reduces peak impact forces, secondary to reduced step length and increased knee flexion at contact (11).

Although it may seem logical that harder running surfaces would generate larger ground reaction forces, this has not been documented. When encountering surfaces of different stiffness, runners typically make individual adjustments in running kinematics that tend to maintain GRFs at a constant level (9). This may be explained to some extent by runners' sensitivity to the shock waves resulting from every heel strike that propagate upward, dynamically loading the musculoskeletal system. There is evidence that when the magnitude of the GRF increases, dynamic loading of the musculoskeletal system increases at five times the rate of the increase in the GRF (29). Research has shown that muscle activity is elicited to minimize soft-tissue vibrations arising from impact forces during running, another sign of runners' sensitivity to dynamic loading (21, 22).

As discussed in Chapter 10, runners generally increase stride length as running speed increases over the slow-to-moderate speed range. Longer strides tend to generate GRFs with larger retarding horizontal components (Figure 12-2). This is one reason that overstriding can be counterproductive. With longer stride lengths and more extended knee angles at contact, muscles crossing the knee also absorb more of the shock that is transmitted upward through the musculoskeletal system, which may translate to additional stress being placed on the knees (7).

●*Since the ground reaction force is an external force acting on the human body, its magnitude and direction affect the body's velocity.*

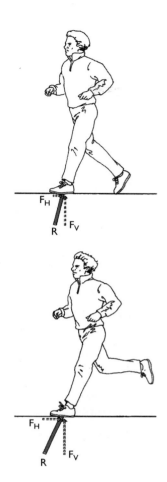

FIGURE 12-2

Use of a longer stride length during running increases the retarding horizontal component (F_H) of the ground reaction force (R).

Since the GRF is an external force acting on the human body, its magnitude and direction have implications for performance in many sporting events. In the high jump, for example, skilled performers are moving with a large horizontal velocity and a slight downwardly directed vertical velocity at the beginning of the stride before takeoff. The GRF reduces the jumper's horizontal velocity and creates an upwardly directed vertical velocity (Figure 12-3). Better jumpers not only enter the takeoff phase of the jump with high horizontal velocities but also effectively use the GRF to convert horizontal velocity to upward vertical velocity (5). Although baseball and softball pitching are often thought of as being primarily upper extremity motions, with GRFs approaching 139% body weight (BW) vertically, 24% BW anteriorly, and 42% BW medially, windmill softball pitchers are commonly seen for lower extremity injuries (14). Maximizing the distance of drives in golf requires generation of large GRFs, with a greater proportion of the GRF on the back foot during the backswing and transfer of this proportion to the front foot during the downswing (17).

Law of Gravitation

Newton's discovery of the law of universal gravitation was one of the most significant contributions to the scientific revolution and is considered by many to mark the beginning of modern science (4). According to legend, Newton's thoughts on gravitation were provoked either by his observation of a falling apple or by his actually being struck on the head by a falling apple. In his writings on the subject, Newton used the example of

FIGURE 12-3

During the high jump takeoff, the horizontal component (F_H) of the ground reaction force (R) decreases the performer's horizontal velocity, and the vertical component (F_V) can contribute to upward vertical velocity.

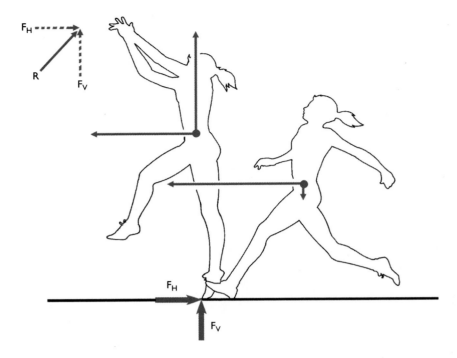

the falling apple to illustrate the principle that every body attracts every other body. Newton's law of gravitation states the following:

> All bodies are attracted to one another with a force proportional to the product of their masses and inversely proportional to the square of the distance between them.

Stated algebraically, the law is the following:

$$F_g = G\frac{m_1 m_2}{d^2}$$

The force of gravitational attraction is F_g, G is a numerical constant, m_1 and m_2 are the masses of the bodies, and d is the distance between the mass centers of the bodies.

For the example of the falling apple, Newton's law of gravitation indicates that just as the earth attracts the apple, the apple attracts the earth, although to a much smaller extent. As the formula for gravitational force shows, the greater the mass of either body, the greater the attractive force between the two. Similarly, the greater the distance between the bodies, the smaller the attractive force between them.

For biomechanical applications, the only gravitational attraction of consequence is that generated by the earth because of its extremely large mass. The rate of gravitational acceleration at which bodies are attracted toward the surface of the earth (9.81 m/s^2) is based on the earth's mass and the distance to the center of the earth.

MECHANICAL BEHAVIOR OF BODIES IN CONTACT

According to Newton's third law of motion, for every action there is an equal and opposite reaction. However, consider the case of a horse hitched to a cart. According to Newton's third law, when the horse exerts a force on the cart to cause forward motion, the cart exerts a backward force of

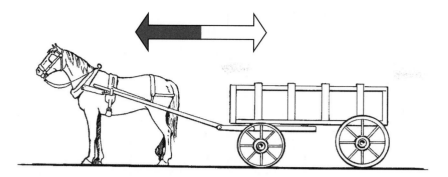

FIGURE 12-4
When a horse attempts to pull a cart forward, the cart exerts an equal and opposite force on the horse, in accordance with Newton's third law.

equal magnitude on the horse (Figure 12-4). Considering the horse and the cart as a single mechanical system, if the two forces are equal in magnitude and opposite in direction, their vector sum is zero. How does the horse-and-cart system achieve forward motion? The answer relates to the presence of another force that acts with a different magnitude on the cart than on the horse: the force of friction.

Friction

Friction is a force that acts at the interface of surfaces in contact in the direction opposite the direction of motion or impending motion. Because friction is a force, it is quantified in units of force (N). The magnitude of the generated friction force determines the relative ease or difficulty of motion for two objects in contact.

Consider the example of a box sitting on a level tabletop (Figure 12-5). The two forces acting on the undisturbed box are its own weight and a reaction force (R) applied by the table. In this situation, the reaction force is equal in magnitude and opposite in direction to the box's weight.

When an extremely small horizontal force is applied to this box, it remains motionless. The box can maintain its static position because the applied force causes the generation of a friction force at the box/table interface that is equal in magnitude and opposite in direction to the small applied force. As the magnitude of the applied force becomes greater and greater, the magnitude of the opposing friction force also increases to a certain critical point. At that point, the friction force present is termed maximum static friction (F_m). If the magnitude of the applied force is increased beyond this value, motion will occur (the box will slide).

Once the box is in motion, an opposing friction force continues to act. The friction force present during motion is referred to as kinetic friction (F_k). Unlike static friction, the magnitude of kinetic friction remains at a constant value that is *less than* the magnitude of maximum static friction. Regardless of the amount of the applied force or the speed of the occurring motion, the kinetic friction force remains the same. Figure 12-6 illustrates the relationship between friction and an applied external force.

What factors determine the amount of applied force needed to move an object? More force is required to move a refrigerator than to move the empty box in which the refrigerator was delivered. More force is also needed to slide the refrigerator across a carpeted floor than to do so across a smooth linoleum floor. Two factors govern the magnitude of the force of maximum static friction or kinetic friction in any situation: the coefficient of friction, represented by the lowercase Greek letter mu (μ), and the normal (perpendicular) reaction force (R):

$$F = \mu R$$

friction
force acting over the area of contact between two surfaces in the direction opposite that of motion or motion tendency

maximum static friction
maximum amount of friction that can be generated between two static surfaces

kinetic friction
constant-magnitude friction generated between two surfaces in contact during motion

coefficient of friction
number that serves as an index of the interaction between two surfaces in contact

normal reaction force
force acting perpendicular to two surfaces in contact

FIGURE 12-5

The magnitude of the friction force changes with increasing amounts of applied force.

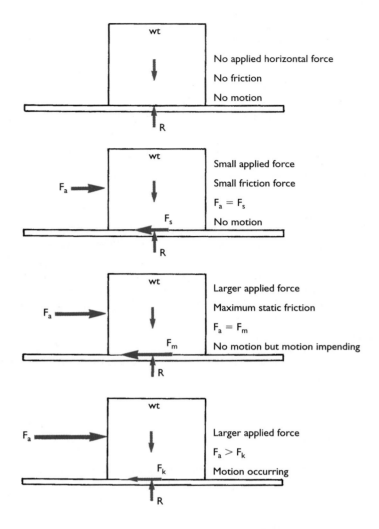

The coefficient of friction is a unitless number indicating the relative ease of sliding, or the amount of mechanical and molecular interaction between two surfaces in contact. Factors influencing the value of μ are the relative roughness and hardness of the surfaces in contact and the type of molecular interaction between the surfaces. The greater the mechanical and molecular interaction, the greater the value of μ. For example, the coefficient of friction between two blocks covered with rough sandpaper is larger than the coefficient of friction between a skate and a smooth surface

FIGURE 12-6

As long as a body is static (unmoving), the magnitude of the friction force developed is equal to that of an applied external force. Once motion is initiated, the magnitude of the friction force remains at a constant level below that of maximum static friction.

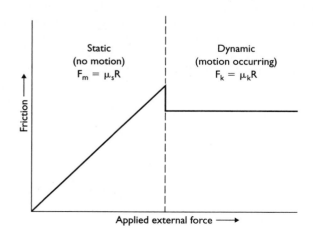

of ice. The coefficient of friction describes the interaction between two surfaces in contact and is not descriptive of either surface alone. The coefficient of friction for the blade of an ice skate in contact with ice is different from that for the blade of the same skate in contact with concrete or wood.

The coefficient of friction between two surfaces assumes one or two different values, depending on whether the bodies in contact are motionless (static) or in motion (kinetic). The two coefficients are known as the *coefficient of static friction* (μ_s) and the *coefficient of kinetic friction* (μ_k). The magnitude of maximum static friction is based on the coefficient of static friction:

$$F_m = \mu_s R$$

The magnitude of the kinetic friction force is based on the coefficient of kinetic friction:

$$F_k = \mu_k R$$

For any two bodies in contact, μ_k is always smaller than μ_s. Kinetic friction coefficients of 0.0071 have been reported for standard ice hockey skates on ice, with new blade designs involving flaring of the blade near the bottom edge lowering the coefficient of friction even further (10). Use of the coefficients of static and kinetic friction is illustrated in Sample Problem 12.2.

•*Because μ_k is always smaller than μ_s, the magnitude of kinetic friction is always less than the magnitude of maximum static friction.*

SAMPLE PROBLEM 12.2

The coefficient of static friction between a sled and the snow is 0.18, with a coefficient of kinetic friction of 0.15. A 250 N boy sits on the 200 N sled. How much force directed parallel to the horizontal surface is required to start the sled in motion? How much force is required to keep the sled in motion?

Known

$$\mu_s = 0.18$$
$$\mu_k = 0.15$$
$$wt = 250\ N + 200\ N$$

Solution

To start the sled in motion, the applied force must exceed the force of maximum static friction:

$$F_m = \mu_s R$$
$$= (0.18)(250\ N + 200\ N)$$
$$= 81\ N$$

The applied force must be greater than 81 N.

To maintain motion, the applied force must equal the force of kinetic friction:

$$F_k = \mu_k R$$
$$= (0.15)(250\ N + 200\ N)$$
$$= 67.5\ N$$

The applied force must be at least 67.5 N.

FIGURE 12-7

As weight increases, the normal reaction force increases.

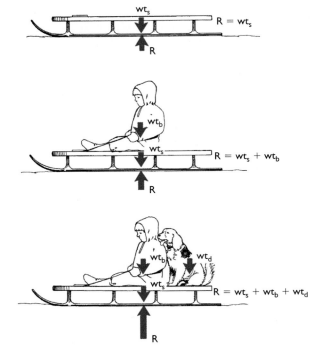

The other factor affecting the magnitude of the friction force generated is the normal reaction force. If weight is the only vertical force acting on a body sitting on a horizontal surface, R is equal in magnitude to the weight. If the object is a football blocking sled with a 100 kg coach standing on it, R is equal to the weight of the sled plus the weight of the coach. Other vertically directed forces such as pushes or pulls can also affect the magnitude of R, which is always equal to the vector sum of all forces or force components acting normal to the surfaces in contact (Figure 12-7).

The magnitude of R can be intentionally altered to increase or decrease the amount of friction present in a particular situation. When a football coach stands on the back of a blocking sled, the normal reaction force exerted by the ground on the sled is increased, with a concurrent increase in the amount of friction generated, making it more difficult for a player to move the sled. Alternatively, if the magnitude of R is decreased, friction is decreased and it is easier to initiate motion.

How can the normal reaction force be decreased? Suppose you need to rearrange the furniture in a room. Is it easier to push or pull an object such as a desk to move it? When a desk is pushed, the force exerted is typically directed diagonally downward. In contrast, force is usually directed diagonally upward when a desk is pulled. The vertical component of the push or pull either adds to or subtracts from the magnitude of the normal reaction force, thus influencing the magnitude of the friction force generated and the relative ease of moving the desk (Figure 12-8).

The amount of friction present between two surfaces can also be changed by altering the coefficient of friction between the surfaces. For example, the use of gloves in sports such as golf and racquetball increases the coefficient of friction between the hand and the grip of the club or racquet. Similarly, lumps of wax applied to a surfboard increase the roughness of the board's surface, thereby increasing the coefficient of friction between the board and the surfer's feet. The application of a thin, smooth coat of wax to the bottom of cross-country skis is designed to decrease the coefficient of friction between the skis and the snow, with different waxes used for various snow conditions.

●It is advantageous to pull with a line of force that is directed slightly upward when moving a heavy object.

●Racquetball and golf gloves are designed to increase the friction between the hand and the racquet or club, as are the grips on the handles of the rackets and clubs themselves.

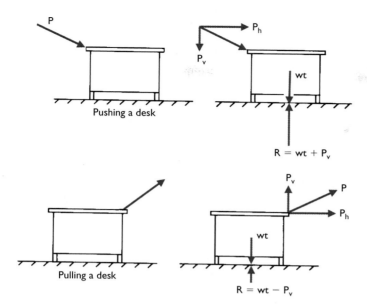

Pushing a desk

$$R = wt + P_v$$

Pulling a desk

$$R = wt - P_v$$

FIGURE 12-8

From a mechanical perspective, it is easier to pull than to push an object such as a desk, since pulling tends to decrease the magnitude of R and F, whereas pushing tends to increase R and F.

A widespread misconception about friction is that greater contact surface area generates more friction. Advertisements often imply that wide-track automobile tires provide better traction (friction) against the road than tires of normal width. However, the only factors known to affect friction are the coefficient of friction and the normal reaction force. Because wide-track tires typically weigh more than normal tires, they do increase friction to the extent that they increase R. However, the same effect can be achieved by carrying bricks or cinder blocks in the trunk of the car, a practice often followed by people who regularly drive on icy roads. Wide-track tires do tend to provide the advantages of increased lateral stability and increased wear, since larger surface area reduces the stress on a properly inflated tire.

Friction exerts an important influence during many daily activities. Walking depends on a proper coefficient of friction between a person's shoes and the supporting surface. If the coefficient of friction is too low, as when a person with smooth-soled shoes walks on a patch of ice, slippage will occur. The bottom of a wet bathtub or shower stall should provide a coefficient of friction with the soles of bare feet that is sufficiently large to prevent slippage.

The amount of friction present between ballet shoes and the dance studio floor must be controlled so that movements involving some amount of sliding or pivoting—such as *glissades*, *assembles*, and *pirouettes*—can be executed smoothly but without slippage. Rosin is often applied to dance floors because it provides a large coefficient of static friction but a significantly smaller coefficient of dynamic friction. This helps to prevent slippage in static situations and allows desired movements to occur freely.

A controversial disagreement occurred between Glenn Allison, a retired professional bowler and member of the American Bowling Congress Hall of Fame, and the American Bowling Congress. The dispute arose over the amount of friction present between Allison's ball and the lanes on which he bowled a perfect score of 300 in three consecutive games. According to the congress, his scores could not be recognized because the lanes he used did not conform to Congress standards for the amount of conditioning oil present (18), which gave Allison the unfair advantage of added ball traction.

The coefficient of friction between a dancer's shoes and the floor must be small enough to allow freedom of motion but large enough to prevent slippage.

●*Rolling friction is influenced by the weight, radius, and deformability of the rolling object, as well as by the coefficient of friction between the two surfaces.*

●*The synovial fluid present at many of the joints of the human body greatly reduces the friction between the articulating bones.*

The magnitude of the rolling friction present between a rolling object, such as a bowling ball or an automobile tire, and a flat surface is approximately one-hundredth to one-thousandth of that present between sliding surfaces. Rolling friction occurs because both the curved and the flat surfaces are slightly deformed during contact. The coefficient of friction between the surfaces in contact, the normal reaction force, and the size of the radius of curvature of the rolling body all influence the magnitude of rolling friction.

The amount of friction present in a sliding or rolling situation is dramatically reduced when a layer of fluid, such as oil or water, intervenes between two surfaces in contact. The presence of synovial fluid serves to reduce the friction, and subsequently the mechanical wear, on the diarthrodial joints of the human body. The coefficient of friction in a total hip prosthesis is approximately 0.01 (23).

Revisiting the question presented earlier about the horse and cart, the force of friction is the determining factor for movement. The system moves forward if the magnitude of the friction force generated by the horse's hooves against the ground exceeds that produced by the wheels of the cart against the ground (Figure 12-9). Because most horses are shod to increase the amount of friction between their hooves and the ground, and most cart wheels are round and smooth to minimize the amount of friction they generate, the horse is usually at an advantage. However, if the horse stands on a slippery surface or if the cart rests in deep sand or is heavily loaded, motion may not be possible.

Momentum

Another factor that affects the outcome of interactions between two bodies is momentum, a mechanical quantity that is particularly important in situations involving collisions. Momentum may be defined generally as the quantity of motion that an object possesses. More specifically, linear momentum is the product of an object's mass and its velocity:

linear momentum
quantity of motion, measured as the product of a body's mass and its velocity

$$M = mv$$

A static object (with zero velocity) has no momentum; that is, its momentum equals zero. A change in a body's momentum may be caused by either a change in the body's mass or a change in its velocity. In most human movement situations, changes in momentum result from changes in velocity. Units of momentum are units of mass multiplied by units of velocity, expressed in terms of kg · m/s. Because velocity is a vector quantity, momentum is also a vector quantity and is subject to the rules of vector composition and resolution.

●*Momentum is a vector quantity.*

When a head-on collision between two objects occurs, there is a tendency for both objects to continue moving in the direction of motion originally possessed by the object with the greatest momentum. If a 90 kg hockey player traveling at 6 m/s to the right collides head-on with an 80 kg player traveling at 7 m/s to the left, the momentum of the first player is the following:

$$
\begin{aligned}
M &= mv \\
&= (90 \text{ kg}) (6 \text{ m/s}) \\
&= 540 \text{ kg} \cdot \text{m/s}
\end{aligned}
$$

The momentum of the second player is expressed as follows:

$$
\begin{aligned}
M &= mv \\
&= (80 \text{ kg}) (7 \text{ m/s}) \\
&= 560 \text{ kg} \cdot \text{m/s}
\end{aligned}
$$

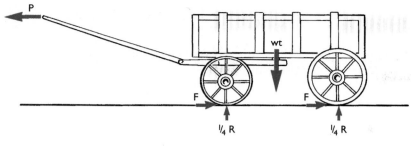

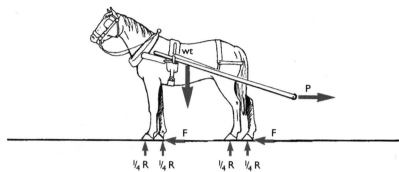

FIGURE 12-9

A horse can pull a cart if the horse's hooves generate more friction than the wheels of the cart.

Since the second player's momentum is greater, both players would tend to continue moving in the direction of the second player's original velocity after the collision. Actual collisions are also affected by the extent to which the players become entangled, by whether one or both players remain on their feet, and by the elasticity of the collision.

Neglecting these other factors that may influence the outcome of the collision, it is possible to calculate the magnitude of the combined velocity of the two hockey players after the collision using a modified statement of Newton's first law of motion (see Sample Problem 12.3). Newton's first law may be restated as the *principle of conservation of momentum:*

> In the absence of external forces, the total momentum of a given system remains constant.

The principle is expressed in equation format as the following:

$$M_1 = M_2$$
$$(mv)_1 = (mv)_2$$

Subscript 1 designates an initial point in time and subscript 2 represents a later time.

Applying this principle to the hypothetical example of the colliding hockey players, the vector sum of the two players' momenta before the collision is equal to their single, combined momentum following the collision (see Sample Problem 12.3). In reality, friction and air resistance are external forces that typically act to reduce the total amount of momentum present.

● *In the absence of external forces, momentum is conserved. However, friction and air resistance are forces that normally act to reduce momentum.*

Impulse

When external forces do act, they change the momentum present in a system predictably. Changes in momentum depend not only on the magnitude of the acting external forces but also on the length of time

SAMPLE PROBLEM 12.3

A 90 kg hockey player traveling with a velocity of 6 m/s collides head-on with an 80 kg player traveling at 7 m/s. If the two players entangle and continue traveling together as a unit following the collision, what is their combined velocity?

Known

$m_1 = 90$ kg
$v_1 = 6$ m/s
$m_2 = 80$ kg
$v_2 = -7$ m/s

m = 90 kg
v = 6 m/s

m = 80 kg
v = 7 m/s

Collision

m = (90 + 80) kg
v = ?

Solution

The law of conservation of momentum may be used to solve the problem, with the two players considered as the total system.

Before collision   After collision

$$m_1v_1 + m_2v_2 = (m_1 + m_2)(v)$$
$$(90 \text{ kg})(6 \text{ m/s}) + (80 \text{ kg})(-7 \text{ m/s}) = (90 \text{ kg} + 80 \text{ kg})(v)$$
$$540 \text{ kg} \cdot \text{m/s} - 560 \text{ kg} \cdot \text{m/s} = (170 \text{ kg})(v)$$
$$-20 \text{ kg} \cdot \text{m/s} = (170 \text{ kg})(v)$$

$v = 0.12$ m/s in the 80 kg player's original direction of travel.

impulse
product of a force and the time interval over which the force acts

over which each force acts. The product of force and time is known as impulse:

$$\text{impulse} = Ft$$

When an impulse acts on a system, the result is a change in the system's total momentum. The relationship between impulse and momentum is derived from Newton's second law:

$$F = ma$$
$$F = m\frac{(v_2 - v_1)}{t}$$
$$Ft = (mv)_2 - (mv)_1$$
$$Ft = \Delta M$$

SAMPLE PROBLEM 12.4

A toboggan race begins with the two crew members pushing the toboggan to get it moving as quickly as possible before they climb in. If crew members apply an average force of 100 N in the direction of motion of the 90 kg toboggan for a period of 7 s before jumping in, what is the toboggan's speed (neglecting friction) at that point?

Known

$$F = 100 \text{ N}$$
$$t = 7 \text{ s}$$
$$m = 90 \text{ kg}$$

100 N

90 kg

Solution

The crew members are applying an impulse to the toboggan to change the toboggan's momentum from zero to a maximum amount. The impulse–momentum relationship may be used to solve the problem.

$$Ft = (mv)_2 - (mv)_1$$
$$(100 \text{ N}) (7 \text{ s}) = (90 \text{ kg}) (v) - (90 \text{ kg}) (0)$$

v = 7.78 m/s in the direction of force application

Subscript 1 designates an initial time and subscript 2 represents a later time. An application of this relationship is presented in Sample Problem 12.4.

Significant changes in an object's momentum may result from a small force acting over a large time interval or from a large force acting over a small time interval. A golf ball rolling across a green gradually loses momentum because its motion is constantly opposed by the force of rolling friction. The momentum of a baseball struck vigorously by a bat also changes because of the large force exerted by the bat during the fraction of a second it is in contact with the ball. It is little surprise that elite sprinters have been shown to develop significantly greater impulse against the starting blocks as compared to well-trained but subelite sprinters (25).

The amount of impulse generated by the human body is often intentionally manipulated. When a vertical jump is performed on a force platform, a graphical display of the vertical GRF across time can be generated (Figure 12-10). Since impulse is the product of force and time, the impulse is the area under the force–time curve. The larger the impulse generated against the floor, the greater the change in the performer's momentum, and the higher the resulting jump. Theoretically, impulse can be increased by increasing either the magnitude of applied force or the time interval over which the force acts. Practically, however, when time of force application against the ground is prolonged during vertical jump execution, the magnitude of the force that can be generated is dramatically reduced, with the ultimate result being a smaller impulse. For performing a maximal vertical jump, the performer must maximize impulse by optimizing the trade-off between applied force magnitude and force duration.

FIGURE 12-10

Force–time histories for **A** high, and **B** low vertical jumps by the same performer. The shaded area represents the impulse generated against the floor during the jump.

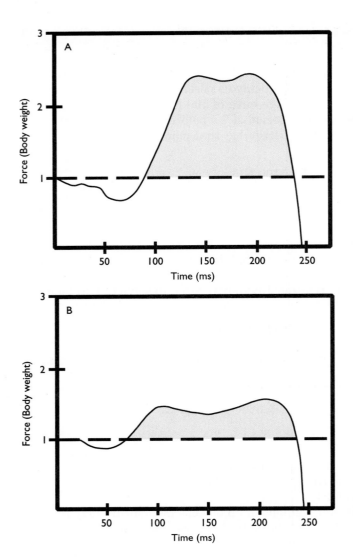

Impulse can also be intentionally manipulated during a landing from a jump (Figure 12-11). A performer who lands rigidly will experience a relatively large GRF sustained over a relatively short time interval. Alternatively, allowing the hip, knee, and ankle joints to undergo flexion during the landing increases the time interval over which the landing force is absorbed, thereby reducing the magnitude of the force sustained. Research has shown that females tend to land in a more erect posture than males, with greater shock absorption occurring in the knees and ankles, and a concomitant greater likelihood of lower-extremity injury (6). One-foot landings also tend to generate higher impact forces and faster loading rates than two-foot landings (27).

It is also useful to manipulate impulse when catching a hard-thrown ball. "Giving" with the ball after it initially contacts the hands or the glove before bringing the ball to a complete stop will prevent the force of the ball from causing the hands to sting. The greater the period is between making initial hand contact with the ball and bringing the ball to a complete stop, the smaller is the magnitude of the force exerted by the ball against the hand, and the smaller is the likelihood of experiencing a sting.

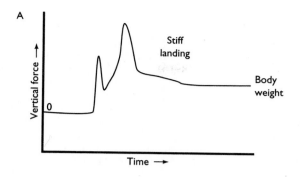

A

Vertical force →

0

Stiff
landing

Body
weight

Time →

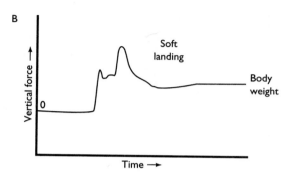

B

Vertical force →

0

Soft
landing

Body
weight

Time →

FIGURE 12-11

Representations of ground
reaction forces during vertical
jump performances: **A.** a
rigid landing, **B.** a landing
with hip, knee, and ankle
flexion occurring. Note the
differences in the magnitudes
and times of the landing
impulses.

Impact

The type of collision that occurs between a struck baseball and a bat is
known as an impact. An impact involves the collision of two bodies over
an extremely small time interval during which the two bodies exert relatively large forces on each other. The behavior of two objects following an
impact depends not only on their collective momentum but also on the
nature of the impact.

For the hypothetical case of a perfectly elastic impact, the relative velocities of the two bodies after impact are the same as their relative velocities
before impact. The impact of a superball with a hard surface approaches
perfect elasticity, because the ball's speed diminishes little during its collision with the surface. At the other end of the range is the perfectly plastic
impact, during which at least one of the bodies in contact deforms and does
not regain its original shape, and the bodies do not separate. This occurs
when modeling clay is dropped on a surface.

Most impacts are neither perfectly elastic nor perfectly plastic, but
somewhere between the two. The coefficient of restitution describes the
relative elasticity of an impact. It is a unitless number between 0 and 1.
The closer the coefficient of restitution is to 1, the more elastic is the
impact; and the closer the coefficient is to 0, the more plastic is the
impact.

The coefficient of restitution governs the relationship between the
relative velocities of two bodies before and after an impact. This relationship, which was originally formulated by Newton, may be stated as
follows:

> When two bodies undergo a direct collision, the difference in their veloci-
> ties immediately after impact is proportional to the difference in their
> velocities immediately before impact.

impact
*collision characterized by the
exchange of a large force during a
small time interval*

perfectly elastic impact
*impact during which the velocity of
the system is conserved*

perfectly plastic impact
*impact resulting in the total loss of
system velocity*

coefficient of restitution
*number that serves as an index of
elasticity for colliding bodies*

"Giving" with the ball during a catch serves to lessen the magnitude of the impact force sustained by the catcher.

This relationship can also be expressed algebraically as the following:

$$-e = \frac{\text{relative velocity after impact}}{\text{relative velocity before impact}}$$

$$-e = \frac{v_1 - v_2}{u_1 - u_2}$$

In this formula, e is the coefficient of restitution, u_1 and u_2 are the velocities of the bodies just before impact, and v_1 and v_2 are the velocities of the bodies immediately after impact (Figure 12-12).

In tennis, the nature of the game depends on the type of impacts between ball and racket and between ball and court. All other conditions being equal, a tighter grip on the racket increases the apparent coefficient of restitution between ball and racket (16). When a pressurized tennis ball is punctured, there is a reduction in the coefficient of restitution between ball and surface of 20% (15). Other factors of influence are racket size, shape, balance, flexibility, string type and tension, and swing kinematics (30).

FIGURE 12-12

The differences in two balls' velocities before impact is proportional to the difference in their velocities after impact. The factor of proportionality is the coefficient of restitution.

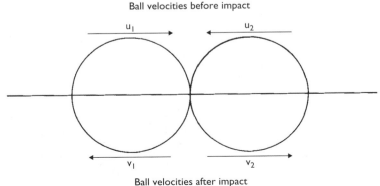

Ball velocities before impact

Ball velocities after impact

$$v_1 - v_2 = -e\,(u_1 - u_2)$$

SAMPLE PROBLEM 12.5

A basketball is dropped from a height of 2 m onto a gymnasium floor. If the coefficient of restitution between ball and floor is 0.9, how high will the ball bounce?

Known

$$h_d = 2 \text{ m}$$
$$e = 0.9$$

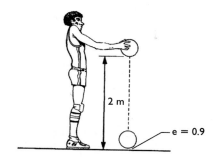

2 m

e = 0.9

Solution

$$e = \sqrt{\frac{h_b}{h_d}}$$

$$0.9 = \sqrt{\frac{h_b}{2 \text{ m}}}$$

$$0.81 = \frac{h_b}{2 \text{ m}}$$

$$\boxed{h_b = 1.6 \text{ m}}$$

The nature of impact between the bat and the ball is also an important factor in the sports of baseball and softball. The hitting surface of the bat is convex, in contrast to the surface of the tennis racquet, which deforms to a concave shape during ball contact. Consequently, hitting a baseball or softball in a direct, rather than a glancing, fashion is of paramount concern. Research has shown that aluminum baseball bats produce significantly higher batted ball speeds than do wood bats, which suggests that the coefficient of restitution between an aluminum bat and baseball is higher than that between a wood bat and baseball (13).

In the case of an impact between a moving body and a stationary one, Newton's law of impact can be simplified because the velocity of the stationary body remains zero. The coefficient of restitution between a ball and a flat, stationary surface onto which the ball is dropped may be approximated using the following formula:

$$e = \sqrt{\frac{h_b}{h_d}}$$

In this equation, e is the coefficient of restitution, h_d is the height from which the ball is dropped, and h_b is the height to which the ball bounces (see Sample Problem 12.5). The coefficient of restitution describes the interaction between two bodies during an impact; it is *not* descriptive of any single object or surface. Dropping a basketball, a golf ball, a racquetball, and a baseball onto several different surfaces demonstrates that some balls bounce higher on certain types of surfaces (Figure 12-13).

The coefficient of restitution is increased by increases in both impact velocity and temperature. In sports such as baseball and tennis, increases in both incoming ball velocity and bat or racket velocity increase the coefficient

Bounce heights of a basketball, golf ball, racquetball, and baseball all dropped onto the same surface from a height of 1 m.

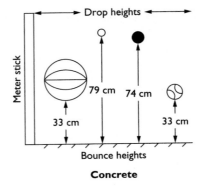

Concrete

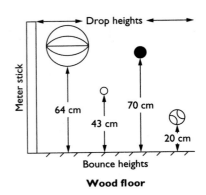

Wood floor

● *Increases in impact velocity and temperature increase the coefficient of restitution.*

of restitution between bat or racket and ball, and contribute to a livelier ball rebound from the striking instrument. In racquetball and squash, where the ball is constantly being deformed against the wall, the ball's thermal energy (temperature) is increased over the course of play. As ball temperature increases, its rebound from both racquet and wall becomes more lively.

WORK, POWER, AND ENERGY RELATIONSHIPS

Work

The word *work* is commonly used in a variety of contexts. A person can speak of "working out" in the weight room, doing "yard work," or "working hard" to prepare for an exam. However, from a mechanical standpoint, work is defined as force applied against a resistance, multiplied by the displacement of the resistance in the direction of the force:

work
in a mechanical context, force multiplied by the displacement of the resistance in the direction of the force

$$W = Fd$$

When a body is moved a given distance as the result of the action of an applied external force, the body has had work performed on it, with the quantity of work equal to the product of the magnitude of the applied force and the distance through which the body was moved. When a force is applied to a body but no net force results because of opposing forces such as friction or the body's own weight, no mechanical work has been done, since there has been no movement of the body.

When the muscles of the human body produce tension resulting in the motion of a body segment, the muscles perform work on the body segment, and the mechanical work performed may be characterized as either positive or negative work, according to the type of muscle action that predominates. When both the net muscle torque and the direction of angular motion at a joint are in the same direction, the work done by the muscles is said to be *positive*. Alternatively, when the net muscle torque and the direction of angular motion at a joint are in opposite directions, the work done by the muscles is considered to be *negative*. Although many movements of the human body involve co-contraction of agonist and antagonist muscle groups, when concentric contraction prevails the work is positive, and when eccentric contraction prevails the work is negative. During an activity such as running on a level surface, the net negative work done by the muscles is equal to the net positive work done by the muscles.

● *Mechanical work should not be confused with caloric expenditure.*

Performing positive mechanical work typically requires greater caloric expenditure than performing the same amount of negative mechanical work. However, no simple relationship between the caloric energy required for performing equal amounts of positive and negative mechanical work has been discovered, and the picture is complicated by the fact that agonist and other muscle groups often co-contract (1, 19).

Units of work are units of force multiplied by units of distance. In the metric system, the common unit of force (N) multiplied by a common unit of distance (m) is termed the *joule* (J).

$$1 \text{ J} = 1 \text{ Nm}$$

Power

Another term used in different contexts is power. In mechanics, power refers to the amount of mechanical work performed in a given time:

$$\text{power} = \frac{\text{work}}{\text{change in time}}$$

$$P = \frac{W}{\Delta t}$$

Using the relationships previously described, power can also be defined as the following:

$$\text{power} = \frac{\text{force} \times \text{distance}}{\text{change in time}}$$

$$P = \frac{Fd}{\Delta t}$$

Because velocity equals the directed distance divided by the change in time, the equation can also be expressed as the following:

$$P = Fv$$

Units of power are units of work divided by units of time. In the metric system, joules divided by seconds are termed *watts* (W):

$$1 \text{ W} = 1 \text{ J/s}$$

In activities such as throwing, jumping, and sprinting and in Olympic weight lifting, the athlete's ability to exert mechanical power or the combination of force and velocity is critical to successful performance. Peak power is strongly associated with maximum isometric strength (26). A problem involving mechanical work and power is shown in Sample Problem 12.6.

Energy

Energy is defined generally as the capacity to do work. Mechanical energy is therefore the capacity to do mechanical work. Units of mechanical energy are the same as units of mechanical work (joules, in the metric system). There are two forms of mechanical energy: kinetic energy and potential energy.

Kinetic energy (KE) is the energy of motion. A body possesses kinetic energy only when in motion. Formally, the kinetic energy of linear motion is defined as one-half of a body's mass multiplied by the square of its velocity:

$$KE = \tfrac{1}{2} mv^2$$

If a body is motionless (v = 0), its kinetic energy is also zero. Because velocity is squared in the expression for kinetic energy, increases in a

power
rate of work production, calculated as work divided by the time during which the work was done

●*The ability to produce mechanical power is critical for athletes competing in explosive track-and-field events.*

kinetic energy
energy of motion, calculated as $\tfrac{1}{2} mv^2$

potential energy
energy by virtue of a body's position or configuration, calculated as the product of weight and height

SAMPLE PROBLEM 12.6

A 580 N person runs up a flight of 30 stairs of riser (height) of 25 cm during a 15 s period. How much mechanical work is done? How much mechanical power is generated?

Known

$$wt\ (F) = 580\ N$$
$$h = 30 \times 25\ cm$$
$$t = 15\ s$$

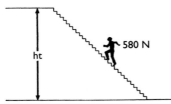

Solution

For mechanical work:

$$W = Fd$$
$$= (580\ N)\ (30 \times 0.25\ m)$$

$$\boxed{W = 4350\ J}$$

For mechanical power:

$$P = \frac{W}{t}$$
$$= \frac{4350\ J}{15\ s}$$

$$\boxed{P = 290\ watts}$$

body's velocity create dramatic increases in its kinetic energy. For example, a 2 kg ball rolling with a velocity of 1 m/s has a kinetic energy of 1 J:

$$KE = \tfrac{1}{2}\,mv^2$$
$$= (0.5)\,(2\ kg)\,(1\ m/s)^2$$
$$= (1\ kg)\,(1\ m^2/s^2)$$
$$= 1\ J$$

If the velocity of the ball is increased to 3 m/s, kinetic energy is significantly increased:

$$KE = \tfrac{1}{2}\,mv^2$$
$$= (0.5)\,(2\ kg)\,(3\ m/s)^2$$
$$= (1\ kg)\,(9\ m^2/s^2)$$
$$= 9\ J$$

The other major category of mechanical energy is potential energy (PE), which is the energy of position. More specifically, potential energy is a body's weight multiplied by its height above a reference surface:

$$PE = wt \cdot h$$
$$PE = ma_g h$$

In the second formula, m represents mass, a_g is the acceleration of gravity, and h is the body's height. The reference surface is usually the floor or the ground, but in special circumstances, it may be defined as another surface.

Because in biomechanical applications the weight of a body is typically fixed, changes in potential energy are usually based on changes in the body's height. For example, when a 50 kg bar is elevated to a height of 1 m, its potential energy at that point is 490.5 J:

$$PE = ma_gh$$
$$= (50 \text{ kg}) (9.81 \text{ m/s}^2) (1 \text{ m})$$
$$= 490.5 \text{ J}$$

Potential energy may also be thought of as stored energy. The term *potential* implies potential for conversion to kinetic energy. A special form of potential energy is called strain energy (SE), or elastic energy. Strain energy may be defined as follows:

$$SE = \tfrac{1}{2} kx^2$$

strain energy
capacity to do work by virtue of a deformed body's return to its original shape

In this formula, k is a spring constant, representing a material's relative stiffness or ability to store energy on deformation, and x is the distance over which the material is deformed. When an object is stretched, bent, or otherwise deformed, it stores this particular form of potential energy for later use. For example, when the muscles and tendons of the human body are stretched, they store strain energy that is released to increase the force of subsequent contraction, as discussed in Chapter 6. During an activity such as a maximal-effort throw, stored energy in stretched musculotendinous units can contribute significantly to the force and power generated and to the resulting velocity of the throw (20). Because they are more extensible than muscle, it is primarily the tendons that store and return elastic energy, with longer tendons performing this function more effectively than shorter ones (3). The Achilles tendon, in particular, stores and returns large amounts of mechanical energy, providing a large component of the mechanical work required for walking (24). Likewise, when the end of a diving board or a trampoline surface is depressed, strain energy is created. Subsequent conversion of the stored energy to kinetic energy enables the surface to return to its original shape and position. The poles used by vaulters store strain energy as they bend, and then release kinetic energy and increase the potential energy of the athlete as they straighten during the performance of the vault (2).

During the pole vault, the bent pole stores strain energy for subsequent release as kinetic energy and heat. Photo courtesy of Chu's Marters, University of Delaware.

Conservation of Mechanical Energy

Consider the changes that occur in the mechanical energy of a ball tossed vertically into the air (Figure 12-14). As the ball gains height, it also gains potential energy (ma_gh). However, since the ball is losing velocity with increasing height because of gravitational acceleration, it is also losing kinetic energy ($\tfrac{1}{2} mv^2$) At the apex of the ball's trajectory (the instant between rising and falling), its height and potential energy are at a maximum value, and its velocity and kinetic energy are zero. As the ball starts to fall, it progressively gains kinetic energy while losing potential energy.

The correlation between the kinetic and potential energies of the vertically tossed ball illustrates a concept that applies to all bodies when the only external force acting is gravity. The concept is known as the *law of conservation of mechanical energy,* which may be stated as follows:

When gravity is the only acting external force, a body's mechanical energy remains constant.

Since the mechanical energy a body possesses is the sum of its potential and kinetic energies, the relationship may also be expressed as the following:

$$(PE + KE) = C$$

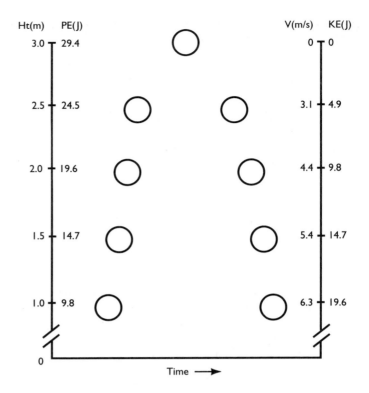

FIGURE 12-14

Height, velocity, potential energy, and kinetic energy change for a 1 kg ball tossed upward from a height of 1 m. Note that PE + KE = C (a constant) throughout the trajectory.

In this formula, C is a constant; that is, it is a number that remains constant throughout the period of time during which gravity is the only external force acting. Sample Problem 12.7 quantitatively illustrates this principle.

Principle of Work and Energy

There is a special relationship between the quantities of mechanical work and mechanical energy. This relationship is described as the *principle of work and energy,* which may be stated as follows:

> The work of a force is equal to the change in energy that it produces in the object acted on.

Algebraically, the principle may be represented thus:

$$W = \Delta KE + \Delta PE + \Delta TE$$

In this formula, KE is kinetic energy, PE is potential energy, and TE is thermal energy (heat). The algebraic statement of the principle of work and energy indicates that the change in the sum of the forms of energy produced by a force is quantitatively equal to the mechanical work done by that force. When a tennis ball is projected into the air by a ball-throwing machine, the mechanical work performed on the ball by the machine results in an increase in the ball's mechanical energy. Prior to projection, the ball's potential energy is based on its weight and height, and its kinetic energy is zero. The ball-throwing machine increases the ball's total mechanical energy by imparting kinetic energy to it. In this situation, the change in the ball's thermal energy is negligible. Sample Problem 12.8 provides a quantitative illustration of the principle of work and energy.

The work–energy relationship is also evident during movements of the human body. For example, the arches in runners' feet act as a mechanical spring to store, and subsequently return, strain energy as they cyclically deform and then regain their resting shapes. The ability of the arches to

• *When gravity is the only acting external force, any change in a body's potential energy necessitates a compensatory change in its kinetic energy.*

SAMPLE PROBLEM 12.7

A 2 kg ball is dropped from a height of 1.5 m. What is its velocity immediately before impact with the floor?

Known

$$m = 2kg$$
$$h = 1.5 \text{ m}$$

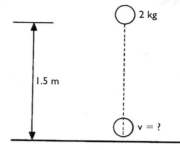

Solution

The principle of the conservation of mechanical energy may be used to solve the problem. The total energy possessed by the ball when it is held at a height of 1.5 m is its potential energy. Immediately before impact, the ball's height (and potential energy) may be assumed to be zero, and 100% of its energy at that point is kinetic.

Total (constant) mechanical energy possessed by the ball:

$$PE + KE = C$$
$$(\text{wt}) (h) + \tfrac{1}{2} mv^2 = C$$
$$(2 \text{ kg}) (9.81 \text{ m/s}^2) (1.5 \text{ m}) + 0 = C$$
$$29.43 \text{ J} = C$$

Velocity of the ball before impact:

$$PE + KE = 29.43 \text{ J}$$

$$(\text{wt}) (h) + \tfrac{1}{2} mv^2 = 29.43 \text{ J}$$

$$(2 \text{ kg}) (9.81 \text{ m/s}^2) (0) + \tfrac{1}{2} (2 \text{ kg}) v^2 = 29.43 \text{ J}$$

$$v^2 = 29.43 \text{ J/kg}$$

$$\boxed{v = 5.42 \text{ m/s}}$$

---

function as a spring reduces the amount of mechanical work that would otherwise be required during running.

Two-joint muscles in the human body also serve to transfer mechanical energy from one joint to another, thereby reducing the mechanical work required of the muscles crossing the second joint during a given movement. For example, during takeoff for a vertical jump, when the hip extensors work concentrically to produce hip extension, if the rectus femoris remains isometrically contracted, a secondary effect is an extensor torque exerted at the knee. In this case, it is the hip extensors that produce the knee extensor torque, since the length of the rectus femoris does not change.

It is important not to confuse the production of mechanical energy or mechanical work by the muscles of the human body with the consumption of chemical energy or caloric expenditure. Factors such as concentric versus eccentric muscular contractions, the transfer of energy between body segments, elastic storage and reuse of energy, and limitations in joint ranges of motion complicate direct quantitative calculation of the

SAMPLE PROBLEM 12.8

How much mechanical work is required to catch a 1.3 kg ball traveling at a velocity of 40 m/s?

Known

$$m = 1.3 \text{ kg}$$
$$v = 40 \text{ m/s}$$

v = 40 m/s
1.3 kg

Solution

The principle of work and energy may be used to calculate the mechanical work required to change the ball's kinetic energy to zero. Assume that the potential energy and thermal energy of the ball do not change:

$$W = \Delta Ke$$
$$= (\tfrac{1}{2} \, mv^2)_2 - (\tfrac{1}{2} \, mv^2)_1$$
$$= 0 - (\tfrac{1}{2})(1.3 \text{ kg})(40 \text{ m/s})^2$$

$$\boxed{W = 1040 \text{ J}}$$

relationship between mechanical and physiological energy estimates (28). Approximately 25% of the energy consumed by the muscles is converted into work, with the remainder changed to heat or used in the body's chemical processes.

Table 12-1 summarizes the formulas used in this chapter.

TABLE 12-1

Formula Summary

DESCRIPTION	FORMULA
Force = (mass)(acceleration)	$F = ma$
Friction = (coefficient of friction)(normal reaction force)	$F = \mu R$
Linear momentum = (mass)(velocity)	$M = mv$
Coefficient of restitution = $\dfrac{\text{relative velocity after impact}}{\text{relative velocity before impact}}$	$-e = \dfrac{v_1 - v_2}{u_1 - u_2}$
Work = (force) (displacement of resistance)	$W = Fd$
Power = $\dfrac{\text{work}}{\text{time}}$	$P = \dfrac{W}{t}$
Power = (force)(velocity)	$P = Fv$
Kinetic energy = $\tfrac{1}{2}$ (mass)(velocity squared)	$KE = \tfrac{1}{2} mv^2$
Potential energy = (weight)(height)	$PE = ma_g h$
Strain energy = $\tfrac{1}{2}$ (spring constant)(deformation squared)	$SE = \tfrac{1}{2} kx^2$
Potential energy + kinetic energy = constant	$PE + KE = C$
Work = change in energy	$W = \Delta KE + \Delta PE + \Delta TE$

SUMMARY

Linear kinetics is the study of the forces associated with linear motion. The interrelationships among many basic kinetic quantities are identified in the physical laws formulated by Sir Isaac Newton.

Friction is a force generated at the interface of two surfaces in contact when there is motion or a tendency for motion of one surface with respect to the other. The magnitudes of maximum static friction and kinetic friction are determined by the coefficient of friction between the two surfaces and by the normal reaction force pressing the two surfaces together. The direction of friction force always opposes the direction of motion or motion tendency.

Other factors that affect the behavior of two bodies in contact when a collision is involved are momentum and elasticity. Linear momentum is the product of an object's mass and its velocity. The total momentum present in a given system remains constant barring the action of external forces. Changes in momentum result from impulses, external forces acting over a time interval. The elasticity of an impact governs the amount of velocity present in the system following the impact. The relative elasticity of two impacting bodies is represented by the coefficient of restitution.

Mechanical work is the product of force and the distance through which the force acts. Mechanical power is the mechanical work done over a time interval. Mechanical energy has two major forms: kinetic and potential. When gravity is the only acting external force, the sum of the kinetic and potential energies possessed by a given body remains constant. Changes in a body's energy are equal to the mechanical work done by an external force.

INTRODUCTORY PROBLEMS

1. How much force must be applied by a kicker to give a stationary 2.5 kg ball an acceleration of 40 m/s^2? (Answer: 100 N)
2. A high jumper with a body weight of 712 N exerts a force of 3 kN against the ground during takeoff. How much force does the ground exert on the high jumper? (Answer: 3 kN)
3. What factors affect the magnitude of friction?
4. If μ_s between a basketball shoe and a court is 0.56, and the normal reaction force acting on the shoe is 350 N, how much horizontal force is required to cause the shoe to slide? (Answer: >196 N)
5. A football player pushes a 670 N blocking sled. The coefficient of static friction between sled and grass is 0.73, and the coefficient of kinetic friction between sled and grass is 0.68.
   a. How much force must the player exert to start the sled in motion?
   b. How much force is required to keep the sled in motion?
   c. Answer the same two questions with a 100 kg coach standing on the back of the sled.
   (Answers: a. >489.1 N; b. 455.6 N; c. >1205.2 N, 1122.7 N)
6. Lineman A has a mass of 100 kg and is traveling with a velocity of 4 m/s when he collides head-on with lineman B, who has a mass of 90 kg and is traveling at 4.5 m/s. If both players remain on their feet, what will happen? (Answer: lineman B will push lineman A backward with a velocity of 0.03 m/s)
7. Two skaters gliding on ice run into each other head-on. If the two skaters hold onto each other and continue to move as a unit after the collision, what will be their resultant velocity? Skater A has a velocity of 5 m/s and a mass of 65 kg. Skater B has a velocity of 6 m/s and a mass of 60 kg. (Answer: v = 0.28 m/s in the direction originally taken by skater B)

8. A ball dropped on a surface from a 2 m height bounces to a height of 0.98 m. What is the coefficient of restitution between ball and surface? (Answer: 0.7)
9. A set of 20 stairs, each of 20 cm height, is ascended by a 700 N man in a period of 1.25 s. Calculate the mechanical work, power, and change in potential energy during the ascent. (Answer: W = 2800 J, P = 2240 W, PE = 2800 J)
10. A pitched ball with a mass of 1 kg reaches a catcher's glove traveling at a velocity of 28 m/s.
    a. How much momentum does the ball have?
    b. How much impulse is required to stop the ball?
    c. If the ball is in contact with the catcher's glove for 0.5 s during the catch, how much average force is applied by the glove?
    (Answers: a. 28 kg · m/s; b. 28 N s; c. 56 N)

ADDITIONAL PROBLEMS

1. Identify three practical examples of each of Newton's laws of motion, and clearly explain how each example illustrates the law.
2. Select one sport or daily activity, and identify the ways in which the amount of friction present between surfaces in contact affects performance outcome.
3. A 2 kg block sitting on a horizontal surface is subjected to a horizontal force of 7.5 N. If the resulting acceleration of the block is 3 m/s^2, what is the magnitude of the friction force opposing the motion of the block? (Answer: 1.5 N)
4. Explain the interrelationships among mechanical work, power, and energy within the context of a specific human motor skill.
5. Explain in what ways mechanical work is and is not related to caloric expenditure. Include in your answer the distinction between positive and negative work and the influence of anthropometric factors.
6. A 108 cm, 0.73 kg golf club is swung for 0.5 s with a constant acceleration of 10 rad/s^2. What is the linear momentum of the club head when it impacts the ball? (Answer: 3.9 kg · m/s)
7. A 6.5 N ball is thrown with an initial velocity of 20 m/s at a 35° angle from a height of 1.5 m.
    a. What is the velocity of the ball if it is caught at a height of 1.5 m?
    b. If the ball is caught at a height of 1.5 m, how much mechanical work is required?
    (Answers: a. 20 m/s; b. 132.5 J)
8. A 50 kg person performs a maximum vertical jump with an initial velocity of 2 m/s.
    a. What is the performer's maximum kinetic energy during the jump?
    b. What is the performer's maximum potential energy during the jump?
    c. What is the performer's minimum kinetic energy during the jump?
    d. How much is the performer's center of mass elevated during the jump?
    (Answers: a. 100 J; b. 100 J; c. 0; d. 20 cm)
9. Using the principle of conservation of mechanical energy, calculate the maximum height achieved by a 7 N ball tossed vertically upward with an initial velocity of 10 m/s. (Answer: 5.1 m)
10. Select one of the following sport activities and speculate about the changes that take place between kinetic and potential forms of mechanical energy.
    a. A single leg support during running
    b. A tennis serve
    c. A pole vault performance
    d. A springboard dive

LABORATORY EXPERIENCES

1. At the *Basic Biomechanics* Online Learning Center (www.mhhe.com/hall6e), go to Student Resources, Chapter 12, Lab Manual, Lab 1, then view Newton's Laws Animation 1 and Animation 2 and Energy Animation 1. Identify the principles that are illustrated, and write explanations of what is demonstrated.

Principle in Newton's Laws Animation 1: _____

Explanation: _____

_____

Principle in Newton's Laws Animation 2: _____

Explanation: _____

_____

Principle in Energy Animation 1: _____

Explanation: _____

_____

2. Following the instructions above, go to the online lab manual and click on Collisions in One Dimension. Play this simulation with all different possible combinations of variable settings. Identify the principle that is illustrated, and write an explanation of what is demonstrated.

Principle: _____

Explanation: _____

_____

_____

_____

3. Have each member of your lab group remove one shoe. Use a spring scale to determine the magnitude of maximum static friction for each shoe on two different surfaces. (Depending on the sensitivity of the spring scale, you may need to load the shoe with weight.) Present your results in a table, and write a paragraph explaining the results.

Shoe	Shoe Weight	Applied Force	μ_s
_____	_____	_____	_____
_____	_____	_____	_____
_____	_____	_____	_____
_____	_____	_____	_____
_____	_____	_____	_____

Explanation: _____

_____

_____

_____

4. Drop five different balls from a height of 2 m on two different surfaces, and carefully observe and record the bounce heights. Calculate the coefficient of restitution for each ball on each surface, and write a paragraph explaining your results.

Ball	Drop Height	Bounce Height	e

5. Using a stopwatch, time each member of your lab group running up a flight of stairs. Use a ruler to measure the height of one stair, then multiply by the number of stairs to calculate the total change in height. Calculate work, power, and change in potential energy for each group member.

Group Member	Wt (N)	Mass (kg)	Time (s)	Av. Vel. (m/s)	Ht Δ (m)	Work (J)	Power (W)	ΔPE (J)

REFERENCES

1. Arampatzis A, Knicker A, Metzler V, and Brüggeman G: Mechanical power in running: a comparison of different approaches, *J Biomech* 33:457, 2000.

2. Arampatzis A, Schade F, and Brüggemann G-P: Effect of the pole-human body interaction on pole vaulting performance, *J Biomech* 37:1353, 2004.

3. Biewener AA and Roberts TJ: Muscle and tendon contributions to force, work, and elastic energy savings: a comparative perspective, *Exerc Sport Sci Rev* 28:99, 2000.

4. Chang YH and Kram R: Metabolic cost of generating horizontal forces during human running, *J Appl Physiol* 86:1657, 1999.

5. Dapena J: Biomechanics of elite high jumpers. In Terauds J et al, eds: *Sports biomechanics,* Del Mar, CA, 1984, Academic Publishers.

6. Decker MJ, Torry MR, Wyland DJ, Sterett WI, and Steadman J: Gender differences in lower extremity kinematics, kinetics and energy absorption during landing, *Clin Biomech* 18:662, 2003.

7. Derrick TR: The effects of knee contact angle on impact forces and accelerations, *Med Sci Sports Exerc* 36:832, 2004.

8. Derrick TR and Mercer JA: Ground/foot impacts: measurement, attenuation, and consequences, *Med Sci Sports Exerc* 36:830, 2004.

9. Dixon SJ, Collop AC, and Batt ME: Surface effects on ground reaction forces and lower extremity kinematics in running, *Med Sci Sports Exerc* 32:1919, 2000.

10. Federolf PA, Mills R, Nigg B: Ice friction of flared ice hockey skate blades, *J Sports Sci* 26:1201, 2008.

11. Gerlach KE, White SC, Burton HW, Dorn JM, Leddy JJ, and Horvath PJ: Kinetic changes with fatigue and relationship to injury in female runners, *Med Sci Sports Exerc* 37:657, 2005.

12. Gottschall JS and Kram R: Ground reaction forces during downhill and uphill running, *J Biomech* 38:445, 2005.

13. Greenwald RM, Penna LH, and Crisco JJ: Differences in batted ball speed with wood and aluminum baseball bats: a batting cage study, *J Appl Biomech* 17:241, 2001.

14. Guido JA Jr, Werner SL, and Meister K: Lower-extremity ground reaction forces in youth windmill softball pitchers, *J Strength Cond Res* 23:1873, 2009.

15. Haake SJ, Carre MJ, and Goodwill SR: The dynamic impact characteristics of tennis balls with tennis rackets, *J Sports Sci* 21:839, 2003.

16. Hatze H: The relationship between the coefficient of restitution and energy losses in tennis rackets, *J Appl Biomech* 9:124, 1993.

17. Hume PA, Keogh J, and Redi D: The role of biomechanics in maximizing distance and accuracy of golf shots, *Sports Med* 35:429, 2005.

18. Kiefer J: Bowling: the great oil debate. In Schrier EW and Allman WF, eds: *Newton at the bat,* New York, 1984, Charles Scribner's Sons.

19. Neptune RR and van den Bogert AJ: Standard mechanical energy analyses do not correlate with muscle work in cycling, *J Biomech* 31:239, 1998.

20. Nigg BM: The role of impact forces and foot pronation: a new paradigm, *Clin J Sport Med* 11:2, 2001.

21. Nigg BM and Wakeling JM: Impact forces and muscle tuning: a new paradigm, *Exerc Sport Sci Rev* 29:37, 2001.

22. Newton RU et al: Influence of load and stretch shortening cycle on the kinematics, kinetics and muscle activation that occurs during explosive upper-body movements, *Eur J Appl Physiol* 75:333, 1997.

23. Saikko VO: A three-axis hip joint simulator for wear and friction studies on total hip prostheses, *Proc Inst Mech Eng* [H] 210:175, 1996.

24. Sawicki GS, Lewis CL, Ferris DP: It pays to have a spring in your step, *Exerc Sport Sci Rev* 37:130, 2009.

25. Slawinski J, Bonnefoy A, Levêque JM, Ontanon G, Riquet A, Dumas R, and Chèze L: Kinematic and kinetic comparisons of elite and well-trained sprinters during sprint start, *J Strength Cond Res* 24:896, 2010.

26. Stone MH, Sanborn K, O'Bryant HS, Hartman M, Stone ME, Proulx C, Ward B, and Hruby J: Maximum strength-power-oerformance relationships in collegiate throwers, *J Strength Cond Res* 17:739, 2003.

27. Tillman MD, Criss RM, Brunt D, and Hass CJ: Landing constraints influence ground reaction forces and lower extremity EMG in female volleyball players, *J Appl Biomech* 20:38, 2004.

28. Van de Walle P, Desloovere K, Truijen S, Gosselink R, Aerts P, and Hallemans A: Age-related changes in mechanical and metabolic energy during typical gait, *Gait Posture* 31:495, 2010.

29. Voloshin A: The influence of walking speed on dynamic loading on the human musculoskeletal system, *Med Sci Sports Exerc* 32:1156, 2000.

30. Wu SK, Gross MT, Prentice WE, and Yu B: Comparison of ball-and-racquet impact force between two tennis backhand stroke techniques, *J Orthop Sports Phys Ther* 31:247, 2001.

ANNOTATED READINGS

Brughelli M and Cronin J: A review of research on the mechanical stiffness in running and jumping: methodology and implications, *Scand J Med Sci Sports* 18:417, 2008.
Discusses the scientific literature related to rate of force development, elastic energy storage and utilization, and athletic performance.

Marín PJ and Rhea MR: Effects of vibration training on muscle power: a meta-analysis, *J Strength Cond Res* 24:871, 2010.
Reviews the scientific literature on fostering increases in muscular power through vibration training.

Moxnes JF and Hausken K: A dynamic model of Nordic diagonal stride skiing, with a literature review of cross country skiing, *Comput Methods Biomech Biomed Engin* 12:531, 2009.
Reviews the scientific literature on Nordic diagonal stride skiing, including the relationships of static and dynamic friction to skier weight, velocity, kicking force angle, and terrain.

Stefani RT: The relative power output and relative lean body mass of World and Olympic male and female champions with implications for gender equity, *J Sports Sci* 24:1329, 2006.
Discusses gender-related differential performance of female and male Olympic and World champions relative to power output applied to the environment.

RELATED WEBSITES

Advanced Medical Technology, Inc.
http://www.amti.com/
Provides information on the AMTI force platforms with reference to ground reaction forces in gait analysis, balance and posture, and other topics.

Answers.com: Mechanical Work
http://www.answers.com/topic/mechanical-work
Lists definitions and examples of mechanical work.

Fear of Physics: What Is Friction?
http://www.fearofphysics.com/Friction/frintro.html
Includes text and illustrations, plus a link to a simulation.

Kistler
http://www.kistler.com
Describes a series of force platforms for measuring ground reaction forces.

Scienceworld: Friction
http://scienceworld.wolfram.com/physics/Friction.html
Includes definitions and examples of friction coefficients.

The Exploratorium's Science of Hockey
http://www.exploratorium.edu/hockey/
Explains scientific concepts related to hockey, including the friction between ice and skate and the mechanics of skating.

The Physics Classroom: Mechanical Energy
http://www.physicsclassroom.com/Class/energy/U5L1d.html
Includes definitions, illustrations, and questions with answers related to work and energy.

KEY TERMS

coefficient of friction	number that serves as an index of the interaction between two surfaces in contact
coefficient of restitution	number that serves as an index of elasticity for colliding bodies
friction	force acting at the area of contact between two surfaces in the direction opposite that of motion or motion tendency
impact	collision characterized by the exchange of a large force during a small time interval
impulse	product of a force and the time interval over which the force acts
kinetic energy	energy of motion calculated as $\frac{1}{2}\,mv^2$
kinetic friction	constant magnitude friction generated between two surfaces in contact during motion
linear momentum	quantity of motion, measured as the product of a body's mass and its velocity
maximum static friction	maximum amount of friction that can be generated between two static surfaces
normal reaction force	force acting perpendicular to two surfaces in contact
perfectly elastic impact	impact during which the velocity of the system is conserved
perfectly plastic impact	impact resulting in the total loss of system velocity
potential energy	energy by virtue of a body's position or configuration, calculated as the product of weight and height
power	rate of work production, calculated as work divided by the time during which the work was done
strain energy	capacity to do work by virtue of a deformed body's return to its original shape
work	in a mechanical context, force multiplied by the displacement of the resistance in the direction of the force

Equilibrium and Human Movement

After completing this chapter, you will be able to:

Define torque, quantify resultant torques and identify the factors that affect resultant joint torques.

Identify the mechanical advantages associated with the different classes of levers and explain the concept of leverage within the human body.

Solve basic quantitative problems using the equations of static equilibrium.

Define center of gravity and explain the significance of center of gravity location in the human body.

Explain how mechanical factors affect a body's stability.

ONLINE LEARNING CENTER RESOURCES

www.mhhe.com/hall6e

Log on to our Online Learning Center (OLC) for access to these additional resources:

- Online Lab Manual
- Flashcards with definitions of chapter key terms
- Chapter objectives
- Chapter lecture PowerPoint presentation
- Self-scoring chapter quiz
- Additional chapter resources
- Web links for study and exploration of chapter-related topics

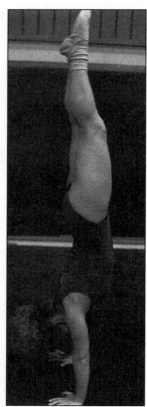

Many athletic skills require mechanical stability.

torque
the rotary effect of a force about an axis of rotation, measured as the product of the force and the perpendicular distance between the force's line of action and the axis

moment arm
shortest (perpendicular) distance between a force's line of action and an axis of rotation

W hy do long jumpers and high jumpers lower their centers of gravity before takeoff? What mechanical factors enable a wheelchair to remain stationary on a graded ramp or a sumo wrestler to resist the attack of his opponent? A body's mechanical stability is based on its resistance to both linear and angular motion. This chapter introduces the kinetics of angular motion, along with the factors that affect mechanical stability.

EQUILIBRIUM

Torque

As discussed in Chapter 3, the rotary effect created by an applied force is known as torque, or *moment of force*. Torque, which may be thought of as *rotary force*, is the angular equivalent of linear force. Algebraically, torque is the product of force and the force's moment arm, or the perpendicular distance from the force's line of action to the axis of rotation:

$$\mathbf{T = Fd_\perp}$$

Thus, both the magnitude of a force and the length of its moment arm equally affect the amount of torque generated (Figure 13-1). Moment arm is also sometimes referred to as *force arm* or *lever arm*.

As may be observed in Figure 13-2, the moment arm is the shortest distance between the force's line of action and the axis of rotation. A force directed through an axis of rotation produces no torque, because the force's moment arm is zero.

Within the human body, the moment arm for a muscle with respect to a joint center is the perpendicular distance between the muscle's line of action and the joint center (Figure 13-3). As a joint moves through a range of motion, there are changes in the moment arms of the muscles crossing the joint. For any given muscle, the moment arm is largest when the angle of pull on the bone is closest to 90°. At the elbow, as the angle of pull moves away from 90° in either direction, the moment arm for the elbow flexors is progressively diminished. Since torque is the product of moment arm and muscle force, changes in moment arm directly affect the joint torque that a muscle generates. For a muscle to generate a constant joint torque during an exercise, it must produce more force as its moment arm decreases.

FIGURE 13-1

Which position of force application is best for opening the swinging door? Experience should verify that position **C** is best.

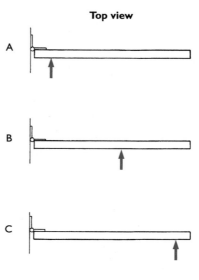

Top view

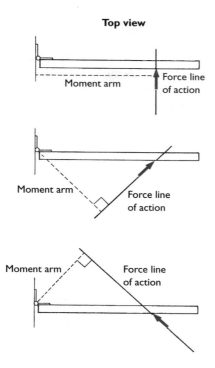

The moment arm of a force is the perpendicular distance from the force's line of action to the axis of rotation (the door hinge).

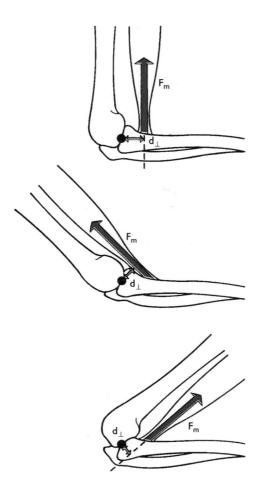

A muscle's moment arm is maximal at a 90° angle of pull. As the line of pull moves away from 90° in either direction, the moment arm becomes progressively smaller.

●It is easiest to initiate rotation when force is applied perpendicularly and as far away as possible from the axis of rotation.

In the sport of rowing, where adjacent crew members traditionally row on opposite sides of the hull, the moment arm between the force applied by the oar and the stern of the boat is a factor affecting performance (Figure 13-4). With the traditional arrangement, the rowers on one side of the boat are positioned farther from the stern than their counterparts on the other side, thus causing a net torque and a resulting lateral oscillation about the stern during rowing (17). The Italian rig eliminates this problem by positioning rowers so that no net torque is produced, assuming that the force produced by each rower with each stroke is nearly the same (Figure 13-4). Italian and German rowers have similarly developed alternative positionings for the eight-member crew (Figure 13-5).

FIGURE 13-4

A. This crew arrangement creates a net torque about the stern of the boat because the sum of the top side oar moment arms $(d_1 + d_2)$ is less than the sum of the bottom side moment arms $(d_3 + d_4)$. **B.** This arrangement eliminates the problem, assuming that all rowers stroke simultaneously and produce equal force, because $(d_1 + d_2) = (d_3 + d_4)$.

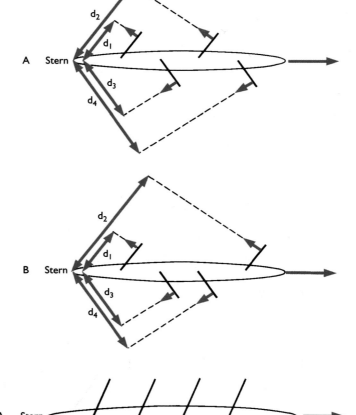

FIGURE 13-5

The Italians and Germans have used alternative positionings for eight-member crews. The torques produced by the oar forces with respect to the stern are balanced in arrangements **B** and **C**, but not in the traditional arrangement shown in **A**.

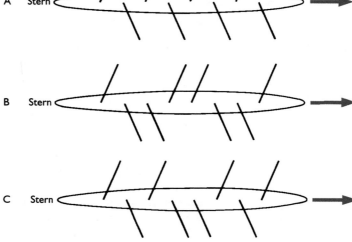

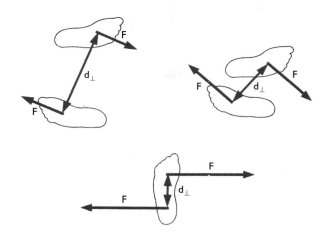

FIGURE 13-6

The wider a dancer's stance, the greater the moment arm for the force couple generated by the feet when a turn is executed. When rotation is initiated from a single-foot stance, the moment arm becomes the distance between the support points of the foot.

Another example of the significance of moment arm length is provided by a dancer's foot placement during preparation for execution of a total body rotation around the vertical axis. When a dancer initiates a turn, the torque producing the turn is provided by equal and oppositely directed forces exerted by the feet against the floor. A pair of equal and opposite forces is known as a force couple. Because the forces in a couple are positioned on opposite sides of the axis of rotation, they produce torque in the same direction. The torque generated by a couple is therefore the sum of the products of each force and its moment arm. Turning from fifth position, with a small distance between the feet, requires greater force production by a dancer than turning at the same rate from fourth position, in which the moment arms of the forces in the couple are longer (Figure 13-6). Significantly more force is required when the torque is generated by a single support foot, for which the moment arm is reduced to the distance between the metatarsals and the calcaneus.

Torque is a vector quantity and is therefore characterized by both magnitude and direction. The magnitude of the torque created by a given force is equal to $Fd_\perp$, and the direction of a torque may be described as clockwise or counterclockwise. As discussed in Chapter 11, the counterclockwise direction is conventionally referred to as the positive (+) direction, and the clockwise direction is regarded as negative (−). The magnitudes of two or more torques acting at a given axis of rotation can be added using the rules of vector composition (see Sample Problem 13.1).

couple
pair of equal, oppositely directed forces that act on opposite sides of an axis of rotation to produce torque

Resultant Joint Torques

The concept of torque is important in the study of human movement, because torque produces movement of the body segments. As discussed in Chapter 6, when a muscle crossing a joint develops tension, it produces a force pulling on the bone to which it attaches, thereby creating torque at the joint the muscle crosses.

Much human movement involves simultaneous tension development in agonist and antagonist muscle groups. The tension in the antagonists controls the velocity of the movement and enhances the stability of the joint at which the movement is occurring. Since antagonist tension development creates torque in the direction opposite that of the torque produced by the agonist, the resulting movement at the joint is a function of the net torque. When net muscle torque and joint movement occur in the same direction, the torque is termed *concentric,* and muscle torque in the direction opposite joint motion is considered to be *eccentric.* Although

• *The product of muscle tension and muscle moment arm produces a torque at the joint crossed by the muscle.*

SAMPLE PROBLEM 13.1

Two children sit on opposite sides of a playground seesaw. If Joey, weighing 200 N, is 1.5 m from the seesaw's axis of rotation, and Susie, weighing 190 N, is 1.6 m from the axis of rotation, which end of the seesaw will drop?

Known

Joey: $wt(F_J) = 200$ N
$d_{\perp J} = 1.5$ m
Susie: $wt(F_S) = 190$ N
$d_{\perp S} = 1.6$ m

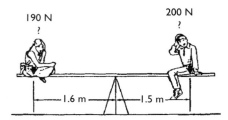

Solution

The seesaw will rotate in the direction of the resultant torque at its axis of rotation. To find the resultant torque, the torques created by both children are summed according to the rules of vector composition. The torque produced by Susie's body weight is in a counterclockwise (positive) direction, and the torque produced by Joey's body weight is in a clockwise (negative) direction.

$$T_a = (F_S)(d_{\perp S}) - (F_J)(d_{\perp J})$$
$$= (190 \text{ N})(1.6 \text{ m}) - (200 \text{ N})(1.5 \text{ m})$$
$$= 304 \text{ N-m} - 300 \text{ N-m}$$
$$= 4 \text{ N-m}$$

The resultant torque is in a positive direction, and Susie's end of the seesaw will fall.

these terms are generally useful descriptors in analysis of muscular function, their application is complicated when two-joint or multijoint muscles are considered, since there may be concentric torque at one joint and eccentric torque at a second joint crossed by the same muscle.

Because directly measuring the forces produced by muscles during the execution of most movement skills is not practical, measurements or estimates of resultant joint torques (joint moments) are often studied to investigate the patterns of muscle contributions. A number of factors, including the weight of the body segments, the motion of the body segments, and the action of external forces, may contribute to net joint torques. Joint torque profiles are typically matched to the requirements of the task at hand and provide at least general estimates of muscle group contribution levels.

To better understand muscle function during running, a number of investigators have studied resultant joint torques at the hip, knee, and ankle throughout the running stride. Figure 13-7 displays representative resultant joint torques and angular velocities for the hip, knee, and ankle during a running stride, as calculated from film and force platform data. In Figure 13-7, when the resultant joint torque curve and the angular velocity curves are on the same side of the zero line, the torque is concentric; the torque is eccentric when the reverse is true. As may be observed from Figure 13-7, both concentric and eccentric torques are present at the lower-extremity joints during running. Interestingly, it has been shown

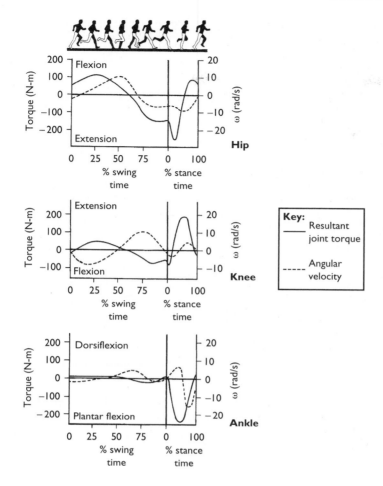

FIGURE 13-7

Representative joint torques
and angular velocity curves
for the lower extremity during
running. Modified from Putnam
CA and Kozey JW: Substantive
issues in running. In Vaughan CL,
ed: *Biomechanics of sport,* Boca
Raton, FL, 1989, CRC Press.

that use of running shoes increases joint torques at the hip, knee, and ankle as compared to running barefoot (6).

Lower-extremity joint torques during cycling at a given power are affected by pedaling rate, seat height, length of the pedal crank arm, and distance from the pedal spindle to the ankle joint. Figure 13-8 shows the changes in average resultant torque at the hip, knee, and ankle joints with changes in pedaling rate at a constant power.

It is widely assumed that the muscular force (and subsequently, joint torque) requirements of resistance exercise increase as the amount of resistance increases. Obese children, who carry extra weight as compared to normal weight children, generate significantly higher torques at the hip, knee, and ankle with every step (16). During resistance exercises, another factor that affects joint torques is the kinematics of the movement. It has been shown, for example, that back squats generate significantly higher extensor torques at the knee as compared to the front squat (2).

Another factor influencing joint torques during exercise is movement speed. When other factors remain constant, increased movement speed is associated with increased resultant joint torques during exercises such as the squat (12). However, increased movement speed during weight training is generally undesirable, because increased speed increases not only the muscle tension required but also the likelihood of incorrect technique and subsequent injury. Acceleration of the load early in the performance of a resistance exercise also generates momentum, which means that the involved muscles need not work so hard throughout the

The torques required at the hip, knee, and ankle during cycling at a given power are influenced by body position and cycle dimensions. Photo courtesy of Steve Allen/Brand X Pictures.

Absolute average joint torques for the hip, knee, and ankle versus pedaling rate during cycling. Modified from Redfield R and Hull ML: On the relation between joint moments and pedalling at constant power in bicycling, J Biomech 19:317, 1986.

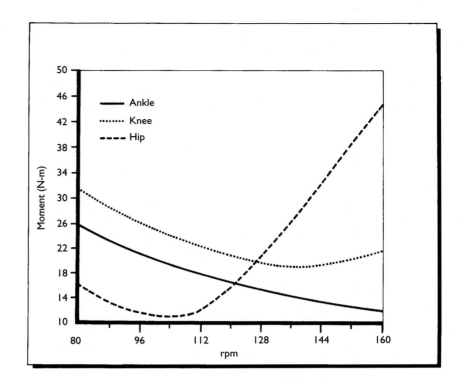

range of motion as would otherwise be the case. For these reasons, it is both safer and more effective to perform exercises at slow, controlled movement speeds.

Levers

When muscles develop tension, pulling on bones to support or move the resistance created by the weight of the body segment(s) and possibly the weight of an added load, the muscle and bone are functioning mechanically as a lever. A lever is a rigid bar that rotates about an axis, or fulcrum. Force applied to the lever moves a resistance. In the human body, the bone acts as the rigid bar; the joint is the axis, or fulcrum; and the muscles apply force. The three relative arrangements of the applied force, resistance, and axis of rotation for a lever are shown in Figure 13-9.

lever
a simple machine consisting of a relatively rigid, barlike body that may be made to rotate about an axis

fulcrum
the point of support, or axis, about which a lever may be made to rotate

Relative locations of the applied force, the resistance, and the fulcrum, or axis of rotation, determine lever classifications.

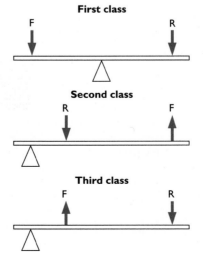

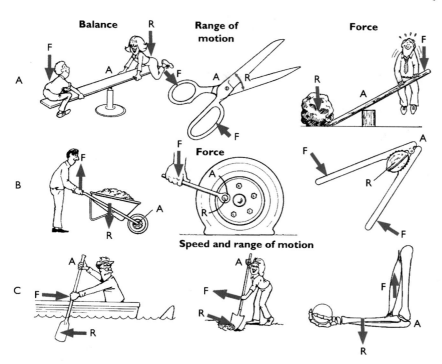

FIGURE 13-10

A. First-class levers.
B. Second-class levers.
C. Third-class levers. Note that the paddle and shovel function as third-class levers only when the top hand does not apply force but serves as a fixed axis of rotation.

With a first-class lever, the applied force and resistance are located on opposite sides of the axis. The playground seesaw is an example of a first-class lever, as are a number of commonly used tools, including scissors, pliers, and crowbars (Figure 13-10). Within the human body, the simultaneous action of agonist and antagonist muscle groups on opposite sides of a joint axis is analogous to the functioning of a first-class lever, with the agonists providing the applied force and the antagonists supplying a resistance force. With a first-class lever, the applied force and resistance may be at equal distances from the axis, or one may be farther away from the axis than the other.

In a second-class lever, the applied force and the resistance are on the same side of the axis, with the resistance closer to the axis. A wheelbarrow, a lug nut wrench, and a nutcracker are examples of second-class levers, although there are no completely analogous examples in the human body (Figure 13-10).

With a third-class lever, the force and the resistance are on the same side of the axis, but the applied force is closer to the axis. A canoe paddle and a shovel can serve as third-class levers (Figure 13-10). Most muscle–bone lever systems of the human body are also of the third class for concentric contractions, with the muscle supplying the applied force and attaching to the bone at a short distance from the joint center compared to the distance at which the resistance supplied by the weight of the body segment or that of a more distal body segment acts (Figure 13-11). As shown in Figure 13-12, however, during eccentric contractions, it is the muscle that supplies the resistance against the applied external force. During eccentric contractions, muscle and bone function as a second-class lever.

A lever system can serve one of two purposes (Figure 13-13). Whenever the moment arm of the applied force is greater than the moment arm of the resistance, the magnitude of the applied force needed to move a given resistance is less than the magnitude of the resistance. Whenever the resistance arm is longer than the force arm, the resistance may be moved through a relatively large distance. The mechanical effectiveness

first-class lever
lever positioned with the applied force and the resistance on opposite sides of the axis of rotation

second-class lever
lever positioned with the resistance between the applied force and the fulcrum

third-class lever
lever positioned with the applied force between the fulcrum and the resistance

FIGURE 13-11

Most levers within the human body are third class. **A.** The biceps at the elbow. **B.** The patellar tendon at the knee. **C.** The medial deltoid at the shoulder.

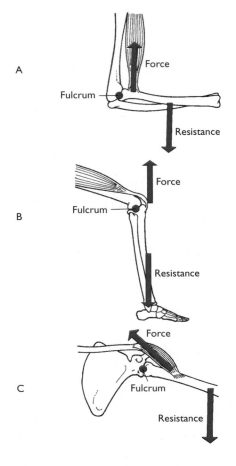

FIGURE 13-12

The elbow flexors contract eccentrically to apply a braking resistance and to control movement speed during the down phase of a curl exercise. In this case, the muscle–bone lever system is second class.

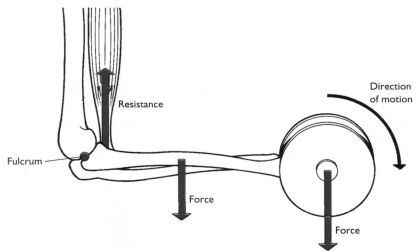

mechanical advantage

ratio of force arm to resistance arm for a given lever

●*The moment arm of an applied force can also be referred to as the* force arm, *and the moment arm of a resistance can be referred to as the* resistance arm.

of a lever for moving a resistance may be expressed quantitatively as its mechanical advantage, which is the ratio of the moment arm of the force to the moment arm of the resistance:

$$\text{mechanical advantage} = \frac{\text{moment arm (force)}}{\text{moment arm (resistance)}}$$

Whenever the moment arm of the force is longer than the moment arm of the resistance, the mechanical advantage ratio reduces to a number that is greater than one, and the magnitude of the applied force required

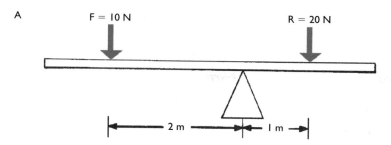

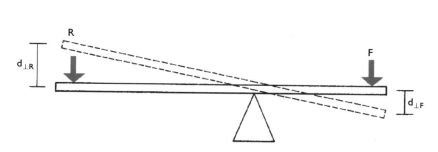

FIGURE 13-13

A. A force can balance a larger resistance when its moment arm is longer than the moment arm of the resistance. **B.** A force can move a resistance through a larger range of motion when the moment arm of the force is shorter than the moment arm of the resistance.

to move the resistance is less than the magnitude of the resistance. The ability to move a resistance with a force that is smaller than the resistance offers a clear advantage when a heavy load must be moved. As shown in Figure 13-10, a wheelbarrow combines second-class leverage with rolling friction to facilitate transporting a load. When removing a lug nut from an automobile wheel, it is helpful to use as long an extension as is practical on the wrench to increase mechanical advantage.

Alternatively, when the mechanical advantage ratio is less than one, a force that is larger than the resistance must be applied to cause motion of the lever. Although this arrangement is less effective in the sense that more force is required, a small movement of the lever at the point of force application moves the resistance through a larger range of motion (see Figure 13-13).

Anatomical Levers

Skilled athletes in many sports intentionally maximize the length of the effective moment arm for force application to maximize the effect of the torque produced by muscles about a joint. During execution of the serve in tennis, expert players not only strike the ball with the arm fully extended but also vigorously rotate the body in the transverse plane, making the spine the axis of rotation and maximizing the length of the anatomical lever delivering the force. The same strategy is employed by accomplished baseball pitchers. As discussed in Chapter 11, the longer the radius of rotation, the greater the linear velocity of the racket head or hand delivering the pitch, and the greater the resultant velocity of the struck or thrown ball.

In the human body, most muscle–bone lever systems are of the third class, and therefore have a mechanical advantage of less than one. Although this arrangement promotes range of motion and angular speed of the body segments, the muscle forces generated must be in excess of the resistance force or forces if positive mechanical work is to be done.

The angle at which a muscle pulls on a bone also affects the mechanical effectiveness of the muscle–bone lever system. The force of muscular tension is resolved into two force components—one perpendicular to

Skilled pitchers often maximize the length of the moment arm between the ball hand and the total-body axis of rotation during the delivery of a pitch to maximize the effect of the torque produced by the muscles. Photo courtesy of Getty Images.

FIGURE 13-14

Muscle force can be resolved into rotary and dislocating components.

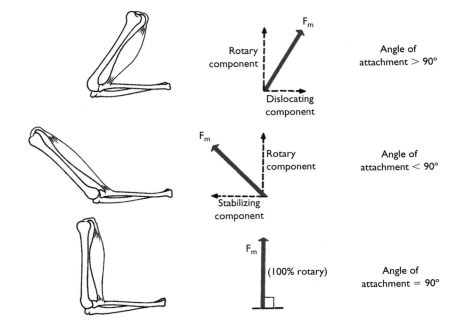

the attached bone and one parallel to the bone (Figure 13-14). As discussed in Chapter 6, only the component of muscle force acting perpendicular to the bone—the rotary component—actually causes the bone to rotate about the joint center. The component of muscle force directed parallel to the bone pulls the bone either away from the joint center (a dislocating component) or toward the joint center (a stabilizing component), depending on whether the angle between the bone and the attached muscle is less than or greater than 90°. The angle of maximum mechanical advantage for any muscle is the angle at which the most rotary force can be produced. At a joint such as the elbow, the relative angle present at the joint is close to the angles of attachment of the elbow flexors. The maximum mechanical advantages for the brachialis, biceps, and brachioradialis occur between angles at the elbow of approximately 75° and 90° (Figure 13-15).

As joint angle and mechanical advantage change, muscle length also changes. Alterations in the lengths of the elbow flexors associated with changes in angle at the elbow are shown in Figure 13-16. These changes affect the amount of tension a muscle can generate, as discussed in Chapter 6. The angle at the elbow at which maximum flexion torque is produced is approximately 80°, with torque capability progressively diminishing as the angle at the elbow changes in either direction (18).

The varying mechanical effectiveness of muscle groups for producing joint rotation with changes in joint angle is the underlying basis for the design of modern variable-resistance strength-training devices. These machines are designed to match the changing torque-generating capability of a muscle group throughout the range of motion at a joint. Machines manufactured by Universal (the Centurion) and Nautilus are examples. Although these machines offer more relative resistance through the extremes of joint range of motion than free weights, the resistance patterns incorporated are not an exact match for average human strength curves.

Isokinetic machines represent another approach to matching torque-generating capability with resistance. These devices are generally designed so that an individual applies force to a lever arm that rotates at a constant

●The force-generating capability of a muscle is affected by muscle length, cross-sectional area, moment arm, angle of attachment, shortening velocity, and state of training.

●Variable-resistance training devices are designed to match the resistance offered to the torque-generating capability of the muscle group as it varies throughout a range of motion.

●The term isokinetic implies constant angular velocity at a joint when applied to exercise machinery.

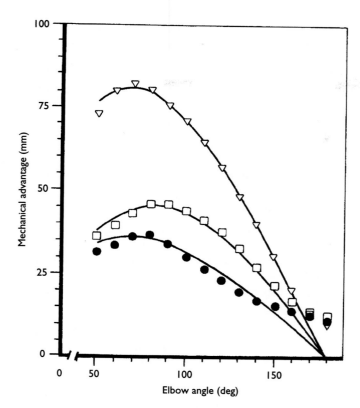

FIGURE 13-15

Mechanical advantage of the brachialis (●), biceps (□), and brachioradialis (▽) as a function of elbow angle. Modified from van Zuylen EJ, van Zelzen A, and van der Gon JJD: A biomechanical model for flexion torques of human arm muscles as a function of elbow angle, J Biomech 21:183, 1988.

angular velocity. If the joint center is aligned with the center of rotation of the lever arm, the body segment rotates with the same (constant) angular velocity of the lever arm. If volitional torque production by the involved muscle group is maximum throughout the range of motion, a maximum matched resistance theoretically is achieved. However, when force is initially

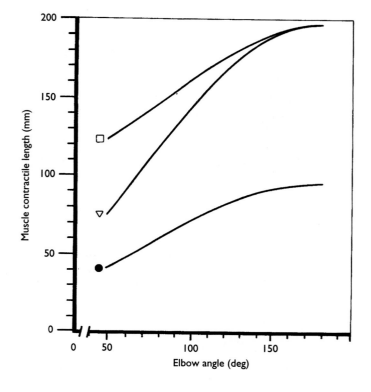

FIGURE 13-16

Contractile length of the brachialis (●), biceps (□), and brachioradialis (▽) as a function of elbow angle. Modified from van Zuylen EJ, van Zelzen A, and van der Gon JJD: A biomechanical model for flexion torques of human arm muscles as a function of elbow angle, J Biomech 21:183, 1988.

A cam in a variable-resistance training machine is designed to match the resistance offered to the mechanical advantage of the muscle.

static equilibrium
a motionless state characterized by $\Sigma F_v = 0$, $\Sigma F_h = 0$, and $\Sigma T = 0$

• *The presence of a net force acting on a body results in acceleration of the body.*

applied to the lever arm of isokinetic machines, acceleration occurs, and the angular velocity of the arm fluctuates until the set rotational speed is reached. Because optimal use of isokinetic resistance machines requires that the user be focused on exerting maximal effort throughout the range of motion, some individuals prefer other modes of resistance training.

Equations of Static Equilibrium

Equilibrium is a state characterized by balanced forces and torques (no net forces and torques). In keeping with Newton's first law, a body in equilibrium is either motionless or moving with a constant velocity. Whenever a body is completely motionless, it is in static equilibrium. Three conditions must be met for a body to be in a state of static equilibrium:

1. The sum of all vertical forces (or force components) acting on the body must be zero.
2. The sum of all horizontal forces (or force components) acting on the body must be zero.
3. The sum of all torques must be zero:

$$\Sigma F_v = 0$$
$$\Sigma F_h = 0$$
$$\Sigma T = 0$$

The capital Greek letter sigma (Σ) means *the sum of,* F_v represents vertical forces, F_h represents horizontal forces, and T is torque. Whenever an object is in a static state, it may be inferred that all three conditions are in effect, since the violation of any one of the three conditions would result in motion of the body. The conditions of static equilibrium are valuable tools for solving problems relating to human movement (see Sample Problems 13.2, 13.3, and 13.4).

SAMPLE PROBLEM 13.2

How much force must be produced by the biceps brachii, attaching at 90° to the radius at 3 cm from the center of rotation at the elbow joint, to support a weight of 70 N held in the hand at a distance of 30 cm from the elbow joint? (Neglect the weight of the forearm and hand, and neglect any action of other muscles.)

Known

$$d_m = 3 \text{ cm}$$
$$wt = 70 \text{ N}$$
$$d_{wt} = 30 \text{ cm}$$

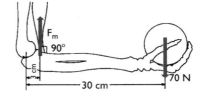

Solution
Since the situation described is static, the sum of the torques acting at the elbow must be equal to zero:

$$\Sigma T_e = 0$$
$$\Sigma T_e = (F_m)(d_m) - (wt)(d_{wt})$$
$$0 = (F_m)(0.03 \text{ m}) - (70 \text{ N})(0.30 \text{ m})$$
$$F_m = \frac{(70 \text{ N})(0.30 \text{ m})}{0.03 \text{ m}}$$

$$\boxed{F_m = 700 \text{ N}}$$

SAMPLE PROBLEM 13.3

Two individuals apply force to opposite sides of a frictionless swinging door. If A applies a 30 N force at a 40° angle 45 cm from the door's hinge and B applies force at a 90° angle 38 cm from the door's hinge, what amount of force is applied by B if the door remains in a static position?

Known

$F_A = 30$ N
$d_{\perp A} = (0.45$ m$)$ (sin 40)
$d_{\perp B} = 0.38$ m

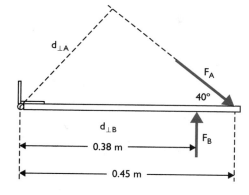

Solution

The equations of static equilibrium are used to solve for F_B. The solution may be found by summing the torques created at the hinge by both forces:

$$\Sigma T_h = 0$$
$$\Sigma T_h = (F_A)(d_{\perp A}) - (F_B)(d_{\perp B})$$
$$0 = (30 \text{ N})(0.45 \text{ m})(\sin 40) - (F_B)(0.38 \text{ m})$$
$$\boxed{F_B = 22.8 \text{ N}}$$

Equations of Dynamic Equilibrium

Bodies in motion are considered to be in a state of dynamic equilibrium, with all acting forces resulting in equal and oppositely directed inertial forces. This general concept was first identified by the French mathematician D'Alembert, and is known as *D'Alembert's principle*. Modified versions of the equations of static equilibrium, which incorporate factors known as *inertia vectors*, describe the conditions of dynamic equilibrium. The equations of dynamic equilibrium may be stated as follows:

$$\Sigma F_x - m\bar{a}_x = 0$$
$$\Sigma F_y - m\bar{a}_y = 0$$
$$\Sigma T_G - \bar{I}\alpha = 0$$

The sums of the horizontal and vertical forces acting on a body are ΣF_x and ΣF_y; $m\bar{a}_x$ and $m\bar{a}_y$ are the products of the body's mass and the horizontal and vertical accelerations of the body's center of mass; ΣT_G is the sum of torques about the body's center of mass, and is the product of the body's moment of inertia about the center of mass and the body's angular acceleration (see Sample Problem 13.5). (The concept of moment of inertia is discussed in Chapter 14.)

A familiar example of the effect of D'Alembert's principle is the change in vertical force experienced when riding in an elevator. As the elevator accelerates upward, an inertial force in the opposite direction is created, and body weight as measured on a scale in the elevator increases. As the elevator accelerates downward, an upwardly directed inertial force decreases body weight as measured on a scale in the elevator. Although

dynamic equilibrium (D'Alembert's principle)
concept indicating a balance between applied forces and inertial forces for a body in motion

SAMPLE PROBLEM 13.4

The quadriceps tendon attaches to the tibia at a 30° angle 4 cm from the joint center at the knee. When an 80 N weight is attached to the ankle 28 cm from the knee joint, how much force is required of the quadriceps to maintain the leg in a horizontal position? What is the magnitude and direction of the reaction force exerted by the femur on the tibia? (Neglect the weight of the leg and the action of other muscles.)

$$wt = 80 \text{ N}$$
$$d_{wt} = 0.28 \text{ m}$$
$$d_F = 0.04 \text{ m}$$

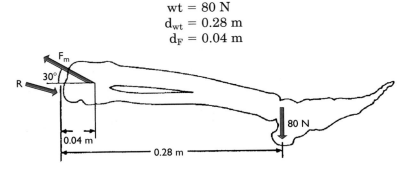

Solution
The equations of static equilibrium can be used to solve for the unknown quantities:

$$\Sigma T_k = 0$$
$$\Sigma T_k = (F_m \sin 30) (d_F) - (wt) (d_{wt})$$
$$0 = (F_m \sin 30) (0.04 \text{ m}) - (80 \text{ N}) (0.28 \text{ m})$$

$$\boxed{F_m = 1120 \text{ N}}$$

The equations of static equilibrium can be used to solve for the vertical and horizontal components of the reaction force exerted by the femur on the tibia. Summation of vertical forces yields the following:

$$\Sigma F_v = 0$$
$$\Sigma F_v = R_v + (F_m \sin 30) - wt$$
$$0 = R_v + 1120 \sin 30 \text{ N} - 80 \text{ N}$$
$$R_v = -480 \text{ N}$$

Summation of horizontal forces yields the following:

$$\Sigma F_h = 0$$
$$\Sigma F_h = R_h - (F_m \cos 30)$$
$$0 = R_h - 1120 \cos 30 \text{ N}$$
$$R_h = 970 \text{ N}$$

The Pythagorean theorem can now be used to find the magnitude of the resultant reaction force:

$$R = \sqrt{(-480 \text{ N})^2 + (970 \text{ N})^2}$$
$$= 1082 \text{ N}$$

The tangent relationship can be used to find the angle of orientation of the resultant reaction force:

$$\tan \alpha = \frac{480 \text{ N}}{970 \text{ N}}$$

$$\alpha = 26.3$$

$$\boxed{R = 1082 \text{ N}, \alpha = 26.3 \text{ degrees}}$$

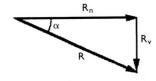

SAMPLE PROBLEM 13.5

A 580 N skydiver in free fall is accelerating at -8.8 m/s^2 rather than -9.81 m/s^2 because of the force of air resistance. How much drag force is acting on the skydiver?

Known

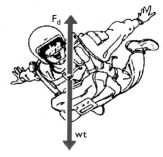

$$\text{wt} = -580 \text{ N}$$
$$a = -8.8 \text{ m/s}^2$$
$$\text{mass} = \frac{580 \text{ N}}{9.81 \text{ m/s}^2} = 59.12 \text{ kg}$$

Solution

Since the skydiver is considered to be in dynamic equilibrium, D'Alembert's principle may be used. All identified forces acting are vertical forces, so the equation of dynamic equilibrium summing the vertical forces to zero is used:

$$\Sigma F_y - \overline{m}a_y = 0$$

Given that $\Sigma F_y = -580 \text{ N} + F_d$, substitute the known information into the equation:

$$-580 \text{ N} + F_d - (59.12 \text{ kg})(-8.8 \text{ m/s}^2) = 0$$

$$\boxed{F_d = 59.7 \text{ N}}$$

---

body mass remains constant, the vertical inertial force changes the magnitude of the reaction force measured on the scale.

CENTER OF GRAVITY

A body's mass is the matter of which it is composed. Associated with every body is a unique point around which the body's mass is equally distributed in all directions. This point is known as the center of mass, or the mass centroid, of the body. In the analysis of bodies subject to gravitational force, the center of mass may also be referred to as the center of gravity (CG), the point about which a body's weight is equally balanced in all directions, or the point about which the sum of torques produced by the weights of the body segments is equal to zero. This definition implies not that the weights positioned on opposite sides of the CG are equal, but that the torques created by the weights on opposite sides of the CG are equal. As illustrated in Figure 13-17, equal weight and equal torque generation on opposite sides of a point can be quite different. The terms *center of mass* and *center of gravity* are more commonly used for biomechanics applications than *mass centroid,* although all three terms refer to exactly the same point. Because the masses of bodies on the earth are subject to gravitational force, the center of gravity is probably the most accurately descriptive of the three to use for biomechanical applications.

The CG of a perfectly symmetrical object of homogeneous density, and therefore homogeneous mass and weight distribution, is at the exact center of the object. For example, the CG of a spherical shot or a solid rubber ball is at its geometric center. If the object is a homogeneous ring, the CG is located in the hollow center of the ring. However, when mass distribution

center of mass
mass centroid
center of gravity
point around which the mass and weight of a body are balanced, no matter how the body is positioned

FIGURE 13-17

The presence of equal torques on opposite sides of an axis of rotation does not necessitate the presence of equal weights on opposite sides of the axis.

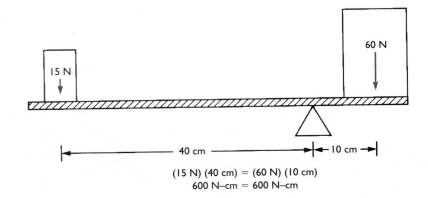

$$(15 \text{ N}) (40 \text{ cm}) = (60 \text{ N}) (10 \text{ cm})$$
$$600 \text{ N–cm} = 600 \text{ N–cm}$$

within an object is not constant, the CG shifts in the direction of greater mass. It is also possible for an object's CG to be located physically outside of the object (Figure 13-18).

Locating the Center of Gravity

The location of the CG for a one-segment object such as a baseball bat, broom, or shovel can be approximately determined using a fulcrum to determine the location of a balance point for the object in three different planes. Because the CG is the point around which the mass of a body is equally distributed, it is also the point around which the body is balanced in all directions.

The location of a body's CG is of interest because, mechanically, a body behaves as though all of its mass were concentrated at the CG. For example, when the human body acts as a projectile, the body's CG follows a parabolic trajectory, regardless of any changes in the configurations of the body segments while in the air. Another implication is that when a weight vector is drawn for an object displayed in a free body diagram, the weight vector acts at the CG. Because the body's mechanical behavior can be traced by following the path of the total-body CG, this factor has been studied as a possible indicator of performance proficiency in several sports.

The path of the CG during takeoff in several of the jumping events is one factor believed to distinguish skilled from less-skilled performance. Research indicates that better Fosbury flop–style high jumpers employ both body lean and body flexion (especially of the support leg) just before takeoff to lower the CG and prolong support foot contact time, thus resulting in increased

FIGURE 13-18

The center of gravity is the single point associated with a body around which the body's weight is equally balanced in all directions.

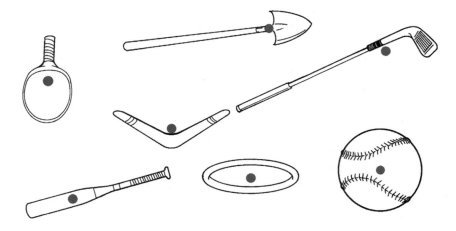

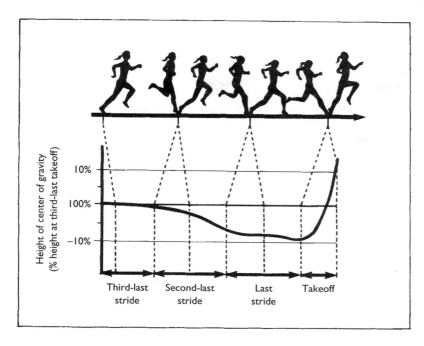

FIGURE 13-19

Height of the athlete's CG during preparation for takeoff in the long jump. Modified from Nixdorf E and Bruggemann P: Zur Absprungvorbereitung beim Weitsprung—Eine biomechanische Untersuchung zum Problem der Korperschwerpunktsenkung, Lehre Leichtathlet, p. 1539, 1983.

takeoff impulse (1). In the long jump, better athletes maintain a normal sprinting stride, with CG height relatively constant, through the second-last step (3). During the last step, however, they markedly lower CG height, then increase CG height going into the jump step (3). Among better pole vaulters, there is progressive elevation of the CG from the third-last step through takeoff. This is partially due to the elevation of the arms as the vaulter prepares to plant the pole. However, research indicates that better vaulters lower their hips during the second-last step, then progressively elevate the hips (and the CG) through takeoff.

The strategy of lowering the CG prior to takeoff enables the athlete to lengthen the vertical path over which the body is accelerated during takeoff, thus facilitating a high vertical velocity at takeoff (Figure 13-19). The speed and angle of takeoff primarily determine the trajectory of the performer's CG during the jump. The only other influencing factor is air resistance, which exerts an extremely small effect on performance in the jumping events.

Locating the Human Body Center of Gravity

Locating the CG for a body containing two or more movable, interconnected segments is more difficult than doing so for a nonsegmented body, because every time the body changes configuration, its weight distribution and CG location are changed. Every time an arm, leg, or finger moves, the CG location as a whole is shifted at least slightly in the direction in which the weight is moved.

Some relatively simple procedures exist for determining the location of the CG of the human body. In the seventeenth century, the Italian mathematician Borelli used a simple balancing procedure for CG location that involved positioning a person on a wooden board (Figure 13-20). A more sophisticated version of this procedure enables calculation of the location of the plane passing through the CG of a person positioned on a reaction board. This procedure requires the use of a scale, a platform of the same height as the weighing surface of the scale, and a rigid board with sharp

The speed and projection angle of an athlete's total-body center of mass largely determine performance outcome in the high jump.

reaction board
specially constructed board for determining the center of gravity location of a body positioned on top of it

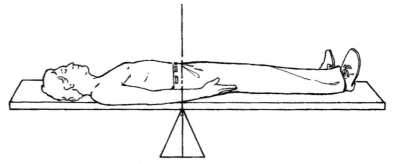

FIGURE 13-20

The relatively crude procedure devised by seventeenth-century mathematician Borelli for approximating the CG location of the human body.

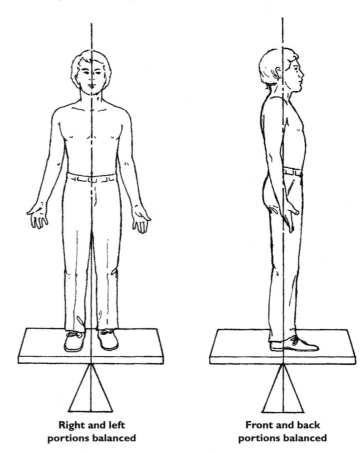

Top and bottom portions balanced

Right and left portions balanced

Front and back portions balanced

• *Location of the CG of the human body is complicated by the fact that its constituents (such as bone, muscle, and fat) have different densities and are unequally distributed throughout the body.*

segmental method
procedure for determining total-body center of mass location based on the masses and center of mass locations of the individual body segments

supports on either end (Figure 13-21). The calculation of the location of the plane containing the CG involves the summation of torques acting about the platform support. Forces creating torques at the support include the person's body weight, the weight of the board, and the reaction force of the scale on the platform (indicated by the reading on the scale). Although the platform also exerts a reaction force on the board, it creates no torque, because the distance of that force from the platform support is zero. Since the reaction board and the subject are in static equilibrium, the sum of the three torques acting at the platform support must be zero, and the distance of the subject's CG plane to the platform may be calculated (see Sample Problem 13.6).

A commonly used procedure for estimating the location of the total body CG from projected film images of the human body is known as the segmental method. This procedure is based on the concept that since the body is composed of individual segments (each with an individual CG),

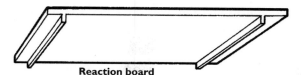

Reaction board

FIGURE 13-21

By summing torques at point a, d (the distance from a to the subject's CG) may be calculated.

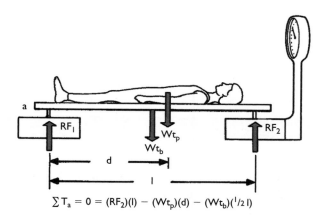

$$\Sigma T_a = 0 = (RF_2)(l) - (Wt_p)(d) - (Wt_b)(\tfrac{1}{2}l)$$

the location of the total-body CG is a function of the locations of the respective segmental CGs. Some body segments, however, are much more massive than others and so have a larger influence on the location of the total-body CG. When the products of each body segment's CG location and its mass are summed and subsequently divided by the sum of all segmental masses (total body mass), the result is the location of the total-body

• *The location of the CG of a multisegmented object is more influenced by the positions of the heavier segments than by those of the lighter segments.*

SAMPLE PROBLEM 13.6

Find the distance from the platform support to the subject's CG, given the following information for the diagram in Figure 13-21:

Known

$$
\begin{aligned}
\text{mass (subject)} &= 73 \text{ kg} \\
\text{mass (board alone)} &= 44 \text{ kg} \\
\text{scale reading} &= 66 \text{ kg} \\
l_b &= 2\text{m}
\end{aligned}
$$

Solution

$$
\begin{aligned}
Wt_p &= (73 \text{ kg})(9.81 \text{ m/s}^2) \\
&= 716.13 \text{ N} \\
Wt_b &= (44 \text{ kg})(9.81 \text{ m/s}^2) \\
&= 431.64 \text{ N} \\
RF_2 &= (66 \text{ kg})(9.81 \text{ m/s}^2) \\
&= 647.46 \text{ N}
\end{aligned}
$$

Use an equation of static equilibrium:

$$
\begin{aligned}
\Sigma T_a = 0 &= (RF_2)(l) - (Wt_p)(d) - (Wt_b)(\tfrac{1}{2}l) \\
0 &= (647.46 \text{ N})(2 \text{ m}) - (716.13 \text{ N})(d) - (431.64 \text{ N})(\tfrac{1}{2})(2 \text{ m})
\end{aligned}
$$

$$d = 1.2 \text{ m}$$

CG. The segmental method uses data for average locations of individual body segment CGs as related to a percentage of segment length:

$$X_{cg} = \Sigma(x_s)(m_s)/\Sigma m_s$$
$$Y_{cg} = \Sigma(y_s)(m_s)/\Sigma m_s$$

•*The segmental method is most commonly implemented through a computer program that reads x,y coordinates of joint centers from a file created by a digitizer.*

In this formula, X_{cg} and Y_{cg} are the coordinates of the total-body CG, x_s and y_s are the coordinates of the individual segment CGs, and m_s is individual segment mass. Thus, the *x*-coordinate of each segment's CG location is identified and multiplied by the mass of that respective segment. The $(x_s)(m_s)$ products for all of the body segments are then summed and subsequently divided by total body mass to yield the *x*-coordinate of the total-body CG location. The same procedure is followed to calculate the *y*-coordinate for total-body CG location (see Sample Problem 13.7).

SAMPLE PROBLEM 13.7

The *x,y*-coordinates of the CGs of the upper arm, forearm, and hand segments are provided on the diagram below. Use the segmental method to find the CG for the entire arm, using the data provided for segment masses from Appendix D.

Known

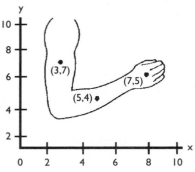

SEGMENT	MASS %	x	(x) (MASS %)	y	(y) (MASS %)
Upper arm	0.45				
Forearm	0.43				
Hand	0.12				
Σ					

Solution

First list the *x*- and *y*-coordinates in their respective columns, and then calculate and insert the product of each coordinate and the mass percentage for each segment into the appropriate columns. Sum the product columns, which yield the *x,y*-coordinates of the total arm CG.

SEGMENT	MASS %	x	(x) (MASS %)	y	(y) (MASS %)
Upper arm	0.45	3	1.35	7	3.15
Forearm	0.43	5	2.15	4	1.72
Hand	0.12	7	0.84	5	0.60
Σ			4.34		5.47

$$x = 4.34$$
$$y = 5.47$$

STABILITY AND BALANCE

A concept closely related to the principles of equilibrium is stability. Stability is defined mechanically as resistance to both linear and angular acceleration, or resistance to disruption of equilibrium. In some circumstances, such as a sumo wrestling contest or the pass protection of a quarterback by an offensive lineman, maximizing stability is desirable. In other situations, an athlete's best strategy is to intentionally minimize stability. Sprinters and swimmers in the preparatory stance before the start of a race intentionally assume a body position allowing them to accelerate quickly and easily at the sound of the starter's pistol. An individual's ability to control equilibrium is known as balance.

Different mechanical factors affect a body's stability. According to Newton's second law of motion ($F = ma$), the more massive an object is, the greater is the force required to produce a given acceleration. Football linemen who are expected to maintain their positions despite the forces exerted on them by opposing linemen are therefore more mechanically stable if they are more massive. In contrast, gymnasts are at a disadvantage with greater body mass, because execution of most gymnastic skills involves disruption of stability.

The greater the amount of friction is between an object and the surface or surfaces it contacts, the greater is the force requirement for initiating or maintaining motion. Toboggans and racing skates are designed so that the friction they generate against the ice will be minimal, enabling a quick disruption of stability at the beginning of a run or race. However, racquetball, golf, and batting gloves are designed to increase the stability of the player's grip on the implement.

Another factor affecting stability is the size of the base of support. This consists of the area enclosed by the outermost edges of the body in contact with the supporting surface or surfaces (Figure 13-22). When the line of action of a body's weight (directed from the CG) moves outside the base of support, a torque is created that tends to cause angular motion of the body, thereby disrupting stability, with the CG falling toward the ground. The larger the base of support is, the lower is the likelihood that this will occur. Martial artists typically assume a wide stance during defensive situations to increase stability. Alternatively, sprinters in the starting blocks maintain a relatively small base of support so that they can quickly disrupt stability

stability
resistance to disruption of equilibrium

balance
a person's ability to control equilibrium

base of support
area bound by the outermost regions of contact between a body and support surface or surfaces

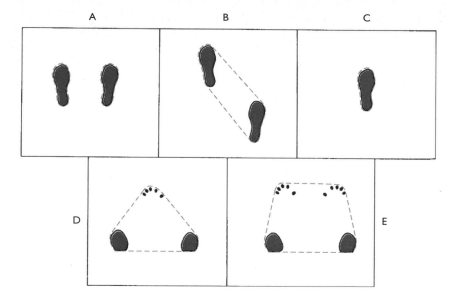

FIGURE 13-22

The base of support for **A** a square stance, **B** an angled stance, **C** a one-foot stance, **D** a three-point stance, and **E** a four-point stance. Areas of contact between body parts and the support surface are shaded. The base of support is the area enclosed by the dashed line.

Performing an arabesque en pointe *requires excellent balance, because lateral movement of the dancer's line of gravity outside the small base of support will result in loss of balance.*

A swimmer on the blocks positions her CG close to the front boundary of her base of support to prepare for forward acceleration.

at the start of the race. Maintaining balance during an *arabesque en pointe,* in which the dancer is balanced on the toes of one foot, requires continual adjustment of CG location through subtle body movements.

The horizontal location of the CG relative to the base of support can also influence stability. The closer the horizontal location of the CG is to the boundary of the base of support, the smaller is the force required to push it outside the base of support, thereby disrupting equilibrium. Athletes in the starting position for a race consequently assume stances that position the CG close to the forward edge of the base of support. Alternatively, if a horizontal force must be sustained, stability is enhanced if the CG is positioned closer to the oncoming force, since the CG can be displaced farther before being moved outside the base of support. Sumo wrestlers lean toward their opponents when being pushed.

The height of the CG relative to the base of support can also affect stability. The higher the positioning of the CG, the greater the potentially disruptive torque created if the body undergoes an angular displacement (Figure 13-23). Athletes often crouch in sport situations when added stability is desirable. A common instructional cue for beginners in many sports is "Bend your knees!" Researchers have proposed a formula for an extrapolated CG position (XcoM) that relates CG height to the base of support in a dynamic situation where a person is walking or running (5). They suggest that to maintain balance, XcoM should remain within the moving base of support. They define XcoM as the vertical position of the CG plus its velocity times a factor of $(l/a_g)^{1/2}$, where l is leg length and a_g is the acceleration of gravity.

Although these principles of stability (summarized in Table 13-1) are generally true, their application to the human body should be made only with the recognition that neuromuscular factors are also influential. Because accidental falls are a significant problem for the growing elderly population, the issue of balance control in this age group is receiving an increasing amount of research attention. Investigators have documented a reduction in anteroposterior CG motion and an increase in mediolateral CG motion in elderly patients with balance disorders as compared to young adults during walking (4). This is of concern because measures of mediolateral sway have been related to the risk of falling (15). Similarly, the ability to vary step width during gait has been shown to be more important for balance control than variations in either step length or step time (9). Researchers hypothesize that difficulty with lateral balance control associated with aging may be related to impaired ability to abduct the leg at the hip with as much strength and speed as needed to maintain dynamic stability (8). Other research with young, healthy adults has shown that rapidly developed, large-magnitude moments at the hip, knee,

FIGURE 13-23

The higher the CG location, the greater the amount of torque its motion creates about the intersection of the line of gravity and the support surface.

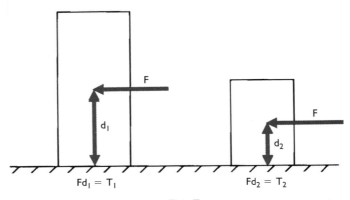

> When other factors are held constant, a body's ability to maintain equilibrium is increased by the following:
> 1. Increasing body mass
> 2. Increasing friction between the body and the surface or surfaces contacted
> 3. Increasing the size of the base of support in the direction of the line of action of an external force
> 4. Horizontally positioning the center of gravity near the edge of the base of support on the side of the oncoming external force
> 5. Vertically positioning the center of gravity as low as possible

TABLE 13-1

Principles of Mechanical Stability

and ankle were required in order to prevent a fall when tripping over an obstacle (11). Both weight-training and aerobic exercise programs can significantly improve postural sway among elderly individuals for whom balance is a concern (7).

Although under normal conditions the size of the base of support is a primary determiner of stability, research shows that a variety of other factors can also limit control of balance. Friction coefficient levels of less than 0.82, decreased resting muscle tension, and impairments in muscle strength, joint movement, balance, gait, hearing, vision, and cognition are all risk factors for falling (10, 13, 14). More research is needed to clarify the application of the principles of stability to human balance.

SUMMARY

Rotary motion is caused by torque, a vector quantity with magnitude and direction. When a muscle develops tension, it produces torque at the joint or joints that it crosses. Rotation of the body segment occurs in the direction of the resultant joint torque.

Mechanically, muscles and bones function as levers. Most joints function as third-class lever systems, well structured for maximizing range of motion and movement speed, but requiring muscle force of greater magnitude than that of the resistance to be overcome. The angle at which a muscle pulls on a bone also affects its mechanical effectiveness, because only the rotary component of muscle force produces joint torque.

When a body is motionless, it is in static equilibrium. The three conditions of static equilibrium are $\Sigma F_v = 0$, $\Sigma F_h = 0$, and $\Sigma T = 0$. A body in motion is in dynamic equilibrium when inertial factors are considered.

The mechanical behavior of a body subject to force or forces is greatly influenced by the location of its center of gravity: the point around which the body's weight is equally balanced in all directions. Different procedures are available for determining center of gravity location.

A body's mechanical stability is its resistance to both linear and angular acceleration. A number of factors influence a body's stability, including mass, friction, center of gravity location, and base of support.

INTRODUCTORY PROBLEMS

1. Why does a force directed through an axis of rotation not cause rotation at the axis?
2. Why does the orientation of a force acting on a body affect the amount of torque it generates at an axis of rotation within the body?
3. A 23 kg boy sits 1.5 m from the axis of rotation of a seesaw. At what distance from the axis of rotation must a 21 kg boy be positioned on the other side of the axis to balance the seesaw? (Answer: 1.6 m)

4. How much force must be produced by the biceps brachii at a perpendicular distance of 3 cm from the axis of rotation at the elbow to support a weight of 200 N at a perpendicular distance of 25 cm from the elbow? (Answer: 1667 N)
5. Two people push on opposite sides of a swinging door. If A exerts a force of 40 N at a perpendicular distance of 20 cm from the hinge and B exerts a force of 30 N at a perpendicular distance of 25 cm from the hinge, what is the resultant torque acting at the hinge, and which way will the door swing? (Answer: $T_h = 0.5$ N-m; in the direction that A pushes)
6. To which lever classes do a golf club, a swinging door, and a broom belong? Explain your answers, including free body diagrams.
7. Is the mechanical advantage of a first-class lever greater than, less than, or equal to one? Explain.
8. Using a diagram, identify the magnitudes of the rotary and stabilizing components of a 100 N muscle force that acts at an angle of 20° to a bone. (Answer: rotary component = 34 N, stabilizing component = 94 N)
9. A 10 kg block sits motionless on a table in spite of an applied horizontal force of 2 N. What are the magnitudes of the reaction force and friction force acting on the block? (Answer: R = 98.1 N, F = 2 N)
10. Given the following data for the reaction board procedure, calculate the distance from the platform support to the subject's CG: $RF_2 = 400$ N, 1 = 2.5 m, wt = 600 N. (Answer: 1.67 m)

ADDITIONAL PROBLEMS

1. For one joint of the lower extremity, explain why eccentric torque occurs during gait.
2. Select one human motor skill with which you are familiar, and sketch a graph showing how you would expect CG height to change during that skill.
3. A 35 N hand and forearm are held at a 45° angle to the vertically oriented humerus. The CG of the forearm and hand is located at a distance of 15 cm from the joint center at the elbow, and the elbow flexor muscles attach at an average distance of 3 cm from the joint center. (Assume that the muscles attach at an angle of 45° to the bones.)
   a. How much force must be exerted by the forearm flexors to maintain this position?
   b. How much force must the forearm flexors exert if a 50 N weight is held in the hand at a distance along the arm of 25 cm? (Answers: a. 175 N; b. 591.7 N)

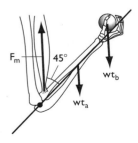

4. A hand exerts a force of 90 N on a scale at 32 cm from the joint center at the elbow. If the triceps attach to the ulna at a 90° angle and at a distance of 3 cm from the elbow joint center, and if the weight of the

forearm and hand is 40 N with the hand/forearm CG located 17 cm from the elbow joint center, how much force is being exerted by the triceps? (Answer: 733.3 N)

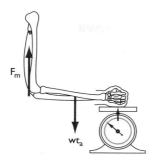

5. A patient rehabilitating a knee injury performs knee extension exercises wearing a 15 N weight boot. Calculate the amount of torque generated at the knee by the weight boot for the four positions shown, given a distance of 0.4 m between the weight boot's CG and the joint center at the knee. (Answers: a. 0; b. 3 N-m; c. 5.2 N-m; d. 6 N-m)

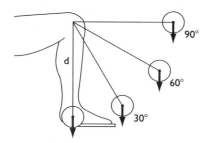

6. A 600 N person picks up a 180 N suitcase positioned so that the suitcase's CG is 20 cm lateral to the location of the person's CG before picking up the suitcase. If the person does not lean to compensate for the added load in any way, where is the combined CG location for the person and suitcase with respect to the person's original CG location? (Answer: Shifted 4.6 cm toward the suitcase)

7. A worker leans over and picks up a 90 N box at a distance of 0.7 m from the axis of rotation in her spine. Neglecting the effect of body weight, how much added force is required of the low back muscles with an average moment arm of 6 cm to stabilize the box in the position shown? (Answer: 1050 N)

8. A man carries a 3 m, 32 N board over his shoulder. If the board extends 1.8 m behind the shoulder and 1.2 m in front of the shoulder, how much force must the man apply vertically downward with his hand that rests on the board 0.2 m in front of the shoulder to stabilize the board in this position? (Assume that the weight of the board is evenly distributed throughout its length.) (Answer: 48 N)

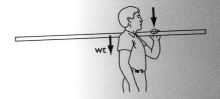

9. A therapist applies a lateral force of 80 N to the forearm at a distance of 25 cm from the axis of rotation at the elbow. The biceps attaches to the radius at a 90° angle and at a distance of 3 cm from the elbow joint center.
   a. How much force is required of the biceps to stabilize the arm in this position?
   b. What is the magnitude of the reaction force exerted by the humerus on the ulna? (Answers: a. 666.7 N; b. 586.7 N)

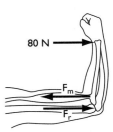

10. Tendon forces T_a and T_b are exerted on the patella. The femur exerts force F on the patella. If the magnitude of T_b is 80 N, what are the magnitudes of T_a and F, if no motion is occurring at the joint? (Answer: T_a = 44.8 N, F = 86.1 N)

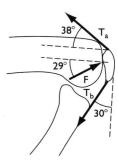

LABORATORY EXPERIENCES

1. Experiment by loosening lug nuts from an automobile tire with a lug wrench, a lug wrench with a short extension on the handle, and a lug wrench with a longer extension on the handle. Write a paragraph explaining your findings, and draw a free body diagram showing the applied force, resistance, and axis of rotation. What provides the resistance?

Explanation:_____

_____

_____

Free Body Diagram:

2. Position a pole across the back of a chair (serving as a fulcrum), and hang a 2 lb weight on one end of the pole. Position a 5 lb weight on the other side of the pole such that the weights are balanced. Measure and record the distances of the two weights from the fulcrum, and write an explanation of your results.

Distance of 5 lb weight from fulcrum: _____

Distance of 2 lb weight from fulcrum: _____

Explanation:_____

_____

_____

3. Perform curl-up exercises under the following conditions: (a) arms folded across the chest, (b) hands behind the neck, and (c) holding a 5 lb weight above the head. Write a paragraph explaining your findings, and draw a free body diagram showing the applied force, resistance, and axis of rotation.

Explanation: _____

_____

_____

_____

Free Body Diagram:

4. Use the reaction board procedure to calculate the sagittal, frontal, and transverse plane positions of the center of gravity of a subject in anatomical position. Repeat the calculations with the subject (a) extending both arms overhead and (b) extending one arm to the right. Present your results in a table and write a paragraph of explanation.

Subject weight: _____

Board weight: _____

Scale reading 1: _____   Scale reading 2: _____   Scale reading 3: _____

   d_1: _____ d_2: _____ d_3: _____

Calculations:

Explanation:_____

_____

_____

_____

_____

5. Using a picture of a person from a magazine or photograph and the anthropometric data from Appendix D, calculate and mark the location of total-body center of gravity using the segmental method. First draw and scale x- and y-axes around the picture. Next, mark the approximate locations of segmental centers of gravity on the picture, using the data from Appendix D. Finally, construct a table using the table in Sample Problem 13.7 as a model.

Segment	Mass %	x	(x) (Mass %)	y	(y) (Mass %)
_____	_____	_____	_____	_____	_____
_____	_____	_____	_____	_____	_____
_____	_____	_____	_____	_____	_____
_____	_____	_____	_____	_____	_____
_____	_____	_____	_____	_____	_____
_____	_____	_____	_____	_____	_____
_____	_____	_____	_____	_____	_____

REFERENCES

1. Dapena J, McDonald C, and Cappaert J: A regression analysis of high jumping technique, *Int J Sport Biomech* 6:246, 1990.
2. Gullett JC, Tillman MD, Gutierrez GM, and Chow JW: A biomechanical comparison of back and front squats in healthy trained individuals, *J Strength Cond Res* 23:284, 2009.
3. Hay JG and Nohara H: The techniques used by elite long jumpers in preparation for take-off, *J Biomech* 23:229, 1990.
4. Hernández A, Silder A, Heiderscheit BC, and Thelen DG: Effect of age on center of mass motion during human walking, *Gait Posture* 30:217, 2009.
5. Hof AL, Gazendam MGJ, and Sinke WE: The condition for dynamic stability, *J Biomech* 38:1, 2005.
6. Kerrigan DC, Franz JR, Keenan GS, Dicharry J, Della Croce U, and Wilder RP: The effect of running shoes on lower extremity joint torques, *PM R* 1:1058, 2009.
7. Messier SP et al: Long-term exercise and its effect on balance in older, osteoarthritic adults: results from the Fitness, Arthritis, and Seniors Trial (FAST), *J Am Geriatr Soc* 48:131, 2000.
8. Mille M-L, Johnson ME, Martinez KM, and Rogers MW: Age-dependent differentces in lateral balance recovery through protective stepping, *Clin Biomech* 20:607, 2005.
9. Owings TM and Grabiner MD: Step width variability, but not step length variability or step time variability, discriminates gait of healthy young and older adults during treadmill locomotion, *J Biomech* 37:935, 2004.
10. Pai YC and Patton J: Center of mass velocity-position predictions for balance control, *J Biomech* 30:347, 1997.
11. Pijnappels M, Bobbert M, and van Dieën JH: How early reactions in the support limb contribute to balance recovery after tripping, *J Biomech* 38:627, 2005.
12. Rahmani A, Viale F, Dalleau G, and Lacour JR: Force velocity and power velocity relationships in squat exercise, *Eur J Appl Physiol* 84:227, 2001.
13. Rietdyk S, Patla AE, Winter DA, Ishac MG, and Little CE: Balance recovery from medio-lateral perturbations of the upper body during standing, *J Biomech* 32:1149, 1999.
14. Robinovitch SN et al: Prevention of falls and fall-related fractures through biomechanics, *Exerc Sport Sci Rev* 28:74, 2000.
15. Rogers MW and Mille M-L: Lateral stability and falls in older people, *Exerc Sports Sci Rev* 31:182, 2003.
16. Shultz SP, Sitler MR, Tierney RT, Hillstrom HJ, and Song J: Effects of pediatric obesity on joint kinematics and kinetics during 2 walking cadences, *Arch Phys Med Rehabil* 90:2146, 2009.
17. Townend MS: *Mathematics in sport,* New York, 1984, John Wiley & Sons.
18. van Zuylen EJ, van Zelzen A, and van der Gon JJD: A biomechanical model for flexion torques of human arm muscles as a function of elbow angle, *J Biomech* 21:183, 1988.

ANNOTATED READINGS

Escamilla RF, Lander JE, and Garhammer J: Biomechanics of powerlifting and weightlifting exercises. In Garrett We Jr. and Kirkendall DT, eds: *Exercise and sport science,* Philadelphia, 2000, Lippincott Williams & Wilkins.
Discusses joint torques and loading during numerous lifting exercises.

Finlayson ML and Peterson EW: Falls, aging, and disability, *Phys Med Rehabil Clin N Am* 21:357, 2010.
Summarizes and compares (1) fall prevalence rates, (2) fall risk factors, (3) consequences of falls, and (4) current knowledge about fall prevention interventions between community-dwelling older adults and people aging with physical disability.

Rogers MW and Mille M-L: Lateral stability and falls in older people, *Exerc Sports Sci Rev* 31:182, 2003.
Discusses age-related changes in specific neuromusculoskeletal factors affecting protective stepping and other balance functions that may precipitate lateral instability and falls.

Winter DA: *Biomechanics and motor control of human movement* (4th ed.), New York, 2009, John Wiley & Sons.
Chapter on kinetics includes useful discussions on interpreting moment of force (torque) curves and the difference between the center of gravity and center of pressure.

RELATED WEBSITES

Advanced Medical Technology, Inc.
http://www.amti.com/
Provides information on the AMTI force platforms, with reference to ground reaction forces in gait analysis, balance and posture, and other topics.
NASA: Center of Gravity
http://www.grc.nasa.gov/WWW/K-12/airplane/cg.html
Official site of the National Aeronautics and Space Administration, providing comprehensive discussion accompanied by slides of the relevance of the CG in the design of airplanes.
NASA: Determining Center of Gravity
http://www.grc.nasa.gov/WWW/K-12/airplane/rktcg.html
Official site of the National Aeronautics and Space Administration, providing comprehensive discussion accompanied by slides of the relevance of the CG in the design of model rockets.
Wikipedia: Center of Gravity
http://en.wikipedia.org/wiki/Center_of_gravity
Includes discussion of relationships among center of gravity, center of mass, and center of buoyancy (Ch. 15).
Wikipedia: Torque
http://en.wikipedia.org/wiki/Torque
Discusses torque as related to other mechanical quantities and to static equilibrium, including many definitions and diagrams.

KEY TERMS

balance	a person's ability to control equilibrium
base of support	area bound by the outermost regions of contact between a body and support surface or surfaces
center of mass **mass centroid** **center of gravity**	point around which the mass and weight of a body are balanced, no matter how the body is positioned
couple	pair of equal, oppositely directed forces that act on opposite sides of an axis of rotation to produce torque
dynamic equilibrium **(D' Alembert's principle)**	concept indicating a balance between applied forces and inertial forces for a body in motion
first-class lever	lever positioned with the applied force and the resistance on opposite sides of the axis of rotation
fulcrum	the point of support, or axis, about which a lever may be made to rotate
lever	a simple machine consisting of a relatively rigid, barlike body that may be made to rotate about an axis
mechanical advantage	ratio of force arm to resistance arm for a given lever
moment arm	shortest (perpendicular) distance between a force's line of action and an axis of rotation
reaction board	specially constructed board for determining the center of gravity location of a body positioned on top of it
second-class lever	lever positioned with the resistance between the applied force and the fulcrum
segmental method	procedure for determining total-body center of mass location based on the masses and center of mass locations of the individual body segments
stability	resistance to disruption of equilibrium
static equilibrium	motionless state characterized by $\Sigma SF_v = 0$, $\Sigma SF_h = 0$, and $\Sigma ST = 0$
third-class lever	lever positioned with the applied force between the fulcrum and the resistance
torque	the rotary effect of a force about an axis of rotation, measured as the product of the force and the perpendicular distance between the force's line of action and the axis

Angular Kinetics
of Human Movement

14

After completing this chapter, you will be able to:

Identify the angular analogues of mass, force, momentum, and impulse.

Explain why changes in the configuration of a rotating airborne body can produce changes in the body's angular velocity.

Identify and provide examples of the angular analogues of Newton's laws of motion.

Define centripetal force and explain where and how it acts.

Solve quantitative problems relating to the factors that cause or modify angular motion.

ONLINE LEARNING CENTER RESOURCES

www.mhhe.com/hall6e

Log on to our Online Learning Center (OLC) for access to these additional resources:

- Online Lab Manual
- Flashcards with definitions of chapter key terms
- Chapter objectives
- Chapter lecture PowerPoint presentation
- Self-scoring chapter quiz
- Additional chapter resources
- Web links for study and exploration of chapter-related topics

W hy do sprinters run with more swing phase flexion at the knee than do distance runners? Why do dancers and ice skaters spin more rapidly when their arms are brought in close to the body? How do cats always land on their feet? In this chapter, we explore more concepts pertaining to angular kinetics, from the perspective of the similarities and differences between linear and angular kinetic quantities.

RESISTANCE TO ANGULAR ACCELERATION

Moment of Inertia

Inertia is a body's tendency to resist acceleration (see Chapter 3). Although inertia itself is a concept rather than a quantity that can be measured in units, a body's inertia is directly proportional to its mass (Figure 14-1). According to Newton's second law, the greater a body's mass, the greater its resistance to linear acceleration. Therefore, mass is a body's inertial characteristic for considerations relative to linear motion.

Resistance to angular acceleration is also a function of a body's mass. The greater the mass, the greater the resistance to angular acceleration. However, the relative ease or difficulty of initiating or halting angular motion depends on an additional factor: the distribution of mass with respect to the axis of rotation.

Consider the baseball bats shown in Figure 14-2. Suppose a player warming up in the on-deck circle adds a weight ring to the bat he is swinging. Will the relative ease of swinging the bat be greater with the weight positioned near the striking end of the bat or with the weight near the bat's grip? Similarly, is it easier to swing a bat held by the grip (the normal hand position) or a bat turned around and held by the barrel?

Experimentation with a baseball bat or some similar object makes it apparent that the more closely concentrated the mass is to the axis of rotation, the easier it is to swing the object. Conversely, the more mass is positioned away from the axis of rotation, the more difficult it is to initiate (or stop) angular motion. Resistance to angular acceleration, therefore, depends not only on the amount of mass possessed by an object but also on the distribution of that mass with respect to the axis of rotation. The inertial property for angular motion must therefore incorporate both factors.

The inertial property for angular motion is moment of inertia, represented as I. Every body is composed of particles of mass, each with its own particular distance from a given axis of rotation. The moment of inertia for a single particle of mass may be represented as the following:

$$I = mr^2$$

In this formula, m is the particle's mass and r is the particle's radius of rotation. The moment of inertia of an entire body is the sum of

• *The more closely mass is distributed to the axis of rotation, the easier it is to initiate or stop angular motion.*

moment of inertia
inertial property for rotating bodies representing resistance to angular acceleration; based on both mass and the distance the mass is distributed from the axis of rotation

FIGURE 14-1

The distribution of mass in a system does not affect its linear momentum.

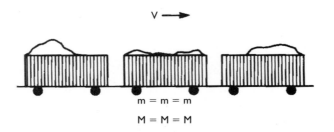

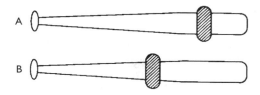

FIGURE 14-2

Although both bats have the same mass, bat A is harder to swing than bat B, because the weight ring on it is positioned farther from the axis of rotation.

the moments of inertia of all the mass particles the object contains (Figure 14-3):

$$I = \Sigma mr^2$$

The distribution of mass with respect to the axis of rotation is more important than the total amount of body mass in determining resistance to angular acceleration, because r is squared. Since r is the distance between a given particle and an axis of rotation, values of r change as the axis of rotation changes. Thus, when a player grips a baseball bat, "choking up" on the bat reduces the bat's moment of inertia with respect to the axis of rotation at the player's wrists, and concomitantly increases the relative ease of swinging the bat. Little League baseball players often unknowingly make use of this concept when swinging bats that are longer and heavier than they can effectively handle. Interestingly, research shows that when baseball players warm up with a weighted bat (with a larger moment of inertia than a regular bat) post-warm-up swing velocity is actually reduced (18).

Within the human body, the distribution of mass with respect to an axis of rotation can dramatically influence the relative ease or difficulty of moving the body limbs. For example, during gait, the distribution of a given leg's mass, and therefore its moment of inertia with respect to the primary axis of rotation at the hip, depends largely on the angle present at the knee. In sprinting, maximum angular acceleration of the legs is desired, and considerably more flexion at the knee is present during the swing phase than while running at slower speeds. This greatly reduces the moment of inertia of the leg with respect to the hip, thus reducing resistance to hip flexion. Runners who have leg morphology involving mass distribution closer to the hip, with more massive thighs and slimmer lower legs than others, have a smaller moment of inertia of the leg with respect to the hip. This is an anthropometric characteristic that is advantageous for sprinters. During walking, in which minimal angular acceleration of the legs is required, flexion at the knee during the swing phase remains relatively small, and the leg's moment of inertia with respect to the hip is relatively large.

During sprinting, extreme flexion at the knee reduces the moment of inertia of the swinging leg.

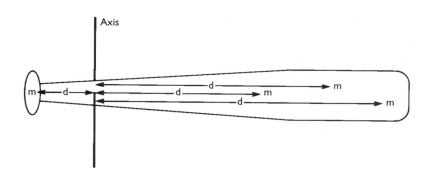

FIGURE 14-3

Moment of inertia is the sum of the products of each particle's mass and radius of rotation squared.

Modern-day golf irons are commonly constructed with the heads bottom weighted, perimeter weighted, or heel and toe weighted. These manipulations of the amount of mass and the distribution of mass within the head of the club are designed to increase club head inertia, thus reducing the tendency of the club to rotate about the shaft during an off-center hit. Research findings indicate that perimeter-weighted club heads perform best for eccentric ball contacts outside the club head center of gravity (CG), with a simple blade head club superior for contacts below the club head CG (16). The most consistent performance, however, was exhibited by a toe-and-bottom-weighted club head, which was second best for all eccentric hits (16). A golfer's individual preference, feel, and experience should ultimately determine selection of club type.

Determining Moment of Inertia

•*The fact that bone, muscle, and fat have different densities and are distributed dissimilarly in individuals complicates efforts to calculate human body segment moments of inertia.*

Assessing moment of inertia for a body with respect to an axis by measuring the distance of each particle of body mass from an axis of rotation and then applying the formula is obviously impractical. In practice, mathematical procedures are used to calculate moment of inertia for bodies of regular geometric shapes and known dimensions. Because the human body is composed of segments that are of irregular shapes and heterogeneous mass distributions, either experimental procedures or mathematical models are used to approximate moment-of-inertia values for individual body segments and for the body as a whole in different positions. Moment of inertia for the human body and its segments has been approximated by using average measurements from cadaver studies, measuring the acceleration of a swinging limb, employing photogrammetric methods, and applying mathematical modeling.

•*Because there are formulas available for calculating the moment of inertia of regularly shaped solids, some investigators have modeled the human body as a composite of various geometric shapes.*

Once moment of inertia for a body of known mass has been assessed, the value may be characterized using the following formula:

$$I = mk^2$$

radius of gyration
distance from the axis of rotation to a point where the body's mass could be concentrated without altering its rotational characteristics

In this formula, I is moment of inertia with respect to an axis, m is total body mass, and k is a distance known as the radius of gyration. The radius of gyration represents the object's mass distribution with respect to a given axis of rotation. It is the distance from the axis of rotation to a point at which the mass of the body can theoretically be concentrated without altering the inertial characteristics of the rotating body. This point is *not* the same as the segmental CG (Figure 14-4). Since the radius of gyration

FIGURE 14-4

Knee angle affects the moment of inertia of the swinging leg with respect to the hip because of changes in the radius of gyration for the lower leg (k_2) and foot (k_3).

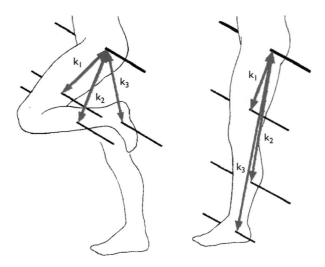

is based on r^2 for individual particles, it is always longer than the radius of rotation, the distance to the segmental CG.

The length of the radius of gyration changes as the axis of rotation changes. As mentioned earlier, it is easier to swing a baseball bat when the bat is grasped by the barrel end rather than by the bat's grip. When the bat is held by the barrel, k is much shorter than when the bat is held properly, since more mass is positioned close to the axis of rotation. Likewise, the radius of gyration for a body segment such as the forearm is greater with respect to the wrist than with respect to the elbow.

The radius of gyration is a useful index of moment of inertia when a given body's resistance to rotation with respect to different axes is discussed. Units of moment of inertia parallel the formula definition of the quantity, and therefore consist of units of mass multiplied by units of length squared (kg · m^2)

Human Body Moment of Inertia

Moment of inertia can only be defined with respect to a specific axis of rotation. The axis of rotation for a body segment in sagittal and frontal plane motions is typically an axis passing through the center of a body segment's proximal joint. When a segment rotates around its own longitudinal axis, its moment of inertia is quite different from its moment of inertia during flexion and extension or abduction and adduction, because its mass distribution, and therefore its moment of inertia, is markedly different with respect to this axis of rotation. Figure 14-5 illustrates the difference in the lengths of the radii of gyration for the forearm with respect to the transverse and longitudinal axes of rotation.

The moment of inertia of the human body as a whole is also different with respect to different axes. When the entire human body rotates free of support, it moves around one of three principal axes: the transverse (or frontal), the anteroposterior (or sagittal), or the longitudinal (or vertical) axis, each of which passes through the total body CG. Moment of inertia with respect to one of these axes is known as a principal moment of inertia.

principal axes
three mutually perpendicular axes passing through the total body center of gravity

principal moment of inertia
total-body moment of inertia relative to one of the principal axes

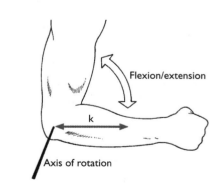
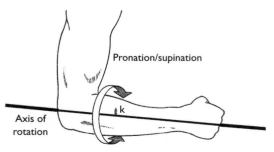

FIGURE 14-5

The radius of gyration (k) of the forearm for flexion/extension movements is much larger than for pronation/supination.

FIGURE 14-6

Principal moments of inertia of the human body in different positions with respect to different principal axes: (1) principal axis; (2) moment of inertia ($kg \cdot m^2$). Modified from Hochmuth G: *Biomechanik sportlicher bewegungen,* Frankfurt, Germany, 1967, Wilhelm Limpart, Verlag.

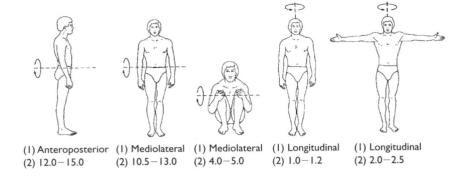

(1) Anteroposterior (1) Mediolateral (1) Mediolateral (1) Longitudinal (1) Longitudinal
(2) 12.0 – 15.0   (2) 10.5 – 13.0   (2) 4.0 – 5.0   (2) 1.0 – 1.2   (2) 2.0 – 2.5

The ratio of muscular strength (the ability of a muscle group to produce torque about a joint) to segmental moments of inertia (resistance to rotation at a joint) is an important contributor to performance capability in gymnastic events. Photo courtesy of Photodisc/Getty Images.

angular momentum

quantity of angular motion possessed by a body; measured as the product of moment of inertia and angular velocity

Figure 14-6 shows quantitative estimates of principal moments of inertia for the human body in several different positions. When the body assumes a tucked position during a somersault, its principal moment of inertia (and resistance to angular motion) about the transverse axis is clearly less than when the body is in anatomical position. Divers performing a somersaulting dive undergo changes in principal moment of inertia about the transverse axis on the order of 15 $kg \cdot m^2$ to 6.5 $kg \cdot m^2$ as the body goes from a layout position to a pike position (7).

As children grow from childhood through adolescence and into adulthood, developmental changes result in changing proportions of body segment lengths, masses, and radii of gyration, all affecting segment moments of inertia. Segment moments of inertia affect resistance to angular rotation, and therefore performance capability, in sports such as gymnastics and diving. Because of smaller moments of inertia, smaller gymnasts have an advantage in performing skills involving whole-body rotations, despite the fact that larger gymnasts may have greater strength and be able to generate more power (1). Several prominent female gymnasts who achieved world-class status during early adolescence faded from the public view before reaching age 20 because of declines in their performance capabilities generally attributed to changes in body proportions with growth.

ANGULAR MOMENTUM

Since moment of inertia is the inertial property for rotational movement, it is an important component of other angular kinetic quantities. As discussed in Chapter 12, the quantity of motion that an object possesses is referred to as its *momentum.* Linear momentum is the product of the linear inertial property (mass) and linear velocity. The quantity of angular motion that a body possesses is likewise known as angular momentum. Angular momentum, represented as H, is the product of the angular inertial property (moment of inertia) and angular velocity:

$$\text{For linear motion: } M = mv$$
$$\text{For angular motion: } H = I\omega$$
$$\text{Or: } H = mk^2\omega$$

Three factors affect the magnitude of a body's angular momentum: (a) its mass (m), (b) the distribution of that mass with respect to the axis of rotation (k), and (c) the angular velocity of the body (ω). If a body has no angular velocity, it has no angular momentum. As mass or angular velocity increases, angular momentum increases proportionally. The factor that most dramatically influences angular momentum is the distribution of mass with respect to the axis of rotation, because angular momentum is proportional to the square of the radius of gyration (see Sample Problem 14.1). Units of angular momentum result from multiplying units of mass, units of length squared, and units of angular velocity, which yields $kg \cdot m^2/s$.

For a multisegmented object such as the human body, angular momentum about a given axis of rotation is the sum of the angular momenta of the individual body segments. During an airborne somersault, the angular momentum of a single segment, such as the lower leg, with respect to the principal axis of rotation passing through the total body CG consists of two components: the local term and the remote term. The local term is based on the segment's angular momentum about its own segmental CG, and the remote term represents the segment's angular momentum about the total body CG. Angular momentum for this segment about a principal axis is the sum of the local term and the remote term:

$$H = I_s \omega_s + mr^2 \omega_g$$

In the local term, I_s is the segment's moment of inertia and ω_s is the segment's angular velocity, both with respect to a transverse axis through

SAMPLE PROBLEM 14.1

Consider a rotating 10 kg body for which k = 0.2 m and ω = 3 rad/s. What is the effect on the body's angular momentum if the mass doubles? The radius of gyration doubles? The angular velocity doubles?

Solution
The body's original angular momentum is the following:

$$H = mk^2\omega$$
$$H = (10 \text{ kg}) (0.2 \text{ m})^2 (3 \text{ rad/s})$$
$$H = 1.2 \text{ kg} \cdot m^2/s$$

With mass doubled:

$$H = mk^2\omega$$
$$H = (20 \text{ kg}) (0.2 \text{ m})^2 (3 \text{ rad/s})$$
$$H = 2.4 \text{ kg} \cdot m^2/s$$

> H is doubled.

With k doubled:

$$H = mk^2\omega$$
$$H = (10 \text{ kg}) (0.4 \text{ m})^2 (3 \text{ rad/s})$$
$$H = 4.8 \text{ kg} \cdot m^2/s$$

> H is quadrupled.

With ω doubled:

$$H = mk^2\omega$$
$$H = (10 \text{ kg}) (0.2 \text{ m})^2 (6 \text{ rad/s})$$
$$H = 2.4 \text{ kg} \cdot m^2/s$$

> H is doubled.

the segment's own CG. In the remote term, m is the segment's mass, r is the distance between the total body and segmental CGs, and ω_g is the angular velocity of the segmental CG about the principal transverse axis (Figure 14-7). The sum of the angular momenta of all the body segments about a principal axis yields the total-body angular momentum about that axis.

During takeoff from a springboard or platform, a competitive diver must attain sufficient linear momentum to reach the necessary height (and safe distance from the board or platform) and sufficient angular momentum to perform the required number of rotations. For multiple-rotation, nontwisting platform dives, the angular momentum generated at takeoff increases as the rotational requirements of the dive increase (6). When a twist is also incorporated into a somersaulting dive, the angular momentum required is further increased. Inclusion of a twist during forward one-and-a-half springboard dives is associated with increased angular momentum at takeoff of 6–19% (17). Adding a somersault while rotating in a tuck rather than a pike position also requires a small increase in angular momentum (14).

Conservation of Angular Momentum

Whenever gravity is the only acting external force, angular momentum is conserved. For angular motion, the principle of conservation of momentum may be stated as follows:

> The total angular momentum of a given system remains constant in the absence of external torques.

Gravitational force acting at a body's CG produces no torque because $d_\perp$ equals zero and so it creates no change in angular momentum.

The principle of conservation of angular momentum is particularly useful in the mechanical analysis of diving, trampolining, and gymnastics events in which the human body undergoes controlled rotations while airborne. In a one-and-a-half front somersault dive, the diver leaves the springboard with a fixed amount of angular momentum. According to the principle of conservation of angular momentum, the amount of angular

●*The magnitude and direction of the angular momentum vector for an airborne performer are established at the instant of takeoff.*

FIGURE 14-7

The angular momentum of the swinging leg is the sum of its local term, $I_s\omega_s$ and its remote term, $mr^2\omega_g$.

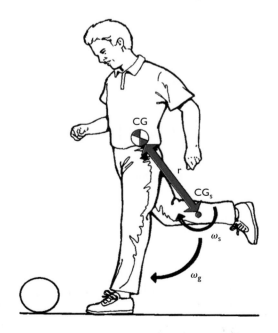

momentum present at the instant of takeoff remains constant throughout the dive. As the diver goes from an extended layout position into a tuck, the radius of gyration is decreased, thus reducing the body's principal moment of inertia about the transverse axis. Because angular momentum remains constant, a compensatory increase in angular velocity must accompany the decrease in moment of inertia (Figure 14-8). The tighter the diver's tuck, the greater the angular velocity. Once the somersault is completed, the diver extends to a full layout position, thereby increasing total-body moment of inertia with respect to the axis of rotation. Again, because angular momentum remains constant, an equivalent decrease in angular velocity occurs. For the diver to appear to enter the water perfectly vertically, minimal angular velocity is desirable. Sample Problem 14.2 quantitatively illustrates this example.

Other examples of conservation of angular momentum occur when an airborne performer has a total-body angular momentum of zero and a forceful movement such as a jump pass or volleyball spike is executed. When a volleyball player performs a spike, moving the hitting arm with a high angular velocity and a large angular momentum, there is a compensatory rotation of the lower body, producing an equal amount of angular momentum in the opposite direction (Figure 14-9). The moment of inertia of the two legs with respect to the hips is much greater than that of the spiking arm with respect to the shoulder. The angular velocity of the legs generated to counter the angular momentum of the swinging arm is therefore much less than the angular velocity of the spiking arm.

Transfer of Angular Momentum

Although angular momentum remains constant in the absence of external torques, transferring angular velocity at least partially from one principal axis of rotation to another is possible. This occurs when a diver changes from a primarily somersaulting rotation to one that is primarily twisting, and vice versa. An airborne performer's angular velocity vector does not necessarily occur in the same direction as the angular momentum vector. It is possible for a body's somersaulting angular momentum and its twisting angular momentum to be altered in midair, though the vector sum of the two (the total angular momentum) remains constant in magnitude and direction.

Researchers have observed several procedures for changing the total-body axis of rotation. Asymmetrical arm movements and rotation of the

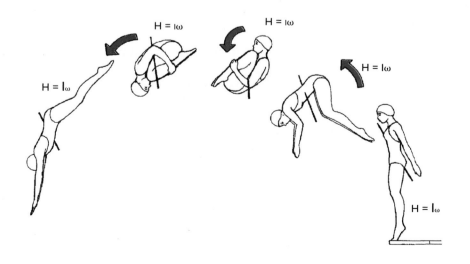

H = Iω

H = Iω

H = Iω

H = Iω

H = Iω

FIGURE 14-8

When angular momentum is conserved, changes in body configuration produce a trade-off between moment of inertia and angular velocity, with a tuck position producing greater angular velocity.

SAMPLE PROBLEM 14.2

A 60 kg diver is positioned so that his radius of gyration is 0.5 m as he leaves the board with an angular velocity of 4 rad/s. What is the diver's angular velocity when he assumes a tuck position, altering his radius of gyration to 0.25 m?

Known

Position 1

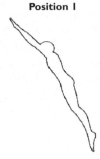

$$m = 60 \text{ kg}$$
$$k = 0.5 \text{ m}$$
$$\omega = 4 \text{ rad/s}$$

$$m = 60 \text{ kg}$$
$$k = 0.25 \text{ m}$$

Position 2

Solution

To find ω, calculate the amount of angular momentum that the diver possesses when he leaves the board, since angular momentum remains constant during the airborne phase of the dive:

Position 1:

$$H = mk^2\omega$$
$$= (60 \text{ kg}) (0.5 \text{ m})^2 (4 \text{ rad/s})$$
$$= 60 \text{ kg} \cdot \text{m}^2/\text{s}$$

Use this constant value for angular momentum to determine ω when $k = 0.25$ m:

Position 2:

$$H = mk^2\omega$$
$$60 \text{ kg} \cdot \text{m}^2/\text{s} = (60 \text{ kg}) (0.25 \text{ m})^2 (\omega)$$

$$\boxed{\omega = 16 \text{ rad/s}}$$

---

hips (termed *hula movement*) can tilt the axis of rotation out of the original plane of motion (Figure 14-10). The less-often-used hula movement can produce tilting of the principal axis of rotation when the body is somersaulting in a piked position. These asymmetrical movements can be used to generate twist and to eliminate twist (20).

Even when total-body angular momentum is zero, generating a twist in midair is possible using skillful manipulation of a body composed of at least two segments. Prompted by the observation that a domestic cat seems always to land on its feet no matter what position it falls from, scientists have studied this apparent contradiction of the principle of conservation of angular momentum (5). Gymnasts and divers can use this procedure, referred to as *cat rotation*, without violating the conservation of angular momentum.

FIGURE 14-9

During the airborne execution of a spike in volleyball, a compensatory rotation of the lower extremity offsets the forcefully swinging arm so that total body angular momentum is conserved.

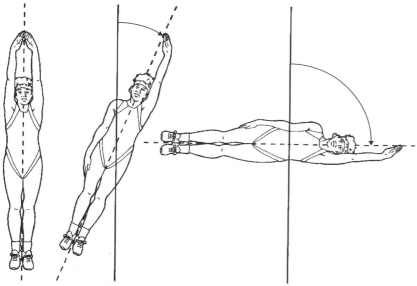

FIGURE 14-10

Asymmetrical positioning of the arms with respect to the axis of angular momentum can shift the axis of rotation.

Cat rotation is basically a two-phase process. It is accomplished most effectively when the two body segments are in a 90° pike position, so that the radius of gyration of one segment is maximal with respect to the longitudinal axis of the other segment (Figure 14-11). The first phase consists of the internally generated rotation of Segment 1 around its longitudinal axis. Because angular momentum is conserved, there is a compensatory rotation of Segment 2 in the opposite direction around the longitudinal axis of Segment 1. However, the resulting rotation is of a relatively small velocity, because k for Segment 2 is relatively large with respect to Axis 1. The second phase of the process consists of rotation of Segment 2 around its longitudinal axis in the same direction originally

A skillful human performer can rotate 180° or more in the air with zero angular momentum because in a piked position there is a large discrepancy between the radii of gyration for the upper and lower extremities with respect to the longitudinal axes of these two major body segments.

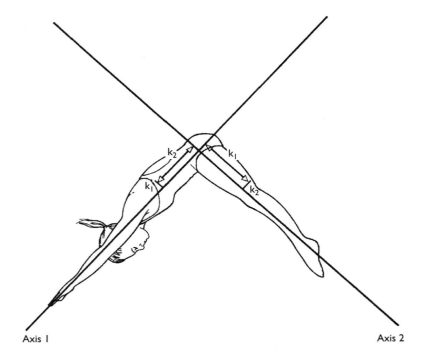

Axis 1   Axis 2

taken by Segment 1. Accompanying this motion is a compensatory rotation of Segment 1 in the opposite direction around Axis 2. Again, angular velocity is relatively small, because k for Segment 1 is relatively large with respect to Axis 2. Using this procedure, a skilled diver can initiate a twist in midair and turn through as much as 450° (5). Cat rotation is performed around the longitudinal axes of the two major body segments. It is easier to initiate rotation about the longitudinal principal axis than about either the transverse or the anteroposterior principal axes, because total-body moment of inertia with respect to the longitudinal axis is much smaller than the total-body moments of inertia with respect to the other two axes.

Change in Angular Momentum

When an external torque does act, it changes the amount of angular momentum present in a system predictably. Just as with changes in linear momentum, changes in angular momentum depend not only on the magnitude and direction of acting external torques but also on the length of the time interval over which each torque acts:

$$\text{linear impulse} = Ft$$
$$\text{angular impulse} = Tt$$

angular impulse

change in angular momentum equal to the product of torque and time interval over which the torque acts

When an angular impulse acts on a system, the result is a change in the total angular momentum of the system. The impulse–momentum relationship for angular quantities may be expressed as the following:

$$Tt = \Delta H$$
$$= (I\omega)_2 - (I\omega)_1$$

As before, the symbols T, t, H, I, and ω represent torque, time, angular momentum, moment of inertia, and angular velocity, respectively, and subscripts 1 and 2 denote initial and second or final points in time. Because angular impulse is the product of torque and time, significant changes in an object's angular momentum may result from the action of a

SAMPLE PROBLEM 14.3

What average amount of force must be applied by the elbow flexors inserting at an average perpendicular distance of 1.5 cm from the axis of rotation at the elbow over a period of 0.3 s to stop the motion of the 3.5 kg arm swinging with an angular velocity of 5 rad/s when k = 20 cm?

Known

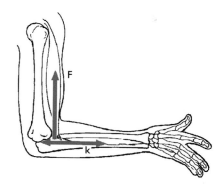

$$d = 0.015 \text{ m}$$
$$t = 0.3 \text{ s}$$
$$m = 3.5 \text{ kg}$$
$$k = 0.20 \text{ m}$$
$$\omega = 5 \text{ rad/s}$$

Solution

The impulse–momentum relationship for angular motion can be used.

$$Tt = \Delta H$$
$$Fdt = (mk^2\omega)_2 - (mk^2\omega)_1$$
$$F(0.015 \text{ m}) (0.3 \text{ s}) = 0 - (3.5 \text{ kg}) (0.20 \text{ m})^2 (5 \text{ rad/s})$$

$$\boxed{F = -155.56 \text{ N}}$$

---

large torque over a small time interval or from the action of a small torque over a large time interval. Since torque is the product of a force's magnitude and the perpendicular distance to the axis of rotation, both of these factors affect angular impulse. The effect of angular impulse on angular momentum is shown in Sample Problem 14.3.

In the throwing events in track and field, the object is to maximize the angular impulse exerted on an implement before release, to maximize its momentum and the ultimate horizontal displacement following release. As discussed in Chapter 11, linear velocity is directly related to angular velocity, with the radius of rotation serving as the factor of proportionality. As long as the moment of inertia (mk^2) of a rotating body remains constant, increased angular momentum translates directly to increased linear momentum when the body is projected. This concept is particularly evident in the hammer throw, in which the athlete first swings the hammer two or three times around the body with the feet planted, and then executes the next three or four whole-body turns while facing the hammer before release. Some hammer throwers perform the first one or two of the whole body turns with the trunk in slight flexion (called *countering with the hips*), thereby enabling a farther reach with the hands (Figure 14-12). This tactic increases the radius of rotation, and thus the moment of inertia of the hammer with respect to the axis of rotation, so that if angular velocity is not reduced, the angular momentum of the thrower/hammer system is increased. For this strategy, the final turns are completed with the entire body leaning away from the hammer, or *countering with the shoulders*. Researchers have suggested that, although the ability to lean forward throughout the turns should increase the angular momentum imparted to the hammer, a natural tendency to protect against excessive spinal stresses

FIGURE 14-12

A hammer thrower must counter the centrifugal force of the hammer to avoid being pulled out of the throwing ring. Countering with the shoulders **A** results in a smaller radius of rotation for the hammer than countering with the hips **B**.

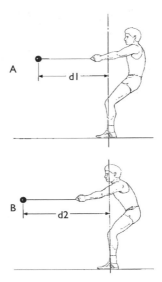

or shoulder strength limitations may prevent the thrower from accomplishing this technique modification (4).

The angular momentum required for the total body rotations executed during aerial skills is primarily derived from the angular impulse created by the reaction force of the support surface during takeoff. During back dives performed from a platform, the major angular impulse is produced during the final weighting of the platform, when the diver comes out of a crouched position through extension at the hip, knee, and ankle joints and executes a vigorous arm swing simultaneously (15). The vertical component of the platform reaction force, acting in front of the diver's CG, creates most of the backward angular momentum required (Figure 14-13).

On a springboard, the position of the fulcrum with respect to the tip of the board can usually be adjusted and can influence performance. Setting the fulcrum farther back from the tip of the board results in greater downward board tip vertical velocity at the beginning of takeoff, which allows the diver more time in contact with the board to generate angular momentum and increased vertical velocity going into the dive (9). Concomitant disadvantages, however, include the requirement of increased hurdle

FIGURE 14-13

The product of the springboard reaction force (F) and its moment arm with respect to the diver's center of gravity ($d_\perp$) creates a torque, which generates the angular impulse that produces the diver's angular momentum at takeoff.

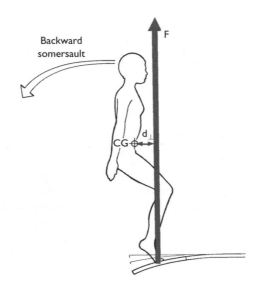

flight duration and the necessity of reversing downward motion from a position of greater flexion at the knees (9). In an optimum reverse dive from a springboard, peak knee extension torque is generated just prior to maximum springboard depression, so that the diver exerts force against a stiffer board (19).

The motions of the body segments during takeoff determine the magnitude and direction of the reaction force generating linear and angular impulses. During both platform and springboard dives, the rotation of the arms at takeoff generally contributes more to angular momentum than the motion of any other segment (6, 13). Highly skilled divers perform the arm swing with the arms fully extended, thus maximizing the moment of inertia of the arms and the angular momentum generated. Less-skilled divers often must use flexion at the elbow to reduce the moment of inertia of the arms about the shoulders so that arm swing can be completed during the time available (13). In contrast to the takeoff during a dive, during the takeoff for aerial somersaults performed from the floor in gymnastics, it is forceful extension of the legs that contributes the most to angular momentum (3). Optimizing performance of aerial somersaults requires generating high linear and angular velocities during the approach, as well as precise timings of body segment motions (11).

Angular impulse produced through the support surface reaction force is also essential for performance of the *tour jeté,* a dance movement that consists of a jump accompanied by a 180° turn, with the dancer landing on the foot opposite the takeoff foot. When the movement is performed properly, the dancer appears to rise straight up and then rotate about the principal vertical axis in the air. In reality, the jump must be executed so that a reaction torque around the dancer's vertical axis is generated by the floor. The extended leg at the initiation of the jump creates a relatively large moment of inertia relative to the axis of rotation, thereby resulting in a relatively low total-body angular velocity. At the peak of the jump, the dancer's legs simultaneously cross the axis of rotation and the arms simultaneously come together overhead, close to the axis of rotation. These movements dramatically reduce moment of inertia, thus increasing angular velocity (12).

Similarly, when a skater performs a double or triple axel in figure skating, angular momentum is generated by the skater's movements and changes in total-body moment of inertia prior to takeoff. Over half of the angular momentum for a double axel is generated during the preparatory glide on one skate going into the jump (2). Most of this angular momentum is contributed by motion of the free leg, which is extended somewhat horizontally to increase total-body moment of inertia around the skater's vertical axis (2). As the skater becomes airborne, both legs are extended vertically, and the arms are tightly crossed. Because angular velocity is controlled primarily by the skater's moment of inertia, tight positioning of the arms and legs close to the axis of rotation is essential for maximizing revolutions while airborne (10).

In overhead sporting movements such as throwing a ball, striking a volleyball, or serving in tennis, the arm functions as what has been described as a "kinetic chain." Accordingly, as the arm moves forcefully forward, angular momentum is progressively transferred from proximal to distal segments. During the overarm throw, the motions of elbow extension and wrist flexion are accelerated by the motions of the trunk and upper arm, as angular momentum is transferred from segment to segment (8). In the performance of the tennis serve, angular momentum is produced by the movements of the trunk, arms, and legs, with a transfer of momentum from the extending lower extremity and rotating trunk to the racket arm, and finally to the racket.

The arm swing during takeoff contributes significantly to the diver's angular momentum.

The surface reaction force is used by the dancer to generate angular momentum during the takeoff of the tour jeté.

ANGULAR ANALOGUES OF NEWTON'S LAWS OF MOTION

Table 14-1 presents linear and angular kinetic quantities in a parallel format. With the many parallels between linear and angular motion, it is not surprising that Newton's laws of motion may also be stated in terms of angular motion. It is necessary to remember that torque and moment of inertia are the angular equivalents of force and mass in substituting terms.

Newton's First Law

The angular version of the first law of motion may be stated as follows:

> A rotating body will maintain a state of constant rotational motion unless acted on by an external torque.

In the analysis of human movement in which mass remains constant throughout, this angular analogue forms the underlying basis for the principle of conservation of angular momentum. Because angular velocity may change to compensate for changes in moment of inertia resulting from alterations in the radius of gyration, the quantity that remains constant in the absence of external torque is angular momentum.

Newton's Second Law

In angular terms, Newton's second law may be stated algebraically and in words as the following:

$$T = I\alpha$$

> A net torque produces angular acceleration of a body that is directly proportional to the magnitude of the torque, in the same direction as the torque, and inversely proportional to the body's moment of inertia.

In accordance with Newton's second law for angular motion, the angular acceleration of the forearm is directly proportional to the magnitude of the net torque at the elbow and in the direction (flexion) of the net torque at the elbow. The greater the moment of inertia is with respect to the axis of rotation at the elbow, the smaller is the resulting angular acceleration (see Sample Problem 14.4).

Newton's Third Law

The law of reaction may be stated in angular form as the following:

> For every torque exerted by one body on another, there is an equal and opposite torque exerted by the second body on the first.

When a baseball player forcefully swings a bat, rotating the mass of the upper body, a torque is created around the player's longitudinal axis. If

TABLE 14-1

Linear and Angular Kinetic Quantities

LINEAR	ANGULAR
mass (m)	moment of inertia (I)
force (F)	torque (T)
momentum (M)	angular momentum (H)
impulse (Ft)	angular impulse (Tt)

SAMPLE PROBLEM 14.4

The knee extensors insert on the tibia at an angle of 308 at a distance of 3 cm from the axis of rotation at the knee. How much force must the knee extensors exert to produce an angular acceleration at the knee of 1 rad/s^2, given a mass of the lower leg and foot of 4.5 kg and k = 23 cm?

Known

$$d = 0.03 \text{ m}$$
$$\alpha = 1 \text{ rad/s}^2$$
$$m = 4.5 \text{ kg}$$
$$k = 0.23 \text{ m}$$

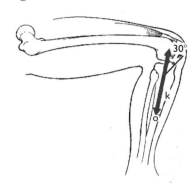

Solution

The angular analogue of Newton's second law of motion may be used to solve the problem:

$$T = I\alpha$$
$$Fd = mk^2\alpha$$
$$(F \sin 30 \text{ N})(0.03 \text{ m}) = (4.5 \text{ kg})(0.23 \text{ m})^2(1 \text{ rad/s}^2)$$

$$\boxed{F = 15.9 \text{ N}}$$

the batter's feet are not firmly planted, the lower body tends to rotate around the longitudinal axis in the opposite direction. However, since the feet usually are planted, the torque generated by the upper body is translated to the ground, where the earth generates a torque of equal magnitude and opposite direction on the cleats of the batter's shoes.

CENTRIPETAL FORCE

Bodies undergoing rotary motion around a fixed axis are also subject to a linear force. When an object attached to a line is whirled around in a circular path and then released, the object flies off on a path that forms a tangent to the circular path it was following at the point at which it was released, since this is the direction it was traveling in at the point of release (Figure 14-14). Centripetal force prevents the rotating body from leaving its circular path while rotation occurs around a fixed axis. The

centripetal force
force directed toward the center of rotation for a body in rotational motion

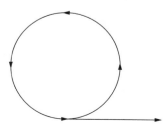

FIGURE 14-14

An object swung in a circle and then released will follow a linear path tangential to the curve at the point of release, since this is the direction of motion at the point of release.

Cyclists and runners lean into a curve to offset the torque created by centripetal force acting on the base of support.

direction of a centripetal force is always toward the center of rotation; this is the reason it is also known as *center-seeking force*. Centripetal force produces the radial component of the acceleration of a body traveling on a curved path (see Chapter 11). The following formula quantifies the magnitude of a centripetal force in terms of the tangential linear velocity of the rotating body:

$$F_c = \frac{mv^2}{r}$$

In this formula, F_c is centripetal force, m is mass, v is the tangential linear velocity of the rotating body at a given point in time, and r is the radius of rotation. Centripetal force may also be defined in terms of angular velocity:

$$F_c = mr\omega^2$$

As is evident from both equations, the speed of rotation is the most influential factor on the magnitude of centripetal force, because centripetal force is proportional to the square of velocity or angular velocity.

When a cyclist rounds a curve, the ground exerts centripetal force on the tires of the cycle. The forces acting on the cycle/cyclist system are weight, friction, and the ground reaction force (Figure 14-15). The horizontal component of the ground reaction force and laterally directed friction provide the centripetal force, which also creates a torque about the cycle/cyclist CG. To prevent rotation toward the outside of the curve, the cyclist must lean to the inside of the curve so that the moment arm of the system's weight relative to the contact point with the ground is large enough to produce an oppositely directed torque of equal magnitude. In the absence of leaning into the curve, the cyclist would have to reduce speed to reduce the magnitude of the ground reaction force, in order to prevent loss of balance.

When rounding a corner in an automobile, there is a sensation of being pushed in the direction of the outside of the curve. What is felt has been referred to as *centrifugal force*. What is actually occurring, however, is that in accordance with Newton's first law, the body's inertia tends to

FIGURE 14-15

Free body diagram of a cyclist on a curve. R_H is centripetal force. When the cyclist is balanced, summing torques at the cyclist's CG, $(R_v)(d_{R_v}) = (R_H)(d_{R_H})$.

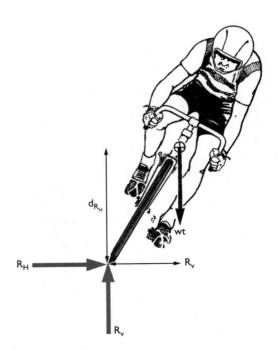

TABLE 14-2

Formula Summary

DESCRIPTION	FORMULA
Moment of Inertia = (mass)(radius of gyration squared)	$I = mk^2$
Angular momentum = (moment of inertia)(angular velocity)	$H = I\omega$
Local term for angular momentum = (segment moment of inertia about segment CG)(segment angular velocity about segment CG)	$H_l = I_s v_s$
Remote term for angular momentum = (segment mass) (distance between total body and segment CGs, squared) (angular velocity of segment about the principal axis)	$H_r = mr^2\omega_g$
Angular impulse = change in angular momentum	$Tt = \Delta H$ $Tt = (mk^2\omega)_2 - (mk^2\omega)_1$
Newton's second law (rotational version)	$T = I\alpha$
Centripetal force $= \dfrac{(mass)\ (velocity\ squared)}{radius\ of\ rotation}$	$F_c = \dfrac{mv^2}{r}$
Centripetal force = (mass)(radius of rotation)(angular velocity squared)	$F_c = mr\omega^2$

cause it to continue traveling on a straight, rather than a curved, path. The car seat, the seat belt, and possibly the car door provide a reaction force that changes the direction of body motion. "Centrifugal force," then, is a fictitious force that might more appropriately be described as the absence of centripetal force acting on an object.

Table 14-2 summarizes the formulas used in this chapter.

SUMMARY

Whereas a body's resistance to linear acceleration is proportional to its mass, resistance to angular acceleration is related to both mass and the distribution of mass with respect to the axis of rotation. Resistance to angular acceleration is known as *moment of inertia*, a quantity that incorporates both the amount of mass and its distribution relative to the center of rotation.

Just as linear momentum is the product of the linear inertial property (mass) and linear velocity, angular momentum is the product of moment of inertia and angular velocity. In the absence of external torques, angular momentum is conserved. An airborne human performer can alter total-body angular velocity by manipulating moment of inertia through changes in body configuration relative to the principal axis around which rotation is occurring. Skilled performers can also alter the axis of rotation and initiate rotation when no angular momentum is present while airborne. The principle of conservation of angular momentum is based on the angular version of Newton's first law of motion. The second and third laws of motion may also be expressed in angular terms by substituting moment of inertia for mass, torque for force, and angular acceleration for linear acceleration.

A linear force that acts on all rotating bodies is centripetal (or a center-seeking) force, which is always directed toward the center of rotation. The magnitude of centripetal force depends on the mass, speed, and radius of rotation of the rotating body.

INTRODUCTORY PROBLEMS

1. If you had to design a model of the human body composed entirely of regular geometric solids, which solid shapes would you choose? Using a straightedge, sketch a model of the human body that incorporates the solid shapes you have selected.

2. Construct a table displaying common units of measure for both linear and angular quantities of the inertial property, momentum, and impulse.

3. Skilled performance of a number of sport skills is characterized by "follow-through." Explain the value of "follow-through" in terms of the concepts discussed in this chapter.

4. Explain the reason the product of body mass and body height squared is a good predictor of body moment of inertia in children.

5. A 1.1 kg racquet has a moment of inertia about a grip axis of rotation of 0.4 kg $\cdot$ m^2. What is its radius of gyration? (Answer: 0.6 m)

6. How much angular impulse must be supplied by the hamstrings to bring a leg swinging at 8 rad/s to a stop, given that the leg's moment of inertia is 0.7 kg $\cdot$ m^2? (Answer: 5.6 kg $\cdot$ m^2/s).

7. Given the following principal transverse axis moments of inertia and angular velocities, calculate the angular momentum of each of the following gymnasts. What body configurations do these moments of inertia represent?

	I_{cg}(kg $\cdot$ m^2)	(rad/s)
A	3.5	20.00
B	7.0	10.00
C	15.0	4.67

(Answers: A = 70 kg $\cdot$ m^2/s; B = 70 kg $\cdot$ m^2/s; C = 70 kg $\cdot$ m^2/s)

8. A volleyball player's 3.7 kg arm moves at an average angular velocity of 15 rad/s during execution of a spike. If the average moment of inertia of the extending arm is 0.45 kg $\cdot$ m^2, what is the average radius of gyration for the arm during the spike? (Answer: 0.35 m)

9. A 50 kg diver in a full layout position, with a total body radius of gyration with respect to her transverse principal axis equal to 0.45 m, leaves a springboard with an angular velocity of 6 rad/s. What is the diver's angular velocity when she assumes a tuck position, reducing her radius of gyration to 0.25 m? (Answer: 19.4 rad/s)

10. If the centripetal force exerted on a swinging tennis racket by a player's hand is 40 N, how much reaction force is exerted on the player by the racket? (Answer: 40 N)

ADDITIONAL PROBLEMS

1. The radius of gyration of the thigh with respect to the transverse axis at the hip is 54% of the segment length. The mass of the thigh is 10.5% of total body mass, and the length of the thigh is 23.2% of total body height. What is the moment of inertia of the thigh with respect to the hip for males of the following body masses and heights?

	MASS (kg)	HEIGHT (m)
A	60	1.6
B	60	1.8
C	70	1.6
D	70	1.8

(Answers: A = 0.25 kg · m², B = 0.32 kg · m², C = 0.30 kg · m², D = 0.37 kg · m²)

2. Select three sport or daily living implements, and explain the ways in which you might modify each implement's moment of inertia with respect to the axis of rotation to adapt it for a person of impaired strength.

3. A 0.68 kg tennis ball is given an angular momentum of $2.72 \cdot 10^{-3} \cdot$ m²/s when struck by a racket. If its radius of gyration is 2 cm, what is its angular velocity? (Answer: 10 rad/s)

4. A 7.27 kg shot makes seven complete revolutions during its 2.5 s flight. If its radius of gyration is 2.54 cm, what is its angular momentum? (Answer: 0.0817 kg · m²/s)

5. What is the resulting angular acceleration of a 1.7 kg forearm and hand when the forearm flexors, attaching 3 cm from the center of rotation at the elbow, produce 10 N of tension, given a 90° angle at the elbow and a forearm and hand radius of gyration of 20 cm? (Answer: 4.41 rad/s²)

6. The patellar tendon attaches to the tibia at a 20° angle 3 cm from the axis of rotation at the knee. If the tension in the tendon is 400 N, what is the resulting acceleration of the 4.2 kg lower leg and foot given a radius of gyration of 25 cm for the lower leg/foot with respect to the axis of rotation at the knee? (Answer: 15.6 rad/s²)

7. A cavewoman swings a 0.75 m sling of negligible weight around her head with a centripetal force of 220 N. What is the initial velocity of a 9 N stone released from the sling? (Answer: 13.4 m/s)

8. A 7.27 kg hammer on a 1 m wire is released with a linear velocity of 28 m/s. What reaction force is exerted on the thrower by the hammer at the instant before release? (Answer: 5.7 kN)

9. Discuss the effect of banking a curve on a racetrack. Construct a free body diagram to assist with your analysis.

10. Using the data in Appendix D, calculate the locations of the radii of gyration of all body segments with respect to the proximal joint center for a 1.7 m tall woman.

LABORATORY EXPERIENCES

1. At the *Basic Biomechanics* Online Learning Center (www.mhhe.com/hall6e), go to Student Center, Chapter 14, Lab Manual, Lab 1, and then view the Angular Momentum Animation of a swinging ball on a rope wrapping around a pole. Answer the following questions.

   a. As the rope winds around the pole, what happens to the angular velocity of the ball?

   _____

   b. As the rope winds around the pole, what happens to the radius of rotation of the ball?

   _____

   c. As the rope winds around the pole, what happens to the angular momentum of the ball?

   Explain: _____

2. View either a video or a live performance of a long jump from the side view. Explain the motions of the jumper's arms and legs in terms of the concepts presented in this chapter.

Contributions of arms:_____

_____

_____

_____

_____

Contributions of legs: _____

_____

_____

_____

_____

3. View either a video or a live performance of a dive incorporating a pike or a somersault from the side view. Explain the motions of the diver's arms and legs in terms of the concepts presented in this chapter.

Contributions of arms:_____

_____

_____

_____

_____

Contributions of legs:_____

_____

_____

_____

_____

4. Stand on a rotating platform with both arms abducted at the shoulders at 90°, and have a partner spin you at a moderate angular velocity. Once the partner has let go, quickly fold your arms across your chest, being careful not to lose your balance. Write a paragraph explaining the change in angular velocity.

Explanation: _____

_____

_____

_____

_____

Is angular momentum in this situation constant? _____

5. Stand on a rotating platform and use a hula-hooping motion of the hips to generate rotation. Explain how total body rotation results.

Explanation:_____

_____

_____

_____

_____

REFERENCES

1. Ackland T, Elliott B, and Tichards J: Growth in body size affects rotational performance in women's gymnastics, *Sports Biomech* 2:163, 2003.
2. Albert WJ and Miller DI: Takeoff characteristics of single and double axel figure skating jumps, *J Appl Biomech* 12:72, 1996.
3. Brüggemann G-P: Biomechanics in gymnastics. In Van Gheluwe B and Atha J: *Current research in sports biomechanics,* Basel, 1987, Karger.
4. Dapena J and McDonald C: A three-dimensional analysis of angular momentum in the hammer throw, *Med Sci Sports Exerc* 21:206, 1989.
5. Frohlich C: The physics of somersaulting and twisting, *Sci Am* 242:154, 1980.
6. Hamill J, Ricard MD, and Golden DM: Angular momentum in multiple rotation nontwisting platform dives, *Int J Sport Biomech* 2:78, 1986.
7. Hay JG: *The biomechanics of sports techniques* (3rd ed.), Englewood Cliffs, NJ, 1985, Prentice Hall.
8. Hirashima M, Yamane K, Nakamura Y, and Ohtsuki T: Kinetic chain of overarm throwing in terms of joint rotations revealed by induced acceleration analysis, *J Biomech* 41:2874, 2008.
9. Jones IC and Miller DI: Influence of fulcrum position on springboard response and takeoff performance in the running approach, *J Appl Biomech* 12:383, 1996.
10. King DL: Performing triple and quadruple figure skating jumps: implications for training, *Can J Appl Physiol* 30:743, 2005.
11. King MA and Yeadon MR: Maximising somersault rotation in tumbling, *J Biomech* 37:471, 2004.
12. Laws K: *The physics of dance,* New York, 1984, Schirmer Books.
13. Miller DI, Jones IC, and Pizzimenti MA: Taking off: Greg Louganis' diving style, *Soma* 2:20, 1988.
14. Miller DI and Sprigings EJ: Factors influencing the performance of springboard dives of increasing difficulty, *J Appl Biomech* 17:217, 2001.
15. Miller DI, et al: Kinetic and kinematic characteristics of 10-m platform performances of elite divers, I: back takeoffs, *Int J Sport Biomech* 5:60, 1989.
16. Nesbit SM et al: A discussion of iron golf club head inertia tensors and their effects on the golfer, *J Appl Biomech* 12:449, 1996.
17. Sanders RH and Wilson BD: Angular momentum requirements of the twisting and nontwisting forward 1½ somersault dive, *Int J Sport Biomech* 3:47, 1988.
18. Southard D and Groomer L: Warm-up with baseball bats of varying moments of inertia: effect on bat velocity and swing pattern, *Res Q Exerc Sport* 74:270, 2003.
19. Sprigings EJ and Miller DI: Optimal knee extension timing in springboard and platform dives from the reverse group, *J Appl Biomech* 20:275, 2004.
20. Yeadon MR: The biomechanics of the human in flight, *Am J Sports Med* 25:575, 1997.

ANNOTATED READINGS

George GS: *Championship gymnastics: Biomechanical techniques for shaping winners,* Carlsbad, CA, 2010, Designs for Wellness Press.
Explains principles of biomechanics related to performance in gymnastics in layman's terms.

Hellström J: Competitive elite golf: a review of the relationships between playing results, technique and physique, *Sports Med* 39:723, 2009.
Reviews the scientific literature related to elite golfer's playing results, technique and physique, including discussion of sequential increase of body segment angular velocities during the swing and moment of inertia of the club head.

Miller DI, Jones IC, and Pizzimenti MA: Taking off: Greg Louganis' diving style, *Soma* 2:20, 1988.
Discusses the linear and angular momentum requirements for performing total body rotations at the world-class level and describes the methods by which these factors may be studied.

Yeadon MR and Mikulcik EC: Stability and control of aerial movements. In Nigg BM, Stefanyshyn D, and Denoth J: *Biomechanics and biology of movement*, Champaign, IL, 2000, Human Kinetics.
Presents a detailed description of strategies and mechanical concepts related to control of rotations and twists during the execution of aerial skills in diving and gymnastics.

RELATED WEBSITES

Centrifugal Force: The False Force
http://regentsprep.org/Regents/physics/phys06/bcentrif/centrif.htm
  Includes description and simulation of a car going around a curve to illustrate the effect of inertia on the contents of the car.
Conceptest on Centripetal Force
http://motor1.physics.wayne.edu/~cinabro/cinabro/education/conceptest15.html
  Poses several questions related to centripetal force with a link to an answer page.
Exploratorium: Angular Momentum
http://www.exploratorium.edu/snacks/momentum_machine/index.html
  Provides description of a simple experiment that illustrates angular momentum.
Angular Momentum: Projects Provided by Students
http://library.thinkquest.org/3042/angular.html
  Includes several practical examples of the effects of angular momentum.

KEY TERMS

angular impulse	change in angular momentum equal to the product of torque and time interval over which the torque acts
angular momentum	quantity of angular motion possessed by a body; measured as the product of moment of inertia and angular velocity
centripetal force	force directed toward the center of rotation for a body in rotational motion
moment of inertia	inertial property for rotating bodies representing resistance to angular acceleration; based on both mass and the distance the mass is distributed from the axis of rotation
principal axes	three mutually perpendicular axes passing through the total body center of gravity
principal moment of inertia	total-body moment of inertia relative to one of the principal axes
radius of gyration	distance from the axis of rotation to a point where the body's mass could be concentrated without altering its rotational characteristics

Human Movement in a Fluid Medium

After completing this chapter, you will be able to:

Explain the ways in which the composition and flow characteristics of a fluid affect fluid forces.

Define *buoyancy* and explain the variables that determine whether a human body will float.

Define *drag*, identify the components of drag, and identify the factors that affect the magnitude of each component.

Define *lift* and explain the ways in which it can be generated.

Discuss the theories regarding propulsion of the human body in swimming.

ONLINE LEARNING CENTER RESOURCES

www.mhhe.com/hall6e

Log on to our Online Learning Center (OLC) for access to these additional resources:

- Online Lab Manual
- Flashcards with definitions of chapter key terms
- Chapter objectives
- Chapter lecture PowerPoint presentation
- Self-scoring chapter quiz
- Additional chapter resources
- Web links for study and exploration of chapter-related topics

The ability to control the action of fluid forces differentiates elite from average swimmers.

W hy are there dimples in a golf ball? Why are some people able to float while others cannot? Why are cyclists, swimmers, downhill skiers, and speed skaters concerned with streamlining their bodies during competition?

Both air and water are fluid mediums that exert forces on bodies moving through them. Some of these forces slow the progress of a moving body; others provide support or propulsion. A general understanding of the actions of fluid forces on human movement activities is an important component of the study of the biomechanics of human movement. This chapter introduces the effects of fluid forces on both human and projectile motion.

THE NATURE OF FLUIDS

fluid
substance that flows when subjected to a shear stress

●*Air and water are fluids that exert forces on the human body.*

●*The velocity of a body relative to a fluid influences the magnitude of the forces exerted by the fluid on the body.*

Although in general conversation the term *fluid* is often used interchangeably with the term *liquid*, from a mechanical perspective, a fluid is any substance that tends to flow or continuously deform when acted on by a shear force. Both gases and liquids are fluids with similar mechanical behaviors.

Relative Motion

Because a fluid is a medium capable of flow, the influence of the fluid on a body moving through it depends not only on the body's velocity but also on the velocity of the fluid. Consider the case of waders standing in the shallow portion of a river with a moderately strong current. If they stand still,

FIGURE 15-1

The relative velocity of a moving body with respect to a fluid is equal to the vector subtraction of the velocity of the wind from the velocity of the body.

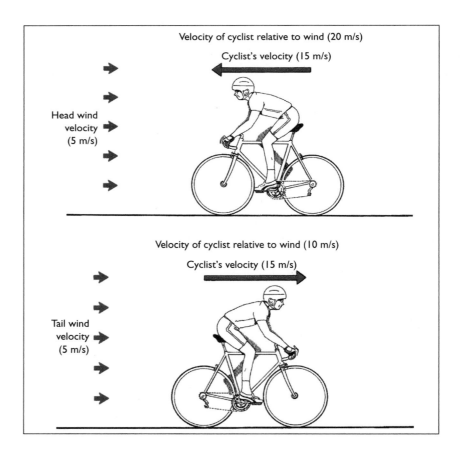

Velocity of cyclist relative to wind (20 m/s)

Cyclist's velocity (15 m/s)

Head wind velocity (5 m/s)

Velocity of cyclist relative to wind (10 m/s)

Cyclist's velocity (15 m/s)

Tail wind velocity (5 m/s)

they feel the force of the current against their legs. If they walk upstream against the current, the current's force against their legs is even stronger. If they walk downstream, the current's force is reduced and perhaps even imperceptible.

When a body moves through a fluid, the relative velocity of the body with respect to the fluid influences the magnitude of the acting forces. If the direction of motion is directly opposite the direction of the fluid flow, the magnitude of the velocity of the moving body relative to the fluid is the algebraic sum of the speeds of the moving body and the fluid (Figure 15-1). If the body moves in the same direction as the surrounding fluid, the magnitude of the body's velocity relative to the fluid is the difference in the speeds of the object and the fluid. In other words, the relative velocity of a body with respect to a fluid is the vector subtraction of the absolute velocity of the fluid from the absolute velocity of the body (see Sample Problem 15.1). Likewise, the relative velocity of a fluid with respect to a

relative velocity
velocity of a body with respect to the velocity of something else, such as the surrounding fluid

SAMPLE PROBLEM 15.1

A sailboat is traveling at an absolute speed of 3 m/s against a 0.5 m/s current and with a 6 m/s tailwind. What is the velocity of the current with respect to the boat? What is the velocity of the wind with respect to the boat?

Known

$$v_b = 3 \text{ m/s} \rightarrow$$
$$v_c = 0.5 \text{ m/s} \leftarrow$$
$$v_w = 6 \text{ m/s} \rightarrow$$

Solution

The velocity of the current with respect to the boat is equal to the vector subtraction of the absolute velocity of the boat from the absolute velocity of the current.

$$v_{c/b} = v_c - v_b$$
$$= (0.5 \text{ m/s} \leftarrow) - (3 \text{ m/s} \rightarrow)$$
$$= (3.5 \text{ m/s} \leftarrow)$$

The velocity of the current with respect to the boat is 3.5 m/s in the direction opposite that of the boat.

The velocity of the wind with respect to the boat is equal to the vector subtraction of the absolute velocity of the boat from the absolute velocity of the wind.

$$v_{w/b} = v_w - v_b$$
$$= (6 \text{ m/s} \rightarrow) - (3 \text{ m/s} \rightarrow)$$
$$= (3 \text{ m/s} \rightarrow)$$

The velocity of the wind with respect to the boat is 3 m/s in the direction in which the boat is sailing.

Laminar flow is characterized by smooth, parallel layers of fluid.

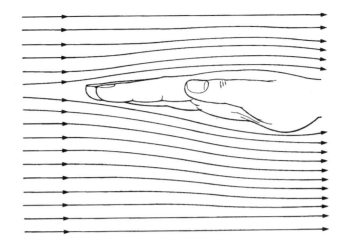

body moving through it is the vector subtraction of the velocity of the body from the velocity of the fluid.

Laminar versus Turbulent Flow

When an object such as a human hand or a canoe paddle moves through water, there is little apparent disturbance of the immediately surrounding water if the relative velocity of the object with respect to the water is low. However, if the relative velocity of motion through the water is sufficiently high, waves and eddies appear.

When an object moves with sufficiently low velocity relative to any fluid medium, the flow of the adjacent fluid is termed laminar flow. Laminar flow is characterized by smooth layers of fluid molecules flowing parallel to one another (Figure 15-2).

When an object moves with sufficiently high velocity relative to a surrounding fluid, the layers of fluid near the surface of the object mix, and the flow is termed *turbulent*. The rougher the surface of the body, the lower the relative velocity at which turbulence is caused. Laminar flow and turbulent flow are distinct categories. If any turbulence is present, the flow is nonlaminar. The nature of the fluid flow surrounding an object can dramatically affect the fluid forces exerted on the object. In the case of the human body during swimming, flow is neither completely laminar nor completely turbulent, but transitional between the two (30).

laminar flow
flow characterized by smooth, parallel layers of fluid

turbulent flow
flow characterized by mixing of adjacent fluid layers

Fluid Properties

Other factors that influence the magnitude of the forces a fluid generates are the fluid's density, specific weight, and viscosity. As discussed in Chapter 3, density (ρ) is defined as mass/volume, and the ratio of weight to volume is known as specific weight (γ). The denser and heavier the fluid medium surrounding a body, the greater the magnitude of the forces the fluid exerts on the body. The property of fluid viscosity involves the internal resistance of a fluid to flow. The greater the extent to which a fluid resists flow under an applied force, the more viscous the fluid is. A thick molasses, for example, is more viscous than a liquid honey, which is more viscous than water. Increased fluid viscosity results in increased forces exerted on bodies exposed to the fluid.

●*Atmospheric pressure and temperature influence a fluid's density, specific weight, and viscosity.*

Atmospheric pressure and temperature influence a fluid's density, specific weight, and viscosity, with more mass concentrated in a given unit of fluid volume at higher atmospheric pressures and lower temperatures.

FLUID*	DENSITY (kg/m³)	SPECIFIC WEIGHT (n/m³)	VISCOSITY (ns/m²)
Air	1.20	11.8	.000018
Water	998	9,790	.0010
Seawater⁺	1,026	10,070	.0014
Ethyl alcohol	799	7,850	.0012
Mercury	13,550.20	133,000.0	.0015

*Fluids are measured at 20°C and standard atmospheric pressure.
⁺10°C, 3.3% salinity.

TABLE 15-1

Approximate Physical Properties of Common Fluids

Because molecular motion in gases increases with temperature, the viscosity of gases also increases. The viscosity of liquids decreases with increased temperature because of a reduction in the cohesive forces among the molecules. The densities, specific weights, and viscosities of common fluids are shown in Table 15-1.

BUOYANCY

Characteristics of the Buoyant Force

Buoyancy is a fluid force that always acts vertically upward. The factors that determine the magnitude of the buoyant force were originally explained by the ancient Greek mathematician Archimedes. Archimedes' principle states that the magnitude of the buoyant force acting on a given body is equal to the weight of the fluid displaced by the body. The latter factor is calculated by multiplying the specific weight of the fluid by the volume of the portion of the body that is surrounded by the fluid. Buoyancy (F_b) is calculated as the product of the displaced volume (V_d) and the fluid's specific weight γ:

$$F_b = V_d \gamma$$

For example, if a water polo ball with a volume of 0.2 m³ is completely submerged in water at 20°C, the buoyant force acting on the ball is equal to the ball's volume multiplied by the specific weight of water at 20°C:

$$\begin{aligned} F_b &= V_d \gamma \\ &= (0.2 \text{ m}^3)(9790 \text{ N/m}^3) \\ &= 1958 \text{ N} \end{aligned}$$

The more dense the surrounding fluid, the greater the magnitude of the buoyant force. Since seawater is more dense than freshwater, a given object's buoyancy is greater in seawater than in freshwater. Because the magnitude of the buoyant force is directly related to the volume of the submerged object, the point at which the buoyant force acts is the object's center of volume, which is also known as the *center of buoyancy*. The center of volume is the point around which a body's volume is equally distributed.

Flotation

The ability of a body to float in a fluid medium depends on the relationship between the body's buoyancy and its weight. When weight and the buoyant force are the only two forces acting on a body and their magnitudes are

Archimedes' principle
physical law stating that the buoyant force acting on a body is equal to the weight of the fluid displaced by the body

center of volume
point around which a body's volume is equally distributed and at which the buoyant force acts

equal, the body floats in a motionless state, in accordance with the principles of static equilibrium. If the magnitude of the weight is greater than that of the buoyant force, the body sinks, moving downward in the direction of the net force.

Most objects float statically in a partially submerged position. The volume of a freely floating object needed to generate a buoyant force equal to the object's weight is the volume that is submerged.

Flotation of the Human Body

In the study of biomechanics, buoyancy is most commonly of interest relative to the flotation of the human body in water. Some individuals cannot float in a motionless position, and others float with little effort. This difference in floatability is a function of body density. Since the density of bone and muscle is greater than the density of fat, individuals who are extremely muscular and have little body fat have higher average body densities than individuals with less muscle, less dense bones, or more body fat. If two individuals have an identical body volume, the one with the higher body density weighs more. Alternatively, if two people have the same body weight, the person with the higher body density has a smaller body volume. For flotation to occur, the body volume must be large enough to create a buoyant force greater than or equal to body weight (see Sample Problem 15.2). Many individuals can float only when holding a large volume of inspired air in the lungs, a tactic that increases body volume without altering body weight.

● *In order for a body to float, the buoyant force it generates must equal or exceed its weight.*

The orientation of the human body as it floats in water is determined by the relative position of the total-body center of gravity relative to the total-body center of volume. The exact locations of the center of gravity and center of volume vary with anthropometric dimensions and body composition. Typically, the center of gravity is inferior to the center of volume due to the relatively large volume and relatively small weight of the lungs. Because weight acts at the center of gravity and buoyancy acts at the center of volume, a torque is created that rotates the body until it is positioned so that these two acting forces are vertically aligned and the torque ceases to exist (Figure 15-3).

● *People who cannot float in swimming pools may float in Utah's Great Salt Lake, in which the density of the water surpasses even that of seawater.*

When beginning swimmers try to float on their back, they typically assume a horizontal body position. Once the swimmer relaxes, the lower end of the body sinks, because of the acting torque. An experienced teacher instructs beginning swimmers to assume a more diagonal position in the water before relaxing into the back float. This position minimizes torque and the concomitant sinking of the lower extremity. Other strategies that a swimmer can use to reduce torque on the body when entering a back float position include extending the arms backward in the water above the head and flexing the knees. Both tactics elevate the location of the center of gravity, positioning it closer to the center of volume.

During swimming with the front crawl stroke, the center of buoyancy is shifted toward the feet when the recovery arm and part of the head are above the surface of the water. At this point in the stroke cycle, the buoyant torque tends to elevate the feet, rather than the reverse (49).

DRAG

Drag is a force caused by the dynamic action of a fluid that acts in the direction of the free-stream fluid flow. Generally, a drag is a *resistance* force: a force that slows the motion of a body moving through a fluid. The

SAMPLE PROBLEM 15.2

When holding a large quantity of inspired air in her lungs, a 22 kg girl has a body volume of 0.025 m^3. Can she float in fresh water if γ equals 9810 N/m^3? Given her body volume, how much could she weigh and still be able to float?

Known

$$m = 22 \text{ kg}$$
$$V = 0.025 \text{ m}^3$$
$$\gamma = 9810 \text{ N/m}^3$$

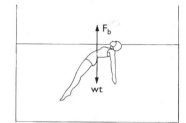

Solution

Two forces are acting on the girl: her weight and the buoyant force. According to the conditions of static equilibrium, the sum of the vertical forces must be equal to zero for the girl to float in a motionless position. If the buoyant force is less than her weight, she will sink, and if the buoyant force is equal to her weight, she will float completely submerged. If the buoyant force is greater than her weight, she will float partly submerged. The magnitude of the buoyant force acting on her total body volume is the product of the volume of displaced fluid (her body volume) and the specific weight of the fluid:

$$F_b = V_\gamma$$
$$= (0.025 \text{ m}^3)(9810 \text{ N/m}^3)$$
$$= 245.52 \text{ N}$$

Her body weight is equal to her body mass multiplied by the acceleration of gravity:

$$wt = (22 \text{ kg})(9.81 \text{ m/s}^2)$$
$$= 215.82 \text{ N}$$

Since the buoyant force is greater than her body weight, the girl will float partly submerged in freshwater.

> Yes, she will float.

To calculate the maximum weight that the girl's body volume can support in freshwater, multiply the body volume by the specific weight of water.

$$wt_{max} = (0.025 \text{ m}^3)(9810 \text{ N/m}^3)$$
$$wt_{max} = 245.25 \text{ N}$$

---

drag force acting on a body in relative motion with respect to a fluid is defined by the following formula:

$$F_D = \tfrac{1}{2} C_D \rho A_p v^2$$

In this formula, F_D is drag force, C_D is the coefficient of drag, ρ is the fluid density, A_p is the projected area of the body or the surface area of the body oriented perpendicular to the fluid flow, and v is the relative velocity

coefficient of drag
unitless number that is an index of a body's ability to generate fluid resistance

FIGURE 15-3

A. A torque is created on a swimmer by body weight (acting at the center of gravity) and the buoyant force (acting at the center of volume). **B.** When the center of gravity and the center of volume are vertically aligned, this torque is eliminated.

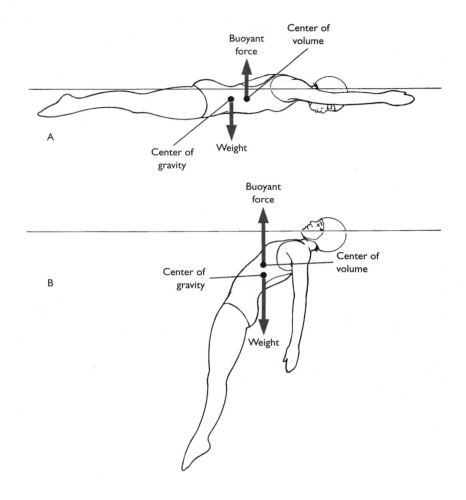

of the body with respect to the fluid. The coefficient of drag is a unitless number that serves as an index of the amount of drag an object can generate. Its size depends on the shape and orientation of a body relative to the fluid flow, with long, streamlined bodies generally having lower coefficients of drag than blunt or irregularly shaped objects. Approximate coefficients of drag for the human body in positions commonly assumed during participation in several sports are shown in Figure 15-4.

The formula for the total drag force demonstrates the exact way in which each of the identified factors affects drag. If the coefficient of drag, the fluid density, and the projected area of the body remain constant, drag increases with the square of the relative velocity of motion. This relationship is referred to as the theoretical square law. According to this law, if cyclists double their speed and other factors remain constant, the drag force opposing them increases fourfold. The effect of drag is more consequential when a body is moving with a high velocity, which occurs in sports such as cycling, speed skating, downhill skiing, the bobsled, and the luge.

In swimming, the drag on a moving body is 500–600 times higher than it would be in the air, with the magnitude of drag varying with the anthropometric characteristics of the individual swimmer, as well as with the stroke used (41). Researchers distinguish between passive drag, which is generated by the swimmer's body size, shape, and position in the water, and active drag, which is associated with the swimming motion. Passive drag is inversely related to a swimmer's buoyancy, which has been found to have a small but important influence on sprint swimming perfor-

theoretical square law
drag increases approximately with the square of velocity when relative velocity is low

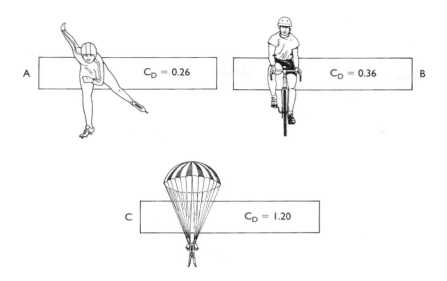

FIGURE 15-4

Approximate coefficients of
drag for the human body.
A. Frontal drag on a speed
skater. **B.** Frontal drag on a
cyclist in touring position.
C. Vertical drag on a
parachutist falling with
the parachute fully opened.
Modified from Roberson JA and
Crowe CT: *Engineering fluid
mechanics* (2nd ed), Boston, 1980,
Houghton Mifflin.

mance (28). Passive drag on male swimmers is also significantly reduced
with shoulder-to-knee and shoulder-to-ankle swimsuits as compared to
briefs (30).

Three forms of resistance contribute to the total drag force. The component of resistance that predominates depends on the nature of the fluid
flow immediately adjacent to the body.

Skin Friction

One component of the total drag is known as skin friction, surface drag, or
viscous drag. This drag is similar to the friction force described in Chapter
12. Skin friction is derived from the sliding contacts between successive
layers of fluid close to the surface of a moving body (Figure 15-5). The
layer of fluid particles immediately adjacent to the moving body is slowed
because of the shear stress the body exerts on the fluid. The next adjacent
layer of fluid particles moves with slightly less speed because of friction
between the adjacent molecules, and the next layer is affected in turn.
The number of layers of affected fluid becomes progressively larger as the
flow moves in the downstream direction along the body. The entire region
within which fluid velocity is diminished because of the shearing resistance caused by the boundary of the moving body is the boundary layer.
The force the body exerts on the fluid in creating the boundary layer

skin friction
surface drag
viscous drag
*resistance derived from friction
between adjacent layers of fluid near
a body moving through the fluid*

boundary layer
*layer of fluid immediately adjacent
to a body*

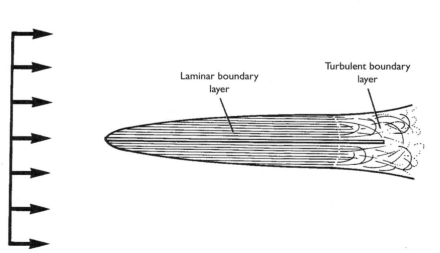

Laminar boundary
layer

Turbulent boundary
layer

FIGURE 15-5

The fluid boundary layer for
a thin, flat plate, shown from
the side view. The laminar
boundary layer gradually
becomes thicker as flow
progresses along the plate.

results in an oppositely directed reaction force exerted by the fluid on the body. This reaction force is known as *skin friction.*

Several factors affect the magnitude of skin friction drag. It increases proportionally with increases in the relative velocity of fluid flow, the surface area of the body over which the flow occurs, the roughness of the body surface, and the viscosity of the fluid. Skin friction is always one component of the total drag force acting on a body moving relative to a fluid, and it is the major form of drag present when the flow is primarily laminar. For front crawl swimming, kayaking, and rowing, skin friction drag predominates at velocities between 1 and 3 m/s (35).

Among these factors, the one that a competitive athlete can readily alter is the relative roughness of the body surface. Athletes can wear tight-fitting clothing composed of a smooth fabric rather than loose-fitting clothing or clothing made of a rough fabric. A 10% reduction of drag occurs when a speed skater wears a smooth spandex suit as opposed to the traditional wool outfit (46). A 6% decrease in air resistance results from cyclists' use of appropriate clothing, including sleeves, tights, and smooth covers over the laces of the shoes (21). Competitive male swimmers and cyclists often shave body hair to reduce skin friction.

The other factor affecting skin friction that athletes can alter in some circumstances is the amount of surface area in contact with the fluid. Carrying an extra passenger such as a coxswain in a rowing event results in a larger wetted surface area of the hull because of the added weight; as a result, skin friction drag is increased.

Form Drag

form drag
profile drag
pressure drag
resistance created by a pressure differential between the lead and rear sides of a body moving through a fluid

A second component of the total drag acting on a body moving through a fluid is form drag, which is also known as profile drag or pressure drag. Form drag is always one component of the drag on a body moving relative to a fluid. When the boundary layer of fluid molecules next to the surface of the moving body is primarily turbulent, form drag predominates. Form drag is the major contributor to overall drag during most human and projectile motion. It is the predominant type of drag for front crawl swimming, kayaking, and rowing at velocities of less than 1 m/s (35).

When a body moves through a fluid medium with sufficient velocity to create a pocket of turbulence behind the body, an imbalance in the pressure surrounding the body—a *pressure differential*—is created (Figure 15-6). At the upstream end of the body where fluid particles meet the body head-on, a zone of relative high pressure is formed. At the downstream end of the body where turbulence is present, a zone of relative low pressure is created. Whenever a pressure differential exists, a force is directed from the region of high pressure to the region of low pressure. For example, a vacuum cleaner creates a suction force because a region of relative low pressure (the relative vacuum) exists inside the machine housing. This force, directed from front to rear of the body in relative motion through a fluid, constitutes form drag.

FIGURE 15-6

Form drag results from the suctionlike force created between the positive pressure zone on a body's leading edge and the negative pressure zone on the trailing edge when turbulence is present.

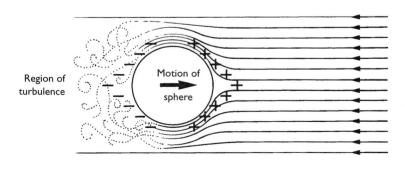

Region of turbulence

Motion of sphere

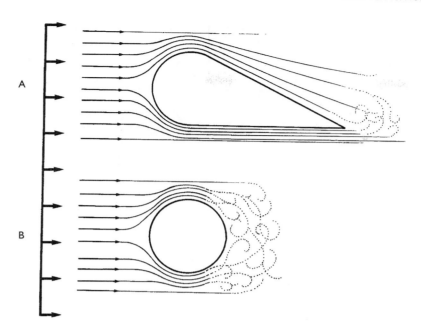

FIGURE 15-7

The effect of streamlining is
a reduction in the turbulence
created at the trailing edge of
a body in a fluid.
A. A streamlined shape.
B. A sphere.

Several factors affect the magnitude of form drag, including the relative velocity of the body with respect to the fluid, the magnitude of the pressure gradient between the front and rear ends of the body, and the size of the surface area that is aligned perpendicular to the flow. Both the size of the pressure gradient and the amount of surface area perpendicular to the fluid flow can be reduced to minimize the effect of form drag on the human body. For example, streamlining the overall shape of the body reduces the magnitude of the pressure gradient. Streamlining minimizes the amount of turbulence created and hence minimizes the negative pressure that is created at the object's rear (Figure 15-7). Assuming a more crouched body position also reduces the body's projected surface area oriented perpendicular to the fluid flow.

Competitive cyclists, skaters, and skiers assume a streamlined body position with the smallest possible area of the body oriented perpendicular to the oncoming airstream. Even though the low-crouched aeroposition assumed by competitive cyclists increases the cyclist's metabolic cost as compared to an upright position, the aerodynamic benefit is an over tenfold reduction in drag (16). Similarly, race cars, yacht hulls, and some cycling helmets are designed with streamlined shapes. The aerodynamic frame and handlebar designs for racing cycles also reduce drag (5, 37).

Streamlining is also an effective way to reduce form drag in the water. The ability to streamline body position during freestyle swimming is a characteristic that distinguishes elite from subelite performers (6). Using a triathlon wet suit can reduce the drag on a competitor swimming at a typical triathlon race pace of 1.25 m/s by as much as 14%, because the buoyant effect of the wet suit results in reduced form drag on the swimmer (11, 43).

The nature of the boundary layer at the surface of a body moving through a fluid can also influence form drag by affecting the pressure gradient between the front and rear ends of the body. When the boundary layer is primarily laminar, the fluid separates from the boundary close to the front end of the body, creating a large turbulent pocket with a large negative pressure and thereby a large form drag (Figure 15-8). In contrast, when the boundary layer is turbulent, the point of flow separation is closer to the rear end of the body, the turbulent pocket created is smaller, and the resulting form drag is smaller.

● *Streamlining helps to minimize form drag.*

FIGURE 15-8

A. Laminar flow results in an early separation of flow from the boundary and a larger drag producing wake as compared to **B**, turbulent boundary flow.

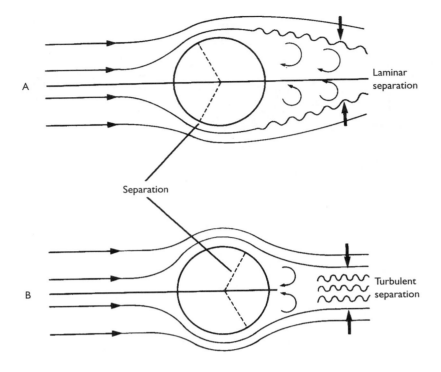

A streamlined cycling helmet.

The nature of the boundary layer depends on the roughness of the body's surface and the body's velocity relative to the flow. As the relative velocity of motion for an object such as a golf ball increases, changes in the acting drag occur (Figure 15-9). As relative velocity increases up to a certain critical point, the theoretical square law is in effect, with drag increasing with the square of velocity. After this critical velocity is reached, the boundary layer becomes more turbulent than laminar, and form drag diminishes because the pocket of reduced pressure on the trailing edge of the ball becomes smaller. As velocity increases further, the effects of skin friction and form drag grow, increasing the total drag. The dimples in a golf ball are carefully engineered to produce a turbulent boundary layer at the ball's surface that reduces form drag on the ball over the range of velocities at which a golf ball travels.

FIGURE 15-9

Drag increases approximately with the square of velocity until there is sufficient relative velocity (v_1) to generate a turbulent boundary layer. As velocity increases beyond this point, form drag decreases. After a second critical relative velocity (v_2) is reached, the drag again increases.

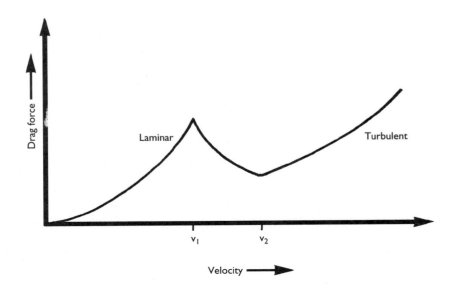

Cyclists drafting to minimize form drag.

Another way in which form drag can be manipulated is through drafting, the process of following closely behind another participant in speed-based sports such as cycling and automobile racing. Drafting provides the advantage of reducing form drag on the follower, since the leader partially shelters the follower's leading edge from increased pressure against the fluid. Depending on the size of the pocket of reduced pressure behind the leader, a suctionlike force may also help to propel the follower forward. In swimming, the optimal drafting distance behind another swimmer in a swimming pool is 0–50 cm from the toes of the lead swimmer (7). Drafting has even been found to improve performance during a long-distance swim, particularly for faster and leaner swimmers (8).

Wave Drag

The third type of drag acts at the interface of two different fluids, for example, at the interface between water and air. Although bodies that are completely submerged in a fluid are not affected by wave drag, this form of drag can be a major contributor to the overall drag acting on a human swimmer, particularly when the swim is done in open water. When a swimmer moves a body segment along, near, or across the air and water interface, a wave is created in the more dense fluid (the water). The reaction force the water exerts on the swimmer constitutes wave drag.

The magnitude of wave drag increases with greater up-and-down motion of the body and increased swimming speed. The height of the bow wave generated in front of a swimmer increases proportionally with swimming velocity, although at a given velocity, skilled swimmers produce smaller waves than less-skilled swimmers, presumably due to better technique (less up-and-down motion) (42). At fast swimming speeds (over 3 m/s), wave drag is generally the largest component of the total drag acting on the swimmer (35). For this reason, competitive swimmers typically propel themselves underwater to eliminate wave drag for a small portion of the race in events in which the rules permit it. One underwater stroke is allowed following the dive or a turn in the breaststroke, and a distance of up to 15 m is allowed underwater after a turn in the backstroke. In most swimming pools, the lane lines are designed to minimize wave action by dissipating moving surface water.

wave drag
resistance created by the generation of waves at the interface between two different fluids, such as air and water

The bow wave generated by a competitive swimmer.

LIFT FORCE

lift
force acting on a body in a fluid in a direction perpendicular to the fluid flow

While drag forces act in the direction of the free-stream fluid flow, another force, known as lift, is generated perpendicular to the fluid flow. Although the name *lift* suggests that this force is directed vertically upward, it may assume any direction, as determined by the direction of the fluid flow and the orientation of the body. The factors affecting the magnitude of lift are basically the same factors that affect the magnitude of drag:

$$F_L = \tfrac{1}{2}\, C_L \rho A_p v^2$$

coefficient of lift
unitless number that is an index of a body's ability to generate lift

In this equation, F_L represents lift force, C_L is the coefficient of lift, ρ is the fluid density, A_p is the surface area against which lift is generated, and v is the relative velocity of a body with respect to a fluid. The factors

The lane lines in modern swimming pools are designed to minimize wave action, enabling faster racing times.

TABLE 15-2

**Factors Affecting the
Magnitudes of Fluid Forces**

FORCE	FACTORS
Buoyant force	Specific weight of the fluid Volume of fluid displaced
Skin friction	Density of the fluid Relative velocity of the fluid Amount of body surface area exposed to the flow Roughness of the body surface Viscosity of the fluid
Form drag	Density of the fluid Relative velocity of the fluid Pressure differential between leading and rear edges of the body Amount of body surface area perpendicular to the flow
Wave drag	Relative velocity of the wave Amount of surface area perpendicular to the wave Viscosity of the fluid
Lift force	Relative velocity of the fluid Density of the fluid Size, shape, and orientation of the body

affecting the magnitudes of the fluid forces discussed are summarized in Table 15-2.

Foil Shape

One way in which lift force may be created is for the shape of the moving body to resemble that of a foil (Figure 15-10). When the fluid stream encounters a foil, the fluid separates, with some flowing over the curved surface and some flowing straight back along the flat surface on the opposite side. The fluid that flows over the curved surface is positively accelerated relative to the fluid flow, creating a region of relative high-velocity flow. The difference in the velocity of flow on the curved side of the foil as opposed to the flat side of the foil creates a pressure difference in the fluid, in accordance with a relationship derived by the Italian scientist Bernoulli. According to the Bernoulli principle, regions of relative high-velocity fluid flow are associated with regions of relative low pressure, and regions of relative low-velocity flow are associated with regions of relative high pressure. When these regions of relative low and high pressure are created on opposite sides of the foil, the result is a lift force directed perpendicular to the foil from the zone of high pressure toward the low-pressure zone.

foil
shape capable of generating lift in the presence of a fluid flow

Bernoulli principle
an expression of the inverse relationship between relative velocity and relative pressure in a fluid flow

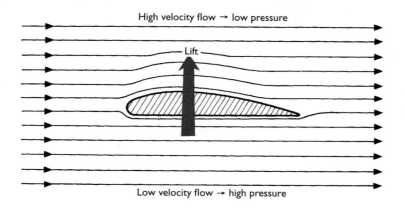

High velocity flow → low pressure

Lift

Low velocity flow → high pressure

FIGURE 15-10

Lift force generated by a foil shape is directed from the region of relative high pressure on the flat side of the foil toward the region of relative low pressure on the curved side of the foil.

Different factors affect the magnitude of the lift force acting on a foil. The greater the velocity of the foil relative to the fluid, the greater the pressure differential and the lift force generated. Other contributing factors are the fluid density and the surface area of the flat side of the foil. As both of these variables increase, lift increases. An additional factor of influence is the *coefficient of lift,* which indicates a body's ability to generate lift based on its shape.

The human hand resembles a foil shape when viewed from a lateral perspective. When a swimmer slices a hand through the water, it generates lift force directed perpendicular to the palm. Synchronized swimmers use a sculling motion, rapidly slicing their hands back and forth, to maneuver their bodies through various positions in the water. The lift force generated by rapid sculling motions enables elite synchronized swimmers to support their bodies in an inverted position with both legs extended completely out of the water.

The semifoil shapes of projectiles such as the discus, javelin, football, boomerang, and frisbee generate some lift force when oriented at appropriate angles with respect to the direction of the fluid flow. Spherical projectiles such as a shot or a ball, however, do not sufficiently resemble a foil and cannot generate lift by virtue of their shape.

angle of attack
angle between the longitudinal axis of a body and the direction of the fluid flow

The angle of orientation of the projectile with respect to the fluid flow—the angle of attack—is an important factor in launching a lift-producing projectile for maximum range (horizontal displacement). A positive angle of attack is necessary to generate a lift force (Figure 15-11). As the angle of attack increases, the amount of surface area exposed perpendicularly to the fluid flow also increases, thereby increasing the amount of form drag acting. With too steep an attack angle, the fluid cannot flow along the curved side of the foil to create lift. Airplanes that assume too steep an ascent can stall and lose altitude until pilots reduce the attack angle of the wings to enable lift (25).

lift/drag ratio
the magnitude of the lift force divided by the magnitude of the total drag force acting on a body at a given time

To maximize the flight distance of a projectile such as the discus or javelin, it is advantageous to maximize lift and minimize drag. Form drag, however, is minimum at an angle of attack of 0°, which is a poor angle for generating lift. The optimum angle of attack for maximizing range is the angle at which the lift/drag ratio is maximum. The largest lift/drag ratio for a discus traveling at a relative velocity of 24 m/s is generated at an angle of attack of 10° (14). For both the discus and the javelin, however, the single most important factor related to distance achieved is release speed (1, 17).

When the projectile is the human body during the performance of a jump, maximizing the effects of lift while minimizing the effects of drag is more complicated. In the ski jump, because of the relatively long period during which the body is airborne, the lift/drag ratio for the human body is particularly important. Research on ski jumping indicates that for optimal performance, ski jumpers should have a flattened body with a large frontal area (for generating lift) and a small body weight (for enabling greater acceleration) during takeoff. The effect of lift is immediate at takeoff, resulting in a higher initial vertical velocity than the jumper generates through impulse against the ramp surface (47). During the first part of the flight, jumpers should assume a small angle of attack to minimize drag (Figure 15-12). During the latter part of the flight, they should increase attack angle up to that of maximum lift. Jumping into a headwind dramatically increases jump length because of the increase in lift acting on the jumper (31).

Magnus Effect

Spinning objects also generate lift. When an object in a fluid medium spins, the boundary layer of fluid molecules adjacent to the object spins

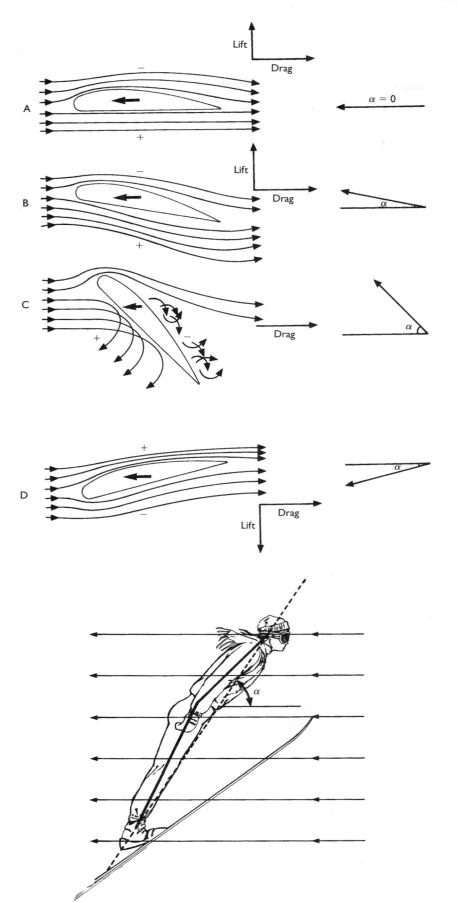

FIGURE 15-11

A. Drag and lift are small because the angle of attack (α) does not create a sufficiently high pressure differential across the top and bottom surfaces of the foil. **B.** An angle of attack that promotes lift. **C.** When the angle of attack is too large, the fluid cannot flow over the curved surface of the foil, and no lift is generated. **D.** When the angle of attack is below the horizontal, lift is created in a downward direction. Modified from Maglischo E: *Swimming faster: A comprehensive guide to the science of swimming*, Palo Alto, CA, 1982, Mayfield.

FIGURE 15-12

The angle of attack is the angle formed between the primary axis of a body and the direction of the fluid flow.

FIGURE 15-13

The relationship between ski jump length and the performer's angle of attack. Modified from Denoth J, Luethi SM, and Gasser HH: Methodological problems in optimization of the flight phase in ski jumping, Int J Sport Biomech 3:404, 1987.

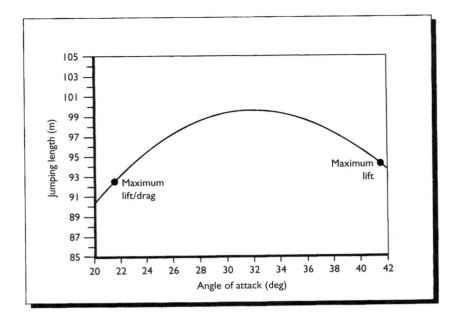

FIGURE 15-14

Magnus force results from the pressure differential created by a spinning body.

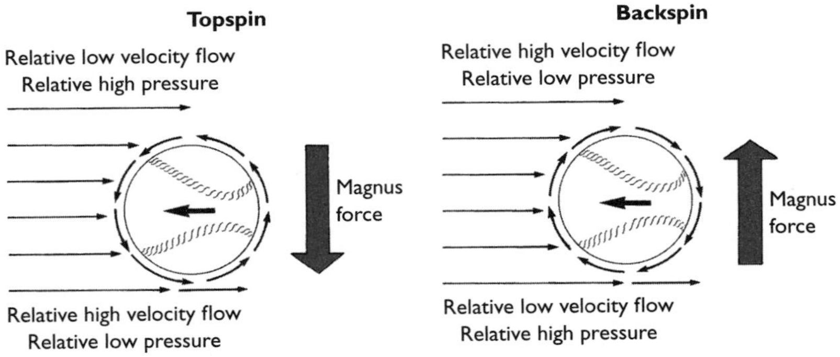

Magnus force
lift force created by spin

Magnus effect
deviation in the trajectory of a spinning object toward the direction of spin, resulting from the Magnus force

● *A ball projected with spin follows a trajectory that curves in the direction of the spin.*

with it. When this happens, the fluid molecules on one side of the spinning body collide head-on with the molecules in the fluid free-stream (Figure 15-14). This creates a region of relative low velocity and high pressure. On the opposite side of the spinning object, the boundary layer moves in the same direction as the fluid flow, thereby creating a zone of relative high velocity and low pressure. The pressure differential creates what is called the Magnus force, a lift force directed from the high-pressure region to the low-pressure region.

Magnus force affects the flight path of a spinning projectile as it travels through the air, causing the path to deviate progressively in the direction of the spin, a deviation known as the Magnus effect. When a tennis ball or table tennis ball is hit with topspin, the ball drops more rapidly than it would without spin, and the ball tends to rebound low and fast, often making it more difficult for the opponent to return the shot. The nap on a tennis ball traps a relatively large boundary layer of air with it as it spins, thereby accentuating the Magnus effect. The Magnus effect can also result from sidespin, as when a pitcher throws a curveball (Figure 15-15). The modern-day version of the curveball is a ball that is intentionally pitched with spin, causing it to follow a curved path in the direction of the spin throughout its flight path.

The extent to which a ball curves, or "breaks," in the horizontal and vertical planes is dependent on the orientation of the spinning ball's axis

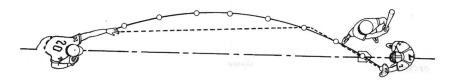

FIGURE 15-15

The trajectory of a ball thrown with sidespin follows a regular curve due to the Magnus effect. The dashed line shows the illusion seen by the players on the field.

of rotation. If the axis of rotation is perfectly vertical, all of the Magnus effect occurs in the horizontal plane. Alternatively, if the axis of rotation is oriented horizontally, the Magnus effect is restricted to the vertical plane. Curveballs thrown by major league pitchers spin as quickly as 27 revolutions per second and deviate horizontally as much as 40 cm over the pitcher-to-batter distance (44).

Soccer players also use the Magnus effect when it is advantageous for a kicked ball to follow a curved path, as may be the case when a player executing a free kick attempts to score. The "banana shot" consists of a kick executed so that the kicker places a lateral spin on the ball, curving it around the wall of defensive players in front of the goal (Figure 15-16).

The Magnus effect is maximal when the axis of spin is perpendicular to the direction of relative fluid velocity. Golf clubs are designed to impart some backspin to the struck ball, thereby creating an upwardly directed Magnus force that increases flight time and flight distance (Figure 15-17).

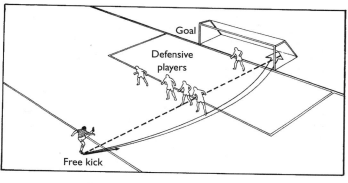

FIGURE 15-16

A banana shot in soccer results from imparting sidespin to the ball.

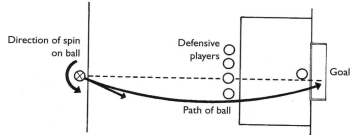

FIGURE 15-17

The loft on a golf club is designed to produce backspin on the ball. A properly hit ball rises because of the Magnus effect.

When a golf ball is hit laterally off-center, a spin about a vertical axis is also produced, causing a laterally deviated Magnus force that causes the ball to deviate from a straight path. When backspin and sidespin have been imparted to the ball, the resultant effect of the Magnus force on the path of the ball depends on the orientation of the ball's resultant axis of rotation to the airstream and on the velocity with which the ball was struck. When a golf ball is struck laterally off-center by a right-handed golfer, the ball unfortunately follows a curved path to one side—commonly known as a hook (to the left) or a slice (to the right).

PROPULSION IN A FLUID MEDIUM

propulsive drag
force acting in the direction of a body's motion

Whereas a headwind slows a runner or cyclist by increasing the acting drag force, a tailwind can actually contribute to forward propulsion. Theoretical calculations indicate that a tailwind of 2 m/s improves running time during a 100 m sprint by approximately 0.18 s (48). A tailwind affects the relative velocity of a body with respect to the air, thereby modifying the resistive drag acting on the body. Thus, a tailwind of a velocity greater than the velocity of the moving body produces a drag force in the direction of motion (Figure 15-18). This force has been termed propulsive drag.

Analyzing the fluid forces acting on a swimmer is more complicated. Resistive drag acts on a swimmer, yet the propulsive forces exerted by the water in reaction to the swimmer's movements are responsible for the swimmer's forward motion through the water. The motions of the body segments during swimming produce a complex combination of drag and lift forces throughout each stroke cycle, and even among elite swimmers, a wide range of kinetic patterns during stroking have been observed. As a result, researchers have proposed several theories regarding the ways in which swimmers propel themselves through the water.

Propulsive Drag Theory

propulsive drag theory
theory attributing propulsion in swimming to propulsive drag on the swimmer

The oldest theory of swimming propulsion is the propulsive drag theory, which was proposed by Counsilman and Silvia (9) and is based on Newton's third law of motion. According to this theory, as a swimmer's hands and arms move backward through the water, the forwardly directed reaction force generated by the water produces propulsion. The theory also suggests that the horizontal components of the downward and backward motion of the foot and the upward and backward motion of the opposite foot generate a forwardly directed reaction force from the water.

FIGURE 15-18

Drag force acting in the same direction as the body's motion may be thought of as propulsive drag because it contributes to the forward velocity of the body.

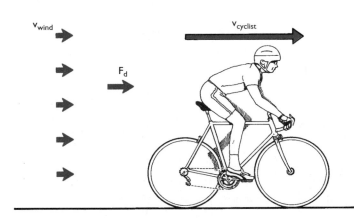

When high-speed movie films of skilled swimmers revealed that swimmers' hands and feet followed a zigzag rather than a straight-back path through the water, the theory was modified. It was suggested that this type of movement pattern enabled the body segments to push against still or slowly moving water instead of water already accelerated backward, thereby creating more propulsive drag. Research shows that holding the hands with the fingers comfortably spread with approximately 12 degrees between them, or a resting hand posture, is optimal for increasing drag against the palm during swimming (26, 29). However, propulsive drag may not be the major contributor to propulsion in swimming.

Propulsive Lift Theory

The propulsive lift theory was proposed by Counsilman in 1971 (4). According to this theory, swimmers use the foillike shape of the hand by employing rapid lateral movements through the water to generate lift. The lift is resisted by downward movement of the hand and by stabilization of the shoulder joint, which translates the forward-directed force to the body, propelling it past the hand. The theory was modified by Firby (13) in 1975, with the suggestion that swimmers use their hands and feet as propellers, constantly changing the pitches of the body segments to use the most effective angle of attack.

A number of investigators have since studied the forces generated by the body segments during swimming. It has been shown that lift does contribute to propulsion and that a combination of lift and drag forces acts throughout a stroke cycle. The relative contributions of lift and drag vary with the stroke performed, the phase within the stroke, and the individual swimmer. For example, lift is the primary force acting during the breaststroke, whereas lift and drag contribute differently to various phases of the front crawl stroke (38). Drag generated by the swimmer's hand is maximal when hand orientation is nearly perpendicular to the flow, and lift is maximal when the hand moves in the direction of either the thumb or the little finger (2).

propulsive lift theory
theory attributing propulsion in swimming at least partially to lift acting on the swimmer

Vortex Generation

Researchers have found a poor correlation between physiological and mechanical approaches to calculating propelling efficiency in swimming (3). This has led to the speculation that some unknown processes may play a role in swimming propulsion, with one possibility being the generation of vortices in the water by the swimmer. Vortex shedding has been found to play a role in the propulsion of both flying and swimming vertebrates and insects (12, 36). The generation of thrust in racing canoe and kayak paddling has also been described in terms of the mechanics of vortex–ring wakes (18). The observation that a swimmer performing the dolphin kick leaves behind a series of bound vortices, or columns of rotating water, has also been made (45). More research is needed to clarify the role of vortex generation in swimming propulsion.

Stroke Technique

Just as running speed is the product of stride length and stride rate, swimming speed is the product of stroke length (SL) and stroke rate (SR). Of the two, SL is more directly related to swimming speed among competitive freestyle swimmers (10, 40). Comparison of male and female swimmers performing at the same competitive distances reveals nearly identical SRs, but longer SLs resulting in higher velocities for the males (34). At slower speeds, skilled freestyle swimmers are able to maintain constant, high levels of SL, with a progressive reduction in SL as exercise intensity increases due to local muscle fatigue (19). The same phenomenon has been observed over the course of distance events, with a general decrease in SL and swimming speed after the first 100 m (22). Research suggests that recreational freestyle swimmers seeking to improve swimming performance should concentrate on applying more force to the water during each stroke to increase SL, as opposed to stroking faster. (50). Interestingly, the contribution of the flutter kick in the front crawl is more related to superior ankle flexibility than to vertical jump power or body size (27). Among backstrokers, although the ability to achieve a high swimming speed is related to SL at submaximal levels, increased speed is achieved through increased SR and decreased SL (20). Better performance in the breaststroke is achieved through highly effective work per stroke, optimal stroke rate, and a glide phase, with less skilled performers tending to use faster stroke rates and less gliding (15, 23, 39).

Another technique variable of importance during freestyle swimming is body roll. In one study, competitive swimmers were found to roll an average of approximately 60° to the nonbreathing side (24). Research shows that body roll in swimming is caused by the turning effect of the fluid forces acting on the swimmer's body. The contribution of body roll is important, since it enables the swimmer to employ the large, powerful muscles of the trunk rather than relying solely on the muscles of the shoulder and arm. It also facilitates the breathing action without any interruption of stroke mechanics (32). Body roll can influence the path of the hand through the water almost as much as the mediolateral motions of the hand relative to the trunk (24). In particular, an increase in body roll has been shown to increase the swimmer's hand speed in the plane perpendicular to the swimming direction, thereby increasing the potential for the hand to develop propulsive lift forces (33). With increasing swimming speed, general body roll decreases, although trunk twist increases, allowing swimmers to benefit from the rolling of the upper trunk, while limiting the increase in the drag of the lower extremity (51).

SUMMARY

The relative velocity of a body with respect to a fluid and the density, specific weight, and viscosity of the fluid affect the magnitudes of fluid forces. The fluid force that enables flotation is buoyancy. The buoyant force acts vertically upward; its point of application is the body's center of volume; and its magnitude is equal to the product of the volume of the displaced fluid and the specific gravity of the fluid. A body floats in a static position only when the magnitude of the buoyant force and body weight are equal and when the center of volume and the center of gravity are vertically aligned.

Drag is a fluid force that acts in the direction of the free-stream fluid flow. Skin friction is a component of drag that is derived from the sliding contacts between successive layers of fluid close to the surface of a moving body. Form drag, another component of the total drag, is caused by a pressure differential between the lead and trailing edges of a body moving with respect to a fluid. Wave drag is created by the formation of waves at the interface between two different fluids, such as water and air.

Lift is a force that can be generated perpendicular to the free-stream fluid flow by a foil-shaped object. Lift is created by a pressure differential in the fluid on opposite sides of a body that results from differences in the velocity of the fluid flow. The lift generated by spin is known as the Magnus force. Propulsion in swimming appears to result from a complex interplay of propulsive drag and lift.

INTRODUCTORY PROBLEMS

For all problems, assume that the specific weight of freshwater equals 9810 N/m^3 and the specific weight of seawater (saltwater) equals 10,070 N/m^3.

1. A boy is swimming with an absolute speed of 1.5 m/s in a river where the speed of the current is 0.5 m/s. What is the velocity of the swimmer with respect to the current when the boy swims directly upstream? Directly downstream? (Answer: 2 m/s in the upstream direction; 1 m/s in the downstream direction)
2. A cyclist is riding at a speed of 14 km/hr into a 16 km/hr headwind. What is the wind velocity relative to the cyclist? What is the cyclist's velocity with respect to the wind? (Answer: 30 km/hr in the direction of the wind; 30 km/hr in the direction of the cyclist)
3. A skier traveling at 5 m/s has a speed of 5.7 m/s relative to a headwind. What is the absolute wind speed? (Answer 0.7 m/s)
4. A 700 N man has a body volume of 0.08 m^3. If submerged in freshwater, will he float? Given his body volume, how much could he weigh and still float? (Answer: Yes; 784.8 N)
5. A racing shell has a volume of 0.38 m^3. When floating in freshwater, how many 700 N people can it support? (Answer: 5)
6. How much body volume must a 60 kg person have to float in freshwater? (Answer: 0.06 m^3)
7. Explain the implications for flotation due to the difference between the specific weight of freshwater and the specific weight of seawater.
8. What strategy can people use to improve their chances of floating in water? Explain your answer.
9. What types of individuals may have a difficult time floating in water? Explain your answer.
10. A beach ball weighing 1 N and with a volume of 0.03 m^3 is held submerged in seawater. How much force must be exerted vertically downward to hold the ball completely submerged? To hold the ball one-half submerged? (Answer: 301.1 N; 150.05 N)

ADDITIONAL PROBLEMS

1. A cyclist riding against a 12 km/hr headwind has a velocity of 28 km/hr with respect to the wind. What is the cyclist's absolute velocity? (Answer: 16 km/hr)
2. A swimmer crossing a river proceeds at an absolute speed of 2.5 m/s on a course oriented at a 45° angle to the 1 m/s current. Given that the absolute velocity of the swimmer is equal to the vector sum of the velocity of the current and the velocity of the swimmer with respect to the current, what is the magnitude and direction of the velocity of the swimmer with respect to the current? (Answer: 3.3 m/s at an angle of 32.6° to the current)

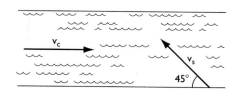

3. What maximum average density can a body possess if it is to float in freshwater? Seawater?

4. A scuba diver carries camera equipment in a cylindrical container that is 45 cm long, 20 cm in diameter, and 22 N in weight. For optimal maneuverability of the container under water, how much should its contents weigh? (Answer 120.36 N)

5. A 50 kg person with a body volume of 0.055 m^3 floats in a motionless position. How much body volume is above the surface in freshwater? In saltwater? (Answer: 0.005 m^3; 0.0063 m^3)

6. A 670 N swimmer oriented horizontally in freshwater has a body volume of 0.07 m^3 and a center of volume located 3 cm superior to the center of gravity.
   a. How much torque does the swimmer's weight generate?
   b. How much torque does the buoyant force acting on the swimmer generate?
   c. What can the swimmer do to counteract the torque and maintain a horizontal position?
   (Answer: 0; 20.6 N-m)

7. Based on your knowledge of the action of fluid forces, speculate as to why a properly thrown boomerang returns to the thrower.

8. Explain the aerodynamic benefits of drafting on a bicycle or in an automobile.

9. What is the practical effect of streamlining? How does streamlining alter the fluid forces acting on a moving body?

10. Explain why a curveball curves. Include a discussion of the aerodynamic role of the seams on the ball.

LABORATORY EXPERIENCES

1. At the *Basic Biomechanics* Online Learning Center (www.mhhe.com/hall6e), go to Student Center, Chapter 15, Lab Manual, Lab 1, then view Fluids Animation 1. Identify the principle illustrated, and write an explanation of what is demonstrated.

Principle: _____

Explanation: _____

_____

_____

_____

_____

2. Slice a hollow ball such as a table tennis ball or a racquet ball in half and float one-half of the ball (concave side up) in a container of water. Gradually add lead shot to the half ball until it floats with the cut edge at the surface of the water. Remove the half ball from the water, then measure its diameter and calculate its volume. Weigh the ball along with the lead shot that was placed in the ball. Using your measurements, calculate the specific weight of the water in the container. Repeat the experiment using water at different temperatures or using different liquids.

Ball diameter: _____ ball volume: _____ weight of shot: _____

Specific weight of water: _____ specific weight of second fluid: _____

Calculations:

3. Position a container of water on a scale and record its weight. Insert your hand, fingers first, into the water until the water line is at the wrist joint. Record the weight registered on the scale. Subtract the original weight of the container from the new weight, divide the difference in half, and add the result to the original weight of the container to arrive at the target weight. Slowly elevate your hand from the water until the target weight is reached. Mark the water line on your hand. What does this line represent? _____

4. Using a stopwatch, time yourself while riding an escalator. Either measure or estimate the length of the escalator and calculate the escalator's speed. Again using a stopwatch, time yourself while carefully running up the escalator and calculate your speed. Calculate your speed relative to the speed of the escalator.

Time riding escalator: _____ time running up escalator: _____

Your speed relative to the elevator's speed: _____

Calculation:

5. Use a variable-speed fan and a spring scale to construct a mock wind tunnel. Position the fan so that it blows vertically upward, and suspend the spring scale from a rigid arm above the fan. This apparatus can be used to test the relative drag on different objects suspended from the scale. Notice that relative drag among different objects may change with fan speed.

Object	Drag

REFERENCES

1. Bartlett R, Müller E, Lindinger S, Brunner F, and Morriss C: Three-dimensional evaluation of the kinematic release parameters for javelin throwers of different skill levels, *J Appl Biomech* 12:58, 1996.
2. Berger MA, deGroot G, and Hollander AP: Hydrodynamic drag and lift forces on human hand/arm models, *J Biomech* 28:125, 1995.
3. Berger MA, Hollander AP, and deGroot G: Technique and energy losses in front crawl swimming, *Med Sci Sports Exerc* 29:1491, 1997.
4. Brown RM and Counsilman JE: The role of lift in propelling swimmers. In Cooper JM, ed: *Biomechanics,* Chicago, 1971, Athletic Institute.
5. Capelli C et al.: Energy cost and efficiency of riding aerodynamic bicycles, *Eur J Appl Physiol* 67:144, 1993.
6. Cappaert JM, Pease DL, and Troup JP: Three-dimensional analysis of the men's 100-m freestyle during the 1992 Olympic games, *J Appl Biomech* 11:103, 1995.
7. Chatard J-C and Wilson B: Drafting distance in swimming, *Med Sci Sports Exerc* 35:1176, 2003.
8. Chollet D, Hue O, Auclair F, Millet G, and Chatard JC: The effects of drafting on stroking variations during swimming in elite male triathletes, *Eur J Appl Physiol* 82:413, 2000.
9. Counsilman JE: *Science of swimming,* Englewood Cliffs, NJ, 1968, Prentice Hall.
10. Craig AB, Jr. et al.: Velocity, stroke rate and distance per stroke during elite swimming competition, *Med Sci Sports Exerc* 17:625, 1985.
11. De Lucas RD, Balikian P, Neiva CM, Greco CC, and Denadai BS: The effects of wet suits on physiological and biomechanical indices during swimming, *J Sci Med Sport* 3:1, 2000.
12. Ellington CP: Unsteady aerodynamics of insect flight, *Symp Soc Exp Biol* 49:109, 1995.
13. Firby H: *Howard Firby on swimming,* London, 1975, Pelham Books.
14. Ganslen RV: *Aerodynamic factors which influence discus flight.* Research report, University of Arkansas.
15. Garland Fritzdorf S, Hibbs A, and Kleshnev V: Analysis of speed, stroke rate, and stroke distance for world-class breaststroke swimming, *J Sports Sci* 27(4):373, 2009.
16. Gnehm P, Reichenback S, Altpeter E, Widmer H, and Hoppeler H: Influence of different racing positions on metabolic cost in elite cyclists, *Med Sci Sports Exerc* 29:818, 1997.
17. Hay JG and Yu B: Critical characteristics of technique in throwing the discus, *J Sports Sci* 13:125, 1995.
18. Jackson PS: Performance prediction for Olympic kayaks, *J Sports Sci* 13:239, 1995.
19. Keskinen KL and Komi PV: Stroking characteristics of front crawl swimming during exercise, *J Appl Biomech* 9:219, 1993.
20. Klentrou PP and Montpetit RR: Energetics of backstroke swimming in males and females, *Med Sci Sports Exerc* 24:371, 1992.
21. Kyle CR and Burke E: Improving the racing bicycle, *Mechanical Engineering* 106:34, 1984.
22. Laffite LP, Vilas-Boas JP, Demarle A, Silva J, Fernandes R, and Billat VL: Changes in physiological and stroke parameters during a maximal 400-m free swimming test in elite swimmers, *Can J Appl Physiol* 29 Suppl:S17, 2004.
23. Leblanc H, Seifert L, and Chollet D: Arm-leg coordination in recreational and competitive breaststroke swimmers, *J Sci Med Sport* 12:352, 2009.
24. Liu Q, Hay JG, and Andrews JG: Body roll and handpath in freestyle swimming: an experimental study, *J Appl Biomech* 9:238, 1993.
25. Maglischo E: *Swimming faster: a comprehensive guide to the science of swimming,* Palo Alto, CA, 1982, Mayfield Publishing.
26. Marinho DA, Barbosa TM, Reis VM, Kjendlie PL, Alves FB, Vilas-Boas JP, Machado L, Silva AJ, and Rouboa AI: Swimming propulsion forces are enhanced by a small finger spread, *J Appl Biomech* 26:87, 2010.

27. McCullough AS, Kraemer WJ, Volek JS, Solomon-Hill GF Jr, Hatfield DL, Vingren JL, Ho JY, Fragala MS, Thomas GA, Häkkinen K, and Maresh CM: Factors affecting flutter kicking speed in women who are competitive and recreational swimmers, *J Strength Cond Res* 23:2130, 2009.

28. McLean SP and Hinrichs RN: Buoyancy, gender, and swimming performance, *J Appl Biomech* 16:248, 2000.

29. Minetti AE, Machtsiras G, and Masters JC: The optimum finger spacing in human swimming, *J Biomech* 42:2188, 2009.

30. Mollendorf JC, Termin AC, Oppenheim E, and Pendergast DR: Effect of swim suit design on passive drag, *Med Sci Sports Exerc* 36:1029, 2004.

31. Müller W: Determinants of ski-jump performance and implications for health, safety and fairness, *Sports Med* 39:85, 2009.

32. Payton CJ, Bartlett RM, Baltzopoulos V, and Coombs R: Upper extremity kinematics and body roll during preferred-side breathing and breath-holding front crawl swimming, *J Sports Sci* 17:689, 1999.

33. Payton CJ, Hay JG, and Mullineaux DR: The effect of body roll on hand speed and hand path in front crawl swimming—a simulation study, *J Appl Biomech* 13:300, 1997.

34. Pelayo P, Sidney M, Kherif T, Chollet D, and Tourny C: Stroking characteristics in freestyle swimming and relationships with anthropometric characteristics, *J Appl Biomech* 12:197, 1996.

35. Pendergast D, Mollendorf J, Zamparo P, Termin A 2nd, Bushnell D, and Paschke D: The influence of drag on human locomotion in water, *Undersea Hyperb Med* 32:45, 2005.

36. Rayner JM: Dynamics of the vortex wakes of flying and swimming vertebrates, *Symp Soc Exp Biol* 49:131, 1995.

37. Richardson RS and Johnson SC: The effect of aerodynamic handlebars on oxygen consumption while cycling at a constant speed, *Ergonomics* 37:859, 1994.

38. Schleihauf RE: A hydrodynamic analysis of swimming propulsion. In Terauds J and Bedingfield E, eds: *Swimming III,* Baltimore, 1979, University Park Press.

39. Seifert L, Leblanc H, Chollet D, and Delignières D: Inter-limb coordination in swimming: effect of speed and skill level, *Hum Mov Sci* 29:103, 2010.

40. Seifert L, Toussaint HM, Alberty M, Schnitzler C, and Chollet D: Arm coordination, power, and swim efficiency in national and regional front crawl swimmers, *Hum Mov Sci* 29:426, 2010.

41. Taïar R, Sagnes P, Henry C, Dufour AB, and Rouard AH: Hydrodynamics optimization in butterfly swimming: position, drag coefficient and performance, *J Biomech* 32:803, 1999.

42. Takamoto M, Ohmichi H, and Miyashita M: Wave height in relation to swimming velocity and proficiency in front crawl stroke. In Winter D et al, eds: *Biomechanics IX-B,* Champaign, IL, 1985, Human Kinetics Publishers.

43. Tomikawa M and Nomura T: Relationships between swim performance, maximal oxygen uptake and peak power output when wearing a wetsuit, *J Sci Med Sport* 12:317, 2009.

44. Townend MS: *Mathematics in sport,* New York, 1984, John Wiley & Sons.

45. Ungerechts BE: On the relevance of rotating water flow for the propulsion in swimming. In Jonsson B, ed: *Biomechanics X-B,* Champaign, IL, 1987, Human Kinetics Publishers.

46. van Ingen Schenau GJ: The influence of air friction in speed skating, *J Biomech* 15:449, 1982.

47. Virmavirta M, Kivekas J, and Komi PV: Take-off aerodynamics in ski jumping, *J Biomech* 34:465, 2001.

48. Ward-Smith AJ: A mathematical analysis of the influence of adverse and favourable winds on sprinting, *J Biomech* 18:351, 1985.

49. Yanai T: Rotational effect of buoyancy in frontcrawl: does it really cause the legs to sink? *J Biomech* 34:235, 2001.

50. Yanai T: Stroke frequency in front crawl: Its mechanical link to the fluid forces required in non-propulsive directions, *J Biomech* 36:53, 2003.

51. Yanai T: Buoyancy is the primary source of generating bodyroll in front-crawl swimming, *J Biomech* 37: 605, 2004.

ANNOTATED READINGS

Barbosa TM, Bragada JA, Reis VM, Marinho DA, Carvalho C, and Silva AJ: Energetics and biomechanics as determining factors of swimming performance: updating the state of the art, J Sci Med Sport 13:262, 2010.
Reviews the scientific literature on the interplay between performance, energetic, and biomechanics in competitive swimming.

Haake SJ: The impact of technology on sporting performance in Olympic sports, 27:1421, 2009.
Discusses the dramatic effects of technological advances in apparel and equipment on sport, comparing the performance statistics for the 100-m sprint, pole vault, javelin, and cycling.

Psycharakis SG and Sanders RH: Body roll in swimming: a review, J Sports Sci 28:229, 2010.
Reviews the scientific literature on effective utilization of body roll in swimming.

Toussaint HM, de Hollander AP, van den Berg C, and Vorontsov AR: Biomechanics of swimming. In Garrett WE and Kirkendall DT: *Exercise and sport science,* Philadelphia, 2000, Lippincott Williams & Wilkins.
Comprehensive review of literature on the effects of drag in swimming and on swimming propulsion techniques and theories.

RELATED WEBSITES

Circulation and the Magnus Effect
http://www.phys.virginia.edu/classes/311/notes/aero/node2.html
Shows a diagram and includes discussion of the Magnus force with practical examples.

NASA: Lift from Pressure
http://www.grc.nasa.gov/WWW/K-12/airplane/right1.html
Provides narrative, definitions of related terms, and slides illustrating lift concepts.

NASA: Lift to Drag Ratio
http://www.grc.nasa.gov/WWW/K-12/airplane/ldrat.html
Provides narrative, definitions of related terms, and slides illustrating lift / drag ratio concepts.

NASA: Relative Velocities
http://www.grc.nasa.gov/WWW/K-12/airplane/move2.html
Provides narrative, definitions of related terms, and slides illustrating relative velocity concepts.

NASA: What Is Drag?
http://www.grc.nasa.gov/WWW/K-12/airplane/drag1.html
Provides narrative, definitions of related terms, and slides illustrating drag concepts.

Tennis: The Magnus Effect
http://wings.avkids.com/Tennis/Book/magnus-01.html
Description and animated drawing of the Magnus effect on a spinning tennis ball.

The Drag Force on a Sphere http://www.ma.iup.edu/MathDept/Projects/CalcDEM-ma/drag/drag0.html
Provides links to pages on the graph of the drag coefficient versus Reynolds number and two models for drag force.

The Open Door Website: Relativity
http://www.saburchill.com/physics/chapters/0083.html
Provides description of a quantitative relative velocity problem with entertaining graphics.

The Physics Classroom: Relative Velocity
http://www.physicsclassroom.com/Class/vectors/U3L1f.html
Includes explanation, graphics, and animations demonstrating relative velocity.

Relative Velocity Applet http://www.math.gatech.edu/~carlen/2507/notes/class-Files/partOne/RelVel.html
Interactive application that enables control of two moving points and provides the ability to graph the absolute and relative motions of the points.

U.S. Centennial of Flight Commission
http://www.centennialofflight.gov/essay/Dictionary/four_forces/DI24.htm
Provides illustrated discussion about the forces acting on an airplane.

KEY TERMS

angle of attack	angle between the longitudinal axis of a body and the direction of the fluid flow
Archimedes' principle	physical law stating that the buoyant force acting on a body is equal to the weight of the fluid displaced by the body
Bernoulli principle	an expression of the inverse relationship between relative velocity and relative pressure in a fluid flow
boundary layer	layer of fluid immediately adjacent to a body
center of volume	point around which a body's volume is equally balanced and at which the buoyant force acts
coefficient of drag	unitless number that is an index of a body's ability to generate fluid resistance
coefficient of lift	unitless number that is an index of a body's ability to generate lift
fluid	substance that flows when subjected to a shear stress
foil	shape capable of generating lift in the presence of a fluid flow
form drag **pressure drag** **profile drag**	resistance created by a pressure differential between the lead and rear sides of a body moving through a fluid
laminar flow	flow characterized by smooth, parallel layers of fluid
lift	force acting on a body in a fluid in a direction perpendicular to the fluid flow
lift/drag ratio	the magnitude of the lift force divided by the magnitude of the total drag force acting on a body at a given time
Magnus effect	deviation in the trajectory of a spinning object toward the direction of spin, resulting from the Magnus force
Magnus force	lift force created by spin
propulsive drag	force acting in the direction of a body's motion
propulsive drag theory	theory attributing propulsion in swimming to propulsive drag on the swimmer
propulsive lift theory	theory attributing propulsion in swimming at least partially to lift acting on the swimmer
relative velocity	velocity of a body with respect to the velocity of something else, such as the surrounding fluid
skin friction **surface drag** **viscous drag**	resistance derived from friction between adjacent layers of fluid near a body moving through the fluid
theoretical square law	drag increases approximately with the square of velocity when relative velocity is low
turbulent flow	flow characterized by mixing of adjacent fluid layers
wave drag	resistance created by the generation of waves at the interface between two different fluids, such as air and water

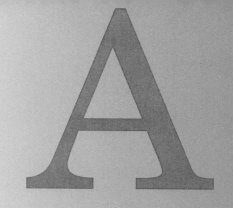

Basic Mathematics and Related Skills

NEGATIVE NUMBERS

Negative numbers are preceded by a minus sign. Although the physical quantities used in biomechanics do not have values that are less than zero in magnitude, the minus sign is often used to indicate the direction opposite the direction regarded as positive. Therefore, it is important to recall the following rules regarding arithmetic operations involving negative numbers:

1. Addition of a negative number yields the same results as subtraction of a positive number of the same magnitude:

$$6 + (-4) = 2$$
$$10 + (-3) = 7$$
$$6 + (-8) = -2$$
$$10 + (-23) = -13$$
$$(-6) + (-3) = -9$$
$$(-10) + (-7) = -17$$

2. Subtraction of a negative number yields the same result as addition of a positive number of the same magnitude:

$$5 - (-7) = 12$$
$$8 - (-6) = 14$$
$$-5 - (-3) = -2$$
$$-8 - (-4) = -4$$
$$-5 - (-12) = 7$$
$$-8 - (-10) = 2$$

3. Multiplication or division of a number by a number of the opposite sign yields a negative result:

$$2 \times (-3) = -6$$
$$(-4) \times 5 = -20$$
$$9 \div (-3) = -3$$
$$(-10) \div 2 = -5$$

4. Multiplication or division of a number by a number of the same sign (positive or negative) yields a positive result:

$$3 \times 4 = 12$$
$$(-3) \times (-2) = 6$$
$$10 \div 5 = 2$$
$$(-15) \div (-3) = 5$$

EXPONENTS

Exponents are superscript numbers that immediately follow a base number, indicating how many times that number is to be self-multiplied to yield the result:

$$5^2 = 5 \times 5$$
$$= 25$$
$$3^2 = 3 \times 3$$
$$= 9$$
$$5^3 = 5 \times 5 \times 5$$
$$= 125$$
$$3^3 = 3 \times 3 \times 3$$
$$= 27$$

SQUARE ROOTS

Taking the square root of a number is the inverse operation of squaring a number (multiplying a number by itself). The square root of a number is the number that yields the original number when multiplied by itself. The square root of 25 is 5, and the square root of 9 is 3. Using mathematics notation, these relationships are expressed as the following:

$$\sqrt{25} = 5$$
$$\sqrt{9} = 3$$

Because -5 multiplied by itself also equals 25, -5 is also a square root of 25. The following notation is sometimes used to indicate that square roots may be either positive or negative:

$$\sqrt{25} = +5$$
$$\sqrt{9} = +3$$

ORDER OF OPERATIONS

When a computation involves more than a single operation, a set of rules must be used to arrive at the correct result. These rules may be summarized as follows:

1. Addition and subtraction are of equal precedence; these operations are carried out from left to right as they occur in an equation:

$$7 - 3 + 5 = 4 + 5$$
$$= 9$$
$$5 + 2 - 1 + 10 = 7 - 1 + 10$$
$$= 6 + 10$$
$$= 16$$

2. Multiplication and division are of equal precedence; these operations are carried out from left to right as they occur in an equation:

$$10 \div 5 \times 4 = 2 \times 4$$
$$= 8$$
$$20 \div 4 \times 3 \div 5 = 5 \times 3 \div 5$$
$$= 15 \div 5$$
$$= 3$$

angle to the length of the hypotenuse. Using the labeled triangle yields the following:

$$\sin \alpha = \frac{\text{opposite}}{\text{hypotenuse}} = \frac{A}{C}$$

$$\sin \beta = \frac{\text{opposite}}{\text{hypotenuse}} = \frac{B}{C}$$

With $A = 3$, $B = 4$, and $C = 5$:

$$\sin \alpha = \frac{A}{C} = \frac{3}{5} = 0.6$$

$$\text{sig } \beta = \frac{B}{C} = \frac{4}{5} = 0.8$$

The cosine (abbreviated *cos*) of an angle is defined as the ratio of the length of the side of the triangle adjacent to the angle to the length of the hypotenuse. Using the labeled triangle yields the following:

$$\cos \alpha = \frac{\text{adjacent}}{\text{hypotenuse}} = \frac{B}{C}$$

$$\cos \beta = \frac{\text{adjacent}}{\text{hypotenuse}} = \frac{A}{C}$$

With $A = 3$, $B = 4$, and $C = 5$:

$$\cos \alpha = \frac{B}{C} = \frac{4}{5} = 0.8$$

$$\cos \beta = \frac{A}{C} = \frac{3}{5} = 0.6$$

The third function, the tangent (abbreviated *tan*) of an angle, is defined as the ratio of the length of the side of the triangle opposite the angle to that of the side adjacent to the angle. Using the labeled triangle yields the following:

$$\tan \alpha = \frac{\text{opposite}}{\text{adjacent}} = \frac{A}{B}$$

$$\tan \beta = \frac{\text{opposite}}{\text{adjacent}} = \frac{B}{A}$$

With $A = 3$, $B = 4$, and $C = 5$:

$$\tan \alpha = \frac{A}{B} = \frac{3}{4} = 0.75$$

$$\tan \beta = \frac{B}{A} = \frac{4}{3} = 1.33$$

Two useful trigonometric relationships are applicable to *all* triangles. The first is known as the law of sines:

> The ratio between the length of any side of a triangle and the angle opposite that side is equal to the ratio between the length of any other side of the triangle and the angle opposite that side.

With respect to the labeled triangle, this may be stated as the following:

$$\frac{A}{\sin \alpha} = \frac{B}{\sin \beta} = \frac{C}{\sin \gamma}$$

A second trigonometric relationship applicable to *all* triangles is the law of cosines:

> The square of the length of any side of a triangle is equal to the sum of the squares of the lengths of the other two sides of the triangle minus two times the product of the lengths of the other two sides and the cosine of the angle opposite the original side.

This relationship yields the following for each of the sides of the labeled triangle:

$$A^2 = B^2 + C^2 - 2BC \cos \alpha$$
$$B^2 = A^2 + C^2 - 2AC \cos \beta$$
$$C^2 = A^2 + B^2 - 2AB \cos \gamma$$

A table of the values of the basic trigonometric functions follows.

Table of Basic Trigonometric Function Values

DEG	SIN	COS	TAN	DEG	SIN	COS	TAN
00	.0000	1.0000	.0000	—	—	—	—
01	.0175	.9998	.0175	46	.7193	.6947	1.0355
02	.0349	.9994	.0349	47	.7314	.6820	1.0723
03	.0523	.9986	.0524	48	.7431	.6691	1.1106
04	.0698	.9976	.0699	49	.7547	.6561	1.1504
05	.0872	.9962	.0875	50	.7660	.6428	1.1918
06	.1045	.9945	.1051	51	.7771	.6293	1.2349
07	.1219	.9925	.1228	52	.7880	.6157	1.2799
08	.1392	.9903	.1405	53	.7986	.6018	1.3270
09	.1564	.9877	.1584	54	.8090	.5878	1.3764
10	.1736	.9848	.1763	55	.8192	.5736	1.4281
11	.1908	.9816	.1944	56	.8290	.5592	1.4826
12	.2079	.9781	.2126	57	.8387	.5446	1.5399
13	.2250	.9744	.2309	58	.8480	.5299	1.6003
14	.2419	.9703	.2493	59	.8572	.5150	1.6643
15	.2588	.9659	.2679	60	.8660	.5000	1.7321
16	.2756	.9613	.2867	61	.8746	.4848	1.8040
17	.2924	.9563	.3057	62	.8829	.4695	1.8807
18	.3090	.9511	.3249	63	.8910	.4540	1.9626
19	.3256	.9455	.3443	64	.8988	.4384	2.0503
20	.3420	.9397	.3640	65	.9063	.4226	2.1445
21	.3584	.9336	.3839	66	.9135	.4067	2.2460
22	.3746	.9272	.4040	67	.9205	.3907	2.3559
23	.3907	.9205	.4245	68	.9279	.3746	2.4751
24	.4067	.9135	.4452	69	.9336	.3584	2.6051
25	.4226	.9063	.4663	70	.9397	.3420	2.7475
26	.4384	.8988	.4877	71	.9456	.3256	2.9042
27	.4540	.8910	.5095	72	.9511	.3090	3.0779
28	.4695	.8829	.5317	73	.9563	.2924	3.2709
29	.4848	.8746	.5543	74	.9613	.2756	3.4874
30	.5000	.8660	.5774	75	.96593	.2588	3.7321
31	.5150	.8572	.6009	76	.9703	.2419	4.0108
32	.5299	.8480	.6249	77	.9744	.2250	4.3315
33	.5446	.8387	.6494	78	.9781	.2079	4.7046
34	.5592	.8290	.6745	79	.9816	.1908	5.1446
35	.5736	.8192	.7002	80	.9848	.1736	5.6713
36	.5878	.8090	.7265	81	.9877	.1564	6.3138
37	.6018	.7986	.7536	82	.9903	.1391	7.1154
38	.6157	.7880	.7813	83	.9925	.1219	8.1443
39	.6293	.7771	.8098	84	.9945	.1045	9.5144
40	.6428	.7660	.8391	85	.99625	.0872	11.4301
41	.6561	.7547	.8693	86	.9976	.0698	14.3007
42	.6691	.7431	.9004	87	.99866	.05239	19.0811
43	.6820	.7314	.9325	88	.9994	.0349	28.6363
44	.6947	.7193	.9657	89	.9998	.0175	57.2900
45	.7071	.7071	1.0000	90	1.0000	.0000	Infinity

Common Units
of Measurement

This appendix contains factors for converting between metric units commonly used in biomechanics and their English system equivalents. In each case, a value expressed in a metric unit can be divided by the conversion factor given to yield the approximate equivalent in an English unit, or a value expressed in an English unit can be multiplied by the conversion factor to find the metric unit equivalent. For example, to convert 100 Newtons to pounds, do the following:

$$\frac{100 \text{ N}}{4.45 \text{ N/lb}} = 22.5 \text{ lb}$$

To convert 100 pounds to Newtons, do the following:

$$(100 \text{ lb})(4.45 \text{ N/lb}) = 445 \text{ N}$$

VARIABLE	METRIC UNIT	← MULTIPLY BY DIVIDE BY →	ENGLISH UNIT
Distance	Centimeters	2.54	Inches
	Meters	0.3048	Feet
	Kilometers	1.609	Miles
Speed	Meters/second	0.447	Miles/hour
Mass	Kilograms	14.59	Slugs
Force	Newtons	4.448	Pounds
Work	Joules	1.355	Foot-pounds
Power	Watts	745.63	Horsepower
Energy	Joules	1.355	Foot-pounds
Linear momentum	Kilogram-meters/second	4.448	Slug-feet/second
Impulse	Newton-seconds	4.448	Pound-seconds
Angular momentum	Kilogram-meters^2/second	1.355	Slug-feet^2/second
Moment of inertia	Kilogram-meters^2	1.355	Slug-feet^2
Torque	Newton-meters	1.355	Foot-pounds

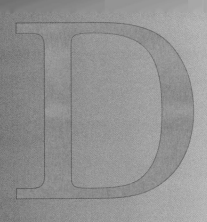

Anthropometric Parameters for the Human Body*

Segment Lengths

SEGMENT	MALES	FEMALES
Head and neck	10.75	10.75
Trunk	30.00	29.00
Upper arm	17.20	17.30
Forearm	15.70	16.00
Hand	5.75	5.75
Thigh	23.20	24.90
Lower leg	24.70	25.70
Foot	4.25	4.25
Segment lengths expressed in percentages of total body height.		

Segment Weights

SEGMENT	MALES	FEMALES
Head	8.26	8.20
Trunk	46.84	45.00
Upper arm	3.25	2.90
Forearm	1.87	1.57
Hand	0.65	0.50
Thigh	10.50	11.75
Lower leg	4.75	5.35
Foot	1.43	1.33
Segment weights expressed in percentages of total body weight		

From Plagenhoef S, Evans FG, and Abdelnour T: Anatomical data for analyzing human motion, Res Q Exerc Sport 54:169, 1983.

*The values reported in these tables represent mean values for limited numbers of individuals as reported in the scientific literature.

Segmental Center of Gravity Locations

SEGMENT	MALES	FEMALES
Head and neck	55.0	55.0
Trunk	63.0	56.9
Upper arm	43.6	45.8
Forearm	43.0	43.4
Hand	46.8	46.8
Thigh	43.3	42.8
Lower leg	43.4	41.9
Foot	50.0	50.0

Segmental center of gravity locations expressed in percentages of segment lengths; measured from the proximal ends of segments.

Segmental Radii of Gyration Measured from Proximal and Distal Segment Ends

SEGMENT	MALES		FEMALES	
	PROXIMAL	DISTAL	PROXIMAL	DISTAL
Upper arm	54.2	64.5	56.4	62.3
Forearm	52.6	54.7	53.0	64.3
Hand	54.9	54.9	54.9	54.9
Thigh	54.0	65.3	53.5	65.8
Lower leg	52.9	64.2	51.4	65.7
Foot	69.0	69.0	69.0	69.0

Segmental radii of gyration expressed in percentages of segment lengths

GLOSSARY

absolute angle	angular orientation of a body segment with respect to a fixed line of reference
acromioclavicular joint	irregular joint between the acromion process of the scapula and the distal clavicle
active insufficiency	limited ability of a two-joint muscle to produce force when joint position places the muscle on slack
active stretching	stretching of muscles, tendons, and ligaments produced by active development of tension in the antagonist muscles
acute loading	application of a single force of sufficient magnitude to cause injury to a biological tissue
agonist	role played by a muscle acting to cause a movement
amenorrhea	cessation of menses
anatomical reference position	erect standing position with all body parts, including the palms of the hands, facing forward; considered the starting position for body segment movements
angle of attack	angle between the longitudinal axis of a body and the direction of the fluid flow
angle of projection	the direction at which a body is projected with respect to the horizontal
angular	involving rotation around a central line or point
angular acceleration	rate of change in angular velocity
angular displacement	change in the angular position or orientation of a line segment
angular impulse	change in angular momentum equal to the product of torque and time interval over which the torque acts
angular momentum	quantity of angular motion possessed by a body; measured as the product of moment of inertia and angular velocity
angular velocity	rate of change in the angular position or orientation of a line segment
anisotropic	exhibiting different mechanical properties in response to loads from different directions
annulus fibrosus	thick, fibrocartilaginous ring that forms the exterior of the intervertebral disc
antagonist	role played by a muscle acting to slow or stop a movement
anteroposterior axis	imaginary line around which frontal plane rotations occur
anthropometric	related to the dimensions and weights of body segments
apex	the highest point in the trajectory of a projectile
appendicular skeleton	bones composing the body appendages
Archimedes' principle	physical law stating that the buoyant force acting on a body is equal to the weight of the fluid displaced by the body
articular capsule	double-layered membrane that surrounds every synovial joint
articular cartilage	protective layer of dense white connective tissue covering the articulating bone surfaces at diarthrodial joints

articular fibrocartilage	soft-tissue discs or menisci that intervene between articulating bones
average	occurring over a designated time interval
axial	directed along the longitudinal axis of a body
axial skeleton	the skull, vertebrae, sternum, and ribs
axis of rotation	imaginary line perpendicular to the plane of rotation and passing through the center of rotation
balance	a person's ability to control equilibrium
ballistic stretching	a series of quick, bouncing-type stretches
base of support	area bound by the outermost regions of contact between a body and support surface or surfaces
bending	asymmetric loading that produces tension on one side of a body's longitudinal axis and compression on the other side
Bernoulli principle	an expression of the inverse relationship between relative velocity and relative pressure in a fluid flow
biomechanics	application of mechanical principles in the study of living organisms
bone atrophy	decrease in bone mass resulting from a predominance of osteoclast activity
bone hypertrophy	increase in bone mass resulting from a predominance of osteoblast activity
boundary layer	layer of fluid immediately adjacent to a body
bursae	sacs secreting synovial fluid internally that lessen friction between soft tissues around joints
cardinal planes	three imaginary perpendicular reference planes that divide the body in half by mass
carpal tunnel syndrome	overuse condition caused by compression of the median nerve in the carpal tunnel and involving numbness, tingling, and pain in the hand
center of gravity	point around which a body's weight is equally balanced, no matter how the body is positioned
center of mass (mass centroid, center of gravity)	point around which the mass and weight of a body are balanced, no matter how the body is positioned
center of volume	point around which a body's volume is equally balanced and at which the buoyant force acts
centripetal force	force directed toward the center of rotation for a body in rotational motion
close-packed position	joint orientation for which the contact between the articulating bone surfaces is maximum
coefficient of drag	unitless number that is an index of a body's ability to generate fluid resistance
coefficient of friction	number that serves as an index of the interaction between two surfaces in contact
coefficient of lift	unitless number that is an index of a body's ability to generate lift
coefficient of restitution	number that serves as an index of elasticity for colliding bodies
collateral ligaments	major ligaments that cross the medial and lateral aspects of the knee
combined loading	simultaneous action of more than one of the pure forms of loading
compression	pressing or squeezing force directed axially through a body
compressive strength	ability to resist pressing or squeezing force
concentric	describing a contraction involving shortening of a muscle
contractile component	muscle property enabling tension development by stimulated muscle fibers
coracoclavicular joint	syndesmosis with the coracoid process of the scapula bound to the inferior clavicle by the coracoclavicular ligament
cortical bone	compact mineralized connective tissue with low porosity that is found in the shafts of long bones
couple	pair of equal, oppositely directed forces that act on opposite sides of an axis of rotation to produce torque

cruciate ligaments	major ligaments that cross each other in connecting the anterior and posterior aspects of the knee
curvilinear	along a curved line
deformation	change in shape
density	mass per unit of volume
dynamic equilibrium (D'Alembert's principle)	concept indicating a balance between applied forces and inertial forces for a body in motion
dynamics	branch of mechanics dealing with systems subject to acceleration
eccentric	describing a contraction involving lengthening of a muscle
electromechanical delay	time between arrival of a neural stimulus and tension development by the muscle
English system	system of weights and measures originally developed in England and used in the United States today
epicondylitis	inflammation and sometimes microrupturing of the collagenous tissues on either the lateral or the medial side of the distal humerus; believed to be an overuse injury
epiphysis	growth center of a bone that produces new bone tissue as part of the normal growth process until it closes during adolescence or early adulthood
extrinsic muscles	muscles with proximal attachments located proximal to the wrist and distal attachments located distal to the wrist
failure	loss of mechanical continuity
fast-twitch fiber	a fiber that reaches peak tension relatively quickly
first-class lever	lever positioned with the applied force and the resistance on opposite sides of the axis of rotation
flat bones	skeletal structures that are largely flat in shape, for example, the scapula
flexion relaxation phenomenon	when the spine is in full flexion, the spinal extensor muscles relax and the flexion torque is supported by the spinal ligaments
fluid	substance that flows when subjected to a shear stress
foil	shape capable of generating lift in the presence of a fluid flow
force	push or pull; the product of mass and acceleration
form drag (profile drag, pressure drag)	resistance created by a pressure differential between the lead and rear sides of a body moving through a fluid
fracture	disruption in the continuity of a bone
free body diagram	sketch that shows a defined system in isolation with all of the force vectors acting on the system
friction	force acting at the area of contact between two surfaces in the direction opposite that of motion or motion tendency
frontal plane	plane in which lateral movements of the body and body segments occur
fulcrum	the point of support, or axis, about which a lever may be made to rotate
general motion	motion involving translation and rotation simultaneously
glenohumeral joint	ball-and-socket joint in which the head of the humerus articulates with the glenoid fossa of the scapula
glenoid labrum	rim of soft tissue located on the periphery of the glenoid fossa that adds stability to the glenohumeral joint
Golgi tendon organ	sensory receptor that inhibits tension development in a muscle and initiates tension development in antagonist muscles
hamstrings	the biceps femoris, semimembranosus, and semitendinosus
humeroradial joint	gliding joint in which the capitellum of the humerus articulates with the proximal end of the radius
humeroulnar joint	hinge joint in which the humeral trochlea articulates with the trochlear fossa of the ulna
iliopsoas	the psoas major and iliacus muscles with a common insertion on the lesser trochanter of the femur

iliotibial band	thick, strong band of tissue connecting the tensor fascia lata to the lateral condyle of the femur and the lateral tuberosity of the tibia
impact	collision characterized by the exchange of a large force during a small time interval
impacted	pressed together by a compressive load
impulse	product of a force and the time interval over which the force acts
inertia	tendency of a body to resist a change in its state of motion
inference	process of forming deductions from available information
initial velocity	vector quantity incorporating both angle and speed of projection
instant center	precisely located center of rotation at a joint at a given instant in time
instantaneous	occurring during a small interval of time
intraabdominal pressure	pressure inside the abdominal cavity; believed to help stiffen the lumbar spine against buckling
intrinsic muscles	muscles with both attachments distal to the wrist
irregular bones	skeletal structures of irregular shapes, for example, the sacrum
isometric	describing a contraction involving no change in muscle length
joint flexibility	a term representing the relative ranges of motion allowed at a joint
joint stability	ability of a joint to resist abnormal displacement of the articulating bones
kinematics	the form, pattern, or sequencing of movement with respect to time
kinesiology	study of human movement
kinetic energy	energy of motion calculated as $\frac{1}{2} mv^2$
kinetic friction	constant magnitude friction generated between two surfaces in contact during motion
kinetics	study of the action of forces
kyphosis	extreme curvature in the thoracic region of the spine
laminar flow	flow characterized by smooth, parallel layers of fluid
laws of constant acceleration	formulas relating displacement, velocity, acceleration, and time when acceleration is unchanging
lever	a relatively rigid object that may be made to rotate about an axis by the application of force
lift	force acting on a body in a fluid in a direction perpendicular to the fluid flow
lift/drag ratio	the magnitude of the lift force divided by the magnitude of the total drag force acting on a body at a given time
ligamentum flavum	yellow ligament that connects the laminae of adjacent vertebrae; distinguished by its elasticity
linear	along a line that may be straight or curved, with all parts of the body moving in the same direction at the same speed
linear acceleration	the rate of change in linear velocity
linear displacement	change in location, or the directed distance from initial to final location
linear momentum	quantity of motion, measured as the product of a body's mass and its velocity
linear velocity	the rate of change in location
long bones	skeletal structures consisting of a long shaft with bulbous ends, for example, the femur
longitudinal axis	imaginary line around which transverse plane rotations occur
loose-packed position	any joint orientation other than the close-packed position
lordosis	extreme curvature in the lumbar region of the spine
Magnus effect	deviation in the trajectory of a spinning object toward the direction of spin, resulting from the Magnus force
Magnus force	lift force created by spin
mass	quantity of matter contained in an object

maximum static friction	maximum amount of friction that can be generated between two static surfaces
mechanical advantage	ratio of force arm to resistance arm for a given lever
mechanics	branch of physics that analyzes the actions of forces on particles and mechanical systems
mediolateral axis	imaginary line around which sagittal plane rotations occur
menisci	cartilaginous discs located between the tibial and femoral condyles
meter	the most common international unit of length, on which the metric system is based
metric system	system of weights and measures used internationally in scientific applications and adopted for daily use by every major country except the United States
moment arm	shortest (perpendicular) distance between a force's line of action and an axis of rotation
moment of inertia	inertial property for rotating bodies representing resistance to angular acceleration; based on both mass and the distance the mass is distributed from the axis of rotation
motion segment	two adjacent vertebrae and the associated soft tissues; the functional unit of the spine
motor unit	a single motor neuron and all fibers it innervates
muscle inhibition	the inability to activate all motor units of a muscle during maximal voluntary contraction
muscle spindle	sensory receptor that provokes reflex contraction in a stretched muscle and inhibits tension development in antagonist muscles
myoelectric activity	electric current or voltage produced by a muscle developing tension
net force	resultant force derived from the composition of two or more forces
neutralizer	role played by a muscle acting to eliminate an unwanted action produced by an agonist
normal reaction force	force acting perpendicular to two surfaces in contact
nucleus pulposus	colloidal gel with a high fluid content, located inside the annulus fibrosus of the intervertebral disc
osteoblasts	specialized bone cells that build new bone tissue
osteoclasts	specialized bone cells that resorb bone tissue
osteopenia	condition of reduced bone mineral density that predisposes the individual to fractures
osteoporosis	a disorder involving decreased bone mass and strength with one or more resulting fractures
parallel elastic component	passive elastic property of muscle derived from the muscle membranes
parallel fiber arrangement	pattern of fibers within a muscle in which the fibers are roughly parallel to the longitudinal axis of the muscle
passive insufficiency	inability of a two-joint muscle to stretch to the extent required to allow full range of motion at all joints crossed
passive stretching	stretching of muscles, tendons, and ligaments produced by a stretching force other than tension in the antagonist muscles
patellofemoral joint	articulation between the patella and the femur
pelvic girdle	the two hip bones plus the sacrum, which can be rotated forward, backward, and laterally to optimize positioning of the hip joint
pennate fiber arrangement	pattern of fibers within a muscle with short fibers attaching to one or more tendons
perfectly elastic impact	impact during which the velocity of the system is conserved
perfectly plastic impact	impact resulting in the total loss of system velocity
periosteum	double-layered membrane covering bone; muscle tendons attach to the outside layer, and the internal layer is a site of osteoblast activity
plantar fascia	thick bands of fascia that cover the plantar aspect of the foot
popliteus	muscle known as the unlocker of the knee because its action is lateral rotation of the femur with respect to the tibia
porous	containing pores or cavities
potential energy	energy by virtue of a body's position or configuration, calculated as the product of weight and height
power	rate of work production, calculated as work divided by the time during which the work was done

pressure	force per unit of area over which a force acts
prestress	stress on the spine created by tension in the resting ligaments
primary spinal curves	curves that are present at birth
principal axes	three mutually perpendicular axes passing through the total-body center of gravity
principal moment of inertia	total-body moment of inertia relative to one of the principal axes
projectile	a body in free fall that is subject only to the forces of gravity and air resistance
projection speed	the magnitude of projection velocity
pronation	combined conditions of dorsiflexion, eversion, and abduction
proprioceptive neuromuscular facilitation	a group of stretching procedures involving alternating contraction and relaxation of the muscles being stretched
propulsive drag	force acting in the direction of a body's motion
propulsive drag theory	theory attributing propulsion in swimming to propulsive drag on the swimmer
propulsive lift theory	theory attributing propulsion in swimming at least partially to lift acting on the swimmer
Q-angle	the angle formed between the anterior superior iliac spine, the center of the patella, and the tibial tuberosity
quadriceps	the rectus femoris, vastus lateralis, vastus medialis, and vastus intermedius
qualitative	involving nonnumeric description of quality
quantitative	involving the use of numbers
radial acceleration	component of acceleration of a body in angular motion directed toward the center of curvature; represents change in direction
radian	unit of angular measure used in angular-linear kinematic quantity conversions; equal to 57.3°
radiocarpal joints	condyloid articulations between the radius and the three carpal bones
radioulnar joints	the proximal and distal radioulnar joints are pivot joints; the middle radioulnar joint is a syndesmosis
radius of gyration	distance from the axis of rotation to a point where the body's mass could be concentrated without altering its rotational characteristics
radius of rotation	distance from the axis of rotation to a point of interest on a rotating body
range	the horizontal displacement of a projectile at landing
range of motion	angle through which a joint moves from anatomical position to the extreme limit of segment motion in a particular direction
reaction board	a specially constructed board for determining the center of gravity location of a body positioned on top of it
reciprocal inhibition	inhibition of tension development in the antagonist muscles resulting from activation of muscle spindles
rectilinear	along a straight line
relative angle	angle at a joint formed between the longitudinal axes of adjacent body segments
relative projection height	the difference between projection height and landing height
relative velocity	velocity of a body with respect to the velocity of something else, such as the surrounding fluid
repetitive loading	repeated application of a subacute load that is usually of relatively low magnitude
resultant	single vector that results from vector composition
retinacula	fibrous bands of fascia
right hand rule	procedure for identifying the direction of an angular motion vector
rotator cuff	band of tendons of the subscapularis, supraspinatus, infraspinatus, and teres minor, which attach to the humeral head
sagittal plane	plane in which forward and backward movements of the body and body segments occur
scalar	physical quantity that is completely described by its magnitude
scapulohumeral rhythm	a regular pattern of scapular rotation that accompanies and facilitates humeral abduction

scoliosis	lateral spinal curvature
secondary spinal curves	the cervical and lumbar curves, which do not develop until the weight of the body begins to be supported in sitting and standing positions
second-class lever	lever positioned with the resistance between the applied force and the fulcrum
segmental method	a procedure for determining total-body center of mass location based on the masses and center of mass locations of the individual body segments
series elastic component	passive elastic property of muscle derived from the tendons
shear	force directed parallel to a surface
short bones	small, cubical skeletal structures, including the carpals and tarsals
skin friction (surface drag, viscous drag)	resistance derived from friction between adjacent layers of fluid near a body moving through the fluid
slow-twitch fiber	a fiber that reaches peak tension relatively slowly
specific weight	weight per unit of volume
spondylolisthesis	complete bilateral fracture of the pars interarticularis, resulting in anterior slippage of the vertebra
spondylolysis	presence of a fracture in the pars interarticularis of the vertebral neural arch
sports medicine	clinical and scientific aspects of sports and exercise
stability	resistance to disruption of equilibrium
stabilizer	role played by a muscle acting to stabilize a body part against some other force
static equilibrium	a motionless state characterized by $\Sigma SF_v = 0$, $\Sigma SF_h = 0$, and $\Sigma ST = 0$
static stretching	maintaining a slow, controlled, sustained stretch over time, usually about 30 seconds
statics	branch of mechanics dealing with systems in a constant state of motion
sternoclavicular joint	modified ball-and-socket joint between the proximal clavicle and the manubrium of the sternum
stiffness	the ratio of stress to strain in a loaded material; that is, the stress divided by the relative amount of change in the structure's shape
strain	amount of deformation divided by the original length of the structure or by the original angular orientation of the structure
strain energy	capacity to do work by virtue of a deformed body's return to its original shape
stress	distribution of force within a body, quantified as force divided by the area over which the force acts
stress fracture	fracture resulting from repeated loading of relatively low magnitude
stress reaction	progressive bone pathology associated with repeated loading
stretch reflex	monosynaptic reflex initiated by stretching of muscle spindles and resulting in immediate development of muscle tension
stretch-shortening cycle	eccentric contraction followed immediately by concentric contraction
summation	building in an additive fashion
supination	combined conditions of plantar flexion, inversion, and adduction
synovial fluid	clear, slightly yellow liquid that provides lubrication inside the articular capsule at synovial joints
system	mechanical system chosen by the analyst for study
tangential acceleration	component of acceleration of a body in angular motion directed along a tangent to the path of motion; represents change in linear speed
tensile strength	ability to resist pulling or stretching force
tension	pulling or stretching force directed axially through a body
tetanus	state of muscle producing sustained maximal tension resulting from repetitive stimulation
theoretical square law	drag increases approximately with the square of velocity when relative velocity is low
third-class lever	lever positioned with the applied force between the fulcrum and the resistance

tibiofemoral joint	dual condyloid articulations between the medial and lateral condyles of the tibia and the femur, composing the main hinge joint of the knee
torque	the rotary effect of a force about an axis of rotation, measured as the product of the force and the perpendicular distance between the force's line of action and the axis
torsion	load producing twisting of a body around its longitudinal axis
trabecular bone	less-compact mineralized connective tissue with high porosity that is found in the ends of long bones and in the vertebrae
trajectory	the flight path of a projectile
transducers	devices that detect signals
translation	linear motion
transverse plane	plane in which horizontal body and body segment movements occur when the body is in an erect standing position
turbulent flow	flow characterized by mixing of adjacent fluid layers
valgus	condition of outward deviation in alignment from the proximal to the distal end of a body segment
varus	condition of inward deviation in alignment from the proximal to the distal end of a body segment
vector	physical quantity that possesses both magnitude and direction
vector composition	process of determining a single vector from two or more vectors by vector addition
vector resolution	operation that replaces a single vector with two perpendicular vectors such that the vector composition of the two perpendicular vectors yields the original vector
viscoelastic	having the ability to stretch or shorten over time
volume	amount of three-dimensional space occupied by a body
wave drag	resistance created by the generation of waves at the interface between two different fluids, such as air and water
weight	gravitational force that the earth exerts on a body
work	in a mechanical context, force multiplied by the displacement of the resistance in the direction of the force
yield point (elastic limit)	point on the load–deformation curve past which deformation is permanent

INDEX